AF698687

SPACE AGE METALS TECHNOLOGY

SOCIETY FOR THE ADVANCEMENT OF
MATERIAL AND PROCESS ENGINEERING

2nd INTERNATIONAL SAMPE METALS AND METALS PROCESSING CONFERENCE

VOLUME 2

SPACE AGE METALS TECHNOLOGY

Dayton Convention Center
Dayton,Ohio
August 2 - 4, 1988

Edited by
F. H. Froes
Ray A. Cull

Additional copies of this publication may be obtained from
Society for the Advancement of Material and Process Engineering

SAMPE Business Office
P.O. Box 2459
Covina, California 91722

Responsbility for the contents and security clearance
of papers published herein rests solely upon the
authors and not upon SAMPE or any of its members

ISBN 0-938994-44-1

© 1988 by Society for the Advancement of Material and Process Engineering

2ND INTERNATIONAL SAMPE METALS AND METALS PROCESSING CONFERENCE
SPONSORED BY: MIDWEST CHAPTER

General Chairman	Ray A. Cull, Dow Corning Corporation
Program Chairman	F. H. Froes, Air Force Wright Aeronautical Lab
Administration Chairman	Karen Sitzman, Metcut Materials Research Group
Finance Chairman	Tony E. Saliba, University of Dayton
Registration Chairman	Susan S. Saliba, University of Dayton Research Inst
Audio Visual Chairman	Robert J. Gran, Universal Technology Corporation
Arrangements Chairman	James T. Johnson, Sinclair Community College
Spouses Program Chairman	Karen Sitzman, Metcut Materials Research Group

INTERNATIONAL OFFICERS

President	Gayle B. Wadsworth
Immediate Past President	Gary W. Valentine
1st Vice President	Jerry Bauer
2nd Vice President	Brian Wilson
3rd Vice President	Bob J. Hunter
Secretary	Roy E. Henrichsen
Treasurer	W. H. Face, Jr.

CHAPTER OFFICERS

Chairman	James T. Johnson
1st Vice Chairman	Tony E. Saliba
2nd Vice Chairman	Arthur K. Behme, Jr.
Secretary	Susan S. Saliba
Treasurer	Robert T. Achard
Sr. International Director	Robert C. Harper
Jr. International Director	Ronald H. Hardy

SESSIONS AND SESSION CHAIRMEN

CASTING

Daniel Eylon, University of Dayton
William J. Barice, Precision Castparts Corporation

ADVANCED ALLOYS - INTERMETALLICS

Young-Won Kim, Metcut Materials Research Group
Alan I. Taub, General Electric Company

RAPID SOLIDIFICATION

Susan D. Kirchoff, Air Force Wright Aeronautical Laboratory
Richard E. Lewis, Lockheed Missiles and Space Company

MELTING

Howard B. Bomberger, TIMET
Paul J. Bania, TIMET

ADVANCED ALLOYS - ALUMINUM

Walter M. Griffith, Air Force Wright Aeronautical Laboratory
Govind R. Chanani, Northrop Corporation

FORMING/MACHINING

John T. Cammett, Metcut Research Associates
Tom Bales, NASA-Langley Research Center

CORROSION/OXIDATION

S. Krishnamurthy, Metcut Materials Research Group
Mohammad Khobaib, University of Dayton Research Institute

ADVANCED ALLOYS - TITANIUM

Tom F. Broderick, Air Force Wright Aeronautical Laboratory
R. Grant Rowe, General Electric Company

FAILURE ANALYSIS

Thomas D. Cooper, Air Force Wright Aeronautical Laboratory
William Bogan, General Electric Company

FRACTURE/FATIGUE

Jim C. Chesnutt, General Electric Company
John E. Hack, Yale University

ADVANCED ALLOYS - MAGNESIUM/BERYLLIUM

Erica Robertson, Air Force Wright Aeronautical Laboratory
Santosh K. Das, Allied-Signal, Inc.

APPLICATIONS AND DESIGN

Andrew M. Johnson, General Electric Company
Jerry E. Fischler, Douglas Aircraft Company

JOINING

William A. Baeslack, Ohio State University
Earl Helder, General Electric Company

SESSIONS AND SESSION CHAIRMEN

METAL MATRIX COMPOSITES

Paul R. Smith, Air Force Wright Aeronautical Laboratory
Gopal Das, Metcut Materials Research Group

APPLICATIONS, DESIGN AND PROCESSING

J. C. Malas, Air Force Wright Aeronautical Laboratory
J. T. Morgan, Air Force Wright Aeronautical Laboratory

PREFACE

The Midwest Chapter of SAMPE is happy to host the Second International SAMPE Metals and Metals Processing Conference in Dayton, Ohio, 2-4 August 1988. As the birthplace of aviation, Dayton is a particularly appropriate location for this conference, though the materials to be discussed are a far cry from the "wood-and-canvas" used so masterfully by the Wright Brothers.

The theme of this conference is "Space Age Metals Technology" reflecting the metals concentration of this meeting. The technical presentations have been selected in close cooperation with the thirty chairmen of the fifteen sessions. The names of these major contributors to the success of this meeting can be found in the summaries of the technical sessions which are located in the following pages of this proceedings. The topics covered have been selected to demonstrate how well the metals community have responded to the challenge from competing materials such as the organic matrix composites. We are pleased to bring together an outstanding group of authors and speakers who will provide the reader and listener with the metals "state-of-the-art."

The technical program is the solid foundation on which a successful conference is built. However, others need also to be recognized including our three internationally known invited speakers Dr. Vince Russo, Dr. Jack Henderson, and Dr. Bart Barthelemy. The program chairmen would like to acknowledge the behind the scenes assistance of SAMPE's Business Office -- Marge Smith, Doris Weaver, and Daun White without whose efforts this conference and publication would not have been possible. We also recognize the local organization committee of Robert J. Gran, James T. Johnson, Joyce A. Martyr, Susan S. Saliba, Tony E. Saliba, and Karen A. Sitzman who gave up many countless hours of their own time to make this conference a success. It is always precarious to pick out one individual to praise, but we choose to take that risk. Miss Karen Sitzman was the "glue" that made the technical program come together and ensured that the papers included in this volume were submitted in a timely manner.

F.H. (Sam) Froes
Technical Chairman

Ray A. Cull
General Chairman

CONTENTS

COPYRIGHT iv

CONFERENCE COMMITTEE & OFFICERS v

SESSIONS AND SESSION CHAIRMEN vi

PREFACE ix

TABLE OF CONTENTS xi

PAPERS AND AUTHORS xii

FOREWORD *by Dr. F. H. Froes* 1

AUTHORS INDEX 530

SUBJECT INDEX 531

PAPERS AND AUTHORS

Page

ADVANCES IN TITANIUM ALLOY CASTING TECHNOLOGY 20

J. K. Thorne and W. J. Barice

EFFECTS OF TREATMENTS ON MECHANICAL PROPERTIES 28

D. Eylon, F. H. Froes and W. J. Barice

MICROSTRUCTURE-PROPERTY RELATIONSHIPS IN ADVANCED NICKEL BASE SUPERALLOY AIRFOIL CASTINGS 37

Alan D. Cetel and David N. Duhl

POTENTIAL FOR NIAl BASED ALLOYS FOR HIGH TEMPERATURE APPLICATIONS 49

K. Vedula

ENGINEERING PROCESSING AND PROPERTIES OF NICKEL ALUMINIDES 62

Vinod K. Sikka

CONTINUAL OPPORTUNITIES FOR SUPERALLOY DEVELOPMENT 76

Keh-Minn Chang

APPLICATIONS OF HIGH TEMPERATURE/HIGH STIFFNESS ALUMINUM ALLOYS 91

P. S. Gilman, S. K. Das, D. Raybould, M. S. Zedalis and J. M. Peltier

RAPID SOLIDIFICATION OF NB-BASE ALLOYS 102

A. B. Gokhale, K. R. Javed, G. J. Abbaschian and R. E. Lewis

RAPID SOLIDIFICATION PROCESSING OF COPPER-BASE ALLOY POWDERS 116

I. E. Anderson

ELECTRON BEAM COLD HEARTH MELTING (EBCHM) OF SPACE AGE METALS 128

Howard R. Harker and Charles H. Entrekin

MELTING TECHNIQUES FOR TITANIUM 135

A. Mitchell and D. W. Tripp

ADVANCED PRODUCTION OF SUPERALLOY AND TITANIUM INGOTS BY COLD HEARTH MELTING 144

H. Stumpp, H. Pannen, H. Ranke, G. Sick and W. Dietrich

DEVELOPMENTS IN ELECTRON BEAM MELTING OF AEROSPACE MATERIALS 157

D. M. Ferris and L. C. Hainz

ALUMINUM-LITHIUM ALLOYS FOR FIGHTER AIRCRAFT APPLICATIONS 166

V. M. Vasey-Glandon, E. A. Walker-Heckman and K. K. Sankaran

LOW DENSITY AL-MG-LI ALLOY FOR AEROSPACE FORGINGS 177

R. D. Schelleng, A. I. Kemppinen and J. H. Weber

EXPERIMENTAL HIGH MODULUS ELEVATED TEMPERATURE Al-Ti BASED ALLOYS BY MECHANICAL ALLOYING 188

P. K. Mirchandani and R. C. Benn

HIGHLY EFFECTIVE HIP PROCESS FOR INTRICATELY-SHAPED PARTS FROM POWDER ALLOYS DIFFICULT TO DEFORM 202

Vsevolod A. Domoratsky, Anatol F. Sanin, Vladimir K. Nikolayev and Lioudmila L. Scheglova

PAPERS AND AUTHORS

	Page
LOCALIZED CORROSION OF GRAPHITE FIBER/6061-T6 ALUMINUM ALLOY METAL-MATRIX COMPOSITES IN AERATED AND DEAERATED SODIUM SULFATE SOLUTIONS L. H. Hirara and R. M. Latanision	213
CORROSION IN THE EARLY STAGES OF CONDENSATION J. H. Payer and S. K. Chawla	225
OXIDATION OF ION-IMPLANTED NICKEL-BASE ALLOYS V. Srinivasan	237
STRESS INDUCED HOT CORROSION OF RENE' 80 S. J. Balsone and M. Khobaib	243
CORROSION/OXIDATION BEHAVIOR OF SURFACE MODIFIED NICKEL M. Khobaib, R. S. Bhattacharya, C. N. Raffoul and A. K. Rai	254
HIGH TEMPERATURE OXIDATION BEHAVIOR OF Ti_3Al ALLOYS M. Khobaib and F. W. Vahldiek	262
TITANIUM REQUIREMENTS FOR CURRENT AND FUTURE MILITARY GAS TURBINE ENGINES E. S. Huron and J. A. Miller	271
Ti-1100: A NEW ELEVATED TEMPERATURE TITANIUM ALLOY Paul J. Bania	286
FAME - AN EXPERT SYSTEM FOR FAILURE ANALYSIS OF METALS R. K. Elsley, W. A. Guthmiller, S. M. El-Soudani, J. D. Young and Russell Henderson	298
HYDROGEN AND ADVANCED AEROSPACE MATERIALS Howard G. Nelson	310
FATIGUE BEHAVIOR OF POWDER METALLURGY Ti-6Al-4V J. W. Bohlen, J. A. Walsh, A. A. Sheinker and G. R. Chanani	320
FATIGUE CRACK GROWTH BEHAVIOR OF A SINGLE CRYSTAL ALLOY AS OBSERVED THROUGH AN IN SITU FATIGUE LOADING STAGE Jack Telesman and Peter Kantzos	334
ADVANCED MAGNESIUM ALLOYS FOR AEROSPACE APPLICATION C. F. Chang, S. K. Das, D. Raybould, R. L. Bye and N. J. Kim	345
A REVIEW OF BERYLLIDES FOR VERY HIGH TEMPERATURE SERVICE J. M. Marder and A. J. Stonehouse	357
THE INFLUENCE OF PROCESSING VARIABLES AND DEFECTS ON THE PERFORMANCE OF ADHESIVELY BONDED JOINTS T. S. Srivatsan, T. A. Place, Raju Mantena, R. F. Gibson and T. S. Sudarshan	368
JOINING OF RAPIDLY SOLIDIFIED POWDER METALLURGY AL-FE-CE ALLOYS W. A. Baeslack III	381

PAPERS AND AUTHORS

	Page
MACHINE VISION FOR CONTROL OF GAS TUNGSTEN ARC WELDING Robert A. Anderson	393
ADAPTIVE GAS TUNGSTEN ARC WELDING SYSTEMS Harvey R. Castner	407
ANALYTICAL MODELING OF 3-D CASTING USING THE CAST3 FINITE ELEMENT CODE Douglas R. Barker, Mark T. Samonds and Sokka M. Doraivelu	421
DEVELOPMENTS AND TRENDS IN ELECTRODEPOSITION J. W. Dini	428
PLASMA APPLICATIONS FOR SPACE AGE METALS M. P. Schlienger	437
PROCESSING AND MECHANICAL PROPERTIES OF LIGHTWEIGHT STRUCTURAL COMPOSITES John J. Lewandowski	450
VARIABLE-POLARITY PLASMA-ARC WELDING OF ALUMINUM-LITHIUM ALLOY 2090 Paul T. Wood and W. A. Baeslack III	458
THE EFFECT OF ELEVATED TEMPERATURE EXPOSURE ON THE PROPERTIES OF SiC/Ti6Al4V COMPOSITES F. W. Wawner and W. J. Whatley	470
INTERACTION ZONE FORMED BETWEEN TiC AND TITANIUM ALUMINIDE M. H. Loretto and D. G. Konitzer	480
FRACTOGRAPHY OF TITANIUM ALUMINIDE METAL MATRIX COMPOSITES Mary Lee Gambone and Kenneth R. Bain	487
THE COMMERCIALIZATION OF METAL-MATRIX COMPOSITES PRODUCED BY INGOT METALLURGY . M. Schuster, M. D. Skibo and F. W. Yep	498
V I D P - A NEW SYSTEM CONCEPT FOR VACUUM MELTING OF SUPERALLOYS, FERROUS AND NON-FERROUS ALLOYS Otto W. Stenzel and Hans-J. Kemmer	504
ENHANCEMENT OF SUPERPLASTIC FORMABILITY IN A HIGH STRENGTH ALUMINUM ALLOY S. P. Agrawal, G. R. Turk & R. Vastava	512

FOREWORD

ADVANCES IN LIGHTWEIGHT METALLIC MATERIALS FOR AEROSPACE APPLICATIONS

F. H. Froes
U.S. Air Force Wright Aeronautical Laboratories
Materials Laboratory, AFWAL/MLLS
Wright-Patterson AFB, OH 45433-6533

Abstract

Development of improved lightweight metallic materials is necessary to enhance the performance capability of aerospace systems. This paper presents recent advances which have been made in the light metals aluminum, titanium, and magnesium. For the three base metals the approaches discussed include ingot metallurgy, powder metallurgy/rapid solidification, castings, and engineered materials including metal matrix composites.

1. INTRODUCTION

Improved material behavior plays a major party in structural weight reduction which in turn leads to improved performance in airframes, engines, and missiles. By far the greatest weight reduction is achieved by a density decrease; which gives a direct weight reduction (Fig. 1). Hence the importance of the low density structural alloys aluminum, titanium, and magnesium. Improvement in material behavior significantly exceeds the contribution to system performance enhancement made by other advanced technologies such as structural concepts, automated design and active load control, and flutter suppression. The increased performance of an engine (as measured by enhanced thrust-to-weight ratio) with the development of improved materials is shown in Fig. 2.

It is certain that organic composite materials will play a significant role in future aerospace systems, however the improvements being made in metallic materials insures that they also will see extensive use. The challenge of the organic composites has forced the metallurgist to reach out and be more innovative. Greater control of chemistry is now possible using techniques such as rapid solidification, and many new and improved processing and fabrication techniques are in various stages of development. It is these advances in chemistry and processing/fabrication that will allow metal components to effectively compete with the organic matrix composites.

The present paper will overview the developments which are being made. The interested reader is referred to more detailed papers in this proceedings for more details.

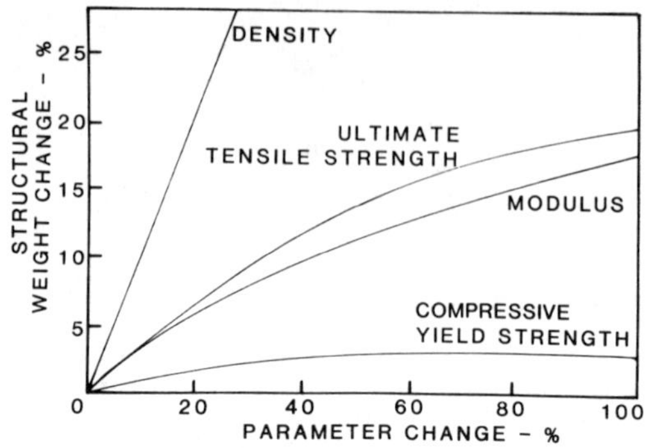

Fig. 1 - Effect of property improvement on structural weight.

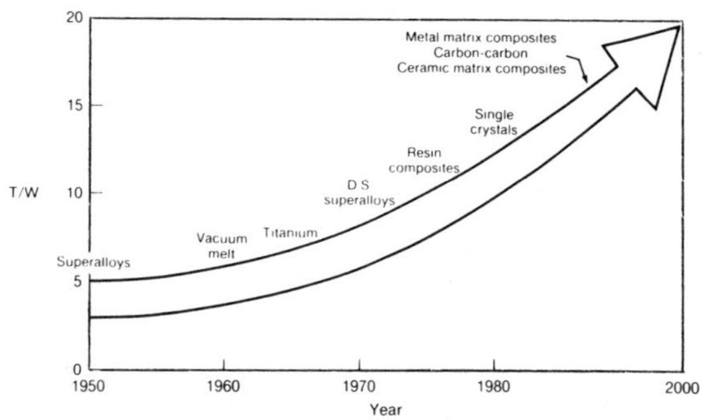

Fig. 2 - Increased engine thrust-to-weight performance with improved materials (courtesy General Electric Company).

2. INGOT METALLURGY (IM)

2.1 Aluminum

The major advance which has occurred in the IM of aluminum alloys is the development of reduced density Al-Li alloys. For every 1% lithium added to aluminum there is a 3% density decrease and a 10% stiffness-to-density increase. These alloys exhibit up to a 10% density decrease compared with conventional IM 2000- and 7000-series alloys at equivalent strength levels. Thus by direct substitution these alloys result in a component weight reduction of up to 10% (Fig. 1). Typical mechanical properties of Al-Li sheet product is shown in Table 1.

In combination with advanced forming techniques, such as superplastic forming (where a combination of slow strain rate and elevated temperature allow elongations of several hundred percent) the Al-Li alloys are strong candidates for extensive use in advanced U.S. Air Force systems such as the advanced fighter and the C-17 transport [1]. A demonstration component on a Boeing aircraft is shown in Fig. 3.

TABLE 1. TYPICAL PROPERTIES OF SEVERAL Al-Li ALLOY SHEET PRODUCTS[a]

Alloy	Grain Structure[b]	Temper	UTS, MPa(ksi) (L)	YS, MPa(ksi) (L)	El., % (L)	Kapp. (L-T)	Kc (L-T)	Source
Lower Strength/Damage Tolerant Tempers								
8090	URx	T8	425(61)	345(50)	6	77	102	RAE
8090	100% Rx	T8	425(61)	325(47)	8	107	165	RAE
2091	100% Rx	T3	425(61)	325(47)	18	100	148	RAE
2091	100% Rx	T8x	440(63)	345(50)	19	---	145	Pechiney
Mid-Strength Tempers								
8090	URx	T6	490(70)	380(55)	5	65	75	RAE
8090	URx	T8	500(72)	425(61)	6	---	---	RAE
2090	URx	T8E50	525(75)	497(71)	5	---	∿65	Alcoa
High Strength Tempers								
2090	URx	T8E41	560(80)	525(75)	5	---	∿55	Alcoa
8091	URx	T8	588(84)	539(77)	5	---	---	Alcan
Corrosion Resistant Temper								
2091	100% Rx	CPH	435(62)	315(45)	14.5	---	140	Pechiney

[a]From Aluminum-Lithium Alloys, Design, Development and Application Update, R. J. Kar, S. P. Agrawal, and W. E. Quist, editors, ASM International, Metals Park, OH (1988).

[b]Rx = recrystallized; URx = unrecrystallized.

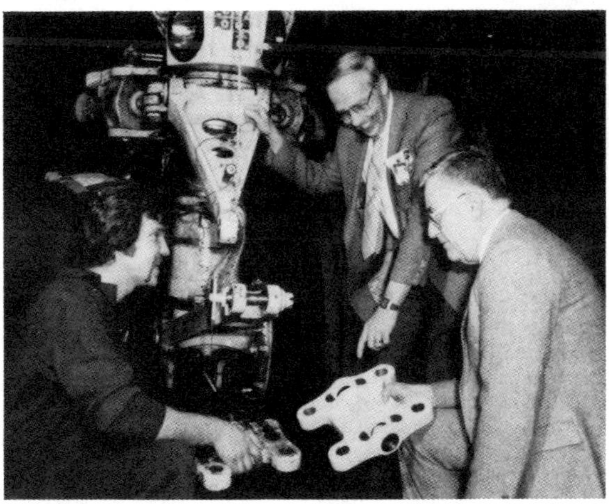

Fig. 3 - Aluminum-lithium fitting (alloy 2090) which has been installed on several Boeing 747 aircraft (courtesy Boeing Company).

The weldability of the 2090 aluminum-lithium alloy has been shown to be comparable to 2219, the high strength aluminum alloy currently used in space-launched fuel tanks (Table 2) [2].

TABLE 2. WELDMENT PROPERTIES[a] [2]

Alloy	Temperature, °C	UTS, MPa	YS, MPa	El., %
2219	RT	285	205	3.0
2090	RT	235	205	5.0
2219	-195	370	205	4.0
2090	-195	325	195	3.5

[a]2219 in -T87 or T81 temper, 2090 in -T8E41 temper.

It is claimed [2] that this makes the alloy a leading candidate for cryogenic tank applications such as the space shuttle external tanks (Fig. 4), and in the recently announced Advanced Launch System (ALS).

Fig. 4 - Space shuttle external tanks, a candidate for Al-Li alloy 2090 applications.

A further advance in IM aluminum alloys is the 7475 controlled toughness alloy developed for sheet and plate applications that require a combination of high strength, superior fracture toughness, and resistance to fatigue crack propagation. This alloy has tighter chemistry control than the current high strength aerospace alloy 7075. Typical fracture toughness values for alloy 7475 plate are approximately 40% greater than alloy 7075 in a similar condition (Table 3); and corrosion resistance fatigue and fatigue crack growth behavior is at least equivalent to current aerospace alloys. Potential applications for the 7475 alloy are in fuselage skins, wing skins, fuselage bulkheads, and wing spars.

2.2 Titanium

Although a number of higher strength alloys have been developed in the past decade (e.g., Ti-10V-2Fe-3Al, Ti-15V-3Al-3Cr-3Zr) [3], the major thrust recently has been to raise the temperature capability of titanium alloys so that they compete with the much heavier nickel-base superalloys. In the United Kingdom the most recent alloy is the IMI-834 Ti-6Al-4Sn-3.5Zr- 0.7Nb- 0.5Mo- 0.35Si- 0.06C which is useful to 600°C (1110°F). A somewhat similar alloy has also been developed in the United States designated Ti-1100 (≡ 600°C) [4].

The next stage in the development of ingot titanium alloys has been to move to the ordered intermetallic alloys such as $Ti_3Al(\alpha)$ and $TiAl(\gamma)$ [5]. Because of their ordered structure these intermetallics exhibit excellent elevated temperature behavior but low room temperature ductility (Table 4).

The aluminides are less dense than conventional titanium alloys, and less than half as dense as the superalloys. These compounds are also of value because of their high stiffness (modulus), an important consideration in many engine components. The only shortcoming in the behavior of the titanium aluminides is their low room temperature ductility, and related fracture toughness and fatigue crack growth rate behavior. A number of programs are currently addressing these areas using innovative processing routes and modified chemistries. The aluminides also perform well in oxidation resistance, the conventional alloys apparently being useful uncoated to a maximum of 600°C (1110°F). A number of engine components have been successfully fabricated from the titanium aluminides (Fig. 5) and have been engine tested successfully. Work is also in progress to use the titanium aluminides on hypersonic vehicles such as the National Aerospace Plane (NASP) [1].

TABLE 3. TYPICAL FRACTURE TOUGHNESS VALUES[a] OF ALUMINUM ALLOYS 7475 AND 7075

Alloy	K_{Ic}, MPa $\sqrt{m}$
7475	55
7075	32

[a]Same temper (T7351) and direction (LT).

3. POWDER METALLURGY/RAPID SOLIDIFICATION (PM/RS)

Production of metallic alloys by the traditional ingot metallurgy (IM) approaches places restrictions on the composition and microstructural scale of the alloys which can be produced. In contrast PM/RS allows extension of the alloying levels which are possible and can lead to much more refined microstructures than are possible using the IM technique. The greatly increased chemistry/microstructure window can lead to enhanced mechanical and physical properties in a variety of metallic systems [6].

3.1 Aluminum

Three new classes of alloys have been explored using the PM/RS approach in the aluminum system.

TABLE 4. PROPERTIES OF ELEVATED TEMPERATURE TITANIUM BASE ALLOYS[a] [5]

Property	Ti-Base	Ti_3Al	TiAl	Superalloys
Density, g/cm^3	4.5	4.1-4.7	3.7	8.3
Modulus, GN/m^2	∿105	110-145	176	206
Maximum Use Temperature, °C	600	815	1040	1095
Ductility[b]	∿20	2-5	1-2	3-5
Ductility[c]	High	5-8	7-12	10-20

[a]Slightly updated.

[b]Room temperature, percent.

[c]Use temperature, percent.

These are the high strength corrosion resistant alloys based on traditional 7000-series alloys, low density Al-Li alloys with increased lithium levels over those possible using the IM approach, and dispersoid strengthened elevated temperature alloys based on low solubility/low diffusion rate additions such as the transition metals (Fe, Ni) and rare earth elements (Ce) [6].

Fig. 5 - High pressure turbine support ring fabricated from the titanium aluminide alloy Ti-14Al-21Nb (left), and superalloy Rene 41 (right) [5].

In the extruded condition the PM/RS alloy CW67 (Al-Zn-Mg-Cu-Zr-Ni) exhibits up to a 30% improvement in strength and as much as a 40% advantage in toughness compared to the high strength IM-7075 alloy. In a stress corrosion cracking environment the PM/RS alloy out performs the IM alloy in the over-aged condition (T73) while exhibiting a much higher strength and fracture toughness (Table 5).

The attractive combination of properties exhibited by CW67 and related PM/RS alloys has resulted in considerable interest in substituting these alloys for existing airframe components which are experiencing costly maintenance problems. Flight tests are already underway on an advanced leading edge flap design (Vought), extruded forward longerons (Warner Robbins Air Logistic Center), and a main landing gear door actuator (Boeing), and a variety of other trials are in the planning process.

A maximum of about 2.7%Li can be added by the IM method limiting the density reduction to a maximum of about 10% compared to conventional IM alloys. In contrast the PM/RS approach since it involves processing directly from the liquid state is not restricted to this maximum value. Alloys developed to date on a semi-commercial basis contain up to 3.6%Li and compare quite favorably with conventional 7075 forgings, but at a 12% reduction in density (Table 6).

The salt spray corrosion rate of PM/RS Al-Li alloys, as determined by weight loss, are much better than the conventional IM-2014-T6 alloy, and the fatigue crack growth rates of the PM/RS material are slower than the 7075-T73 alloy. Since there are problems still to be resolved with IM Al-Li forgings it is possible that the first applications for PM/RS Al-Li alloys will be in this product form.

TABLE 5. HIGH STRENGTH ALUMINUM ALLOYS

Property	IM 7075-T73[a]	PM/RS CW67-T7[b]
Yield Strength, MPa	435	580
Ultimate Tensile Strength, MPa	505	615
Elongation, %	13	12
Fracture Toughness, MPa $\sqrt{m}$	32	47
SCC Resistance, MPa	290	310

[a]Plate.

[b]Extrusions.

TABLE 6. TOUGHNESS AND TENSILE PROPERTIES OF CONVENTIONAL IM AND PM Al-Li ALLOYS[a]

Alloy	YS, MPa	UTS, MPa	El., %	K_q, MPa$\sqrt{m}$
7075-T73	435	505	13	32
PM/RS Al-Li	455	595	9	25

[a]Longitudinal properties, IM forging, PM/RS extrusion.

The elevated temperature dispersion strengthened aluminum alloys are the third class of PM/RS aluminum alloys being developed. These alloys are impossible to produce using conventional IM techniques. Use of this class of aluminum alloys would result in replacement of titanium in many moderate temperature applications (to 300°C [570°F] and perhaps beyond). Al-Fe-Ce alloys exhibit strength capabilities about double those of the highest strength IM-2219 alloy to 230°C (445°F). The elevated temperature stability is also good with room temperature strength being retained even after exposures of 100 hr at 300°C (570°F). This Al-Fe-Ce PM/RS class of alloys also have good SCC resistance in addition to attractive general corrosion and exfoliation resistance. A comparison of alloys in this dispersion strengthened category is shown in Table 7.

The elevated temperature strength capability of this generic class of dispersion strengthened alloy is shown in comparison with aluminum-based materials also designed for elevated temperature use (discussed later in this article) in (Fig. 6).

An alternative to PM/RS is to produce the alloys directly from the vapor [7]. Using electron beam melting methods (Fig. 7) the process is now at the stage where aluminum alloy slabs on the 50-100 kg (110-220 lb) scale up to 44 mm (1.75 in.) thick can be produced at a deposition rate of 6 mm/hr (0.25 in./hr) [7] (Fig. 8). The strength of a vapor deposit Al-7.5Cr-1.5Fe alloy is significantly higher than the strength of PM/RS alloys up to at least 250°C (480°F), a result of the fine microstructure produced by the vapor deposition technique. Above about 300°C (570°F) both types of alloys have similar strength levels.

TABLE 7. TENSILE PROPERTIES OF ELEVATED TEMPERATURE PM/RS ALUMINUM ALLOYS AT 315°C (570°F)[a] [6]

Alloy	UTS, MPa	YS, MPa	El., %
Al-Fe-Ce	270	225	7
Al-Fe-Mo	235	210	10
Al-Fe-V-Si	310	300	7
Al-Zr-Cr-Mn	235	215	--

[a]Longitudinal.

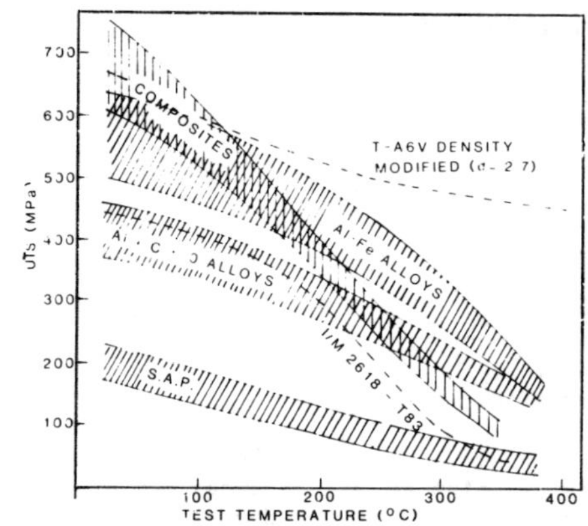

Fig. 6 - Elevated temperature strength capability of aluminum-based materials (courtesy Cegedur Pechiney).

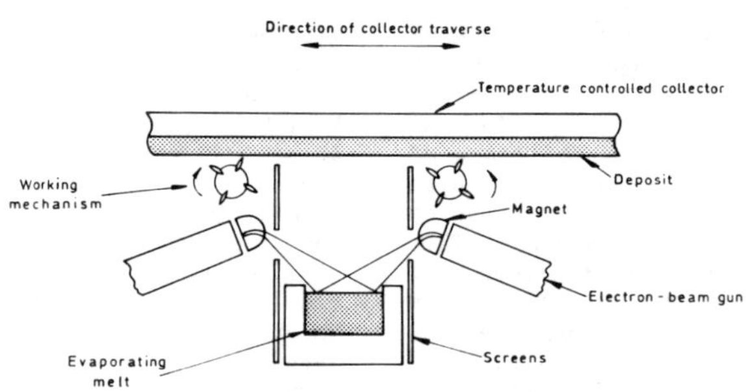

Fig. 7 - Schematic of electron beam vapor deposition process [7].

Fig. 8 - Slab of aluminum alloy produced using electron beam vapor deposition process [7].

Another alternative to PM/RS is mechanical alloying (MA) [8] in which heavy working of powder particles results in intimate alloying by a process of repeated welding, fracturing and rewelding. This process has been mainly explored to date in order to produce oxide dispersion strengthened (ODS). The mechanical property combination which can be obtained in a MA Al-base alloy is shown in Table 8.

3.2 Magnesium

Conventional magnesium alloys (the majority of which are used in the as-cast condition) exhibit lower than desirable strength, ductility, and creep behavior, and the non-protective oxide skin can lead to severe corrosion problems. Using PM/RS magnesium alloys containing rare earth (Y, Nd) additions have been developed which have strength and ductility

TABLE 8. ROOM TEMPERATURE MECHANICAL PROPERTIES OF MECHANICALLY ALLOYED ALUMINUM ALLOYS[a] [9]

Process	Alloy	UTS, MPa	YS, MPa	El., %	K_{Ic}, MPa$\sqrt{m}$	Density, g/cm^3
MA	Al-Li-Mg-O-C	450	510	10	45	2.55
Conventional IM	7075-T73	435	505	13	32	2.81

[a]Longitudinal.

The MA Al-Li alloys exhibit minimal degradation of properties when stressed in the transverse direction and are characterized by excellent general corrosion resistance which is 100x better than that exhibited by the IM-7053-T73 alloy. Inco Alloys International now has a full-scale production facility in place to MA aluminum alloys in Pittsboro, North Carolina.

combinations equivalent to high strength aluminum alloys and are five times more corrosion resistant than the most resistant conventional magnesium alloys. The specific strength and corrosion behavior of the PM/RS alloys is shown in Figs. 9 and 10 in comparison with conventional magnesium and a common aluminum alloy. The technology of PM/RS magnesium based alloys is much less mature than that of PM/RS aluminum-based alloys and applications have yet to be defined.

10, 11], although other programs have addressed high strength and low density alloys [3]. The use of PM/RS dispersion strengthened

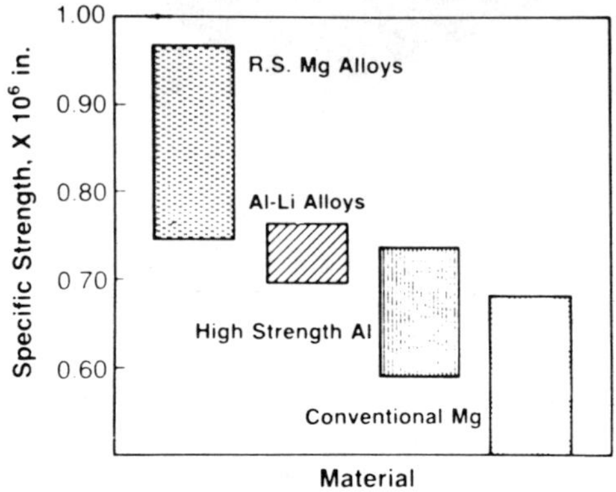

Fig. 9 - Specific strength of RS magnesium alloys in comparison with competing materials (courtesy Allied-Signal, Inc.).

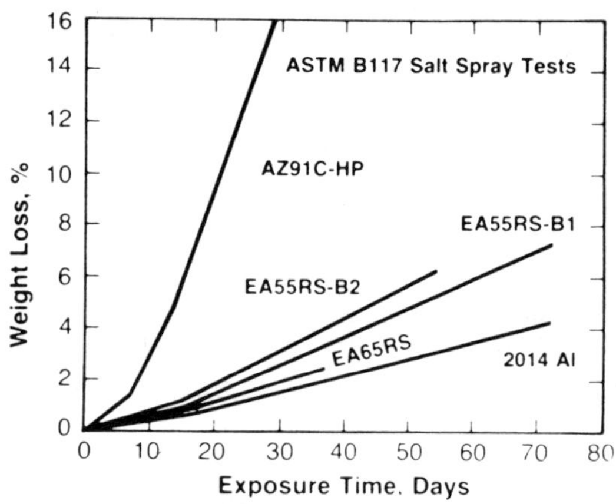

Fig. 10 - Salt spray corrosion behavior of RS magnesium alloys in comparison with competing materials (courtesy Allied-Signal, Inc.).

3.3 Titanium

To date the major concentration on PM/RS alloys has been to enhance the elevated temperature capability through dispersion hardening [3,

alloys with temperature capabilities beyond those of conventional IM titanium alloys (i.e., >600°C [>1110°F]) would lead to considerable weight savings in advanced jet engines. However, to date this activity remains in the research base, although results to date indicate the promise of this alloy class (Fig. 11).

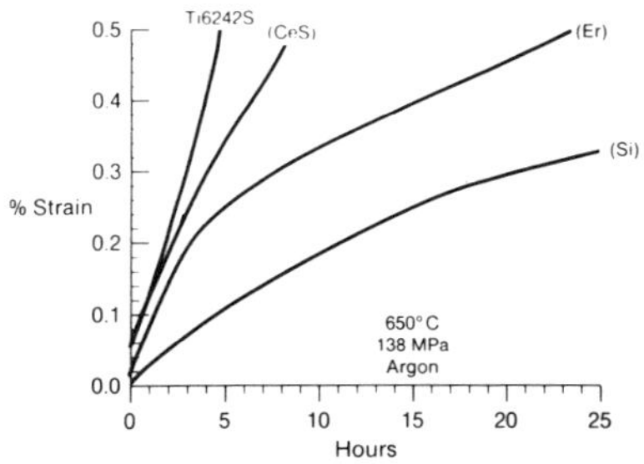

Fig. 11 - Creep behavior of dispersion strengthened titanium alloys (courtesy General Electric Company).

Other work has evaluated the potential of the titanium aluminide intermetallics using the PM/RS approach [10, 11]. Here the possible advantages of using the PM/RS method are a refined grain structure, gettering of oxygen from the matrix resulting in an improvement in ductility, and rare earth oxide dispersions to enhancing the high temperature creep performance. However, although there are indications that PM/RS may help in the processing of the Ti_3Al [10, 11] much further optimization is required.

4. CASTINGS

Of all the processing options available castings offer the lowest cost potential. However, traditionally castings have been considered to be inferior to cast and wrought (IM) material because of the inability to control the basic microstructure, and as a result of the porosity inherent to the casting process. Recently for both the aluminum and titanium systems the introduction of hot isostatic pressing (HIP'ing) to close porosity, and innovative processing, has greatly enhanced mechanical properties. Increased use will also result from elimination of the casting factor which is generally in the range of 25-33%. This factor is the ratio by which casting mechanical properties are reduced because of the wide variation in properties that can be obtained both within a casting and from one casting design to another [12, 13).

4.1 Aluminum

Recently intensive work on the aluminum casting alloy 357 culminated in a "casting factor free" design [12] for the air inlet duct of the F-16 fighter. By controlling uniformity through tighter chemistry and the use of metallographic controls (i.e., a check on the structure of the casting) a product with sufficiently reproducible properties was developed to allow use with no casting factor (Table 9).

TABLE 9. ALUMINUM 357 CASTING ALLOY, TENTATIVE DESIGN ALLOWABLES [12][a]

UTS, MPa	YS, MPa	El., %
350	295	3

[a]MIL-Handbook 5, B values.

Large, reliable, high strength aluminum alloy castings can now be produced resulting in replacement of costly assemblies traditionally fabricated from numerous machined parts [2]. Thin walls can be cast (0.25-0.33 cm [0.10-0.13 in.]), with tolerances of ±0.08 cm (±0.03 in.) with adjacent heavy wall sections for the attachment of mating components. Components being produced by this method include the Harpoon anti-ship missile fuel tank, fuselage section of the AGM-86B air-launched cruise missile, the forward and aft fairings for the B-1B air-launched cruise missiles, the F-14 canopy frame, and complex engine nacelle hinge structures for the Boeing 737.

4.2 Titanium

Initiation related mechanical properties particularly fatigue, have traditionally been the major drawback for use of titanium castings in demanding applications. However, the use of HIP'ing combined with innovative heat treatments or thermochemical processing (TCP, use of hydrogen as a temporary alloying element) has raised the level of fatigue behavior of titanium castings to at least that of IM levels [3]. This, in combination with increased size and shape-making capabilities, should ensure greatly increased use of titanium castings in both engine and airframe applications. The shipment of titanium castings has increased by a factor of 3 over the last decade, to a current level exceeding 1M lb/annum [3, 13].

An example of a large investment cast titanium fan frame is shown in Fig. 12. Further use will come from elimination of the casting factor. Already a number of integrally bladed fan components have been used in a non-manrated engine (no pilot) without a casting factor being used.

Fig. 12 - Today's largest production titanium investment castings CF680C2 aircraft engine fan frame. The 130 cm (52 in.) diameter part is produced from the Ti-6Al-4V alloy by PCC.

A recent advance is the casting of titanium alloys other than the conventional Ti-6Al-4V alloy [13]. The Ti-3Al-8V-6Cr-4Zr-4Mo alloy (38-6-44 or Beta C) exhibited excellent tensile properties as shown in Table 10.

TABLE 10. BETA C TITANIUM CASTING, TENSILE PROPERTIES

YS, MPa	UTS, MPa	El., %	RA, %
1170	1275	7	15

Equally impressive fatigue behavior was observed in this alloy with an endurance limit 85% above the average value of 600MPa (85 ksi) typical of the Ti-6Al-4V alloy. Further advances in titanium castings will come when alloys are developed which are designed specifically as casting alloys, rather than just using "off-the-shelf" wrought alloys.

4.3 Magnesium

The poor corrosion resistance of magnesium arises from its position in the electromotive series and the presence of undesirable metallic impurities, particularly nickel, iron, copper, and to an extent silicon. This latter fact has spurred the development of a new series of high purity (HP) corrosion resistant magnesium alloys. The salt spray corrosion resistance of alloys such as AZ91HP (Mg-9Al-1Zn) is better than the commonly used 380 die cast aluminum alloy and carbon steel. This new corrosion resistant high purity magnesium alloy has the same mechanical properties as normal AZ91 since the major alloying additions are identical. It is already seeing use in non-aerospace applications such as automotive parts and computer peripheral components and should see widespread use in those situations where paint or elaborate coatings are either undesirable or prohibitively expensive.

5. ENGINEERED MATERIALS

The success exhibited by organic matrix composites in which an organic matrix is combined with an elemental (e.g., graphite) or ceramic (e.g., SiC) fiber has led to parallel efforts on metal matrix composites (MMC).

5.1 Aluminum

Aluminum-base alloys have been used to develop hybrid engineered materials such as MMC, DISPAL (mechanically alloyed dispersion strengthened aluminum), and ARALL (aramid aluminum laminates). A comparison of the elevated temperature behavior of a variety of aluminum-based materials is shown in Fig. 6.

By dispersing ceramic fibers (whiskers) through an aluminum matrix increased stiffness, higher strengths, and better resistance to creep than unreinforced materials can be obtained [14]. Depending on the intended application, the aluminum-based MMC contains from 15-45 vol% SiC reinforcement. Generally, the whiskers have an average diameter of 0.6 µm (2.5×10^{-5} in.) and lengths ranging from 10-80 µm ($4\text{--}30 \times 10^{-4}$ in.) with tensile strengths up to 7000MPa (1000 ksi). Properties of aluminum MMC produced using a powder metallurgy blending approach are shown in Table 11.

TABLE 11. TYPICAL LONGITUDINAL PROPERTIES OF EXTRUDED ROD [14]

Material	UTS, MPa	Ys, MPa	El., %	E, GPa
6061-T6	290	255	17	70
6061-T6 (20% SiC)	585	440	4	120
6061-T6 (30% SiC)	795	570	2	140

DURAL metal matrix composites consist of an aluminum matrix, such as 2014, reinforced with a ceramic particulate like SiC, at about the 15 vol% level. Using a low cost casting approach the particulate are dispersed uniformily in the matrix (Fig. 13) [15].

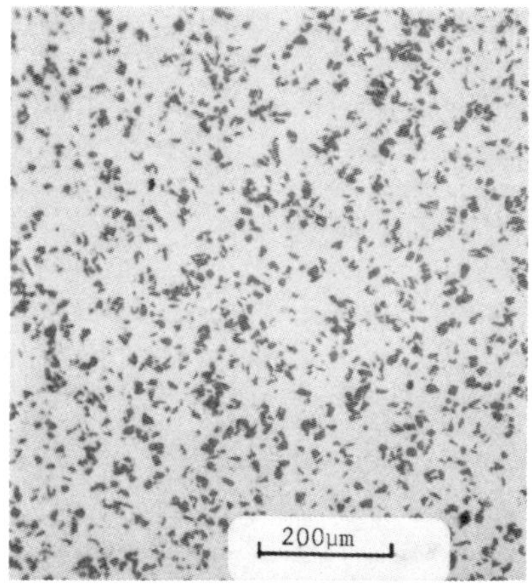

Fig. 13 - Optical photomicrograph of DURAL 15 vol% SiC in 2014 aluminum matrix, transverse extrusion.

The specific stiffness (stiffness divided by density) of DURAL SiC-Al is close to that of quasi-isotropic graphite-epoxy, at the same specific strength level, and much higher than competing monolithic metals (Fig. 14) [15].

Tensile properties of DURAL book castings exhibit higher strength and stiffness than unreinforced material but with a reduction in ductility (Table 12). Very recent results on direct chill (DC) cast material has higher ductility values [15].

At a cost projected to be a few dollars per pound, DURAL should be attractive for both non-aerospace and aerospace applications. The latter with use of high strength 2000- or 7000-series alloys. Generally the casting approach offers the opportunity to manufacture ceramic fiber reinforced MMC at reduced cost compared to powder metallurgy or foil interleaving techniques [16]. The challenge, which now appears to be an achieveable goal, is to combine this with acceptable mechanical properties.

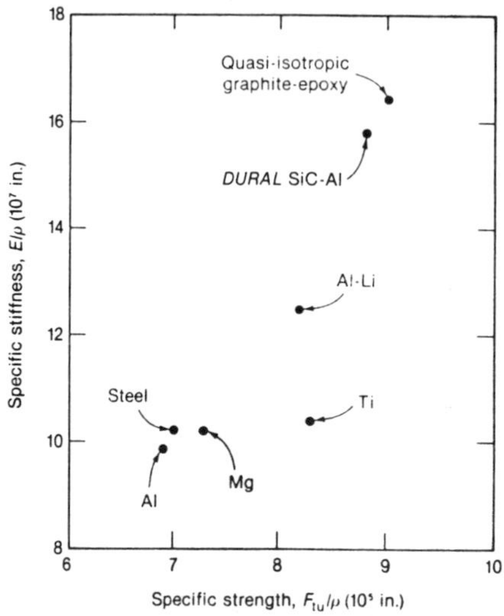

Fig. 14 - Specific stiffness/ strength of various structural materials [15].

TABLE 12. TENSILE PROPERTIES OF 6061-T6 Al-SiC I-BEAM EXTRUSIONS [15]

Vol% SiC	UTS, ksi	YS, ksi	El., %	E, x 10^6 psi
0	43.5	39.0	10.5	9.6
10	49.4	46.5	3.8	11.7
20	51.1	48.6	2.3	14.0

The DISPAL material is dispersion hardened by reaction milling with carbon black to produce a 12% volume fraction of Al_4C_3 dispersoids [17]. This material has excellent elevated temperature stability; for example, a 260°C (500°F)/100 hr soak causes essentially no loss in strength. The material exhibits ultimate tensile strengths of 400MPa (60 ksi) and 180MPa (25 ksi) at room temperature and 500°C (930°F), respectively. The disadvantages of this material at the present time appear to be low room temperature strength, a decrease in ductility as the temperature increases, and the expense of the DISPAL which makes it comparable in cost with titanium alloys.

Very recently a new approach to production of MMC has been developed [18]. In the XD (XD is a trademark of Martin Marietta Corporation) approach an in-situ precipitation technique is used for the formation of the reinforcement directly in the matrix of interest. Key features claimed by the inventors include: thermodynamically stable reinforcements; clean particle/matrix interfaces; single-crystal reinforcement; size, chemistry, and shape control; uniform distribution; compatibility with conventional ingot processing techniques; and effective grain refinement.

Using XD technology it is possible to simultaneously introduce a number of different reinforcements in a given matrix, thus allowing design of the microstructure to meet a specific material need, e.g., fine particles for strength and whiskers for creep reduction. Thus, XD technology is a very useful technique for achieving "designer" or "engineered" microstructures. Such flexibility at the microstructure level now makes possible the concept of concurrent design, i.e., the simultaneous design of the material and component.

XD technology has been applied to a number of systems in addition to aluminum, including NiAl, TiAl, and Cu. Significant property improvements have been noted in the systems studied so far. These include higher strength and creep resistance, reduced CTE, improved strength at high temperatures, and increased modulus. Furthermore, in the difficult to fabricate alloys such as the Ti-aluminides it appears that the fine microstructures resulting from XD processing aid in fabricability. Thus, although still in the early stages of development, XD technology appears to be a powerful tool in manipulating microstructure and improving properties and thereby enhancing system performance. A typical microstructure of XD aluminum is shown in Fig. 15.

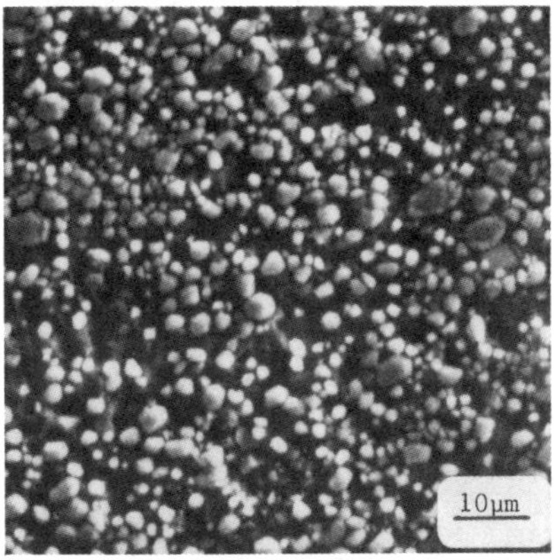

Fig. 15 - XD Al-2024 with 15 vol% TiB_2.

Typical extruded parts produced using this technique are shown in Fig. 16.

ARALL laminates are claimed to be a high strength composite material which handles like aluminum sheet. The system is based on the concept of bonding high strength aramid fibers embedded in a special resin between thin sheets of high strength aluminum alloys (Fig. 17). The primary benefit is a significant enhancement in fatigue and fatigue crack growth properties compared with monolithic aluminum materials. In fatigue tests ARALL has shown 100-1000 times fatigue life compared to monolithic aluminum. Tensile properties also are claimed to exceed conventional aluminum by 15-30%. The material is 15% less dense than monolithic aluminum and the outer aluminum layers give ARALL laminates protection against impact damage and moisture, both major concerns with normal organic composites. The outstanding fatigue behavior of ARALL makes it a strong candidate for fatigue dominated sheet applications such as are typically found in lower wing skins, fuselage, and tail skin locations. The first ARALL part to fly will be on the underwing of a Fokker F-50 aircraft (Fig. 18).

Fig. 16 - Typical extruded parts produced from XD aluminum.

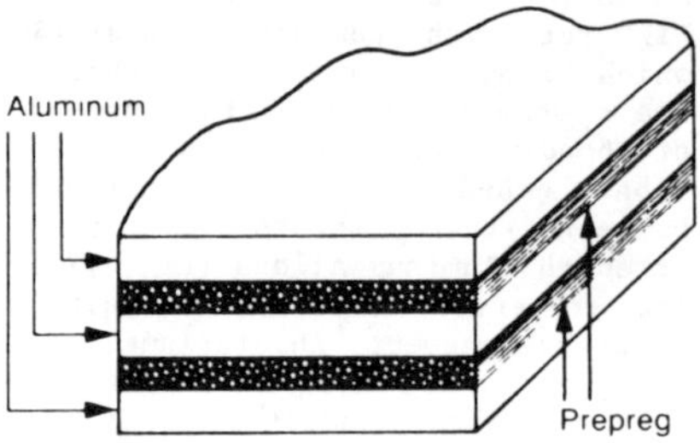

Fig. 17 - Schematic of ARALL laminate (courtesy Alcoa).

Fig. 18 - ARALL experimental component on Fokker F-50 aircraft.

5.2 Titanium

Titanium alloys reinforced with continuous ceramic composites also enhance the strength and modulus (Fig. 19) [3]. Here a major concern is the reaction zone formed between the fiber and the matrix which can strongly influence properties such as fatigue [3]. Typical properties of titanium MMC are shown in Table 13.

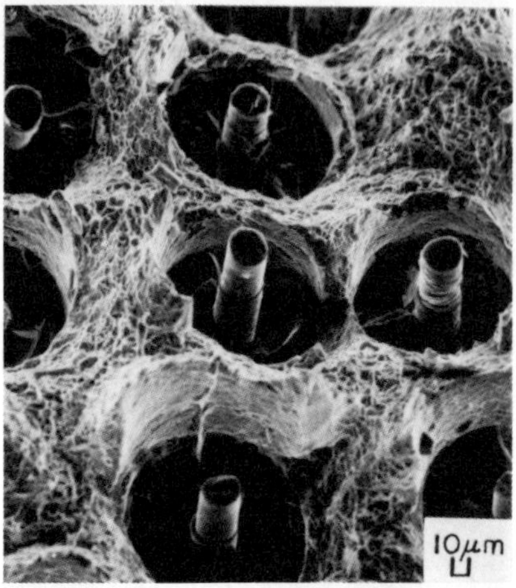

Fig. 19 - Scanning electron microscope micrograph of the fracture surface of a continuous silicon carbide reinforced Ti-6Al-4V matrix, x2000 magnification [3].

TABLE 13. TYPICAL ROOM TEMPERATURE TENSILE PROPERTIES OF TITANIUM METAL MATRIX COMPOSITES [3]

System	UTS, MPa (L)	UTS, MPa (T)	Modulus, GPa (L)
Ti-6Al-4V	890	890	120
B_4C-B/ Ti-6Al-4V	1055	310	205
SCS-6/ Ti-6Al-4V	1455	340	240

5.3 Magnesium

Reinforcement of a magnesium alloy with discontinuous fibers can enhance strength and modulus (Table 14) [18].

TABLE 14. MECHANICAL PROPERTIES OF REINFORCED ZK60A(Mg-5.5Zn-0.5Zn) EXTRUDED ROD[a]

Reinforcement[b]	Volume Fraction, %	YS, MPa	UTS, MPa	El., GPa	E, %
SiC_p	15	330	420	4.7	68
SiC_p	20	370	455	3.9	74
SiC_w	15	450	570	2.0	83
B_4C_p	20	405	490	2.0	83
Baseline, PM	--	260	325	15.0	44

[a]T5 temper.

[b]p = particulate; w = whisker.

The significant increase in strength and modulus compared to the baseline non-reinforced material should lead to use in components requiring these characteristics especially in an environment where corrosion is not a problem (i.e., space).

6. CONCLUSIONS

The challenge presented by the organic matrix composites has accelerated development of improved lightweight metallic materials based on aluminum, titanium, and magnesium. This includes ingot metallurgy, powder metallurgy/rapid solidification, castings, and engineered materials including metal matrix composites.

7. ACKNOWLEDGMENTS

The author would like to acknowledge the assistance of Miss Karen A. Sitzman and Ms Janice S. Halldorson for manuscript preparation. In addition, helpful discussions and material from the following is gratefully noted: P. J. Bania, W. J. Barice, R. C. Benn, P. A. Blenkinsop, H. B. Bomberger, J. Brupbaker, W. Cebulak, L. Christodoulou, S. K. Das, D. Eylon, R. Gardiner, C. L. Harmsworth, A. Johnson, S. Kampe, W. R. Mohn, J. Pickens, W. E. Quist, R. D. Schelleng, D. M. Schuster, A. D. Smith, B. W. Smith, P. R. Smith, and J. Thorne.

8. REFERENCES

1. Phillips, E. H., "McDonnell Emphasizes Thermoplastics in Advanced Composite Research," Aviation Week and Space Technology, February 22, 1988, pp. 57-61.

2. ALCOA Technology Report to the Aerospace Industry, Vol. 7 (April 1988)

3. Froes, F. H., Eylon, D., and Bomberger, H. B., Ed., Titanium Technology: Present Status and Future Trends, Titanium Development Association, 1985.

4. Bania, P. J., J. Metals, 40, No. 3, 20-22 (March 1988).

5. Lipsitt, H. A., "Titanium Aluminides - An Overview," High-Temperature Ordered Intermetallic Alloys, Vol. 39, Ed. C. C. Koch, C. T. Liu, and N. S. Stoloff, Materials Research Society, 1985, pp. 351-364.

6. Froes, F. H., Kim, Y-W., and Hehmann, F., "Rapid Solidification

of Al, Mg and Ti," J. Metals, 39, No. 8, 14-21 (August 1987).

7. Bickerdike, R. L., Clark, D., Hughes, G., McConnell, M. C., Mair, W. N., Partridge, P. G., and Viney, B. W., "The Preparation and Properties of Vapour Deposited Al-Cr-Fe Alloys," Rapidly Solidified Materials, Ed. Peter W. Lee and Robert S. Carbonara, American Society for Metals, 1985, pp. 145-152.

8. Sundaresan, R. and Froes, F. H., "Mechanical Alloying," J. Metals, 39, No. 8, 22-27 (August 1987).

9. Narayanan, G. H., Graham, W. H., Quist, W. E., Wingert, A. L., and Ronald, T. M. F., "Important Considerations in the Production of P/M Aluminum-Lithium Alloys," Processing of Structural Metals by Rapid Solidification, Ed. F. H. Froes and S. J. Savage, ASM International, 1987, pp. 321-336.

10. Rowe, R. G. and Froes, F. H., "Titanium Rapid Solidification - Alloys and Processes," Processing of Structural Metals by Rapid Solidification, Ed. F. H. Froes and S. J. Savage, ASM International, 1987, pp. 163-173.

11. Rowe, R. G., Taub, A. I., and Froes, F. H., "Developments in Rapidly Solidified Nickel and Titanium Intermetallics," presented at the ASM International European Symposium on Advanced Materials and Processing Techniques for Structural Applications, Paris, France, 7-9 September 1987.

12. Harmsworth, C. L., "Damage Tolerant Design Without a Casting Factor," Mechanical Properties of Aluminum Castings, American Foundrymen's Society, Inc., 1987, pp. 83-94.

13. Eylon, D., Barice, W. J., and Froes, F. H., "Beta Titanium Alloy Castings for Demanding Fatigue Applications," Technology for Premium Quality Castings, Ed. E. M. Dunn and D. R. Durham, The Metallurgical Society, 1988, in print.

14. Mohn, W. R. and Vukobratovich, D., "Engineered Metal Matrix Composites for Precision Optical Systems," SAMPE Journal, 26-35 (January/February 1988).

15. Schuster, D. M., private communication, 28 Apr 88, DURAL Aluminum Composites Corporation, San Diego, CA.

16. Verma, S. K. and Dorcic, J. L., "Squeezing Production Costs from Metal-Ceramic Composites," Ad. Mat. and Proc., 133, No. 5, 48-51 (May 1988).

17. Jangg, G., "Dispersion Strengthened Al-Al_4C_3 Materials," Radex-Rundschau, 2/3, (1986).

18. Christodoulou, L., Kampe, S., and Brupbaker, J., private communication, 29 Apr 88, Martin Marietta Laboratories, Baltimore, MD.

19. Pickens, J., private communication, 8 Mar 88, Advanced Composite Materials Corporation, Greer, SC.

9. BIOGRAPHY

F. H. (Sam) Froes, U.S. Air Force Wright Aeronautical Laboratories, Materials Laboratory, Structural Metals Branch (AFWAL/MLLS), Wright-Patterson AFB, OH.

Dr. Froes has been conducting basic and applied research and development programs on structural metals, with emphasis on titanium, for almost 20 years. He received his B.S. in physical metallurgy from Liverpool University, England and subsequently was awarded M.S. and Ph.D. degrees in the same field at Sheffield University, England. He worked for Crucible Steel Company for 11 years in a number of positions culminating with an assignment as manager of the titanium program. In 1978 he joined the U.S. Air Force Materials Laboratory where he is currently Branch Chief of the Structural Metals Branch, Metals and Ceramics Division.

Dr. Froes has authored more than 250 technical publications and has edited five technical books including Titanium Technology: Present Status and Future Trends published by the Titanium Development Association. He is currently Vice-Chairman of the Titanium Committee of The Metallurgical Society and as such will become Chairman of this committee and Co-chairman of the World Titanium Conference when it is held in the United States in 1992. He is the Technical Program Chairman of the present SAMPE Metals Conference. Dr. Froes teaches courses on physical metallurgy, powder metallurgy, rapid solidification, and light metals technology at both the University of Dayton and Wright State University, and for ASM International. Dr. Froes has been an invited technical speaker in several parts of the world including the Soviet Union, India, and Israel. In 1986 he was elected a Fellow of ASM International for his contribution to titanium science and technology and in 1987 received the Ben Gurion medal for his contribution to light metals technology.

2nd International SAMPE Metals Conference
August 2-4, 1988

ADVANCES IN TITANIUM ALLOY CASTING TECHNOLOGY

J. K. Thorne and W. J. Barice
Precision Castparts Corp.
Portland, Oregon, U.S.A.,

Abstract

The state-of-the-art for titanium investment castings has advanced significantly in the last five years. Development of enabling technologies relative to size capability, chemical milling, ceramic cores, and mechanical property improvement has resulted in large, complex, critically stressed aerospace components for static and rotating applications.

1. INTRODUCTION

Recent advances in foundry technology have enabled titanium investment castings to be applied to larger, more complex and critical applications. These technologies include large part size capability, HIP, precision chemical milling, preformed cores, alternate alloys and heat treatments, and rigorous quality and process controls.

Casting programs are near production status for critical airframe structure and net-shape engine fan and compressor rotors. Many others are in development.

Future advancements will be toward wide-spread application of castings as engine airfoil components. Blades and vanes, both separately and integrally cast, and frames with airfoil features, will be supplied near net shape with complex cored passages. Higher temperature service, for both structural and rotating components, will be possible for titanium castings via the newer alloys and titanium-based intermetallics. More airframe critical structures, will be supplied via net-shape castings.

2. AIRFOIL COMPONENTS

Titanium casting technology has advanced in recent years and several fully engine-qualified components, both non-rotating and rotating, are entering production. Procedures for achieving, measuring, and maintaining airfoil dimensional contours and tolerances have been developed. Chemical polishing techniques to achieve as-supplied surface finishes of 40 RMS or better have also been developed.

Rigorous quality controls have been "built in" to the process and confirmed via comprehensive destructive and non-destructive testing.

Figure 1 shows three (3) Ti-6Al-2Sn-4Zr-6Mo alloy integrally-cast rotors 20 cm (8 in.) O.D., for a small gas turbine engine. The airfoil surfaces are supplied net-shape, and machining is required only on the hub and outer diameter of the blade tips.

Comprehensive engine tests have confirmed performance and full scale production is being initiated. These precision cast rotors offer an economic alternative to the conventional method of machining the integral airfoils from a forged block.

Both individually cast vanes and integrally cast vane assemblies are present state-of-the-art. Figure 2 shows a Ti 6Al-4V alloy integrally cast stator for an advanced gas turbine engine application. This one-piece component, has 70 airfoil vanes, is about 91 cm. (36 in.) in diameter and weighs 14 kg. (31 lbs. No machining of the airfoil surfaces is required. Compared to a conventionally fabricated stator assembly, this component economically offers airfoil design flexibility and weight reduction. It is engine qualified and bill-of-material.

The two (2) preceeding examples have shown cast titanium solid airfoils. Hollow airfoils are the next step in the state-of-the-art. They require special preformed core technology. Although not yet in production, hollow individually cast non-rotating airfoils, such as hollow inlet guide vanes, will be in service within one year.

Cast hollow rotating airfoils, Figure 3, are in early development and will require extensive testing and qualification for production utilization. The complex pedestal supports between the airfoil surfaces of these blades make preformed cores a necessity. Design flexibility for wall thickness variation, complex, wide chord shapes, stiffness with low rotating mass, and the potential ease of manufacture as a casting are the primary drivers. Cast hollow airfoil integral components, such as blisks with hollow blades, are the next logical technology advancement.

Fig. 1 - Integrally-Cast Ti-6Al-2Sn-4Zr-6Mo Alloy Rotors

Fig. 2 - Ti 6Al-4V Integrally Cast Stator

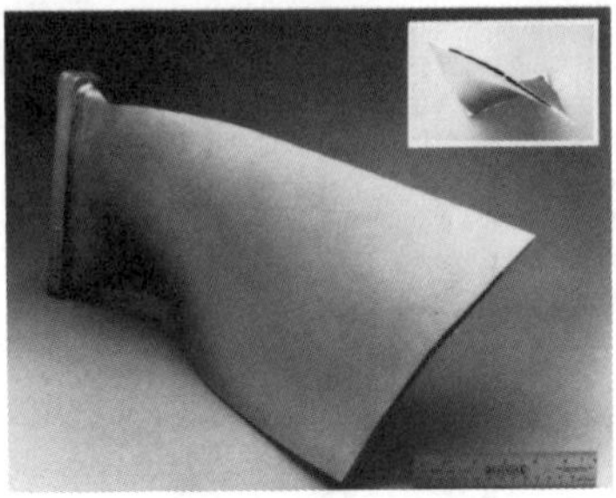

Fig. 3 - Hollow Ti 6Al-4V Cast Fan Blade of Generic Design For Development Program, 56 cm. (22 in.) Long, 5 kg. (12 lbs.)

3. ENGINE STRUCTURAL COMPONENTS

Since the incorporation of HIP in the mid-70's, cast titanium structural components have become widely specified for aircraft gas turbine engines. The state-of-the-art has steadily advanced toward more complex and larger components. Fig. 4 shows a cast actuation ring for the augmentor section on an advanced production engine, with the photo-insert showing the complex cross-section of this ring. The alternative manufacturing technique would include extensive ring-rolling, machining and fabrication.

The fan frame shown in Fig. 5, presently the world's largest production titanium investment casting, was formerly produced via machining and fabrication of 96 separate components. It is produced at more than 30 pieces per month. Compressor cases are also widely specified in cast titanium and range from relatively small cases for helicopter gas turbines to the largest for commercial aircraft engines.

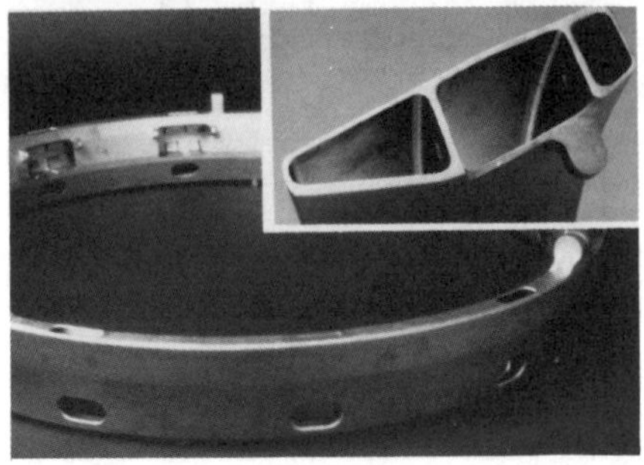

Fig. 4 - Ti 6Al-4V Actuation Ring, 102 cm (40 in.) Dia., 14 kg. (30 lbs.). Insert Shows Ring Cross-Section

Fig. 5 - Ti 6Al-4V Fan Frame 132 cm. (52 in.) Dia., 182 kg. (400 lbs.)

Fig. 6 shows a typical compressor case. The bosses for the variable vanes are cast near-net shape.

Titanium castings are being utilized in critical structures. Fig. 7 shows engine mounts for attachment of a large commercial gas turbine engine to the airframe. These castings are near-net shape and have relatively large section thicknesses.

Fig. 6 - Ti 6Al-2Sn-4Zr-2Mo Compressor Case, 71 cm. (32 in.) Dia., 70 kg. (154 lbs.)

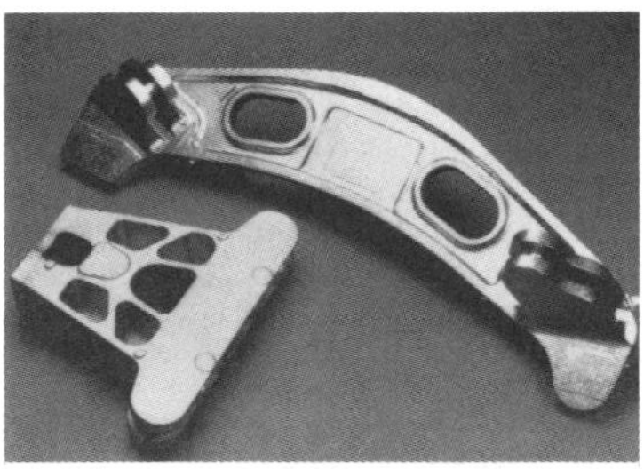

Fig. 7 - Ti 6A-4V Engine Mounts, Largest Is 124 cm. (35 in.) Long, 1.0 in. Max. Thickness, 17 kg. (37 lbs.)

4. AIRFRAME STRUCTURAL COMPONENTS

Although widely used in aircraft engines, titanium investment castings are not routinely specified for critical airframe structure due to "casting factor" design penalties and the absence of widely-accepted design-allowables. This is rapidly changing, due to advancements in foundry processing and controls (1). Several major U.S. organizations have cast material design-allowables programs underway for the Ti 6Al-4V alloy. The interest is driven by the economic benefits of near-net shape, and titanium's excellent compatibility with graphite-composite materials. As a result, airframe structural applications are emerging in both commercial and military aircraft. Figures 8 and 9 show Ti 6Al-4V alloy structural airframe components for a new transport aircraft. Fig. 8 shows one of several longerons, or box-like structural members, in an aircraft nacelle. These are supplied as net-shape castings which are end-welded together to form a 190 cm. (75 in.) long part.

Fig. 9 shows a Ti 6A-4V prototype cascade casting. The integrally-cast airfoil vanes direct the reverse thrust of the engine. Several different cascade components are near term for production utilization.

Critical airframe structures such as wing ribs, and missile body components, are presently under design consideration.

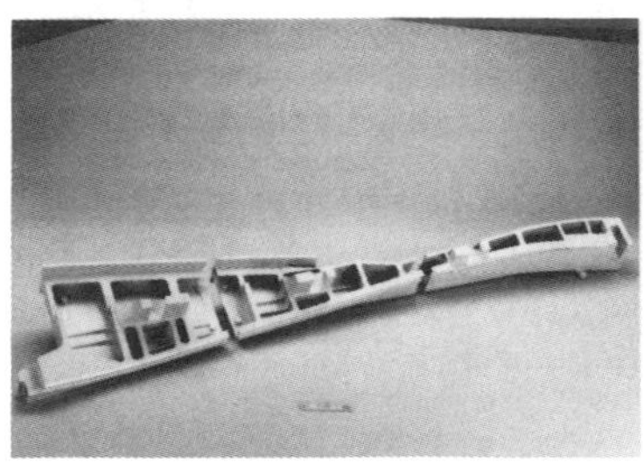

Fig. 8 - Ti 6Al-4V Longeron Sections. 3 Pcs., 63 cm. (25 in.) Long, 10 kg. (23 lbs.) Ea., End-Welded To Overall 190 cm. (75 in.) Length

Fig. 9 - Ti 6Al-4V Cascade 63 cm. (25 in.) Long, 15 kg. (33 lbs.)

5. MECHANICAL PROPERTIES

All of the titanium castings shown are hot isostatically pressed. HIP has become a routine process step in the manufacture of titanium castings for aerospace service. The mechanical property benefits of HIP, particularly in reduction of fatigue life variability, have been well documented. (2) In addition to HIP, some type of vacuum or inert atmosphere heat treatment is utilized in the finishing process. For Ti 6Al-4V alloy, a 1350°F or 1550°F anneal is commonly used to assure elimination of internal stress, providing distortion-free final machining and more consistent tensile and fracture toughness properties. Recent metallurgical development emphasis has been to investigate alternative heat treatments, such as solution treat and age (STA), to achieve microstructural refinement and mechanical properties improvements. For Ti 6Al-4V alloys, solutioning at slightly below or above the beta transus followed by rapid gas quench and age has been investigated. Microstructural characteristics after STA are smaller alpha plate- lets, broken-up structure, and reduced grain boundary alpha phase. (3)

STA heat treatment of cast beta alloys typically shows a much finer microstructure than the near alpha alloys and requires less dependency on the cooling rate between the solution and age temperatures.

Table I summarizes the room temperature tensile properties of cast-plus-HIP titanium alloys. Both mean and minimum values are shown on data for cast-to-size test bars; data for bars excised from parts is dependent on specific geometry.

While alloys Ti 6Al-4V and Ti 6Al-2Sn-4Zr-2Mo exhibit good tensile properties in the annealed condition, the STA heat treatments extend the strength limits of these alloys even further into the regime of wrought titanium and steel alloys. The beta alloys, combined with the proper heat treatment, exhibit tensile strengths up to 195 ksi. The ductility at the higher strength levels is lower but, in most cases, it is still above minimum acceptable levels. It should be noted that the 1880°F STA of Ti 6Al-4V, the 1800°F STA of TI 6Al-2Sn-4Zr-2Mo, and the STA heat treatments of the beta alloys are relatively new. Alternative STA age cycles will be developed to increase ductility, at some compromise in strength.

The beta alloys, such as Ti 3Al-8V-6Cr-4Mo-4Zr (Beta C), are reaching production status as an attractive alternative to Ti 6Al-4V. The good castability, easeof heat treatment, and "deep hardening" characteristics cause these alloys to have excellent producibility in complex and thick-sectioned components. Fig. 10 compares the room temperature specific strengths of cast-plus-HIP, heat-treated Ti 6Al-4V, Ti 6Al-2Sn- 4Zr-2Mo, Ti 6Al-2Sn-4Zr-6Mo, and several beta alloys to those of selected iron, nickel, and aluminum base alloys.

ALLOY	HEAT TREATMENT	UTS ksi (MPa)	0.2% YS ksi (MPa)	ELONG (%)	RA (%)	
Ti-6Al-4V	1350 or 1550F	135 (931)	124 (855)	10	15	mean
	Anneal	130 (896)	120 (827)	6	10	min
	1750F/1h/GFC	141 (972)	131 (903)	8	14	mean
	+1150F/2h	136 (938)	124 (855)	5	8	min
	1880F/0.5h/GFC	149 (1027)	134 (924)	8	14	mean
	+1150F/2h	140 (965)	125 (862)	5	10	min
Ti-6Al-2Sn-4Zr-2Mo	1550F Anneal	146 (1007)	132 (910)	10	21	mean
		139 (958)	124 (855)	6	15	min
	1750/1h/GFC	148 (1020)	132 (910)	11	17	mean
	+1100F/8h	142 (979)	126 (869)	7	11	min
	1800F/1h/GFC	158 (1089)	138 (952)	8	13	mean
	+1100F/8h	152 (1048)	133 (917)	7	10	min
Ti-6Al-2Sn-4Zr-6Mo	1735F/1h/GFC	195 (1345)	184 (1269)	2	3	mean
	+1100F/8h	185 (1276)	177 (1220)	1	1	min
Ti-3Al-8V-6Cr-4Mo-4Zr	1725F/1h/GFC+	193 (1331)	180 (1241)	7	12	mean
	850/12+1000/12	181 (1248)	164 (1131)	5	4	min
Ti-15V-3Al-3Cr-3Sn	1550F/1h/GFC	163 (1124)	151 (1041)	7	13	mean
	+990F/16h	149 (1027)	137 (945)	2	7	min
WROUGHT ALLOYS						
Ti-6Al-4V	Mill Annealed	140 (965)	127 (876)	13	25	
Aluminum 7075	T6	83 (572)	73 (503)	11	16	
Stainless Steel 17-4PH	STA	160 (1103)	145 (1000)	15	55	
Nickel IN718	STA	194 (1338)	167 (1151)	24	35	

TABLE I - Room Temperature Tensile Properties Of Titanium And Other Alloys

As mentioned, the advent of HIP has considerably improved the high cycle fatigue response of cast titanium alloys by healing internal defects. Fig. 11 compares the room temperature high cycle fatigue strength of several cast-plus-HIP heat treated titanium alloys and various wrought alloys. STA heat treated Ti 3Al-8V-6Cr-4Mo-4Zr (Beta C) is particularly promising from an HCF standpoint. The HCF strength of cast-plus-HIP annealed Ti 6Al-4V is lower than the mill annealed counterpart; the STA heat treatment significantly increases the HCF strength, surpassing that of wrought mill annealed material.

Titanium cast-parts for continuous service at 1100°F and higher are presently of interest in gas turbine engine and missile applications. In response, development programs are underway to characterize the latest high temperature titanium alloys in the cast-plus-HIP and heat-treated condition. Intermetallics are also of great interest and much work has been done on titanium aluminides, both alpha-two and gamma. The alpha-two alloy is in semi-production status for castings. Substantial progress can be expected on both the high temperature titanium alloys and on the intermetallics as castings in the next several years.

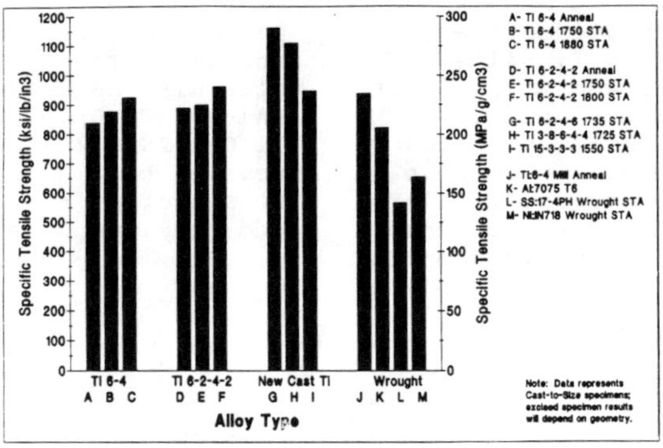

Fig. 10 - Room Temperature Specific Strength Of Titanium Castings And Wrought Alloys

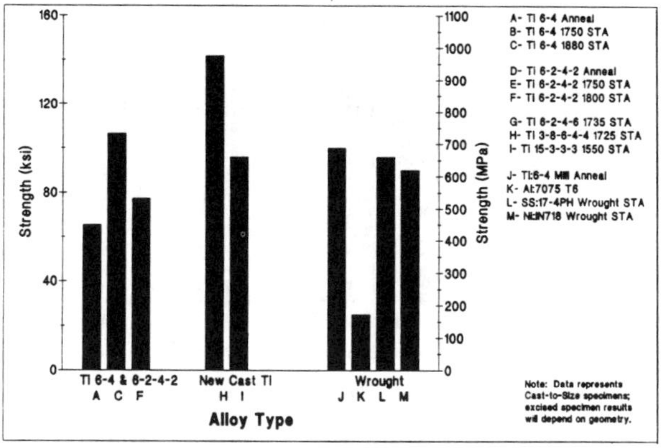

Fig. 11 - Room Temperature Fatigue Strength Of Titanium Casting And Wrought Alloys (At 5×10^6 Cycles, R = 0.1, K_t = 1.0)

6. CONCLUSION

In less than 15 years, titanium precision investment castings have moved from a position of relative insignificance in the overall titanium industry to a broad base of complex, net-shape applications. This is only the beginning. Castings are now at the application forefront for the most critical service requirements such as engine rotating components and primary airframe structure. Design flexibility, net-shape economics, and use of alternate alloys and heat treatments are the cause and this trend will continue.

7. REFERENCES

1. R. Tisler, "Fatigue And Fracture Characteristics Of Ti 6Al-4V HIP'd Investment Castings", 1986 Inter. Conf. On Titanium Pdts. And Applns., San Francisco, Ca., Titanium Development Assoc., Vol. I, pg. 23-41.

2. D. Eylon, F. H. Froes, and R.W. Gardiner, "Developments In Titanium Alloy Casting Technology", J. Of Metals, Vol. 35, No. 2, Feb. 1983, pg. 35-47.

3. D. Eylon, W. J. Barice, and F. H. Froes, "Microstructure Modification Of Ti 6Al-4V Castings", 17th National SAMPE Tech. Conf., Oct. 22-24, 1985, SAMPE, Covina, Ca. pg. 585-595.

8. BIOGRAPHY

<u>Dr. John K. Thorne</u> is a graduate of Cornell University with a B. Met. E. and M.B.A. He completed the Ph.D. in Metallurgical Engineering in 1968 at the University of Michigan. He has been involved with titanium investment casting technology since 1976. As PCC's Technical Director for the Titanium Business Operation, his responsibilities center on advancement of applications and manufacturing technologies for titanium investment castings.

<u>William J. Barice</u> received his BS in Aeronautical Engineering from Embry Riddle University in 1966. He is presently responsible for R&D and Lab Management at PCC's Technical Center. He has more than 20 years experience in manufacturing and evaluation of nickel and titanium base alloy cast components for aerospace applications.

2nd International SAMPE Metals Conference
August 2-4, 1988

EFFECTS OF TREATMENTS ON MECHANICAL PROPERTIES OF TITANIUM ALLOY CASTINGS

D. Eylon[a], F. H. Froes[b], and W. J. Barice[c]

[a]University of Dayton
Graduate Materials Engineering, KL407
Dayton, OH 45469-0001

[b]U.S. Air Force Wright Aeronautical Laboratories
Materials Laboratory, AFWAL/MLLS
Wright-Patterson AFB, OH 45433-6533

[c]Precision Castparts Corp.
4600 S.E. Harney Drive
Portland, OR 97206-0898

Abstract

Refinement of the alpha phase is one of the most effective methods for improving tensile and fatigue properties of alpha+beta titanium alloys. However, microstructure refinement in titanium alloy net-shape products, like castings, requires different approaches than those used for wrought material since work is not applied during net-shape processing. Refinement and modification of the alpha structure in Ti-6Al-4V castings by eight (8) different methods is discussed in this paper and fatigue life and tensile properties are presented and correlated to the modified microstructures. These methods are divided into two major categories; (1) heat treatment approach and (2) temporary alloying with hydrogen. All discussed methods resulted in fatigue life and tensile properties equal or superior to anneal wrought data of the same alloy.

1. INTRODUCTION

Of the major titanium net-shape technologies such as powder metallurgy (PM), casting, and precision forging, casting is by far the most developed and the most widely used. The annual shipment of titanium castings in the United States has increased more than three-fold between 1978 and 1986 [1] making it the fastest growing titanium technology. This is largely due to the fact that hot isostatically pressed (HIP'd), titanium alloy castings display microstructures and properties typical of wrought beta processed products [2, 3].

It is interesting to note that casting is not only the most developed and widely used net-shape technology for metals, but is also the oldest known net-shape technology. It originated at the beginning of the copper age in Asia Minor, Mesopotamia, and the Nile Valley as early as the 5th millennium B.C. [4]. It is even more surprising to note that the lost-wax technique, which is the heart of present day aerospace oriented titanium investment casting technology, was practiced in the Middle East in the 4th millennium B.C. [5].

In recent years efforts have been made to improve the behavior of cast titanium alloy products so that they will qualify for use in critical aerospace applications. This was done by increasing the casting integrity through better control of the melting process, mold materials, alloy chemistry, and by applying HIP'ing [2, 3]. In addition, a substantial amount of work has been done to improve the microstructure of castings for mechanical property improvements with a special emphasis on tensile and high cycle fatigue characteristics.

The objective of this paper is to review and compare the various methods employed in recent years to modify the microstructure and improve properties of titanium alloy cast parts.

2. EXPERIMENTAL

All methods reviewed are related to the modification of Ti-6Al-4V castings. However, the same methods can be successfully applied to treating other alpha+beta or near-alpha alloys. It also should be noted that these methods can be applied to other net-shapes such as PM [6, 7], or even to wrought products. The main effort was in eliminating the continuous grain boundary alpha phase, the large alpha plate colonies, and the individual long alpha plates, typical of slowly cooled titanium alloy castings (Figure 1a). This was accomplished by either solution treatment or temporary alloying with hydrogen. The details of these methods are listed in Table 1; additional information on the processes is available in the references which are also listed in this table. All material evaluated in these works was produced by the investment casting method and was treated after a HIP'ing step.

Typical microstructures of the ABST, BST, BUS, HTH, and CST methods (the full titles of the abbreviations are in Table 1) are shown in Figures 1b-1f. Most tensile and fatigue tests were performed on 5 mm diameter x 22 mm long smooth bar specimens. High cycle fatigue tests were done at R = 0.1, frequency = 5 Hz with triangular waveform.

3. RESULTS AND DISCUSSION

By comparing the cast+HIP microstructure in Figure 1a with some of the treated microstructures (Figures 1b-1f) it can be seen that these treatments were effective in eliminating the large alpha plate colonies and the continuous and thick grain boundary alpha phase. Only the alpha+beta solution treated condition (ABST) (Figure 1b) retained some of the cast+HIP features (Figure 1a) and as a result showed the least improvement in both tensile and fatigue behavior.

All seven reported treatments resulted in improved strength when compared to both cast+HIP and wrought beta annealed baseline conditions (Table 2). The solution treatments (ABST, BST, BUS, and GTEC) resulted in a 6% to 10% increase in strength over cast+HIP level. The hydrogen treatments (CST, HTH, and TCT) resulted in a 10% to 17% increase. This is probably a result of the higher level of refinement of the alpha phase obtained by the hydrogen treatments.

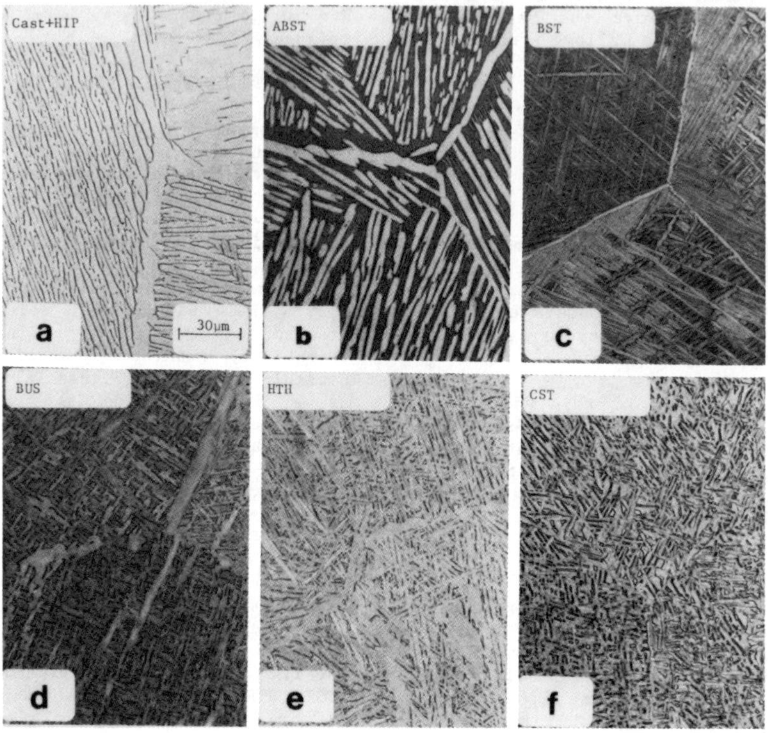

Figure 1 - The microstructure of Ti-6Al-4V castings (original magnifications 600x) (a) Cast+HIP, (b) ABST (alpha+beta solution treatment) treated, (c) BST (beta solution treatment) treated, (d) BUS (broken-up structure) treated, (e) HTH (high temperature hydrogenation) treated, and (f) CST (constitutional solution treatment) treated. (See Table 1 for process details.)

TABLE 1. METHODS FOR MODIFYING THE MICROSTRUCTURE OF ALPHA+BETA TITANIUM ALLOY[a] NET-SHAPE PRODUCTS

Method designation	Solution treatment, °C (°F)/hr	Hydrogenation temperature, °C (°F)	Intermediate treatment, °C (°F)	Dehydrogenation temperature, °C (°F)	Annealing or aging treatment, °C (°F)/hr	Applied to	References
BUS (Broken-up structure)	1040 (1900)/½	--	--	--	815 (1550)/24	Cast, PM, IM	8-11
GTEC (Garrett Turbine Engine Company)	1050 (1925)/½	--	--	--	815 (1550)/½ +705 (1300)/2	Cast	12
BST (Beta solution treatment)	1040 (1900)/½ + GFC[b]	--	--	--	540 (1000)/8	Cast, IM	13
ABST (Alpha-beta solution treatment)	955 (1750)/1 + GFC[b]	--	--	--	540 (1000)/8	Cast, IM	13
HVC (Hydrovac process)	--	650 (1200)	870 (1600) (glass encapsulated)	760 (1400)	--	PM	14, 15
TCT (Thermo-chemical treatment)	1040 (1900)/½	600 (1100)	Cool to RT[c]	760 (1400)	--	Cast, PM, IM	15-17
CST[d] (Constitutional solution treatment)	--	870 (1600)	No intermediate step (continuous process)	815 (1500)	--	Cast	18
HTH (High temperature hydrogenation)	--	900 (1650)	Cool to RT[c]	700 (1300)	--	Cast, PM, IM	19

[a]Most data applies to Ti-6Al-4V. Beta transus temperature approximately 995°C (1825°F).

[b]GFC = gas fan cooled (150°C/min.).

[c]RT = room temperature.

[d]CST is a tradename of Howmet Corporation.

TABLE 2. TENSILE PROPERTIES OF HIP'd AND TREATED Ti-6Al-4V CAST COUPONS

Material condition	YS, MPa	YS, ksi	UTS, MPa	UTS, ksi	El., %	RA, %	UTS increase above cast+HIP, %	References
Cast+HIP	870	126	960	139	10	18	0	11, 13
BUS[a]	940	136	1040	151	8	12	9	11, 13
GTEC[a]	940	136	1025	149	8	11	7	12
BST[a]	930	135	1055	153	9	15	10	13
ABST[a]	930	135	1020	148	8	12	6	13
TCT[a]	1060	153	1120	163	6	9	17	15, 20
CST[a]	985	143	1055	153	8	15	10	18
HTH[a]	1060	153	1100	160	8	15	15	19
Wrought beta anneal	860	125	955	139	9	21	0	21

[a]See Table 1 for process designation and details.

The fatigue life scatterbands of cast and cast+HIP Ti-6Al-4V are compared to that of the wrought annealed alloy in Figure 2. It can be seen that untreated cast+HIP data are well below that of wrought annealed data. The fatigue results of ABST, BST, BUS, CST, GTEC, and HTH treated castings (all HIP'd) are compared in Figure 3 to wrought annealed scatterband data. All treatments resulted in improved fatigue life equal or above levels in the upper scatterband region of the wrought annealed material. The fatigue results from material given hydrogen treatments (CST and HTH) resulted in the highest improvement in fatigue strength. This can also be interpreted on the basis of the highest level of alpha phase refinement and the elimination of the continuous grain boundary alpha phase as evident from Figure 1.

To study the effect of microstructural refinement on fatigue, the ratios S of fatigue strength at 5×10^6 cycles to ultimate tensile strength were calculated in Table 3 and compared to cast+HIP material [13]. All six reported methods displayed higher ratio than the 0.55 typical of cast+HIP material. This is a strong indication that the improvement in fatigue strength is not only the result of increased tensile strength (Table 2), but also the result of microstructure refinement. The fatigue ratio of the modified conditions varies from 0.61 for ABST treated castings to as high as 0.73 for CST treated material. The higher fatigue ratio of the CST treatment is considered to be the result of the higher level of alpha structure refinement and the total elimination of grain boundary alpha phase (Figure 1f).

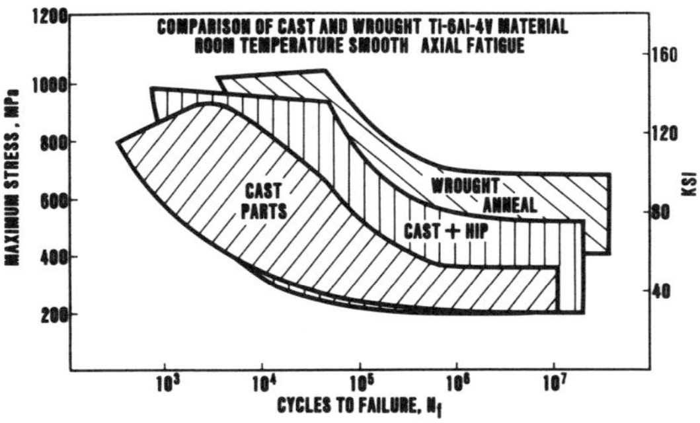

Figure 2 - Comparison of fatigue life scatterbands of Ti-6Al-4V cast, cast+HIP, and wrought annealed products.

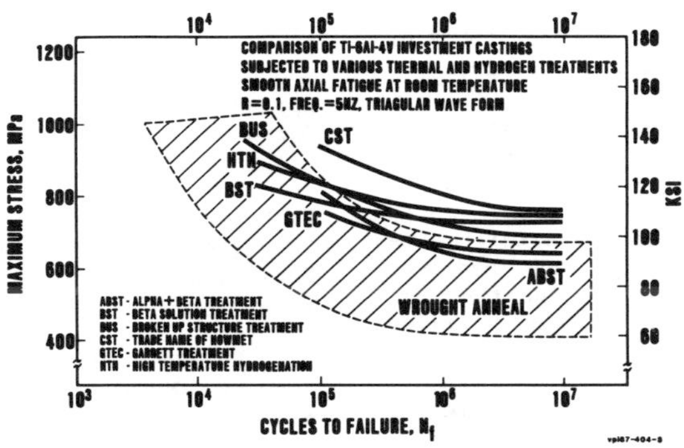

Figure 3 - Smooth axial fatigue life of Ti-6Al-4V HIP'd castings subjected to various thermal and hydrogen treatments (listed in Table 1) compared to the wrought annealed data scatterband.

TABLE 3. RATIO S OF FATIGUE STRENGTH AT 5 x 10^6 CYCLES TO TENSILE STRENGTH

Condition	UTS, MPa	σ_f, MPa	S σ_f/UTS
ABST	1020	620	0.61
BST	1055	730	0.69
BUS	1040	690	0.66
GTEC	1025	650	0.63
CST	1055	770	0.73
HTH	1100	745	0.68
Cast+HIP [13]	958	530	0.55

4. SUMMARY AND CONCLUSIONS

1. Eight methods for modification and refinement of the alpha phase structure in Ti-6Al-4V net-shape castings were reviewed and compared in this paper. These methods were divided into solution treatment and hydrogen treatment approaches.

2. All methods resulted in tensile strengths above cast+HIP or beta annealed wrought material.

3. Six methods were compared in fatigue to wrought annealed alloy. All were found to have fatigue strengths equal or superior to wrought annealed material. The temporary alloying with hydrogen methods were found to be the most effective for improving fatigue life.

4. The improved tensile strength was correlated to the refinement of the alpha phase. The improved fatigue strength was also correlated to this refinement of alpha phase, and in addition was found to relate to the elimination of alpha plate colonies, and the elimination of continuous grain boundary alpha phase.

5. ACKNOWLEDGMENTS

The authors wish to acknowledge Miss Karen A. Sitzman for manuscript preparation.

6. REFERENCES

1. "Titanium 1986, Statistical Review 1978-1986," Annual Report of the Titanium Development Association, Dayton, OH (1987).

2. D. Eylon, F. H. Froes, and R. W. Gardiner: "Developments in Titanium Alloy Casting Technology," Journal of Metals, Vol. 35, No. 2 (February 1983), pp. 35-47; also in Titanium Technology: Present Status and Future Trends, edited by F. H. Froes, D. Eylon, and H. B. Bomberger, Titanium Development Association, Dayton, OH (1985), pp. 35-47.

3. D. Eylon and F. H. Froes: "Titanium Casting - A Review," in Titanium Net Shape Technologies, edited by F. H. Froes and D. Eylon, The Metallurgical Society, Warrendale, PA (1984), pp. 155-178.

4. R. F. Tylecote: A History of Metallurgy, Chapter 2, The Metals Society, London (1979), pp. 5-13.

5. P. Bar-Adon: "The Cave of the Treasure," Israel Exploration Society, Jerusalem (1980), Appendix D by R. Potszkin and K. Bar-Avi.

6. F. H. Froes, D. Eylon, G. E. Eichelman, and H. M. Burte: "Developments in Titanium Powder Metallurgy," Journal of Metals, Vol. 32, No. 2 (1980), pp. 47-54.

7. F. H. Froes and D. Eylon: "Powder Metallurgy of Titanium Alloys - A Review," in Titanium, Science and Technology, Vol. 1, edited by G. Lutjering, U. Zwicker, and W. Bunk, DGM, Oberursel, West Germany (1985), pp. 267-286; also in Powder Metallurgy International (PMI), Vol. 17, No. 4 (1985), pp. 163-167 and continued in Vol. 17, No. 5 (1985), pp. 235-238; also in Titanium Technology: Present Status and Future Trends, edited by F. H. Froes, D. Eylon, and H. B. Bomberger, Titanium Development Association, Dayton, OH (1985), pp. 49-59.

8. D. Eylon and F. H. Froes: "Method for Refining Microstructures of Cast Titanium Articles," U.S. Patent 4,482,398 (November 13, 1984).

9. D. Eylon and F. H. Froes: "Method for Refining Microstructures of Prealloyed Powder Metallurgy Titanium Articles," U.S. Patent 4,534,808 (August 13, 1985).

10. D. Eylon and F. H. Froes: "Method for Refining Microstructure of Blended Elemental Powder Metallurgy Titanium Articles," U.S. Patent 4,536,234 (August 20, 1985).

11. D. Eylon, F. H. Froes, and L. Levin: "Effect of Hot Isostatic Pressing and Heat Treatment on Fatigue Properties of Ti-6Al-4V Castings," in <u>Titanium, Science and Technology</u>, Vol. 1, edited by G. Lutjering, U. Zwicker, and W. Bunk, DGM, Oberursel, West Germany (1985), pp. 179-186.

12. D. L. Ruckle and P. P. Millan: "Method for Heat Treating Cast Titanium Articles to Improve Their Mechanical Properties," U.S. Patent 4,631,092 (December 23, 1986).

13. D. Eylon, W. J. Barice, and F. H. Froes: "Microstructure Modification of Ti-6Al-4V Castings," in <u>Overcoming Material Boundaries</u>, Vol. 17, SAMPE, Covina, CA (1985), pp. 585-595.

14. W. R. Kerr, P. R. Smith, M. E. Rosenblum, F. J. Gurney, Y. R. Mahajan, and L. R. Bidwell: "Hydrogen as an Alloying Element in Titanium (Hydrovac)," in <u>Titanium '80, Science and Technology</u>, edited by H. Kimura and O. Izumi, The Metallurgical Society, Warrendale, PA (1980), pp. 2477-2486.

15. R. G. Vogt, F. H. Froes, D. Eylon, and L. Levin: "Thermo-Chemical Treatment (TCT) of Titanium Alloy Net Shapes," in <u>Titanium Net Shape Technologies</u>, edited by F. H. Froes and D. Eylon, The Metallurgical Society, Warrendale, PA (1984), pp. 145-154.

16. L. Levin, R. G. Vogt, D. Eylon, and F. H. Froes: "Method for Refining Microstructures of Titanium Alloy Castings," U.S. Patent 4,612,066 (September 16, 1986).

17. L. Levin, R. G. Vogt, D. Eylon, and F. H. Froes: "Method for Refining Microstructures of Prealloyed Titanium Powder Compacted Articles," U.S. Patent 4,655,855 (April 7, 1987).

18. R. J. Smickley and L. E. Dardi: "Microstructure Refinement of Cast Titanium," U.S. Patent 4,505,764 (March 19, 1985).

19. C. F. Yolton, D. Eylon, and F. H. Froes: "Microstructure Modification of Titanium Alloy Products by Temporary Alloying with Hydrogen," presented at the Sixth World Conference on Titanium, Cannes, France (June 1988), and to be published in the proceedings.

20. L. Levin, R. G. Vogt, D. Eylon, and F. H. Froes: "Fatigue Resistance Improvement of Ti-6Al-4V by Thermo-Chemical Treatment," in <u>Titanium, Science and Technology</u>, Vol. 4, edited by G. Lutjering, U. Zwicker, and W. Bunk, DGM, Oberursel, West Germany (1985), pp. 2107-2114.

21. G. R. Yoder, L. A. Cooley, and T. W. Crooker: "Fatigue Crack Propagation Resistance of Beta-Annealed Ti-6Al-4V Alloys of Differing Interstitial Oxygen Content," Met. Trans. A, Vol. 9A (1978), pp. 1413-1420.

7. BIOGRAPHY

<u>Daniel Eylon</u> is a professor of metallurgy at the University of Dayton, Graduate Materials Engineering Department. For the past 15 years he was also associated with the U.S. Air Force Materials Laboratory doing research in the area of aerospace materials and powder metallurgy.

<u>F. H. (Sam) Froes</u> received his Ph.D. in physical metallurgy from the University of Sheffield,

England. He is currently Branch Chief, U.S. Air Force Materials Laboratory, Structural Metals Branch, Wright-Patterson AFB, Ohio. His responsibilities include the planning of research and applications related to programs on the light metals aluminum, titanium, magnesium, and beryllium. He is an adjunct professor at the University of Dayton and Wright State University and is Technical Session Chairman of this conference.

William J. Barice received his B.S. in Aeronautical Engineering from Embry Riddle University in 1966. He is presently responsible for R&D and Lab Management at PCC's Technical Center. He has more than 20 years experience in manufacturing and evaluation of nickel and titanium-base alloy cast components for aerospace applications.

2nd International SAMPE Metals Conference
August 2-4, 1988

MICROSTRUCTURE-PROPERTY RELATIONSHIPS IN ADVANCED NICKEL BASE SUPERALLOY AIRFOIL CASTINGS

Alan D. Cetel and David N. Duhl
Pratt & Whitney, East Hartford, CT 06108

Abstract

Microstructural - property relationships in nickel-base superalloys have been of major interest to metallurgists since the initial utilization of these high temperature materials almost forty years ago. Progess has been made in understanding these relationships in both polycrystalline and single crystal alloys. A review of this work will be presented with particular emphasis on the behavior of the advanced high strength single crystal cast airfoil alloys. These alloys contain high volume fractions of the ordered intermetallic Ni_3Al γ' phase which accounts for their unique high temperature strength. The role of γ' size and morphology in controlling the properties of high strength superalloys will be emphasized.

Post casting heat treatment, in particular temperature and cooling rate, can strongly influence the size and distribution of the γ' strengthening particles which in turn affect creep and tensile strength. It is possible to define a heat treatment which produces an optimum γ' size and morphology for an alloy, which maximizes its creep and tensile properties.

Secondary microstructural features including shrinkage porosity and carbides, which are typically present in cast components, significantly affect fatigue life. Efforts have been directed at minimizing or eliminating these potential crack initiation sites. Refractory rich Topologically Close Packed (TCP) and other secondary phase instabilities which can precipitate in heavily alloyed compositions during long time exposures, have been found to adversely affect low temperature fatigue and high temperature creep strength. Alloy development efforts have focused on designing compositions to be microstructurally stable avoiding the formation of these deleterious phases.

1. INTRODUCTION

The stringent demands placed on turbine airfoils in modern gas turbine engines has led to the development of a class of cast, high strength, nickel-base superalloys which are required to operate at ever increasing metal temperatures. To meet these accelerating demands on high temperature properties, significant advances in alloy compositional design and casting technology have led to the development of a series of gamma prime strengthened, directionally solidified alloys. These advanced, columnar grain and single crystal alloys contain high volume fractions of the ordered intermetallic Ni_3Al phase, referred to as gamma prime (γ'), which accounts for their unique high temperature strength.

While some of these high strength alloys were developed in the 1960's and have been in service for more than 15 years, a clear understanding of the relationship between their microstructure and properties has begun to emerge only recently (1-6). To attain the maximum strength potential of these advanced alloys, microstructural control is essential. This paper will review what has been learned about the microstructural property relationships in these alloys over the past two decades.

2. EFFECT OF GAMMA PRIME MORPHOLOGY ON PROPERTIES

2.1 Creep Strength

Since the advanced high strength nickel-base alloys owe their exceptional high temperature properties to the high volume fraction of the ordered γ' phase they contain, it should not be surprising that controlling precipitate distribution and morphology can profoundly affect properties. Post casting processing of these alloys, especially solution heat treatment, can radically affect microstructure. Although directionally solidified alloys had been routinely solution heat treated since their inception in the early 1960's, the role of this process in achieving increases in creep strength was not well understood.

The high strength alloys typically contain 55-70 volume percent of the γ' precipitates which in the as-cast condition are coarse (.4-1.0μ) and irrequlary shaped cuboidal particles (Figure 1). Solution heat treatment dissolves this coarse γ' and upon cooling reprecipitates it in a finer more uniform distribution (Figure 2).

The highly alloyed nickel-base alloys solidify dendritically and due to chemical segregation affects across the dendrites, higher concentrations of the γ' forming elements such as aluminum and titanium are present in interdendritic areas than in the dendrite core. This results in the γ' solvus (temperature at which γ' first precipitates upon cooling) being lower in the core region than interdendritically. For a high volume fraction γ' single crystal alloy, such as PWA 1480, some solutioning is evident as low as 1260°C, while full solutioning occurs at 1290°C (Figure 3). Although PWA 1480 was the subject of many of the Pratt & Whitney (P&W) studies presented in this paper, the conclusions drawn with PWA 1480 apply equally well to a wide range of high strength alloys.

Varying the cooling rate from the solution heat treatment temperature can significantly affect γ' particle size as rapid rates do not allow sufficient time for the particles to coarsen as they precipitate upon cooling below the γ' solvus temperature. Increasing the cooling rate from the solution heat treatment temperature for PWA 1480 from 30 to 120°C/minute results in an average particle size refinement of more than 30% (Figure 4). By controlling both the solution heat treatment temperature and cooling rate, both the volume fraction of fine particles as well as their size can be controlled. Heat treating an alloy close to its γ' solvus temperature, completely solutioning its coarse γ' particles, can produce consistently high elevated temperature creep-rupture strength.

For an alloy such as PWA 1480 where all the coarse γ' is solutioned at 1290°C (Figure 5) rupture lives at 980°C/248 MPa of 80 hours are achieved. Reducing the solution heat treatment temperature by 30°C, however, can result in a 45% reduction in rupture life (Figure 5). Work done at P&W has demonstrated that rupture life is directly related to the volume fraction of coarse γ' taken into solution (Figure 6).

While undersolutioning can reduce creep strength, an even greater reduction can result from heat treating an alloy above its incipient melting temperature (Figure 5).

Severe incipient melting (Figure 7) can reduce creep lives at 980°C by more than 60% and thus must be avoided by proper temperature control. Melted regions which do

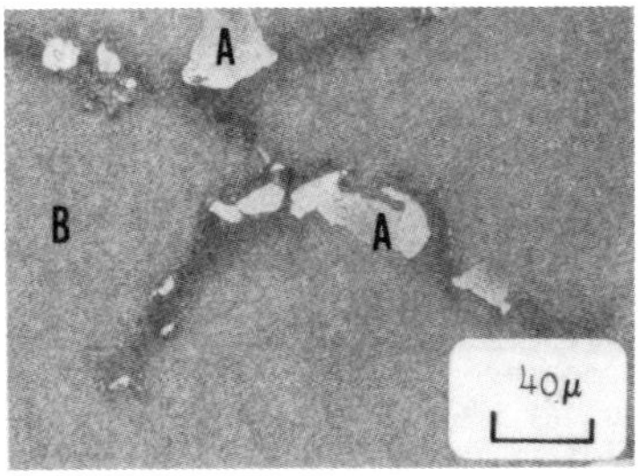

GAMMA/GAMMA PRIME EUTECTICS (A) AND UNRESOLVED GAMMA PRIME PRECIPITATE (B) IN GAMMA MATRIX

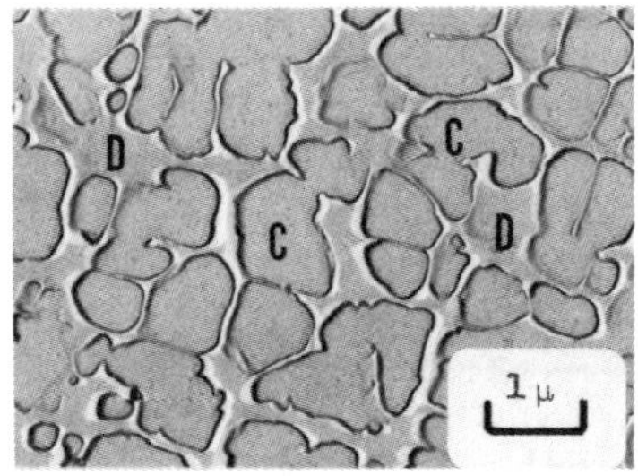

COARSE GAMMA PRIME PRECIPITATES (C) IN GAMMA MATRIX (D)

FIGURE 1. TYPICAL AS-CAST MICROSTRUCTURE OF HIGH STRENGTH NICKEL-BASE SUPERALLOY

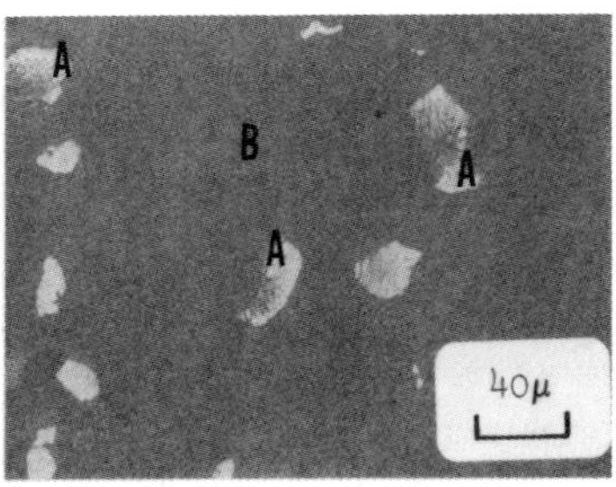

GAMMA/GAMMA PRIME EUTECTICS (A) AND UNRESOLVED GAMMA PRIME PRECIPITATE (B) IN GAMMA MATRIX

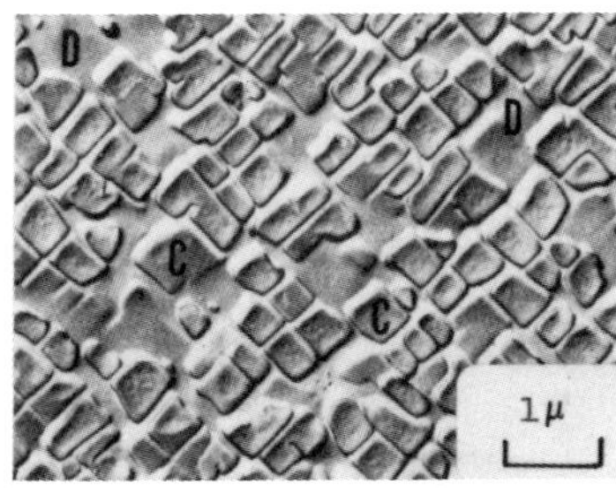

FINE GAMMA PRIME PRECIPITATES (C) IN GAMMA MATRIX (D)

FIGURE 2. TYPICAL FULLY SOLUTIONED MICROSTRUCTURE OF HIGH STRENGTH NICKEL-BASE SUPERALLOY

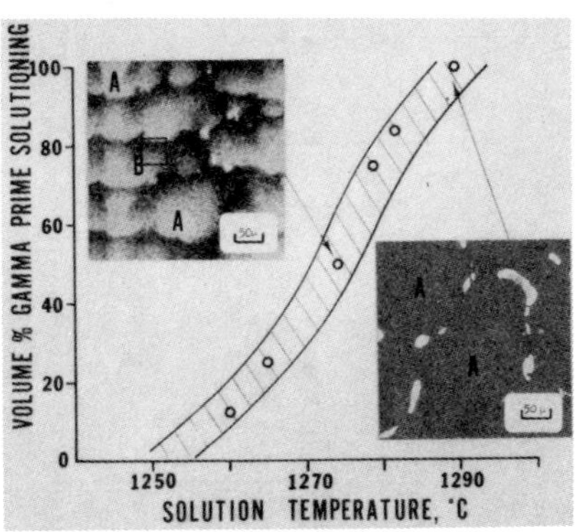

FIGURE 3. AMOUNT OF GAMMA PRIME SOLUTIONED IS RELATED TO HEAT TREATMENT TEMPERATURE, SOLUTIONED REGION (A), COARSE NON SOLUTIONED REGION (B)

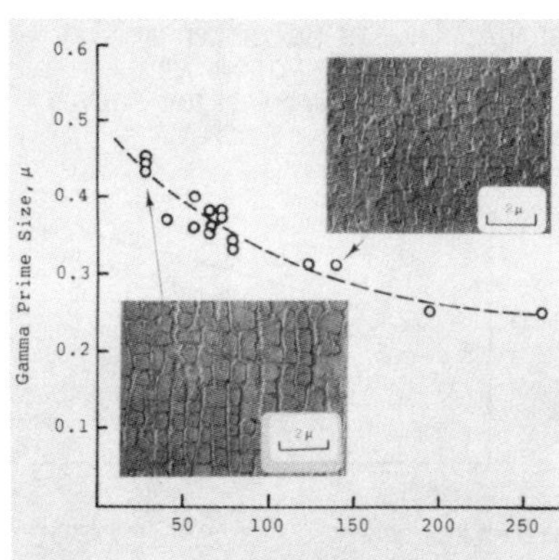

FIGURE 4. RAPID COOLING AFTER HEAT TREATMENT REFINES GAMMA PRIME SIZE

not retain the microstructure of the alloy solidified under controlled conditions, and can contain voids, are sites for early crack initiation.

In addition to achieving complete γ' solutioning, the size and morphology of the resulting particles must also be carefully controlled to achieve the full strength potential of the alloy. Work done by P&W (6), ONERA (7) and NASA (8) has shown that an optimum γ' size must be produced to maximize creep strength in the 760 - 1100°C temperature range. Caron and Kahn (7) have shown that for CMSX-2 the optimum γ' size is .35 - 60μ (Figure 8). Work performed by Nathal et al (8) indicates that the optimum γ' size for an alloy is

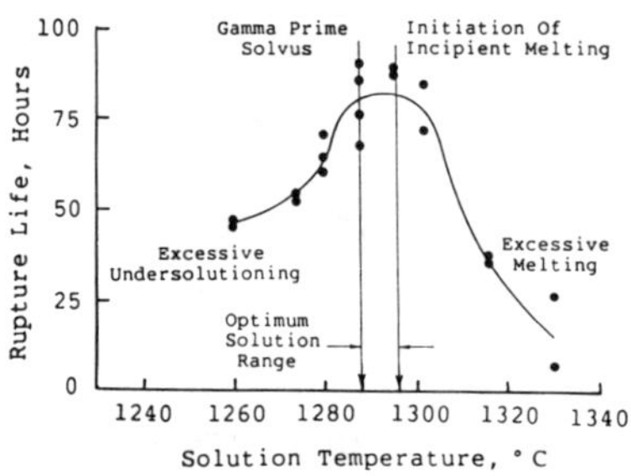

FIGURE 5. CREEP-RUPTURE STRENGTH AT 980°C/248MPa IS MAXIMIZED BY SOLUTION HEAT TREATMENT AT OR SLIGHTLY ABOVE GAMMA PRIME SOLVUS TEMPERATURE

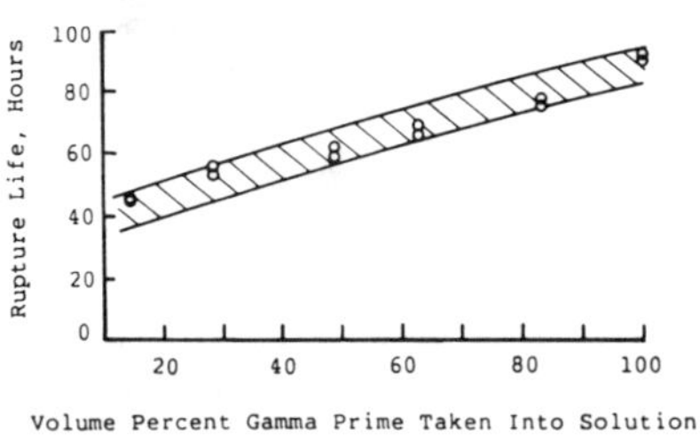

FIGURE 6. CREEP-RUPTURE STRENGTH OF PWA 1480 AT 980°C/248MPa IS DIRECTLY RELATED TO AMOUNT OF GAMMA PRIME SOLUTIONED

dependent on the lattice mismatch between the γ and γ' phases (Figure 9), which is composition dependent. Alloys such as PWA 1480 and CMSX-2 which display an intermediate level of mismatch thus exhibit maximum creep lives over a limited range of γ' sizes.

To achieve the optimum size and morphology of the γ' particles for alloys such as CMSX-2, Caron & Kahn (9) have found that post solution heat treatment cycles between 1050 and 1100°C are effective. These heat treatment cycles coarsen the

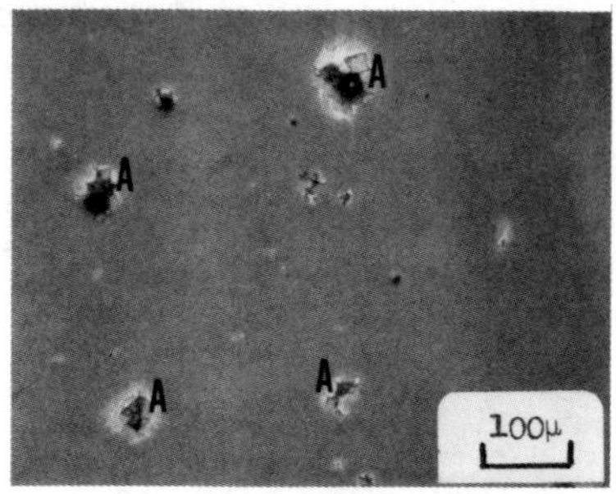

FIGURE 7. SOLUTION HEAT TREATING TOO FAR ABOVE GAMMA PRIME SOLVUS CAN LEAD TO EXCESSIVE INCIPIENT MELTING (A)

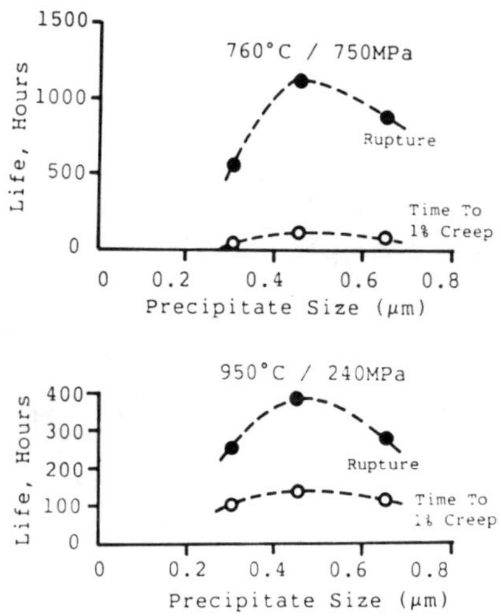

FIGURE 8. CREEP-RUPTURE STRENGTH IS MAXIMIZED BY HEAT TREATING TO PRODUCES AN OPTIMUM GAMMA PRIME SIZE (FROM REF. 7)

very fine γ' particles (≤.3μ), that result during cooling from the solution heat treatment temperature, to an optimum size of ≈.5μ. These heat treatments have also been shown to produce a more uniform array of cuboidal particles which provides an additional benefit to creep strength. Work done on PWA 1480 has shown that by achieving a rapid (≥65°C/minute) cooling rate from the solution heat treatment temperature, and employing a 1080°C post solution heat treatment, an optimum γ' size (.35 - .45μ) and distribution can be achieved (Figure 10) which maximizes creep strength.

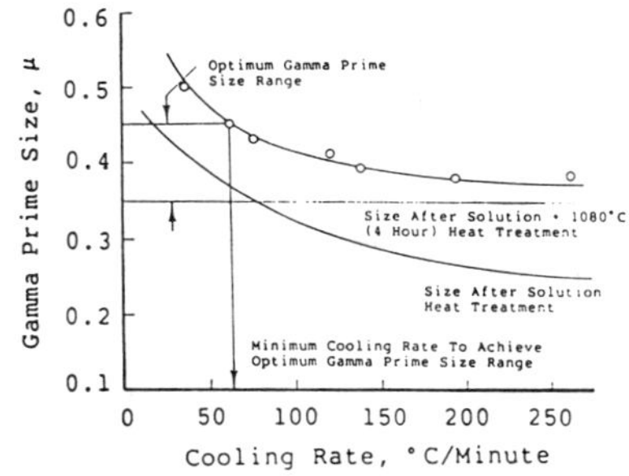

FIGURE 10. OPTIMUM GAMMA PRIME SIZE ACHIEVED BY RAPID COOL FROM SOLUTION TEMPERATURE COMBINED WITH POST SOLUTION HEAT TREATMENT

Testing at 980°C (Figure 11) has shown that rupture lives greater than 75 hours can be consistently attained by achieving cooling rates in excess of 65°C/minute. Cooling at rates greater than this, however, are not effective in further increasing rupture life, since the optimum γ' size range is attained by rapid cooling at 65°C/minute and faster cooling reduces the γ' size below the optimum needed for maximum creep strength.

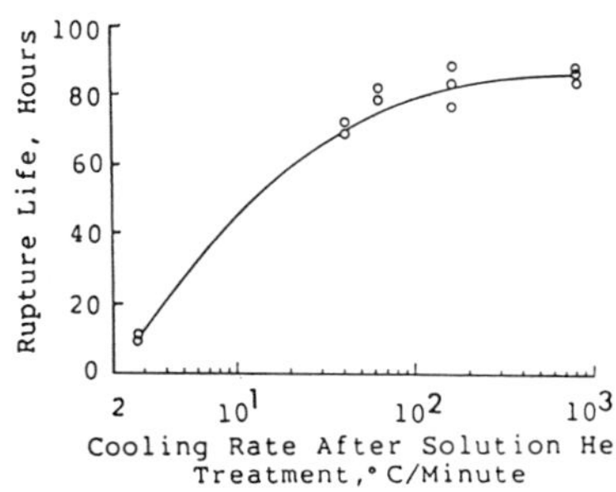

FIGURE 11. CREEP-RUPTURE STRENGTH AT 980°C/248MPa IS SIGNIFICANTLY INCREASED BY RAPID COOLING FROM THE SOLUTION HEAT TREATMENT TEMPERATURE

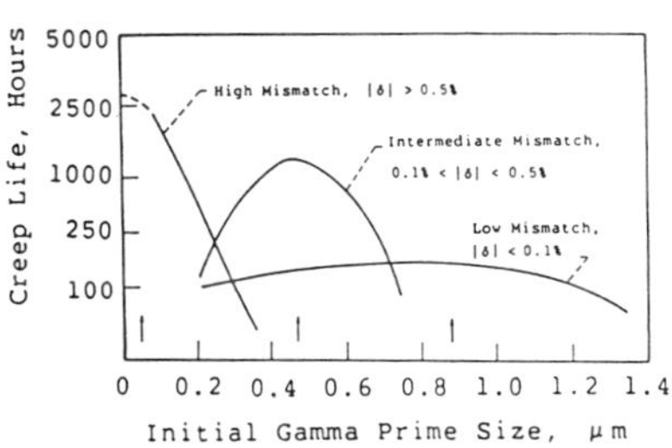

FIGURE 9. OPTIMUM GAMMA PRIME SIZE TO MAXIMIZE CREEP STRENGTH IS DEPENDENT ON MISMATCH BETWEEN GAMMA AND GAMMA PRIME PHASES (FROM REF. 8)

2.2 Intermediate Temperature Yield Strength

The yield strength of cast,high volume fraction γ' nickel-base alloys is relatively insensitive to temperatures below 750°C. Turbine blade attachments operate at intermediate temperatures around 600°C where the yield strength is a parameter of interest. At temperatures below 750°C, the yield strength of advanced cast nickel-base alloys have been found to be significantly affected by post casting heat treatment conditions. Studies conducted at P&W have shown that the 593°C yield strength is maximized by completely solutioning all the coarse γ' precipitate in the alloy (Figure 12). Solutioning below the γ' solvus temperature results in excessive coarse γ' which adversely affects yield strength. Heat treating significantly above the incipient melting temperature degrades yield strength through the creation of locally melted areas (Figure 13).

A study conducted on the effect of cooling rate from the solution heat treatment temperature has shown that 593°C yield strength increases continuously with cooling rate (Figure 14). The yield strength of PWA 1480 at 593°C is raised by more than 250 MPa through an increase in cooling rate from 3 to 65°C/minute. This rate increase refines γ' size as previously shown in Figure 10 and results in increased yield strength due to refinement of the γ' size (Figure 15). As opposed to the results previously presented on the effect of γ' size on creep-rupture strength, intermediate temperature yield strength is inversely related to γ' size with no peak in strength being evident. Reducing γ' size from .8μ (3°C/minute) to .3μ (550°C/minute) by increased cooling rates results in an improvement of more than 350 MPa in the yield strength of PWA 1480 at 593°C.

Further increases in the yield strength of many advanced alloys have been achieved with low temperature (700 - 870°C) aging heat treatments. These aging heat treatments, applied after post solution heat treatments (1050 - 1100°C) employed to maximize creep-rupture strength, result in the additional precipitation of a small amount of ultra fine γ' particles (≤.1μ in size) between the existing larger γ' particles (Figure 16). This precipitation of ultra fine γ' can result in a significant increase in yield strength at 593°C (Figure 17).

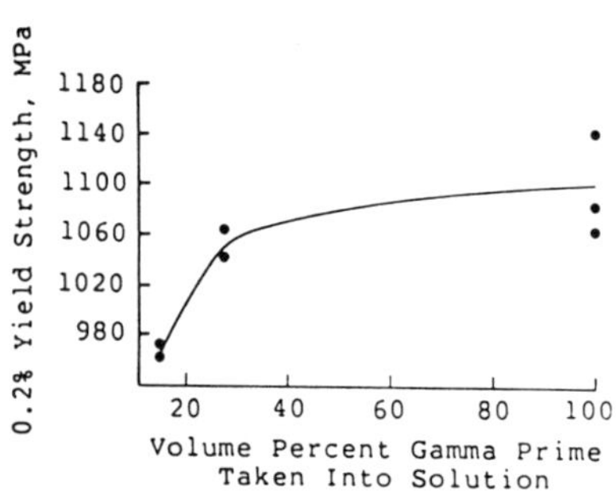

FIGURE 12. YIELD STRENGTH AT 593°C IS RELATED TO GAMMA PRIME SOLUTIONED

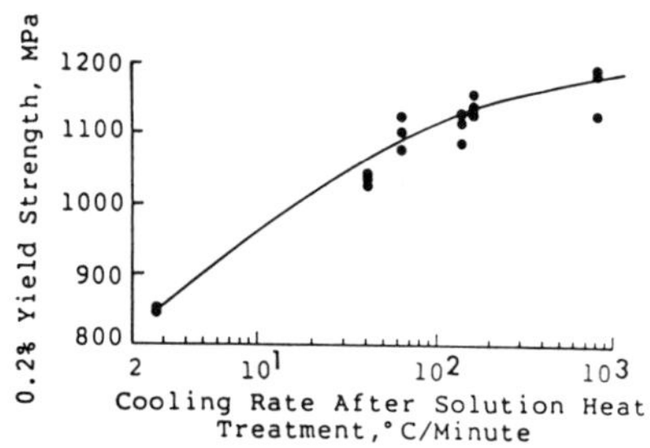

FIGURE 14. YIELD STRENGTH AT 593°C IS SIGNIFICANTLY INCREASED BY RAPID COOLING AFTER SOLUTION HEAT TREATMENT

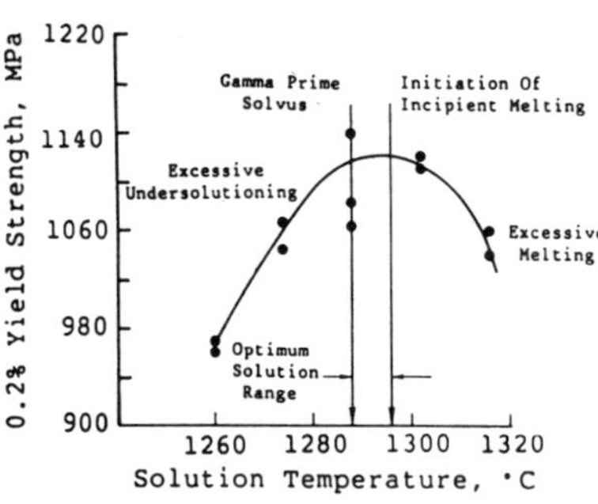

FIGURE 13. YIELD STRENGTH OF PWA 1480 AT 593°C IS MAXIMIZED BY SOLUTION HEAT TREATING AT OR SLIGHTLY ABOVE THE GAMMA PRIME SOLVUS TEMPERATURE

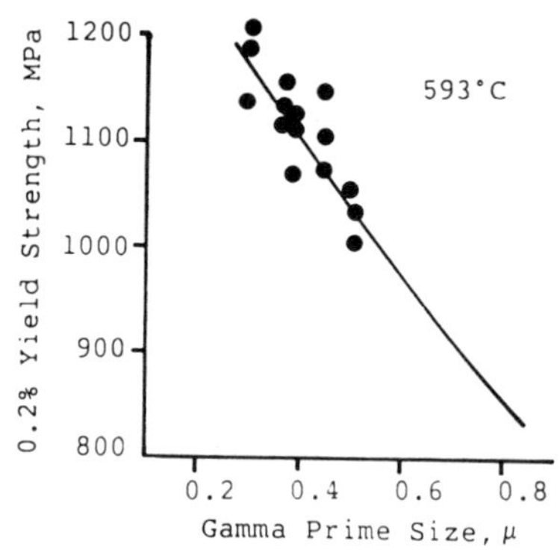

FIGURE 15. YIELD STRENGTH AT 593°C IS SIGNIFICANTLY INCREASED BY REFINING GAMMA PRIME SIZE

2.3 Effect of Post Heat Treatment Thermal Exposure

Although a uniform distribution of fine cuboidal γ' particles can be achieved in the as-heat treated condition, any additional elevated temperature processing or service exposure will tend to coarsen the γ' precipitates. The coarsening is temperature and time dependent and can cause significant growth of the γ' precipitates at exposure temperatures exceeding 1100°C (Figure 18). Excessive γ' coarsening produced by such elevated temperature thermal exposures can degrade yield strength (Figure 19). Severe thermal exposures can also degrade creep-rupture strength. Creep-rupture life at 871°C can be reduced by more than 30% through exposures which produce γ' sizes greater than .7μ (Figure 20).

The effects of thermal exposure on γ' morphology discussed here have all been based on unstressed aging. Work by Nathal (8) and Tien (10) has shown, however, that thermal exposures conducted under stress can completely change γ' morphology (Figure 21). Directional coarsening of particles occurs with lamellar or rafts produced which are arranged perpendicular to the applied tensile stress axis during creep exposure. The degree of alignment of the rafts was found to be related to the uniformity of the initial γ' particle distribution. The thickness of the γ' rafts, which develops during creep exposure, was shown to be very close to the initial γ' particle size.

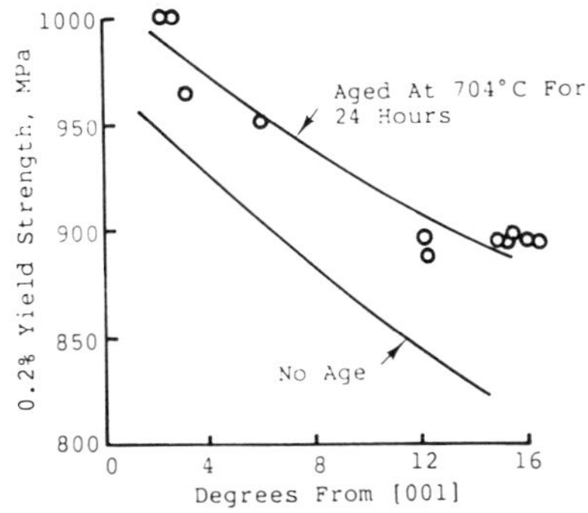

FIGURE 17. A LOW TEMPERATURE AGING HEAT TREATMENT SIGNIFICANTLY INCREASES YIELD STRENGTH AT 593°C

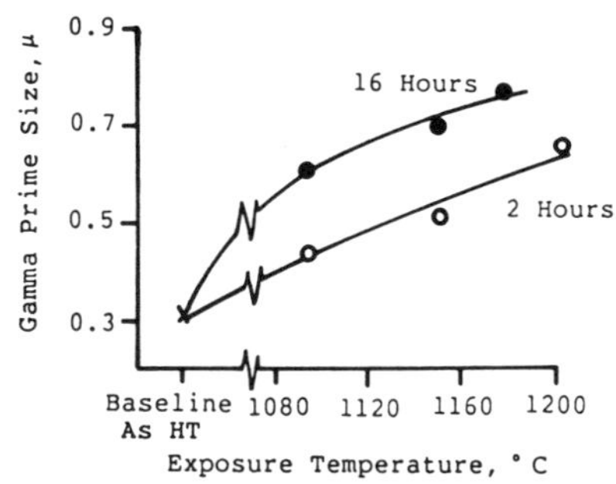

FIGURE 18. ELEVATED TEMPERATURE CAN SIGNIFICANTLY COARSEN GAMMA PRIME SIZE

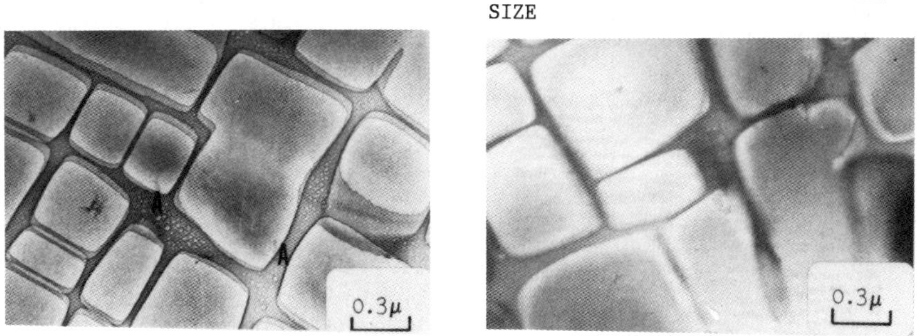

704°C (1300°F) FOR 24 HOURS **NO AGE**

FIGURE 16. LOW TEMPERATURE (LESS THAN OR EQUAL TO 870°C) AGING HEAT TREATMENT CAN PRECIPITATE AN ADDITIONAL SMALL VOLUME FRACTION OF ULTRA FINE GAMMA PRIME PRECIPITATES (A)

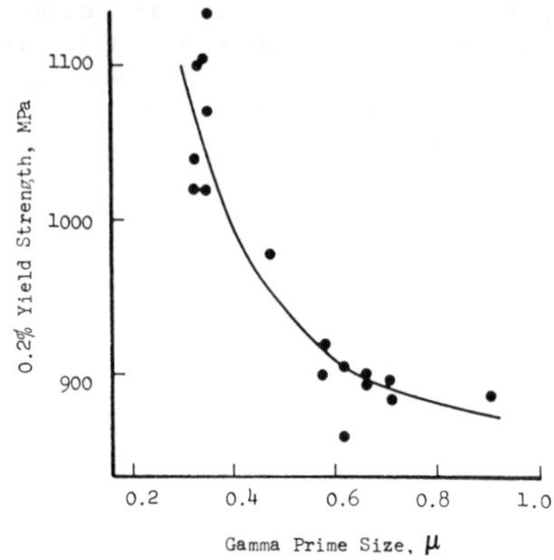

FIGURE 19. YIELD STRENGTH AT 593°C IS REDUCED BY THERMAL EXPOSURE WHICH SIGNIFICANTLY COARSENS GAMMA PRIME SIZE

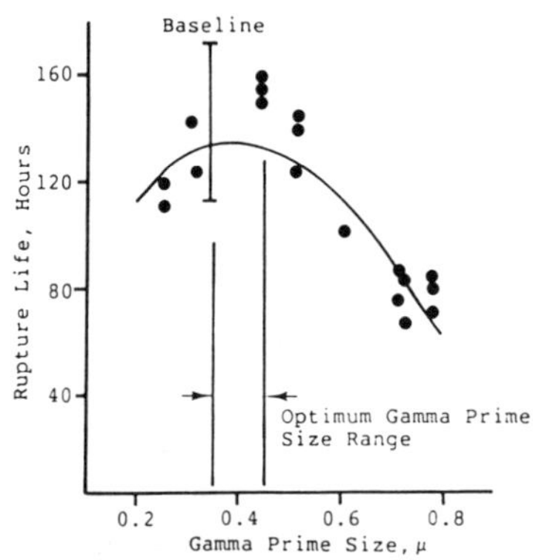

FIGURE 20. CREEP-RUPTURE LIFE AT 871°C AND 483 MPa IS REDUCED BY THERMAL EXPOSURE WHICH COARSENS GAMMA PRIME SIZE SIGNIFICANTLY BEYOND 0.45μ.

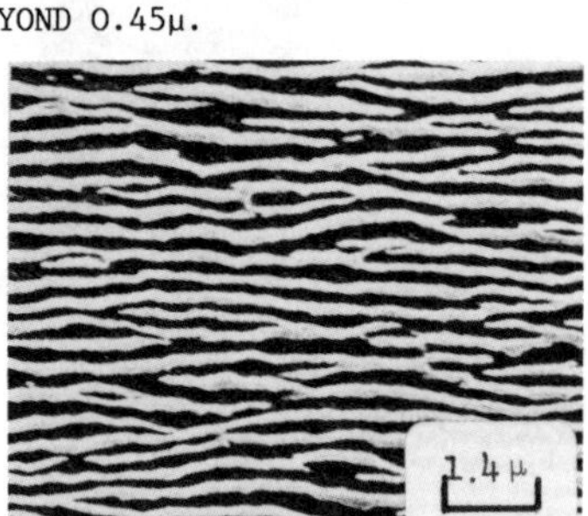

FIGURE 21. DIRECTIONAL COARSENING (RAFTING) PRODUCED IN EXPERIMENTAL ALLOY THROUGH CREEP-EXPOSURE AT 1024°C/207 MPa

3. EFFECT OF SECONDARY MICRO-STRUCTURAL FEATURES ON PROPERTIES

3.1 Porosity and Carbides

Advanced nickel-base superalloys commonly contain microstructural features such as casting shrinkage porosity and carbides (Figure 22) which can act as fatigue initiation sites. The larger the size of these pores or carbides the greater the stress concentration they present and the greater their impact on fatigue life (11). Polycrystalline and columnar grain alloys contain carbon additions to help improve grain-boundary strength and ductility. While carbon additions are beneficial for gain boundary ductility, the large carbides that form can adversely affect fatigue life. Both low and high cycle fatigue cracking have been observed to initiate at large (length greater than .05mm) carbides (Figure 23) present in these alloys (11). When polycrystalline alloys were cast in single crystal form it was determined that carbides did not impart any beneficial strengthening effects in the absence of grain boundaries, and thus could be eliminated (11). Producing essentially carbon-free single crystal alloys led to significant improvements in fatigue life as large carbide colonies were no longer present to initiate fatigue cracks (Figure 24).

Microporosity which is present in various amounts in all castings results from the shrinkage liquid metal undergoes as it solidifies. This leaves small voids unless additional liquid metal is supplied to fill these areas before they freeze. Cracks in both low and high cycle fatigue can initiate at pores as shown by scanning electron microscope (SEM) photos in Figure 25. In carbon-free alloys, pores have been determined to be the prime sites for fatigue crack initiation. In alloys containing carbides, which are of similar size to the pores they contain, fatigue cracking from both types of sites is likely. Distributions of porosity vary widely between castings and within a casting. Fatigue testing at P&W has shown that in extreme cases life can vary by more than 10X between relatively sound castings and those containing large amounts of shrinkage, although still at levels below detectibility limits by x-ray or other nondestructive inpection techniques.

Since some degree of porosity will likely be present in all components cast under even the most carefully controlled conditions, elimination of porosity can best be accomplished through post casting hot isostatic pressing (HIP) processing. This process collapses and closes pores through the simultaneous application of heat and hydrostatic pressure to a part in a pressure vessel under an inert gas atmosphere. By completely eliminating all pores in a carbide free single crystal alloy, fatigue life improvements greater than 4X have been demonstrated (Figure 26) even when compared to sound castings containing fine pores (≤.05mm diameter). This life improvement is a result of the elimination of all common defect sites in the alloy making fatigue initiation much more difficult.

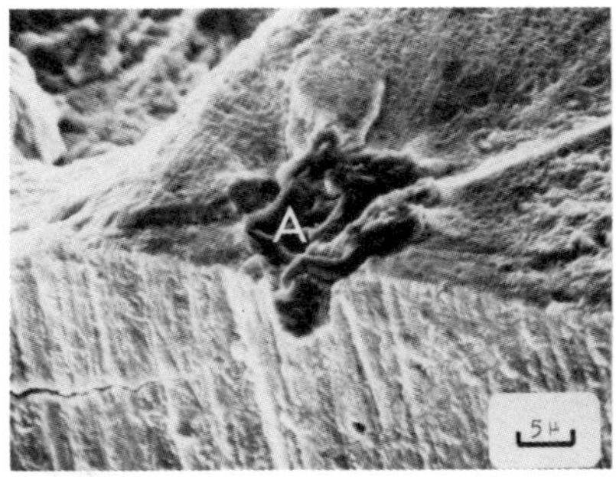

FIGURE 23. LARGE CARBIDES (A) CAN ACT AS FATIGUE INITIATION SITES

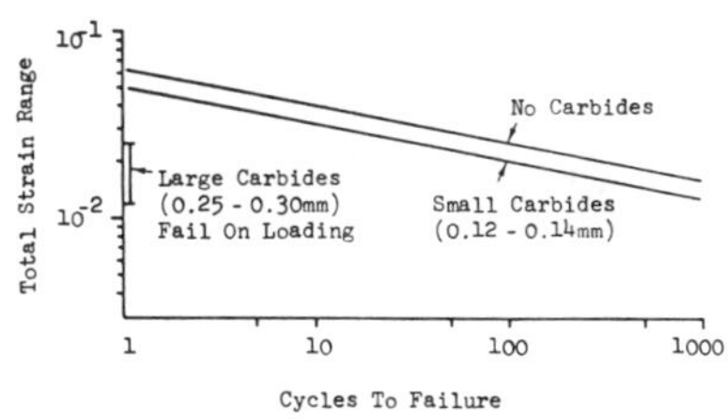

FIGURE 24. FATIGUE LIFE OF MAR-M200 IS SIGNIFICANTLY IMPROVED AT ROOM TEMPERATURE BY ELIMINATING CARBIDES (FROM REF. 11)

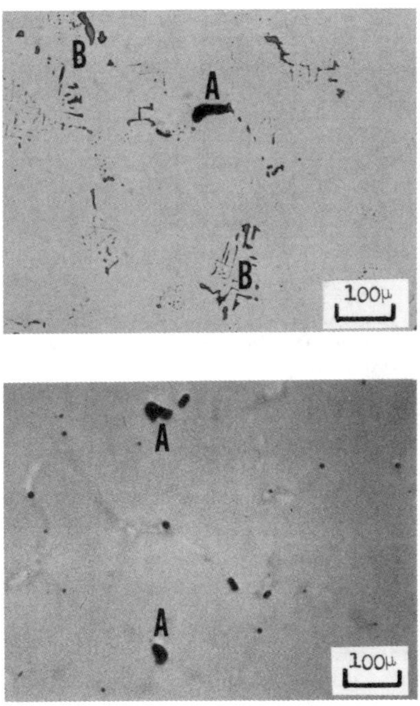

FIGURE 22. COLUMNAR GRAIN CASTINGS (TOP) TYPICALLY CONTAIN POROSITY (A) AND CARBIDES (B) WHILE CARBON-FREE SINGLE CRYSTAL CASTINGS (BOTTOM) TYPICALLY CONTAIN POROSITY (A) ONLY WHICH CAN ACT AS FATIGUE INITIATION SITES

3.2 Phase Instabilities

Modern superalloy compositions which are highly alloyed, containing large amounts of the refractory elements chromium, molybdenium, tungsten and rhenium, have been designed to maintain microstructural stability during long-time service operation. However, unless strict restrictions are placed on their compositions a series of TCP or other refractory rich secondary phases can precipitate in these alloys when they are thermally exposed for long times at temperatures greater than 760°C. These phases which include sigma (σ)., mu (μ), alpha tungsten and

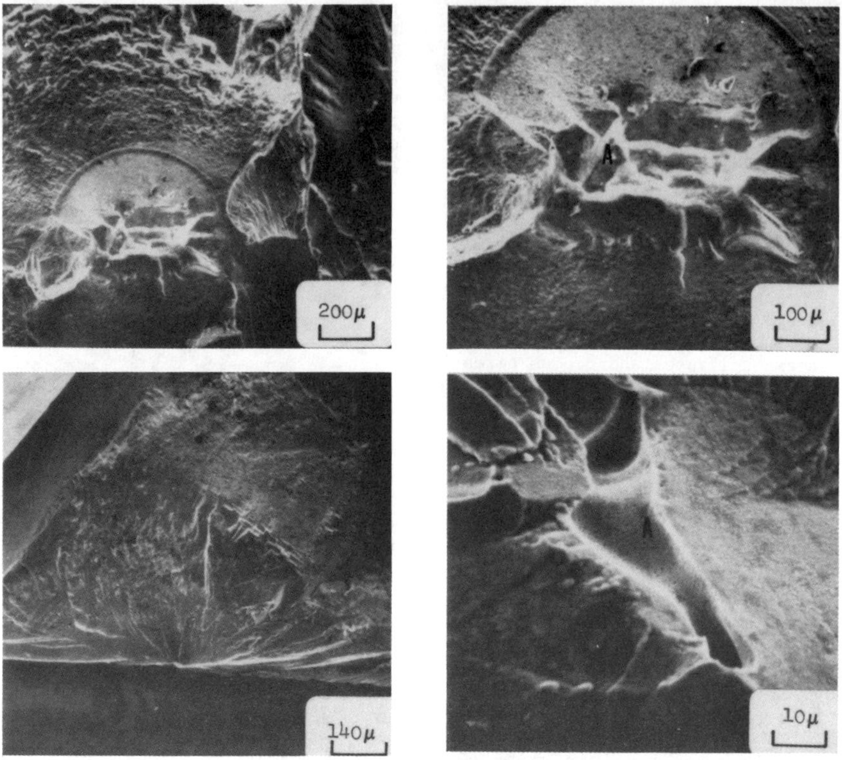

FIGURE 25. LOW CYCLE (TOP PHOTOGRAPHS) AND HIGH CYCLE FATIGUE (BOTTOM PHOTOGRAPHS) CRACKING INITIATES AT MICROPOROSITY (A)

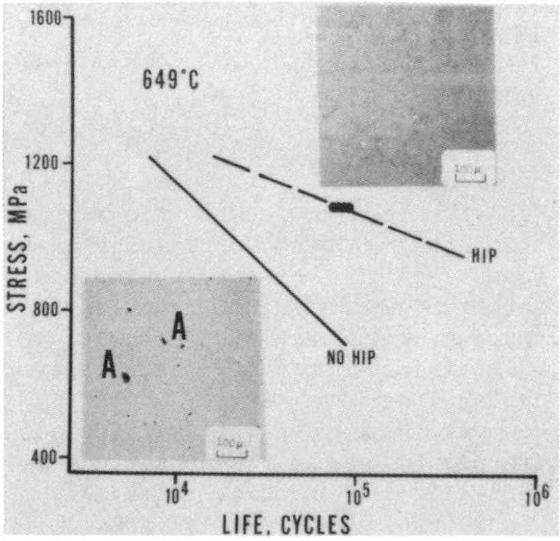

FIGURE 26. HOT ISOSTATIC PRESSING (HIP) SIGNIFICANTLY IMPROVES LOW CYCLE FATIGUE LIFE AT 649°C BY ELIMINATING POROSITY (A)

alpha chromium, commonly precipitate in acicular plate-like morphologies (Figure 27).

Data is available from a variety of sources (12 and 13) showing that these instabilities are to be avoided as they can degrade both high temperature creep strength as well as lower temperature fatigue strength in long-time service. The platelike morphology, incoherency and hardness of these phases make them prime sites for low temperature crack initiation, adversely affecting fatigue capability. The high refractory content of the instabilities depletes the nickel- rich matrix of its most important solid solution strengthening elements, significantly reducing the high temperature creep capability of the alloy.

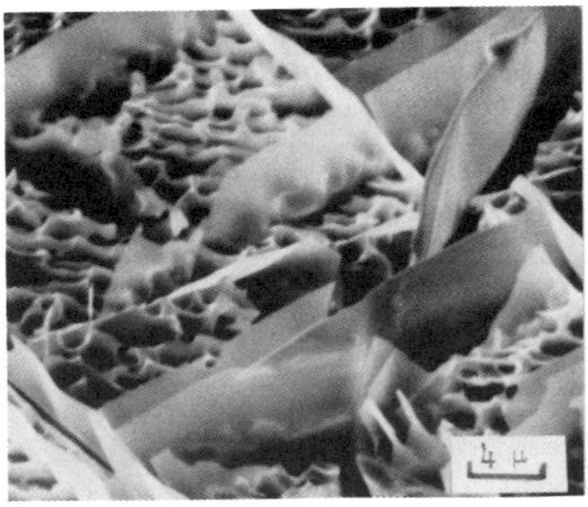

FIGURE 27. TCP AND OTHER SECONDARY REFRACTORY RICH PHASES PRECIPITATE IN PLATELET MORPHOLOGY AS SHOWN IN DEEP ETCHED SCANNING ELECTRON MICROSCOPE (SEM) PHOTOGRAPH

The detrimental effect of σ phase on creep-rupture strength was first observed more than 25 years ago in tests conducted on the poly-crystalline alloys IN100 and Astroloy (13). Creep- rupture strength of Astroloy showed a significant drop off, (Figure 28), when testing was conducted in the 816-871°C range where it was determined that σ phase precipitation occurred at the most rapid rate. The lower the stress and longer the test duration in this critical temperature range the greater the reduction in strength exhibited by the unstable alloy. The longer exposure enables a larger amount of the σ phase to precipitate during the test further weakening the alloy.

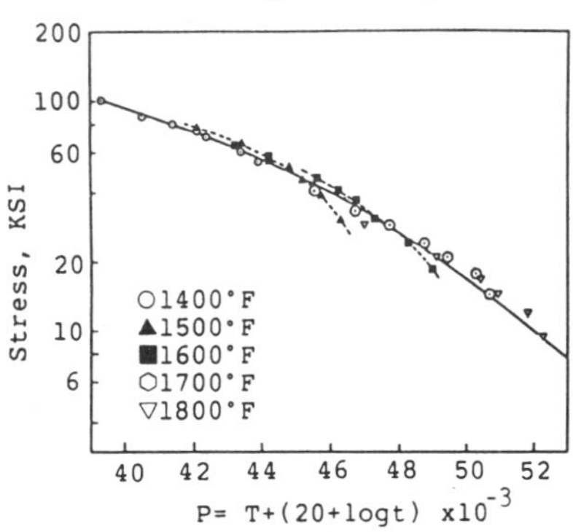

FIGURE 28. CREEP-RUPTURE STRENGTH OF ASTROLOY IS DEGRADED BY σ FORMING DURING LONG TIME TESTS IN 816°-871°C TEMPERATURE RANGE (FROM REF. 13)

A similar degradation in creep-rupture strength has been found in testing of a high tungsten experimental alloy. The creep-rupture strength of a particular unstable heat which was observed to form large volume fractions of alpha tungsten and mu phases (Figure 29) was shown to be significantly lower than that for data obtained on stable heats of the alloy in testing at 980°C and 110 MPa (Figure 30). The large volume of tungsten rich phases which had precipitated during the creep-rupture tests in the unstable material significantly depleted the alloy of its prime solid solution strengthening agent thus causing the marked reduction in creep-rupture strength of this heat.

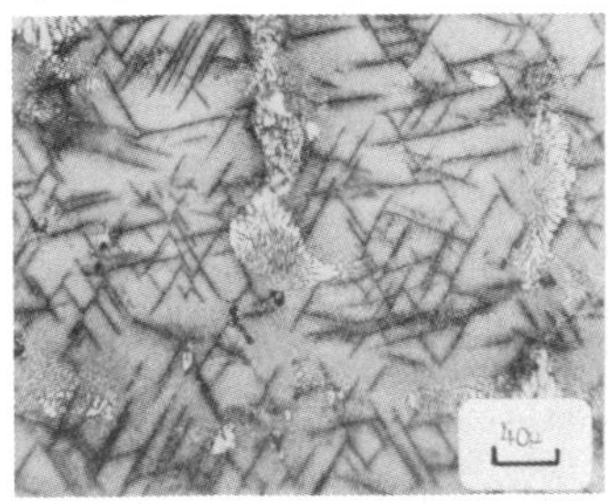

FIGURE 29. ALPHA TUNGSTEN AND MU ACICULAR PHASE PRECIPITATION IN UNSTABLE HEAT OF EXPERIMENTAL ALLOY AFTER EXPOSURE AT 980°C

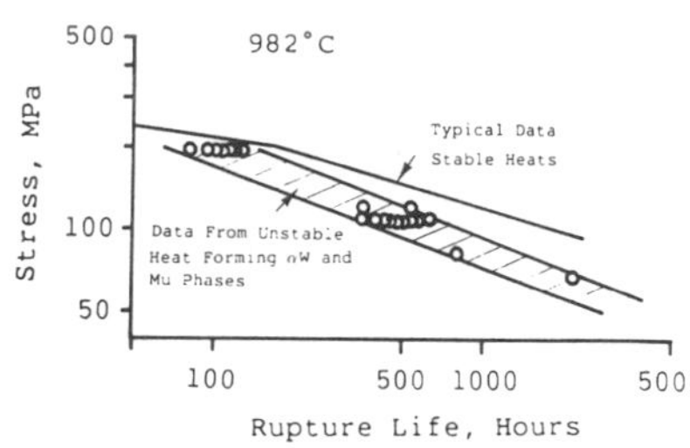

FIGURE 30. CREEP-RUPTURE LIFE AT 982°C IS SIGNIFICANTLY REDUCED BY PRECIPITATION OF ALPHA TUNGSTEN AND MU PHASES IN AN UNSTABLE HEAT OF AN EXPERIMENTAL ALLOY

4. CONCLUSIONS

1. γ' precipitate morphology is significantly affected by solution heat treatment temperature and cooling rate.
2. Full solutioning and refinement of the coarse as cast γ' precipitates significantly enhances creep rupture and yield strength.
3. An optimum γ' size range and distribution exists for most high strength alloys which maximizes elevated temperature creep strength.
4. The optimum size range is achieved through proper control of the cooling rate from the solution heat treatment temperature in conjunction with an intermediate temperature (1050-1100°C) post solution heat treatment.
5. Yield strength at 593°C is maximized by achieving a fine γ' size through rapid cooling from the solution heat treatment temperature. Further increases in yield strength can be attained by employing a low temperature age (700 - 870°C) following an intermediate temperature post solution heat tratment which precipitates ultra fine γ' particles.
6. Severe thermal exposures (temperatures greater than 1100°C) after heat treatment which produce excessive γ' coarsening can degrade creep and tensile strength.
7. Minimizing or eliminating large carbides and shrinkage porosity in castings which can act as crack initiation sites will significantly raise the fatigue capability of an alloy under both low and high cycle fatigue conditions. Producing carbon free single crystal alloys and then HIPing after casting will eliminate both major fatigue initiation sites.
8. High strength superalloys must be designed to maintain microstructural stability and be free from secondary phase precipitation during long time service operation. Refractory rich phases, such as sigma, mu, alpha tungsten and alpha chromium, which can precipitate as acicular plates in unstable alloys have been shown to adversely affect low temperature fatigue strength and high temperature creep strength.

5. ACKNOWLEDGEMENTS

The authors wish to thank Mr. R. L. Perkins for his metallographic assistance and preparation of the technical illustrations for this paper.

6. REFERENCES

1. "Copley, S.M. and Kear, B.H., TMS-AIME, 239, 977, (1967)."
2. "Copley, S.M. and Kear, B.H., TMS-AIME, 239, 984, (1967)."
3. "Donachie, M.J., Ed., Superalloy Source Book, ASM, 1984, pp 102-111."
4. "Leverant, G.R. and Kear, B.H., Met. Trans. 1, 491-498, (1970)."
5. "Jackson, J.J., et al, Met. Trans., Vol. 8A No.10, pp. 1615-1620, (1977)."
6. "Gell, M., and Duhl, D.N., Advanced High Temperature Alloys, ASM, 1985, pp. 41-49."
7. "Khan, T., Meeting on High Temperature Alloys for Gas Turbine and Other Applications, Liege, Belgium, Oct. 6-9, 1986, ONERA, Tp 1986-150."
8. "Nathal, M.V., Met. Trans., 18A, pp. 1961-1970, (1987)."
9. "Caron, P., and Khan, T., Materials Science and Engineering, 61, pp. 173-184, (1983)."
10. "Tien, J.K. and Copley, S.M., Met. Trans., 2, pp. 543-557, (1971)."
11. "Gell, M. and Leverant, G.R., Met. Trans., 242, p. 1869, (1968)."
12. "Ross, E.W., Recent Research on IN100 92nd AIME Annual Meeting, Dallas, Texas (1963)."
13. "Barker, J.F., General Electric Co. FPD/TIS, Report R65FPD51 (1965)."

2nd International SAMPE Metals Conference
August 2-4, 1988

POTENTIAL FOR NIAl BASED ALLOYS FOR HIGH TEMPERATURE APPLICATIONS

K. Vedula
Associate Professor
Department of Materials Science and Engineering
Case Western Reserve University, Cleveland, Ohio 44106

Abstract

The B2 nickel aluminide, NiAl, is of potential use for high temperature structural applications because of its high melting point, low density and oxidation resistance. This paper presents an overview of recent attempts to improve the low temperature ductility as well as the high temperature strength of this alloy, including results of recent work at Case Western Reserve University. Room temperature tensile ductility of 2.5% can be obtained in a stoichiometric binary alloy by careful control of the composition and processing conditions. Alloying additions to improve high temperature creep resistance are most effective when fine distributions of ternary intermetallic phases can be obtained. A promising direction for exploiting the advantages of this alloy may be as a matrix for composite materials.

1. INTRODUCTION

The Ni-Al phase diagram, shown in Figure 1, reveals a monotectic with about a 15at% range in solid solubility and a congruent maximum in melting temperature of $\sim$ 1913 K. This phase diagram includes a modification of the Ni_5Al_3 phase field based on recent work by Khadkikar and Vedula[1,2]. The B2 crystal structure of the NiAl phase is illustrated in Figure 2. The wide range of stoichiometry in this phase is accomodated by the formation of defect structures[3,4]. For Ni-rich alloys, anti-structure defects are predominant where excess Ni atoms fill Al sites. For Al-rich alloys, excess Al atoms, generally, do not form anti-structure defects; instead these atoms fill normal Al sites and the predominant point defects are vacant Ni sites. However, the point defect structure of NiAl at stoichiometry is believed to be composed of triple defects, consisting of two vacancies on Ni sites plus a Ni atom on an Al site.

Although the composition dependence of the type and number of point defects in NiAl alloys at low temperatures has been explored in some detail, their temperature dependence has received less attention. Point defects must play an important role in the mechanical deformation of these alloys. They can contribute to defect hardening at low temperatures and to diffusion processes at elevated temperatures. Furthermore, any attempts at solid solution alloying will most certainly change the type and number of point defects, thereby impacting on mechanical behavior of these alloys.

2. SINGLE CRYSTAL DEFORMATION

Single crystals of NiAl are reported to be ductile at room temperature as well as at elevated temperatures. The details of operative slip systems have been studied to some extent and the behavior is shown to be a very sensitive function of orientation and of temperature[5,6,7].

Studies of near-stoichiometric single crystal alloys have shown that <100> oriented crystals are more difficult to deform than other orientations, making this alloy very anistropic. In compression, all orientations possess some ductility irrespective of the temperature. However, due to the anisotropy of slip, the non-<100> crystals are much more ductile at lower temperatures than the <100> oriented crystals. At 673 K and above, all orientations possess considerable ductility even when tested in tension. There is general agreement that for non-<100> oriented crystals, deformation involves {100} <001>, and {110} <001> slip systems at all temperatures. The behavior of <100> oriented crystals is less clear and conflicting data exists; at or below room temperature, <111> slip directions with {123} and {112} slip planes have been observed; while high temperature deformation has been reported to be the result of slip in <111>, <110> and <100> directions. In addition, neither antiphase boundaries nor superdislocations are found in annealed or deformed material, since <100> dislocations do not create APBs.

These observations have important implications for polycrystalline deformation. If only <100> slip is operative at low temperatures, there are insufficient slip systems for general deformation in a polycrystalline alloy.

3. POLYCRYSTALLINE MATERIAL

3.1 Low Temperature Deformation

The development of NiAl alloys has been seriously hampered by room temperature brittleness of polycrystalline alloys.

One of the earliest attempts to judge the potential of polycrystalline B2 NiAl alloys for use as high temperature structural materials was conducted by Westbrook [8]. Through microhardness testing, he demonstrated the strong dependence of the strength of NiAl on both composition and temperature. A typical hardness-composition curve at room temperature is reproduced in Figure 3, which reveals that a minimum in hardness exists at nominally 50 at% Al. Westbrook interpreted this behavior to mean that vacant atom sites (Al-rich alloys) as well as substitutional atoms (Al-poor alloys) provide strengthening effects.

Although some earlier reports have indicated compressive ductilities at room temperature [5,9,10], only one report by Rozner and Wasilewski [11] has reported 2% room temperature tensile ductility in a polycrystalline stoichiometric alloy.

The number of slip systems which operate if only the <100> slip occurs are insufficient for von Mises' criterion for polycrystalline deformation. Much higher stress levels are required in order for other slip systems to be activated. This is possible in compression, since fracture initiation and propagation can be delayed until these higher stress levels are reached. Compressive ductility is, therefore, observed. In tension, however, no ductility is observed presumably because fracture intervenes at lower stresses than needed to activate slip in the harder orientations.

On the other hand, if <111> slip is possible (as in the case of disordered BCC alloys as well as some other B2 compounds), there are enough number of slip systems of the type {110} <111>, to satisfy von Mises' criterion and lead to polycrystalline tensile ductility. Based on this premise, many of the current efforts at obtaining room temperature ductility in B2 NiAl are concentrating on encouraging slip of the <111> type in these alloys through alloying additions. This approach cannot, however, guarantee ductility since the fracture stress may still be lower than the stress needed for activation of these <111> slip systems.

Grain refinement has been proposed as another possible approach towards improving polycrystalline ductility in these alloys [12]. This is based on the argument that polycrystalline fracture stress increases more rapidly with decreasing grain size than does yield stress. Hence, fracture can be delayed sufficiently in finer grain sized materials to allow for the onset of some plastic deformation. However the required grain size for this to happen is probably very fine and this approach has not been demonstrated as feasible.

Intrigue in the published results of Rozner and Wasilewski [11] and the failure of recent efforts to reproduce their results led to a more recent investigation of binary B2 NiAl alloys by Hahn and Vedula(13). The objective was to attempt to duplicate their results and see if it is, in fact, possible to obtain significant tensile ductility at room temperature in polycrystalline NiAl.

Four compositions near the stoichiometric composition were selected, since there is ample evidence that small deviations from stoichiometry can significantly alter the deformation behavior of these alloys. Compositions selected were Ni-47at% Al, Ni-49at% Al, Ni-50at% Al and Ni-51at% Al. Selected compositions were cast in the form of 1.25" diameters, 4" long cylindrical ingots at General Electric Co., Evendale. The cast billets were homogenized in argon atmosphere for 24 hours at 1473 K. The casting route was preferred over a powder approach, since large levels of interstitial contamination are typically observed in P/M alloys(14). These impurities are suspected to greatly influence the mechanical properties of these materials, even more so than they do in conventional BCC metals and alloys.

The castings were placed in mild steel cans and extruded by Brush Wellman Inc. at Cleveland. The extrusion temperature is believed to be critical since the material tested in the work by Rozner and Wasilewski(11) was thermomechanically processed and since substructure due to prior deformation may play a vital role in the room temperature deformation of these alloys.

Tensile specimens were prepared by centreless grinding 2" lengths of the extrusions to the dimensions illustrated in Figure 4. These tensile specimens were mechanically polished and electropolished in a solution of 10% perchloric acid in methanol prior to testing. Tensile tests were carried out in air at temperatures ranging from room temperature to 1273 K, using an Instron Universal Testing machine. The test temperatures were obtained within ±2 K accuracy using an electric resistance furnace. All specimens were held for 30 minutes at test temperature before testing at an initial strain rate of 1.4×10^{-3}/sec (cross head velocity being kept constant).

The chemical analysis of these extrusions showed that they were very close to the nominal compositions and that the level of interstitial contamination was, in fact, low. The actual compositions were: 47.0 at% Al, 49.4 at% Al, 50.3 at% Al, and 51.5 at% Al with an accuracy of within 0.2 at% Al. The levels of carbon and oxygen were analyzed for the 50.3 at% Al alloy and were found to be 50 and 82 ppm respectively.

The other castings are presumed to contain similar levels of oxygen and carbon. The grain sizes of the four alloys are in the range of 11 to 16 μm for all 4 alloys. A typical longitudinal optical microstructure of the 50.3 at% Al alloy is illustrated in Figure 5 in order to emphasize the uniformity of the grain morphology and size. The grain size in the longitudinal orientation is very similar to the grain size in the transverse direction, suggesting that the alloy was recrystallized either during or immediately following the extrusion process.

The engineering stress - engineering plastic strain plots for all the four alloys tested are shown in Figures 6a, b, c and d for the lower temperatures. Definite, although limited, discontinuous yields and plastic deformation are observed in the nominally stoichiometric alloy at 298 K and at 473 K. This alloy containing 50.3 at% Al showed tensile plastic elongations of 2.5% at room temperature, 4.5% at 473 K, 31% at 673 K and 47% at 873 K. The other three alloys were brittle with no significant plastic elongations at room temperature and at 473 K. They showed the beginnings of plastic deformation at 673 K and were very plastic at 873 K.

The fracture surfaces of specimens for the nominally stoichiometric alloy tested at these four temperatures are illustrated in Figure 7. This alloy exhibits a predominantly intergranular fracture mode with some transgranular fracture at room temperature. The extent of transgranular fracture increases at 473 K and becomes the dominant fracture mode at 673 K. Increases in ductility are therefore associated with an increased tendency towards transgranular fracture. The fracture surface of the specimen tested at 873 K exhibits typical ductile fracture showing dimples and evidence of microvoid coalescence.

The completely brittle fracture surfaces of the off-stoichiometric alloys tested are also predominantly intergranular. At higher temperatures, where ductility is observed in these alloys as well, the fracture becomes transgranular.

The precise mechanism by which ductility is made possible in the nominally stoichiometric alloys is still under investigation. Transmission electron microscopy is the major technique being used. The substructures of the alloys in the extruded condition are being compared with the substructures of the deformed specimens of the stoichiometric alloy.

Texture in the alloys is also being investigated as a potential source of the observed ductility. A recent characterization of extrusions of several aluminides, reported by Khadkikar and Vedula[15], show these aluminide extrusions to be strongly textured as a result of the recrystallization associated with hot extrusion. For the Ni-47 at% Al alloy which was cast and extruded at 1073 K, and a Ni-45 at% Al alloy powder-extruded at 1400 K, they have reported a texture with <111> as the major component in the as-extruded condition in both cases. <211> and <110> form the minor components in the powder-extruded alloy, whereas <110> is the only minor component in the cast and extruded alloy. A sharper texture with increased amount of <111>, at the expense of <110> and <211> is produced in the powder-extruded alloy by annealing at 1400 K for 24 hours. Increased amounts of both <111> and <211> over <110> are observed in the cast and extruded NiAl after annealing.

3.2 Elevated Temperature Deformation

The strength of NiAl decreases with temperature, as do regular metals. A drastic decrease in strength is accompanied by a sudden increase in the ductility at about 0.4 Tm as illustrated in Figure 8[13]. The reasons for this brittle to ductile transition are related to temperature dependence of slip behavior and the onset of diffusion assisted process.

The dislocation configurations in an extruded, stoichiometric NiAl alloy show evidence for the existence of glissile dislocations with Burger's vectors of <100>, <110> and <111>[16]. The additional two vectors provide enough independent slip systems at high temperatures to satisfy von Mises criterion for general plastic flow in a polycrystalline aggregrate. These additional slip systems coupled with climb mechanism which play an important role at elevated temperatures make NiAl ductile at elevated temperatures.

Elastic modulus is another key factor in the design of high temperature structural parts and some measurements for this alloy[17] show high values of modulus at low temperatures with a gradual decrease as temperature increases.

The behavior of NiAl alloys at slow strain rates and high temperatures, studied by Vandervoort et al[18] and Yang and Dodd[19], is illustrated in Figure 9. Although the stoichiometric alloy is the weakest at low temperatures, it is the strongest alloy at high temperatures. The stress exponents are typically between 3 and 5, suggesting combinations of dislocation climb controlled creep and viscous glide controlled creep. These studies indicate an intriguing result that fine grain size alloys are stronger than large grain size alloys between 1073 and 1273 K.

Whittenberger[20] recently studied alloys of NiAl ranging from 43.9 to 52.7 at% Al in compressive creep from 1200 to 1400 K and did not find any major differences in the creep behavior of these alloys. The deformation could be described by a single stress exponent (around 5) and a single activation energy, at low temperatures and fast strain rates. The deformation is found to be independent of grain size and appears to be controlled by subgrain formation. However, at higher temperatures and slower strain rates, diffusional creep seems to dominate.

In spite of the beneficial effect of ordering on the creep resistance, the phase NiAl is much too soft at high temperatures in comparison with commercial superalloys[21]. Alloying additions are needed to modify properties. However, information on the phase diagrams for ternary additions is scarce[22]. Ternary additions have been shown to result in substantial improvements in the high temperature compressive creep of NiAl alloys by Pathare and Vedula[14,23,24]. Several ternary elements have been studied, including those which have large solubilities (Cr, Mn, Ti, Co) and those which have limited solubilities (Nb, Ta, Hf, Mo, W, Zr, B) in the binary alloys. The results of compressive creep studies of solid solution ternary alloys at 1300 K have suggested that deformation mechanisms in ternary alloys at 1300 K are similar to those in binary alloys. Dislocation climb controlled creep with a stress exponent of about 5 at high strain rates and diffusional creep at lower strain rates with a stress exponent closer to 1 are typically observed.

Ternary alloys which result in precipitation of second phases have shown the best creep resistance at 1300 K. In particular, small amounts of addition of Ta and Nb have been shown to cause substantial improvements in compressive creep behavior. Precipitates of ternary intermetallic phases have been observed and are believed to cause strengthening by dislocation-particle interactions. These ternary intermetallics have been carefully characterized and are typically the Heusler type compounds (Ni_2AlTa, Ni_2AlNb) and the hexagonal compounds (NiAlTa, NiAlNb). Figure 10 compares the compressive creep behavior of binary NiAl alloy with ternary alloys containing different amounts of Ta and processed in different ways. The cast alloy with 5 at% Ta shows the best creep resistance, due to a combination of the larger volume fraction of precipitates as well as the larger grain size as illustrated in the microstructure in Figure 11. Ternary alloys with second phase strengthening, demonstrate a stress exponent closer to 3, possibly indicating viscous glide control.

4. PROCESSING CONSIDERATIONS

Conventional casting of these alloys in large sizes is believed to be difficult because of segregation of elements, brittleness of the alloy as well as sensitivity of mechanical properties to compositional deviations. Secondary processing is usually essential in order to obtain uniform microstructures. Hot extrusion of small castings has been shown to break up the cast structure and produce fine grained recrystallized microstructures(13).

Powder processing by hot extrusion of canned powders is a common technique used for obtaining fine grained microstructures. Vacuum hot pressing as well as hot isostatic pressing have also been demonstrated to be effective(14). The hot extrusion process has the advantage that it can break up the prior particle boundaries and in addition, preferred orientations can be developed by control of processing conditions.

Rapid solidification by melt spinning has also been attempted(25). The observed results are not particularly encouraging since the grain sizes are not refined to any large extent and no metastable structures result. Room temperature bend tests of these ribbons give results which are quite similar to those observed in P/M extruded materials.

5. POTENTIAL AS MATRIX MATERIALS FOR COMPOSITE APPLICATIONS

Although the high temperature creep resistance of NiAl alloys has not yet been shown to be significantly superior to conventional superalloys, it is possible to use these alloys as matrices for composite materials in order to take advantage of their low densities, excellent oxidation resistance as well as high melting points. Several research groups are actively pursuing this direction. One report of composites based on NiAl will serve to illustrate the promise of this approach. Viswanadham and Whittenberger(26) have studied particulate composites of TiB_2 in NiAl. The composites were made by XD synthesis and hot pressed to full density with the average size of the TiB_2 particles being about 1 μm. Compression testing of these composites in air showed that the flow strength of NiAl at 1300 K is increased by a factor of 3 for 20 vol% TiB_2 and by a factor of about 3.5 for 30 vol% TiB_2. Post-deformation TEM analysis showed that the primary feature of the dislocation substructure of deformed NiAl was the formation of very well-defined low-angle boundaries, while that of the composites consisted of a very high density of tangled dislocations, loops and a few irregular boundaries. These observations suggest that TiB_2 particles enhance strength of NiAl by stabilizing a completely different dislocation substructure than that of NiAl.

No deleterious chemical interactions between the NiAl matrix and the TiB_2 particles were observed. This shows that the materials are chemically compatable and that the aluminide is an effective barrier to oxygen. The latter is particularly critical since diborides are notorious for their poor oxidation resistance.

6. SUMMARY

Efforts to exploit NiAl based alloys reveal encouraging trends. At the low temperature end, with proper control of composition and processing techniques, significant room temperature tensile ductility can be expected. At the high temperature end, creep resistance can be substantially improved by second phase precipitates as well as by particulate additions.

7. ACKNOWLEDGEMENTS

The author acknowledges the support of several agencies and private industry for different portions of the results presented in this paper. These include NASA Lewis Research Center, Cleveland; Oak Ridge National Laboratories through their DOE-ECUT program, ALCOA Technical Center, General Electric Co., Evandale and Brush Wellman Inc., Cleveland. It is needless to say that the actual results were obtained through the tireless work of several graduate students, to whom the author is grateful.

8. REFERENCES

1. Khadkikar, P.S., Ph.D. Thesis, Case Western Reserve University, 1988.
2. Khadkikar, P.S. and Vedula, K, paper submitted to Met. Trans. for publication.
3. Georgopoulos, P. and Cohen, J.B., Scripta Met. Vol. 11, p. 147, 1977.
4. Georgopoulos, P. and Cohen, J.B., Acta Met. Vol. 29, p. 1535, 1981.
5. Ball, A. and Smallman, R.E., Acta Met. Vol. 14, p. 1349, 1966.
6. Ball, A and Smallmann, R.E., Acta Met. Vol. 14, p. 1517, 1966.
7. Wasilewski, R.J., et al. Trans. Met. Soc. AIME, Vol. 239, p. 1357, 1967.
8. Westbrook, J.H., J. Electrochem. Soc. Vol. 103, p. 54, 1956.
9. Pascoe, R.T. and Newey, C.W.A., Met. Sci. J. Vol. 2, p. 138, 1968.
10. Pascoe, R.T. and Newey, C.W.A., Phys. Stat. Sol. Vol. 29, p. 357, 1968.
11. Rozner, A.G. and Wasilewski, R.J., J. Inst. Metals, Vol. 94, p. 169, 1966.
12. Schulson, E.M. and Barker, D.R., Scripta Met. Vol. 17, p. 519, 1983.
13. Hahn, K.H. and Vedula, K., paper submitted to Scripta Met. May 1988.
14. Pathare, V.M., Ph.D. Thesis, Case Western Reserve University, 1987.
15. Khadkikar, P.S., Michal, G.M. and Vedula, K., paper submitted to Met. Trans., April 1988.
16. Lloyd, C.H. and Loretto, H.M., Phy. Stat. Sol. Vol. 14, p. 1517, 1966.
17. Wolfenden, A., NASA Report #WAG3-314, December 1985.

18. Vandervoort, R.R., et al. Trans. ASM, Vol. 59, p. 930, 1966.
19. Yang, W., et al. Met. Trans. Vol. 3, p. 2049, 1972.
20. Whittenberger, J.D., J. of Mat. Sci. Vol. 22, p. 394, 1987a.
21. Sauthoff, G., Z. Metallkde, Bd. 77, p. 10, 1986.
22. Nash, P., MRS Symposia Proceedings, High Temperature Ordered Alloys, Vol. 39, p. 423, 1984.
23. Pathare, V.M., Michal,G.M. and Vedula, K. Scripta Met., Vol. 21, p. 283, 1987.
24. Vedula, K., et al., Modern Developments in Powder Metallurgy, Vol. 16, p. 727, 1984.
25. Gaydosh, D.J. and Crimp, M.A. MRS Symposia Proceedings, High Temperature Ordered Alloys, Vol. 39, p. 429, 1984.
26. Viswanadham, R.K. and Whittenberger, J.D., MRS Symposia Proceedings, Reno, Spring, 1988.

9. BIOGRAPHY

Professor Krishna Vedula has been with the Department of Materials Science and Engineering at Case Western Reserve University since 1980. He obtained his Ph.D. from Michigan Technological University in 1980, his M.S. from Drexel University in 1969 and his B. Tech from I.I.T. Bombay in 1967. Professor Vedula has been actively involved with research in intermetallic compounds, powder metallurgy and high temperature oxidation.

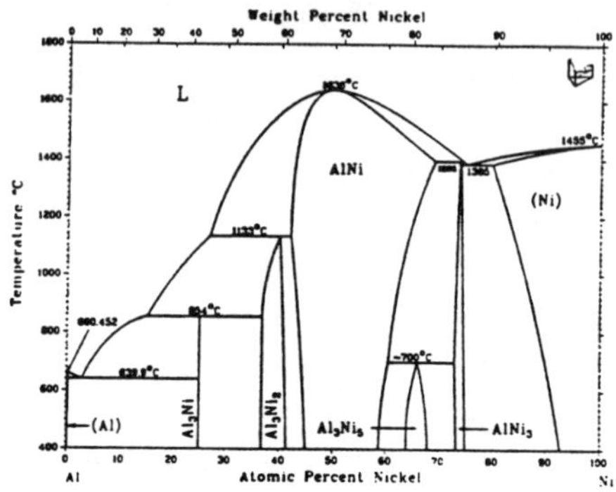

Fig. 1. Al-Ni phase diagram (Ref. 1)

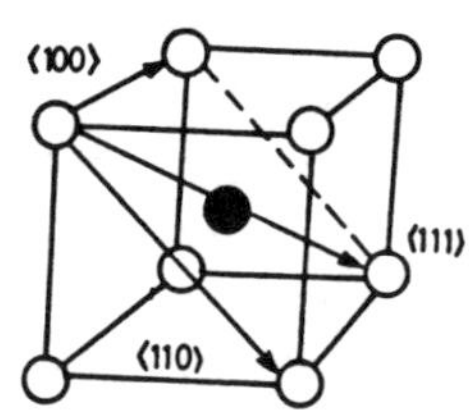

Fig. 2. B2 crystal structure

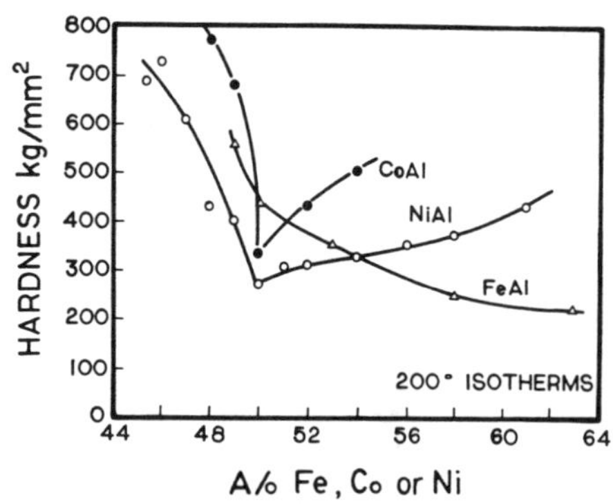

Fig. 3. Room temperature hardness of B2 compounds (Ref. 8)

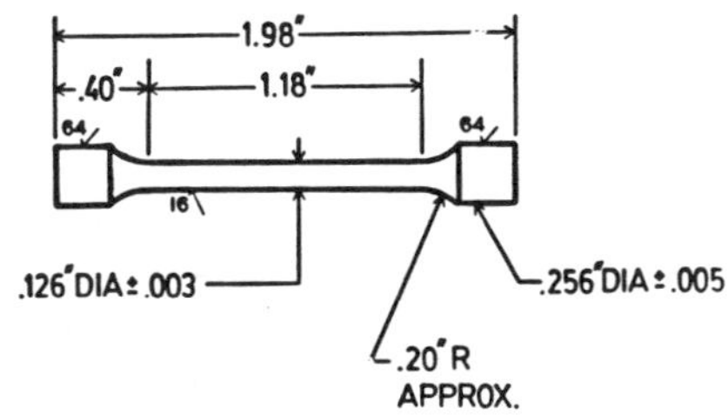

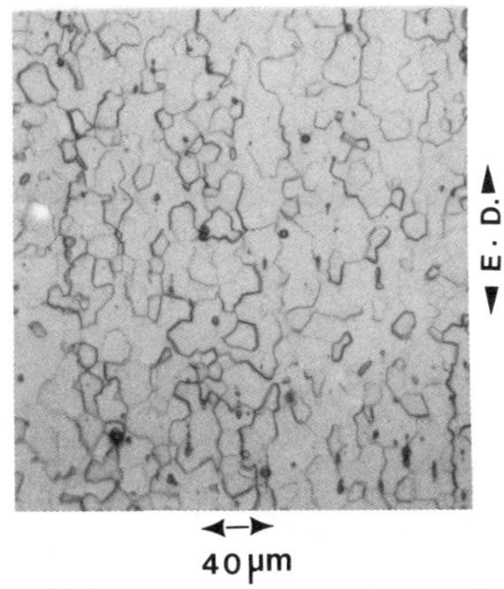

Fig. 4. Specimen geometry for tensile testing (Ref. 13)

Fig. 5. Microstructure of longitudinal section of 50.3 at% Al alloy (Ref. 13)

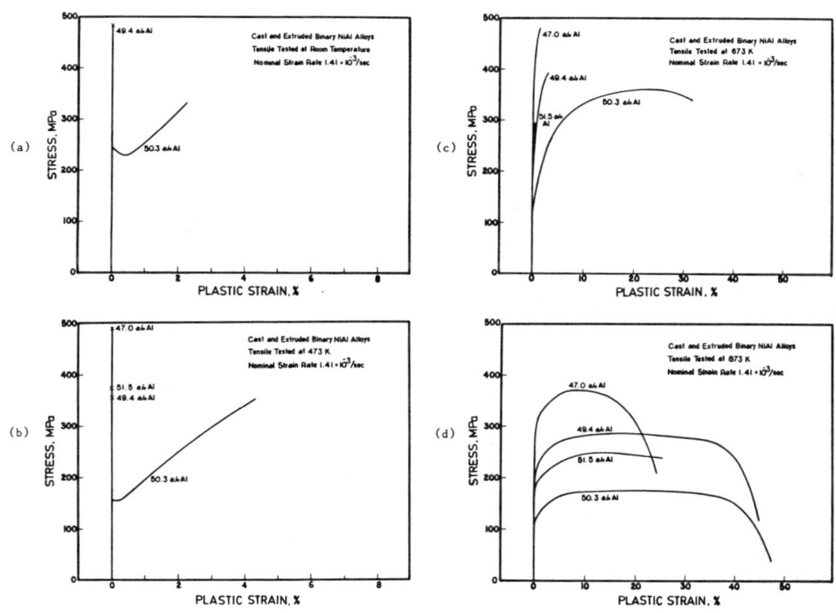

Fig. 6. Engineering stress - engineering plastic strain plots for 4 binary NiAl alloys tested at (a) room temperature (b) 473 K (c) 673 K (d) 873 K (Ref. 13)

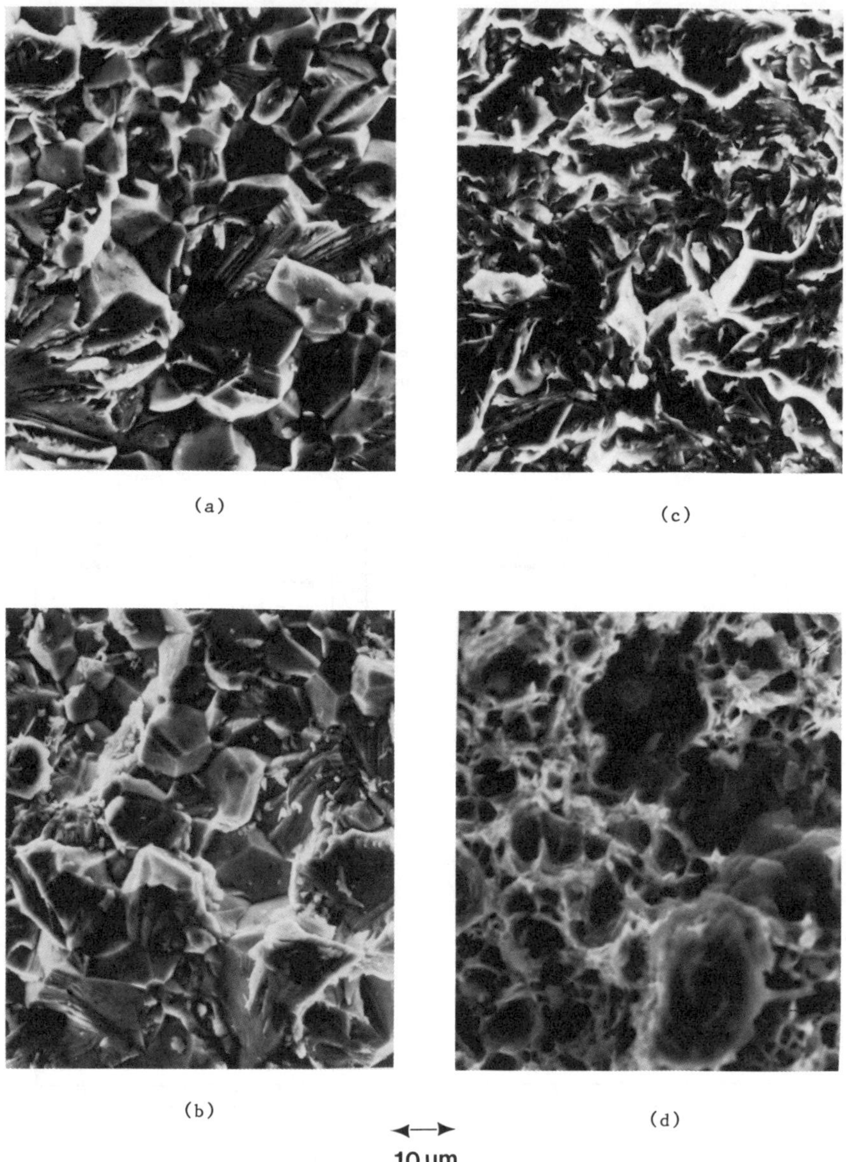

Fig. 7. Fracture surfaces of 50.3 at% Al alloy tensile tested at (a) room temperature (b) 473 K (c) 673 K (d) 873 K (Ref. 13)

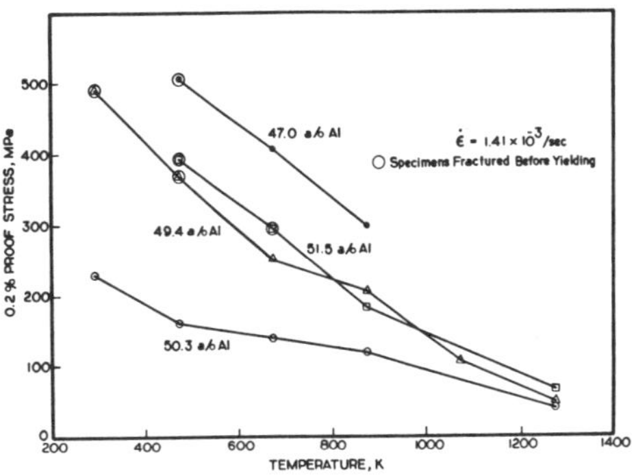

Fig. 8. Temperature dependence of 0.2% proof stress of NiAl alloys (Ref. 13)

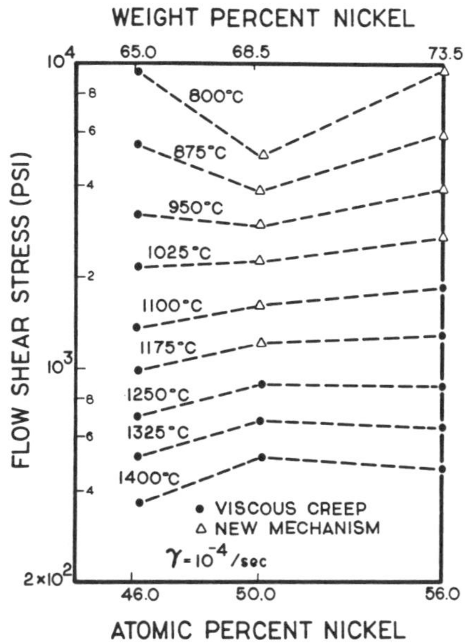

Fig. 9. Temperature dependence of slow strain rate high temperature behavior of NiAl alloys (Ref. 19)

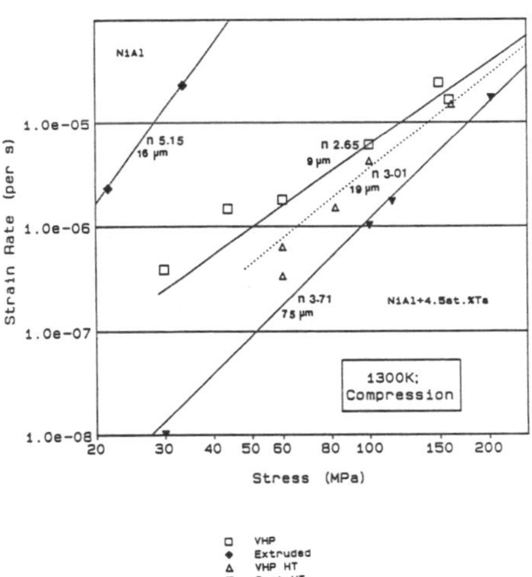

Fig. 10. Compressive creep behavior of NiAl-Ta alloys at 1300 K (Ref. 14)

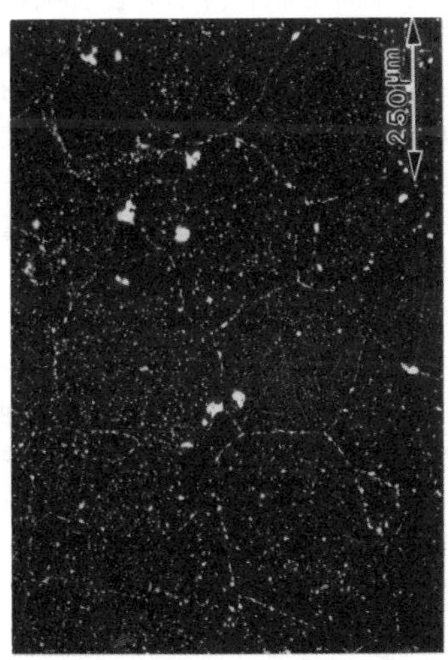

Fig. 11. Ternary intermetallic phases in cast NiAl-5 at% Ta alloy (Ref. 14)

2nd International SAMPE Metals Conference
August 2-4, 1988

ENGINEERING PROCESSING AND PROPERTIES OF NICKEL ALUMINIDES*

Vinod K. Sikka
Oak Ridge National Laboratory
P.O. Box 2008
Oak Ridge, TN 37831-6083

Abstract

Ordered intermetallic compounds of iron, nickel, and titanium are materials recently under development for structural applications. Among these, Ni_3Al has been made reasonably ductile by the addition of small amounts of boron. Further additions of zirconium and chromium have been utilized for enhancement of high temperature strength and intermediate temperature ductility. Nickel aluminide alloys based on Ni_3Al are near commercialization This paper describes the melting, processing, mechanical properties, physical properties, corrosion, and weldability of these alloys. Applications for nickel aluminides have been identified. Potential suppliers who have recently licensed the nickel aluminide technology from Oak Ridge National Laboratory (ORNL) are also listed.

1. INTRODUCTION

There has recently been a great interest in the development and application of intermetallic compounds. Among various intermetallics, nickel aluminides (Ni_3Al-based) are near commercialization because of significant ductility improvements[1-4] by additions of up to approximately 200 ppm by weight of boron in slightly nickel-rich compositions, the increase in yield strength with test temperature up to 750°C, and high temperature oxidation resistance. This paper will present alloy compositions near commercialization, status of processing techniques for various product forms, mechanical properties, physical properties,

*Research sponsored by the Office of Energy Utilization Research, Energy Conversion and Utilization Technologies (ECUT) Program, U.S. Department of Energy, under contract DE-AC05-84OR21400 with Martin Marietta Energy Systems, Inc.

corrosion data, weldability, and machinability. Applications for nickel aluminides are identified along with the status of licenses signed by various companies for these materials from ORNL.

2. ALLOY COMPOSITIONS

The compositions of three of the Ni_3Al alloys under investigation at ORNL in conjunction with industry are listed in Table 1, and their optical microstructures are compared in Fig. 1. The IC50 alloy has an ordered structure (γ'-phase) from room temperature to its melting point. The IC218 and IC221 alloys have primarily ordered structures (γ'-phase) with small amounts of the disordered structure (γ-phase) at room temperature. The fraction of disordered phase increases at high temperatures ($\geqslant$ 1000°C).

3. MELTING TECHNIQUES

Because of their high aluminum content, nickel aluminides are easily adaptable to air melting. Aluminum is an excellent deoxidizer; it removes oxygen from the melt and forms Al_2O_3 which protects the underlying molten metal from the atmosphere. The recovery of alloying elements of boron, aluminum, and zirconium is $\geqslant$ 95%. Air melting has been used by ORNL and several commercial melters for heats approaching 900 kg. Chemical compositions achieved in air-melted heats are presented in Table 2. Note that not only can the aim chemistry be fairly well met, but the oxygen levels are generally very low. All melting studies

TABLE 1

Composition of Three Nickel Aluminide Intermetallics Being Investigated at ORNL

	Weight Percent		
Element	IC50	IC218	IC221
Al	11.3	8.5	8.5
Cr	--	7.8	7.8
Zr	0.6	0.8	1.7
B	0.02	0.02	0.02
Ni	Bal	Bal.	Bal.

(a) (b) (c)

Fig. 1. Optical microstructure of the three nickel aluminide-base alloys, (a) IC50, (b) IC218, and (c) IC221. Note the presence of disordered γ phase in IC218 and IC221 alloys. The IC50 alloy is fully ordered γ'.

except one have been conducted using virgin materials. In one study, a 250-kg heat was successfully remelted. In addition to air melting, argon and vacuum-induction-melting techniques have been used for heats up to 250 kg. Secondary melting techniques used successfully include electroslag remelting (ESR) and vacuum-arc double electrode remelting (VADER). The ESR ingots were produced commercially in two sizes, 100- and 200-mm diam. The ESR process produces an excellent ingot surface and solidification grain structure, which improves plastic flow during the forging and extruding processes. The VADER process was developed to produce fine-grain cast ingots of superalloys, and it did work for nickel aluminide alloys. The chemical analyses of VADER ingots are also included in Table 2.

Based on the melt data available, it is believed that the existing commercial techniques will be directly applicable to the scale-up of nickel aluminide alloys.

4. PROCESSING TECHNOLOGY

The processing technology[5-8] for nickel aluminide has been explored by conventional techniques, innovative near-net-shape techniques, and by powder metallurgy and superplastic forming. All techniques were attempted to identify the most economical methods for a given application. Brief descriptions of the products of various technologies are presented here.

The air-melted nickel aluminides can be cast in sand, metal, or graphite molds. Most of the castings (extrusion billets and shaped castings) at ORNL have been produced in graphite molds. The 250-kg air-melted heats of IC50 and IC218 have been centrifugally cast by Combustion Engineering Incorporated into tube hollows 125-mm OD × 25-mm wall × 3000-mm length. The OD and ID machining

Table 2

Comparison of Chemical Analyses with Aim Composition for IC50 and IC218 Nickel Aluminide Alloys Melted By Various Processes

	IC50				IC218		
Element	Specification	250-kg Heat Combustion Engineering	250-kg Heat Steel Casting Engineering	VADER 250-kg Heat Special Metals	Specification	250-kg Heat Combustion Engineering	ESR 50-kg Heat Cabot Corporation
Al	11.3	11.35	11.17	11.42	8.5	9.18	9.27
Cr	---	---	---	---	7.8	7.92	7.94
Hf[a]	1.88	---	---	1.79	--	---	--
Zr	0.6	0.64	0.51	0.02	0.8	1.01	1.13
B	0.02	0.017	0.0019	0.020	0.02	0.026	0.019
Ni	Bal.	87.7	Bal.	86.7	Bal.	81.2	Bal.
O		0.0035	0.0014	0.019		0.0030	0.0023
N		0.0019	0.0008	0.0009		0.004	0.0005
C		0.025	0.01	0.004		0.029	0.010
S		0.001	0.0008	0.0002		0.003	0.0003
Fe		---	---	0.03		0.41	0.06
Si		0.03	0.05	---		0.03	--

[a]Hf substituted for Zr in VADER heat.

of a 610-mm-long piece of the centrifugal cast tube hollow has been completed. This tube is currently being cold pilgered into a range of tube sizes. Cast ingots can be successfully extruded in a range of sizes if they are first canned in stainless steel. The stainless steel cans add to the cost of nickel aluminide alloys. More recently, slightly modified compositions to those listed in Table 1 have been successfully extruded by simply wrapping them in a 0.50-mm-thick carbon steel sheet. Although it has not yet been demonstrated, plasma coating of billets may also yield good results. A thin section of up to 6-mm thick can be hot rolled from the ingots using 1.5-mm-thick stainless steel cover sheets. Hot extrusions and rolling have been carried out at 1100 and 1200°C. The alloys can also be cold rolled with intermediate anneals at 1100°C. The cold rolling has been most successful when the starting thickness is around 6 mm. Nickel aluminides will be used in the as-cast condition in various shapes, and we believe that these shapes can be made using existing foundry practice.

5. INNOVATIVE NEAR-NET-SHAPE PROCESSING TECHNIQUES

These techniques include direct casting of liquid metal into sheet and bar product. The sheet produced by the direct casting process ranges in thicknesses from 1.0 to 1.3 mm. At present, sheet can be cast into 300-mm width and continuously coiled (see Fig. 2). The thickness of the sheet was uniform within ±3% except 20-mm edges on both sides, which were much thinner (approximately -10%) than the mid section. The as-cast sheet possesses excellent ductility as demonstrated by the fabrication of a tray with tight bends in Fig. 3. The as-cast sheet could be cold rolled up to 50% without intermediate annealing. Post-cold work annealing at 1100°C for 1 h produced equiaxed fine-grain structure. This sheet was isotropic in strength and ductility properties.

Fig. 2. Photograph of direct cast sheet of IC50 and IC218 alloys. The large coil of IC50 weighs 740 kg.

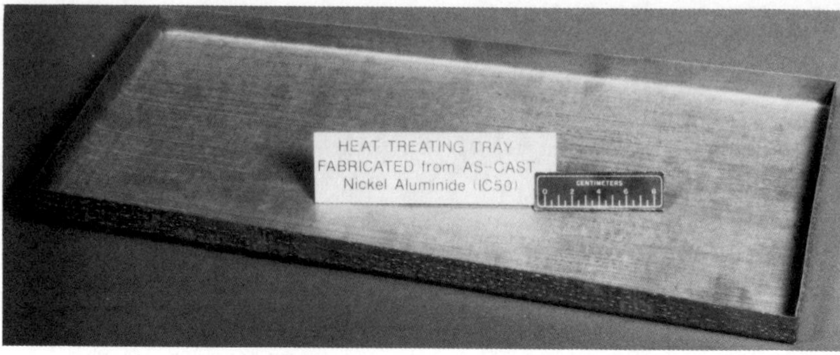

Fig. 3. Photograph of a heat-treating tray fabricated from direct cast sheet of IC50. The tight bends in this tray demonstrate good ductility of direct cast IC50 sheet.

Direct casting is probably the most economical method for producing thin sheet of nickel aluminide alloys.

The direct bar casting from liquid metal was performed at Steel Casting Engineering. The casting is carried out through a mold that is directly connected to the induction-melting furnace. This casting is pulled out continuously in a horizontal plane. Nickel aluminides have been successfully cast into 8- and 12-mm-diam rods. As with direct cast sheet, the cast bar can also be coiled (see Fig. 4). The as-cast bar was swaged up to 60% prior to annealing at 1100°C and further swaging of 60% was possible. Two 60% reductions refined the columnar structure to equiaxed structure. This direct casting method for bar is expected to be the most economical process for the subsequent production of wire.

6. POWDER METALLURGY AND SUPERPLASTIC FORMING

Inert gas atomized powder of nickel aluminides can be commercially produced by Homogeneous Metals and the Powder Product Division of Cytemp Specialty Steel Division. Several processes used for the consolidation of powder are

1. extrusion of powder,
2. hot isostatic pressing (HIP),
3. HIP followed by extrusion,
4. consolidation under atmospheric pressure (CAP) using glass molds,
5. CAP followed by extrusion,
6. rapid omnidirectional consolidation (ROC), and
7. injection molding.

All consolidation techniques except CAP and injection molding produce fully dense material. The CAP process showed some closed porosity that was eliminated by extrusion. Other processes such as HIP or ROC, which were not tried, may also be possible. The injection-molded

Fig. 4. Photograph showing the coil of direct cast 12-mm bar of IC50. The direct casting was carried out at Steel Casting Engineering, Orange, California.

parts contained approximately 30% porosity, corresponding to the volume of the binder used. The porosity of injection-molded parts could be eliminated by the ROC or HIP process.

The most significant observation from powder processing was the superplastic behavior of nickel aluminide alloys. This is demonstrated by a series of tensile tests at a strain rate of 8.3×10^{-4}/s and at test temperatures from 25 to 1100°C shown pictorially in Fig. 5. Note 638% elongation was observed at 1100°C. The superplastic flow behavior observed during tensile testing was confirmed by the fabrication of a prototype turbine disk at Ladish Company using isothermal forging at 1100°C and a strain rate of

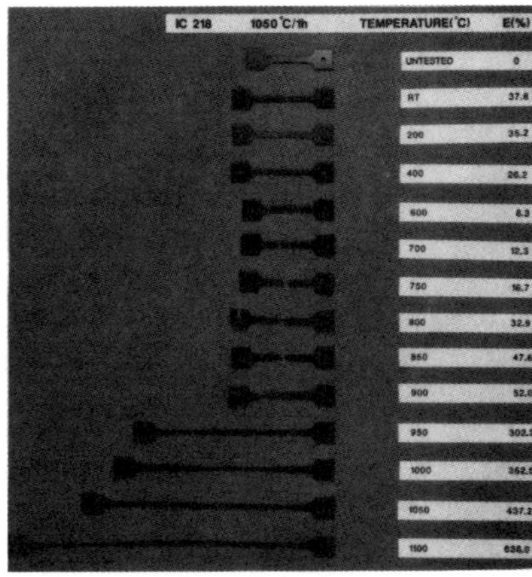

Fig. 5. Effect of test temperature on tensile ductility in air of IC218 alloy at a strain rate of 8.3×10^{-4}/s. Superplasticity occurred at temperatures ≥ 950°C at this strain rate.

8.3×10^{-3}/s (see Fig. 6).

7. MECHANICAL PROPERTIES

Tensile properties of IC50 alloy in the cast and wrought conditions are compared in Fig. 7. Data in these figures show the improvement in strength and ductility achieved by thermomechanical processing.

These figures also show the spread in the data for several specimens tested from the same heat and the spread in data between different heats and product forms. Nickel aluminide work hardens rapidly resulting in large increases in yield and tensile strength with cold work levels of 20 to 40%. Tensile properties similar to those presented in Fig. 7 are also available for IC218 and IC221 alloys. Compared to IC50, IC218 and IC221 show much less increase in strength with increasing test

Fig. 6. Prototype turbine disks produced by isothermal deformation of powder-extruded bar of IC218 at 1100°C and 8.3×10^{-3}/s. Forging was carried out at Ladish Company, Cudahy, Wisconsin.

temperature. The less positive temperature dependence probably results from disordered phase precipitate strengthening in IC218 and IC221 at lower test temperatures.

Creep-rupture testing of IC50, IC218, and IC221 is in progress in both wrought and cast conditions. The 10^2- and 10^3-h rupture strength from parametric plots of the data available thus far are compared in Table 3. This table shows a dramatic increase in rupture strength of large-grain castings as compared to fine-grained wrought material.

Fatigue and crack growth testing of nickel aluminide alloys is under way at Rensselaer Polytechnic Institute.(10-12) An important observation from this study is that the fatigue crack growth rates of ordered alloys are an order of magnitude lower than conventional alloys. Furthermore, the 10^7 cycle fatigue life of ordered alloys is above their yield strength.

8. PHYSICAL PROPERTIES

Thermal expansion, thermal conductivity, and electrical resistivity data of Ni_3Al have been studied at ORNL and reported by Williams, et al.(13-14) Data for the above mentioned properties are listed in Table 4. Room temperature values of thermal expansion and thermal conductivity of nickel aluminides with and without ternary elements are compared in Table 5. Note that the thermal expansion of nickel aluminide alloys is the same as commercial nickel-base alloys. The thermal conductivity of binary nickel aluminide is approximately twice that of the conventional alloys. The effect of chromium addition is expected to be similar to that presented above for iron.

9. AQUEOUS CORROSION BEHAVIOR

The aqueous corrosion behavior of IC50, IC218, and IC221 was recently studied by Buchanan et al.(15) under a grant from the Nickel Development Institute of Canada. Summary results of these tests

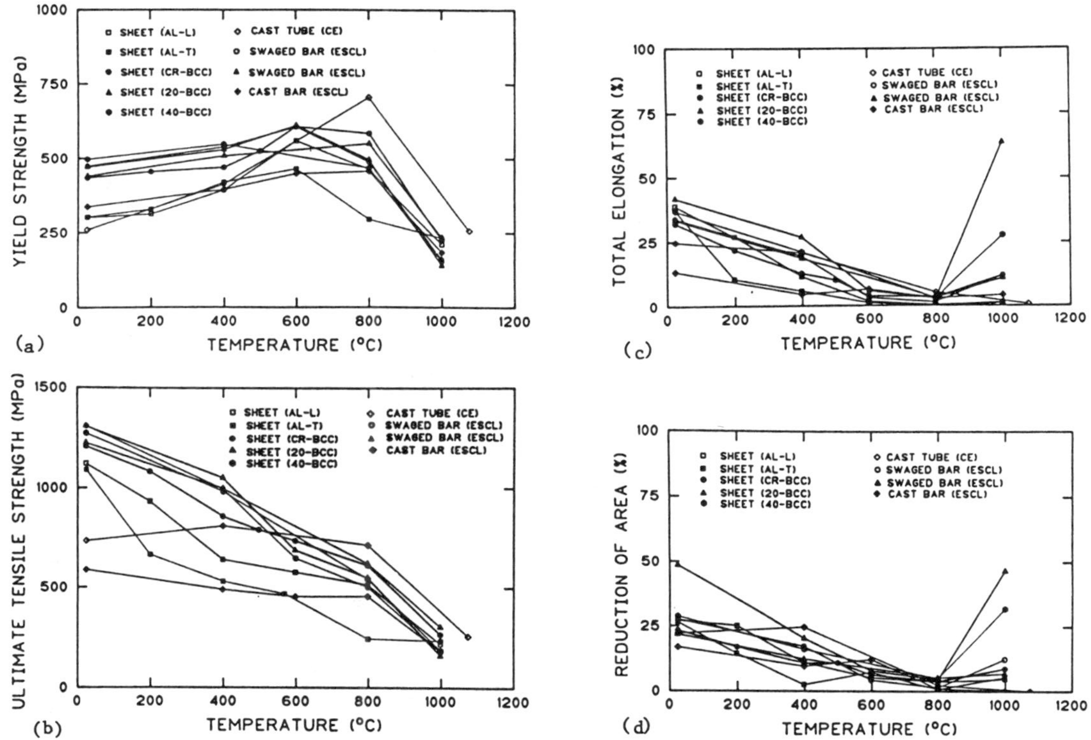

Fig. 7. Tensile properties of nickel aluminide (IC50) fabricated by several different processes. All tests were in air and at a strain rate of 8.3×10^{-4}/s. (a) yield, (b) tensile, (c) total elongation, and (d) reduction of area.

Table 3

Creep Rupture Strength of Three Nickel Aluminides As Affected By Processing

Temperature (°C)	Time To Rupture (h)	Stress (MPa)					
		IC50	IC218			IC221	
		Electroslag		Wrought			
		Cast	Cast	7–10 μm	14–16 μm	Cast	Wrought
649	1000	---	---	427	---	---	655
732	100	310	586	255	538	614	379
732	1000	207	379	138	276	393	207
816	100	172	290	93	186	296	138
816	1000	124	193	48	103	193	69
982	100	37	69	---	23	67	16
982	1000	14	26	---	---	24	---

Table 4

Physical Property Values for Ni_3Al

Temperature (K)	$Ni_{0.76}Al_{0.24}$		
	Electrical Resistivity (10^{-8} Ωm)	Thermal Conductivity (W/m·K)	Thermal Expansion Coefficient[a] (10^{-6} K^{-1})
4.2	9.05	---	--
300	51.3	21.37	12.5
400	58.89	24.21	13.2
500	64.17	26.35	13.8
600	68.60	28.30	14.3
700	72.60	30.14	14.8
800	76.52	31.82	15.4
900	80.41	33.13	16.1
1000	84.24	32.70	16.8

[a]$(1/L_{300})\ dL/dT$.

Table 5

Physical Property Comparison at Room Temperature of Nickel Aluminides With Conventional Nickel Base Alloys

Material	Room Temperature Values	
	Thermal Expansion Coefficient (10^{-6} K^{-1})	Thermal Conductivity W/m·K
Inconel 625	13.3	9.8
Inconel 718	13.6	10.9
Inconel X-750	14.0	11.9
Ni-24 at. % Al	13.3	21.4
Ni-20 at. % Al—10 at. % Fe	13.3	14.2

at room temperature and 95°C are presented in Tables 6 and 7.
The literature values of room temperature corrosion data for type 304 stainless steel, nickel-base alloys 76Ni-16Cr-7Fe (Inconel 600) and 54Ni-15Cr-16Mo + Fe,W (Hastelloy C), and pure nickel and aluminum are also included for comparison in Table 6. At room temperature it appears that when the corrosion behavior of pure aluminum in a given medium was better, the nickel aluminides tended to behave like aluminum; and when the corrosion behavior of pure nickel in a given medium was better, the nickel aluminide tended to behave like nickel. From Table 6, we can observe that, at room temperature, nickel aluminides appear to be as good or better than type 304, Inconel 600, and possibly Hastelloy C in the following media at concentrations listed in the table: hydrofluoric acid, phosphoric acid, oxalic acid, acetic acid, sodium chloride, sodium hydroxide, and ammonium hydroxide solutions. At 95°C (Table 7) nickel aluminides produced good corrosion resistance in organic acids (oxalic and acetic), sodium chloride solutions, and the bases (sodium hydroxide and ammonium hydroxide solutions). In terms of localized corrosion susceptibility, all three nickel aluminide alloys underwent pitting corrosion initiation within the 200-h exposure to 3.5 wt % sodium chloride solution.

Initial studies have indicated nickel aluminide alloys to be resistant to stress corrosion cracking in sodium chloride solution. Detailed studies are currently under way at the National Bureau of Standards.

10. HIGH TEMPERATURE OXIDATION

Because of their high aluminum content, nickel aluminides show excellent high temperature oxidation resistance. The protective oxide is alpha alumina[16] and forms in the temperature range of 1000 to 1100°C. For chromium-containing alloys, oxide consists of Al_2O_3-Cr_2O_3. Average thickness of 2-3 μm is the optimum protective oxide for these alloys.

11. WELDABILITY

Nickel aluminides can be welded using electron beam and gas tungsten arc techniques. Procedures for these techniques are currently under development.

12. MACHINABILITY

Nickel aluminides can be easily machined by electrodischarge machining. However, the regular machining is difficult using conventional cutting tools because of the high work-hardening rate and protective oxide film formation during casting and heat treating.

13. APPLICATIONS

Because of the above-mentioned properties, nickel aluminides have been identified for a range of applications. These include gas, water, and steam turbines because of their excellent cavitation,

Table 6

Comparisons of Average Penetration Rates (mils/year), 25°C

Solution (Room Temp)	Concentration		Avg. of 200-h continuous and discontinuous exposures			From NACE Corrosion Date Survey (1974)*				
	Molar	Weight % (w/o)	IC-50	IC-218	IC-221	Nickel	Aluminum	Type 304 SS	Ni-Cr-Fe 76-16-7	Ni-Cr-Mo 54-15-16 + Fe,W
Sulfuric Acid	1 18	9 96	2 17	4 19	3 15	<20 >50	- -	20-50 <2	<20 20-50 at 80 w/o	- -
Hydrochloric Acid	1 12	4 37	5 196	5 223	3 765	>50 at 10 w/o >50	>50 at 10 w/o >50	>50 at 10 w/o >50	>50 at 10 w/o >50	- -
Nitric Acid	1 8 16	6 40 70	302 >8025 78	273(>1600) 1 2	671 6 3	>50 at 10w/o >50 >50	<2 >50 <20	<2 <2 <2	<20 at 10 w/o <20 20-50	<2 at 10 w/o <2 -
Hydrofluoric Acid	1	2	3	3	3	<20 at 10% w/o	>50 at 10 w/o	>50 at 10 w/o	<2 at 10 w/o	-
Phosphoric Acid	1	10	2	2	3	<2	<2	<2	<20	-
Oxalic Acid	1	10	1	1	1	20-50	<20	<20	<20	<20
Acetic Acid	1	6	2	<1	<1	<20 at 10 w/o	<20 at 10 w/o	<2	<20 at 10 w/o	<2 at 10 w/o
Sodium Chloride	0.6	3.5	1	1	1	<2 at 10 w/o	<2 at 10 w/o	<20 at 10 w.o	<2 at 10 w/o	<2 at 10 w/o
Sodium Hydroxide	10	30	<1	<1	<1	<2	>50	<20	<2	-
Ammonium Hydroxide	10	18	4	<1	<1	<2	<2	<2	<2	<2
Ferric chloride	1.2	20	>3990	>4587	993	>50	>50	>50	20-50	<20

*Values given only as <2, <20, 20-50, or >50.
- indicates no data.

Table 7

200-Hour Mass Loss and Average Penetration Rate
Continuous Exposure, 95°C

Solution (Room Temp)	Concentration		IC-50		IC-218		IC-221	
	Molar	Weight % (w/o)	Mass Loss in 200-h (g/cm²)	Average Penetration Rate (mpy)	Mass Loss in 200h (g/cm²)	Average Penetration Rate (mpy)	Mass Loss in 200h (g/cm²)	Average Penetration Rate (mpy)
Sulfuric Acid	1 18	9 96	$>2.33\times10^{-1}$ (a) 3.20×10^{-2}	>490 (a) 67	1.44×10^{-1} 8.39×10^{-3}	302 18	$>4.12\times10^{-1}$ (a) 9.36×10^{-3}	>867 (a) 20
Hydrochloric Acid	1 12	4 37	2.43×10^{-1} $>2.40\times10^{-1}$ (a)	510 >505 (a)	$>2.50\times10^{-1}$ (a) $>2.42\times10^{-1}$ (a)	>526 (a) >509 (a)	5.25×10^{-1} $>6.35\times10^{-1}$ (a)	1101 >1337 (a)
Nitric Acid	1 8 16	6 40 70	$>2.43\times10^{-1}$ (a) $>2.27\times10^{-1}$ (a) $>2.45\times10^{-1}$ (a)	>511 (a) >477 (a) >514 (a)	$>2.52\times10^{-1}$ (a) 2.12×10^{-2} 2.83×10^{-2}	>530 (a) 45 59	1.47×10^{-1} 2.68×10^{-2} 9.42×10^{-2}	308 56 198
Hydrofluoric Acid	1	2	$>2.55\times10^{-1}$ (a)	>534 (a)	$>2.34\times10^{-1}$ (a)	>49[illegible] (a)	1.91×10^{-1}	402
Phosphoric Acid	1	10	$>2.26\times10^{-1}$ (a)	>475 (a)	$>2.46\times10^{-1}$ (a)	>517 (a)	3.58×10^{-1}	752
Oxalic Acid	1	10	-7.63×10^{-3} (b)	(b)	1.25×10^{-3}	3	1.70×10^{-3}	4
Acetic Acid	1	6	1.94×10^{-3}	4	1.93×10^{-3}	4	1.97×10^{-3}	4
Sodium Chloride	0.6	3.5	7.56×10^{-5}	0.2	-7.82×10^{-5} (b)	(b)	-9.68×10^{-5} (b)	(b)
Sodium Hydroxide	10	30	-8.15×10^{-5} (b)	(b)	-1.60×10^{-4} (b)	(b)	-9.47×10^{-5} (b)	(b)
Ammonium Hydroxide	10	18	-1.40×10^{-4} (b)	(b)	-1.60×10^{-4} (b)	(b)	-1.88×10^{-4} (b)	(b)
Ferric chloride	1.2	20	$>2.49\times10^{-1}$ (a)	>524 (a)	$>2.3\times10^{-1}$ (a)	>484 (a)	$>5.38\times10^{-1}$ (a)	>1132 (a)

(a) Total dissolution in <200h
(b) Very small mass gain

erosion, and oxidation resistance; aircraft fasteners because of lower density and ease by which the desired strength is achieved; heating elements because of better oxidation resistance and three to five times longer life in simulated tests; automotive turbochargers because of higher fatigue resistance and lower density; pistons and valves because of better wear resistance and capability of developing thermal barrier by high-temperature oxidation treatment; bellows for expansion joints for corrosive environments because of good aqueous corrosion resistance; tooling because of high-temperature strength and wear resistance developed through preoxidation; and permanent molds because of the ability to develop a thermal barrier coating by high-temperature oxidation.

14. COMMERCIAL SUPPLIERS

The nickel aluminide alloy technology has been transferred from ORNL to at least three commercial users and producers through licensing arrangements. These include Armada Research Corporation in Hamburg, Michigan, Armco Inc. in Middletown, Ohio, and Metallamics in Traverse City, Michigan. For further information on alloys, readers can contact ORNL, Armada, Armco, or Metallamics.

15. SUMMARY

Ductilization effect of small addition of boron to ordered nickel aluminides observed in the small experimental heats were confirmed for large heats in the laboratory and heats melted commercially. Air induction was shown to be the best method for melting large heats. Electroslag remelting is probably the best method for secondary processing of air-melted material. Nickel aluminides can be cast by all conventional techniques and near-net-shape processes. The secondary processing of large ingots is most successful by extrusion. Consistent success of hot extrusion requires surface protection, which can be achieved by wrapping with mild steel sheet. Grain-refined ingots or extruded bar can be hot forged over a range of temperature from 1000–1200°C. Cast and hot processed ingots can be cold rolled to various levels with intermediate anneals at 1100°C. Near-net-shape cast sheet and bar can be successfully processed by cold rolling or cold swaging and wire drawing.

Nickel aluminides can be processed by all powder metallurgy processes available including superplastic forming.

Tensile, creep, fatigue and crack growth rate data are starting to become available for these alloys. Cold working is the most effective method of strengthening the alloys up to 800°C. Properties of these alloys are sensitive to grain size at both low and high temperature.

For low temperature ($\leq$700°C) applications, fine-grain microstructure is desirable for high strength properties. At high temperatures ($\geq$700°C), large-grain wrought or cast structure are most desirable because of high creep strength.

Physical property data of nickel aluminides show that base alloy has twice the thermal conductivity of nickel-base alloys. Alloys containing chromium (IC218 and IC221) are expected to have approximately 30% higher thermal conductivity than nickel-base alloys. Thermal expansion behavior of nickel aluminides is the same as nickel-base alloys.

At room temperature, corrosion resistance of nickel aluminides appears as good or better than type 304, Inconel 600, and possibly Hastelloy C in concentrations listed in the paper for hydrofluoric acid, phosphoric acid, oxalic acid, acetic acid, sodium chloride, sodium hydroxide, and ammonium hydroxide solutions. At 95°C, nickel aluminides have good corrosion resistance in organic acids (oxalic and acetic), sodium chloride, and the bases (sodium hydroxide and ammonium hydroxide solutions). High temperature oxidation in the temperature range of 1000-1100°C forms a protective alpha alumina on the surface.

Nickel aluminides can be welded by electron beam and gas tungsten arc welding techniques. Several applications for these alloys have been identified, along with potential suppliers who have recently licensed the nickel aluminides technology from ORNL.

16. ACKNOWLEDGMENTS

The author wishes to extend his special thanks to Armco Research and Technology, Carpenter Technology Corporation, Steel Casting Engineering, Ladish Company, Powder Products Division of Cytemp Specialty Steel, Homogeneous Metals, Inc., Metallamics, Haynes International, Special Metals Corporation, Combustion Engineering Incorporated, AMAX Special Metals Corporation, and Wyman-Gordon Company for their participation; K. S. Blakely, C. E. Dunn, C. W. Holland, Jr., C. R. Howell, and J. D. Vought for their melting processing efforts at ORNL; R. H. Baldwin for plotting of test data; J. R. Weir, Jr. for his valuable comments; T. K. Roche and J. R. Weir, Jr. for internal paper review; and M. L. Atchley and F. C. Stooksbury for preparing the manuscript.

17. REFERENCES

1. Liu, C. T. and Stiegler, J. O., Science, 226, 636 (November 1984).
2. Aoki, K. and Izumi, O., Nippon Kinzoku Gakkaishi 43, 1190 (1979).
3. Liu, C. T., White, C. L., and Horton, J. A., Acta Metall., 33(2), 213 (1985).
4. Liu, C. T. and Sikka, V. K.,

J. Met. 38, 19 (1986 .

5. Sikka, V. K. and Loria, E. A., "Deformation Behavior of Powder Metallurgy Ni_3Al-Hf-B," Proceedings of CIM's 25th Annual Conference of Metallurgists, 1986, p.293.
6. Sikka, V. K., Liu, C. T., and Loria, E. A., Enhanced Properties of Structural Metals Via Rapid Solidification, ed. F. H. Froes, American Society for Metals, Metals Park, Ohio, 1987, p. 417.
7. Sikka, V. K., Proceedings of Mater. Res. Soc., Vol. 81, Materials Research Society, 1987, p. 487.
8. Sikka, V. K., First ASM Europe Technical Conference, cosponsored by SFM, Paris, France, 1987, to be published.
9. Sikka, V. K., Advanced Materials and Manufacturing Processes, 1988, to be published.
10. Stoloff, N. S., Fuchs, G. E., Kuruvilla, A. K., and Choe, S. J., p. 247 in MRS Proc. Vol. 81, ed. N. S. Stoloff, C. C. Koch, C. T. Liu, and O. Izumi, 1987.
11. Chang, K. M., Huang, S. C., and Taub, A. I., General Electric Company, Report 86CRD202, October 1986.
12. Stoloff, N. S., submitted to Int. Met. Rev., 1988.
13. Williams, R. K., Graves, R. S., Weaver, F. J., and McElroy, D. L., Proceedings of Mater. Res. Soc., Vol. 39, Materials Research Society, 1985, p. 505.
14. Williams, R. K., Graves, R. S., and Weaver, F. J., J. Applied Physics 61(4), 1486 (February 1987).
15. Buchanan, R. A. et al., "Aqueous Corrosion Characteristics of Nickel Aluminides," final report to the Nickel Development Institute, Toronto, Ontario, April 1988.
16. Cathcart, J. V., "Corrosion of Ni and Ni-Fe Aluminides," presentation made at workshop on nickel aluminides, Oak Ridge, TN, May 13, 1986.

"The submitted manuscript has been authored by a contractor of the U.S. Government under contract No. DE-AC05-84OR21400. Accordingly, the U.S. Government retains a nonexclusive, royalty-free license to publish or reproduce the published form of this contribution, or allow others to do so, for U.S. Government purposes."

2nd International SAMPE Metals Conference
August 2-4, 1988

CONTINUAL OPPORTUNITIES FOR SUPERALLOY DEVELOPMENT

Keh-Minn Chang
GE Corporate Research and Development
Schenectady, New York

Abstract

Superalloys have played a significant and unique role in the development of the aircraft turbine engine. The entire hot-end components of current engines are totally dependent on superalloys. Although the modern engine has been designed to operate at a maximum gas temperature above the melting point of superalloys, engine hardware serviced at intermediate temperatures still relies upon the superior high-temperature properties of superalloys to meet the specific requirements. The opportunity appears constantly for superalloy development when new processing techniques or new design methods are employed. Two examples of a new superalloy approach are to be demonstrated: the high-strength alloy with defect tolerance and the casting alloy with weldability.

The application of defect-tolerance requirements to the turbine disks has introduced a new challenge to superalloy designers. Powder metallurgy (P/M) superalloy compositions currently available seem to satisfy the strength requirements at elevated temperature, but none of them can provide the adequate defect tolerance evaluated by the fracture mechanics methodology. A new direction for the disk alloy development would emphasize the crack-growth resistance at high temperatures, especially under the time-dependent conditions.

Structural casting can reduce drastically the number of component parts and the machining hours. More and more supporting engine hardware, like frames and cases of various stages, is made through the investment-casting technique. New casting alloys are urgently needed to have the improved temperature capability without sacrificing the processing characters, such as castability and weldability.

1. INTRODUCTION

Current research effort has completely neglected the development of new superalloys. Since the early 1980s when single-crystal superalloy was introduced into aircraft engines, many people have questioned the need of the future development of superalloys. In spite of the progress in temperature capability of superalloy single crystals in the recent years, the major research effort on high-temperature materials has been shifted from superalloys to others, such as intermetallic alloys, refractory metals, and ceramic composites.

A misleading concept resulting from such a change is that all potential superalloy compositions have been explored, and the opportunity for inventing a new alloy with improved properties is no longer feasible.

In a modern aircraft engine, superalloys make up more than 30% in weight for different components that cover the entire hot section with the service temperature from 500° to 1100°C. Within this wide temperature range, engine components are made from different alloys through different processes into various forms. Each component has its own unique combination of duty requirements that are included in the alloy specification. For example, there are components, such as turbine blades, which must be extremely creep-resistant, and components, such as turbine disks, which require high-tensile strength and long fatigue life. The optimum property of components can only be achieved by proper alloy composition and adequate processing.

For the past two decades the improvement of superalloy property was predominately attributable to new metal processing techniques (1). New methods, like directional solidification (DS), single-crystal, powder metallurgy (P/M), large structural casting, etc., had surpassed alloy development as the prime forward drive to enhance the superalloy performance. In the "age of processing," the research efforts in alloy development were limited and concentrated only on the application of high temperature, such as single crystals (2,3). In contrast, there is no new alloy designed specifically for intermediate temperature, high-stress applications that has been commercialized in the past twenty years.

As the processing techniques become more and more mature, the idea of adapting an old alloy composition by advanced processing is facing too many drawbacks. The alloy properties cannot attain a significant improvement until a major commitment is made to alloy development. Therefore, many opportunities of superalloy development reappear today, especially in the area where alloy design advancement was lagging in the past.

Before demonstrating some successful examples of recent superalloy development, a brief review on the metallurgy of high-strength superalloys will provide a better understanding on the basic principles served as the foundation for these alloy developments. A more detailed survey about this subject is available in the literature (4-6).

2. METALLURGY OF HIGH-STRENGTH SUPERALLOYS

Precipitation-hardened nickel-base superalloys have been used exclusively for high-stress intermediate temperature applications in turbine engines. In addition to their excellent environmental resistance, these alloys exhibit an attractive alloy strength extending to high temperatures. Figure 1 shows schematically the typical relationship between alloy strength and temperature of a high-strength superalloy. The tensile strength, in the level of 1350 MPa or above, remains approximately constant with increasing temperature up to ~ 0.5 to 0.6 T_m, melting point. The temperature at which the alloy strength becomes temperature-sensitive can decrease somewhat for the rupture strength. The overall property of these alloys stands for the best material performance available at present.

As seen in Table 1, the composition of high-strength superalloys is relatively complicated; but their microstructure is simple when compared with titanium or with steel alloys. These alloys can be produced in the

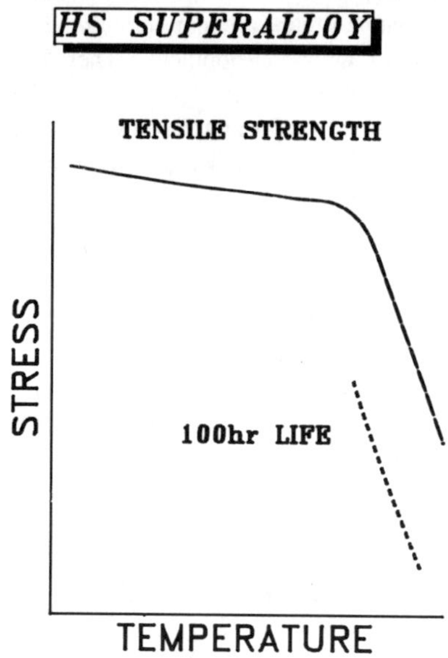

Figure 1 Schematic plot of the relationship between alloy strength and temperature for high strength superalloys.

cast, wrought and powder forms, with each process either improving specific properties or resulting in economic advantage for certain applications. The major constituent in their microstructure consists of a certain volume fraction of coherent precipitation phases adequately dispersed in a solid-solution, face-centered-cubic matrix. The precipitate content can be from 20% to 60% in volume. Appropriate control on the composition and the morphology of the precipitation phase is the key factor in attaining the excellent mechanical properties.

Precipitation-hardening elements generally employed in nickel-base superalloys include four alloying elements: Al, Ti, Nb, and Ta. This class of elements has a common feature in the nickel-base alloys: their solubility increases with temperature and reaches to a maximum at the eutectic temperature. The intermetallic phases forming the eutectic with the fcc matrix have a nominal chemistry form of Ni_3X, X is the hardening element, and their crystal lattice consists of ordered closed packing planes

Table 1 Chemical Composition of High-Strength Superalloys

	Nominal Composition, wt%													
Alloy	Ni	Cr	Co	Fe	Mo	W	Al	Ti	Nb	Ta	Hf	Zr	C	B	ppt v%
							Cast & Wrought							
IN 901	Bal	12.5	-	36.0	5.7	-	0.2	2.8	-	-	-	-	0.050	0.015	31.5
IN 718	Bal	19.0	-	18.5	3.0	-	0.5	0.9	5.1	-	-	-	0.040	-	31.9
Waspaloy	Bal	19.5	13.5	-	4.3	-	1.3	3.0	-	-	-	-	0.050	0.006	29.0
Rene'41	Bal	19.0	11.0	-	10.0	-	1.5	3.1	-	-	-	-	0.060	0.006	33.1
Astroloy	Bal	15.0	17.0	-	5.0	-	4.0	3.5	-	-	-	-	0.060	0.015	48.5
							Powder Metallurgy							
AP 1	Bal	15.0	17.0	-	5.0	-	4.0	3.5	-	-	-	0.05	0.020	0.030	48.5
MERL 76	Bal	12.5	18.5	-	3.2	-	5.0	4.4	1.4	-	0.40	0.06	0.020	0.020	61.1
AF2-1DA	Bal	12.0	10.0	-	2.8	6.5	4.6	2.8	-	1.5	-	-	0.040	0.010	52.1
Rene'95	Bal	13.0	8.0	-	3.5	3.5	3.5	2.5	3.5	-	-	0.05	0.060	0.010	49.0
AF 115	Bal	15.0	10.7	-	2.8	5.9	3.8	3.9	1.7	-	0.75	0.05	0.050	0.020	55.3

Table 2
Phase Relationship of Precipitation Hardening Elements With Nickel

Alloying Element	Al	Ti	Nb	Ta
Ni_3X Crystal Structure	$L1_2$	DO_{24}	DO_a	DO_{22}
Ordering on the closed packing	triangle	triangle	rectangle	rectangle
Eutectic Temperature (°C)	1385	1304	1282	1360
Maximum Solubility (at%)	20.2	13.7	12.6	14.0
Solubility at 700°C (at%)	10.3	9.0	3.3	<3.0

similar to {111} planes in fcc lattice. The data extracted from binary phase diagrams (7) are listed in Table 2.

In a precipitation-hardening superalloy, the precipitate phases can be completely dissolved in an fcc matrix by annealing above the temperature at which the total content of hardening elements equals the solubility. This temperature is called the precipitate solvus temperature. A supersaturated solid solution results, and a subsequent aging treatment can develop a uniform distribution of fine precipitate particles. The hardening element atoms will segregate together by diffusion and will form the ordered structure on the closed packing planes, {111}, of fcc matrix. Depending upon the final equilibrium phase, the ordering pattern on {111} planes is either in triangle (for Al and Ti) or in rectangle (for Nb and Ta). The triangularly ordered fcc precipitates are designated as γ', which has a $L1_2$ crystal lattice. The precipitates with a rectangular fcc ordering are called γ'' having a DO_{22} crystal structure. Figure 2 shows the unit cell and the closed packing planes of two precipitate crystals.

Without forming a high-energy interface, both types of precipitates can homogeneously nucleate throughout the alloy matrix. After reaching to the adequate size, the alloy is fully hardened and maintains its superior strength to elevated temperatures.

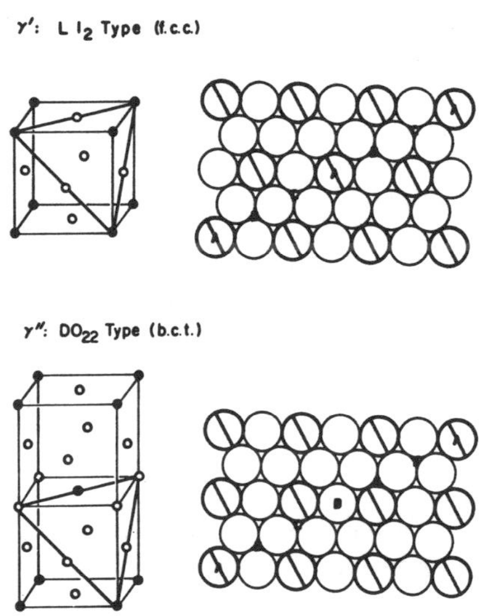

Figure 2 Schematic plots of crystal unit cell and closed packing plane for strengthening precipitate phases in superalloys.

The coherency strain caused by the lattice misfit between alloy matrix and precipitate phase is relatively large (> 0.5%) for γ'' precipitates, which results in a plate-like morphology. In contrast, the γ' precipitates with a low coherency strain (< 0.3%) are usually in a spherical or a cuboidal shape. In the equivalent volume fraction, the γ'' precipitates can strengthen the superalloy more effectively than the γ' precipitates.

In addition to hardening the alloy, the precipitates have also been employed with thermomechanical processing (TMP) to control the alloy microstructure. The existence of precipitate particles during the hot working process can prevent the grain growth and retain the residual strains. One practical example is the directly aged 718 alloy (DA 718), whose tensile and fatigue strengths are enhanced over those of conventional 718 alloy by TMP (8). If the superalloy with a high level of precipitate content is processed through powder metallurgy, the precipitate particles can offer the alloy an advantage of superplastic deformation under the specific forging condition (9).

3. DEVELOPMENT OF CRACK GROWTH RESISTANT ALLOYS

Low-cycle fatigue has been traditionally one of the major concerns for high-strength superalloys in turbine engine applications. The concern became a critical issue when the P/M superalloys were employed (9). The problem was partly caused by the cleanliness of powder batches and was partly attributable to the high operating stress designed for the P/M superalloys. In spite of the superior alloy strength, these alloys had limited fatigue life in some advanced designs. A new methodology based on linear elastic fracture mechanics (LEFM) has been adapted to low-cycle fatigue design so that the component life in turbine engines can be well managed.

LEFM methodology permits a quantitative characterization of fatigue crack propagation in a component under the cyclic stress, $\Delta\sigma$. The crack growth rate, da/dN, of a given crack of length, a, is a function of the cyclic range of the applied stress intensity, ΔK, which has a general form as follows: $\Delta K = \Delta\sigma \times \sqrt{a} \times f$, where f is a geometric factor. The component life depends upon two important material parameters: the maximum allowable defect size and the crack-growth rate under the operation conditions.

For every alloy there exists a fatigue threshold, ΔK_{th}, the limit below which crack growth will not occur. The criterion to employ LEFM methodology is then determined by the threshold crack length, a*, a function of $(\Delta K_{th}/\Delta\sigma)^2$. P/M superalloys exhibit an alloy strength higher than other conventional cast and wrought superalloys, and thus suggest a higher design stress. Furthermore, a high-strength alloy usually shows a low fatigue threshold, ΔK_{th}. As a result, a* for P/M superalloys becomes so small that it is not possible to be satisfactorily detected by the current nondestructive testing (NDT). As the metal processing engineers are trying to produce clean powders of high-strength superalloys, superalloy metallurgists must accept the challenge to understand the fatigue crack propagation (FCP) behavior in P/M superalloys at elevated temperatures and to develop new alloys with excellent crack growth resistance.

Figure 3 shows the crack-growth rate of a typical high-strength superalloy, IN 718, as a function of fatigue frequency at different temperatures under a fixed-cyclic stress intensity, ΔK (10). Two regimes of FCP behavior are clearly observed. At 400°C,

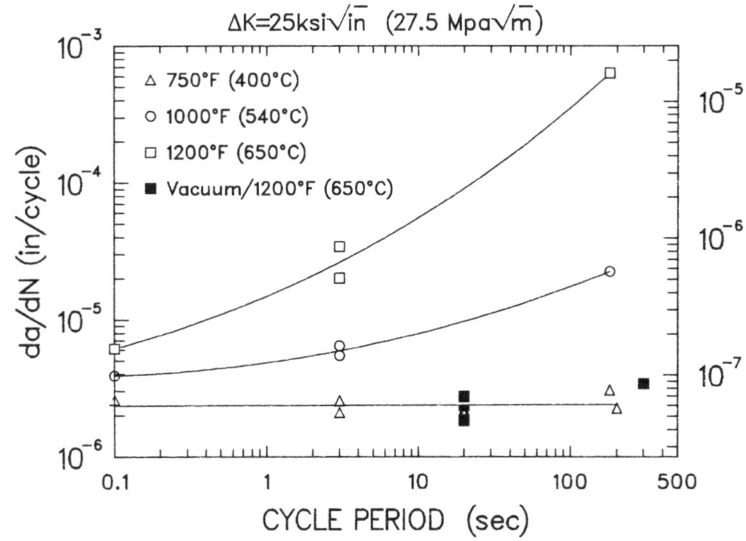

Figure 3 The relationship between crack growth rate and period of fatigue cycles in IN 718 alloy.

the alloy shows a constant da/dN under a given ΔK. The crack growth is completely controlled by the cyclic stress intensity, independent of period and wave form of fatigue cycle. In this cycle-dependent regime, crack-growth rate is believed to be insensitive to the variation of microstructure and alloy chemistry. As the temperature increases, the crack growth starts to accelerate substantially even if ΔK is kept at the same value. For a given temperature, da/dN increases dramatically with the cycle period. Such a time-dependent behavior is aggravated by increasing temperature.

The fractography on the fracture surface suggests that the change of fracture mode corresponds to the transition from the cycle-dependent to time-dependent regime. As seen in Figure 4, transgranular fracture consisting of regular striations is

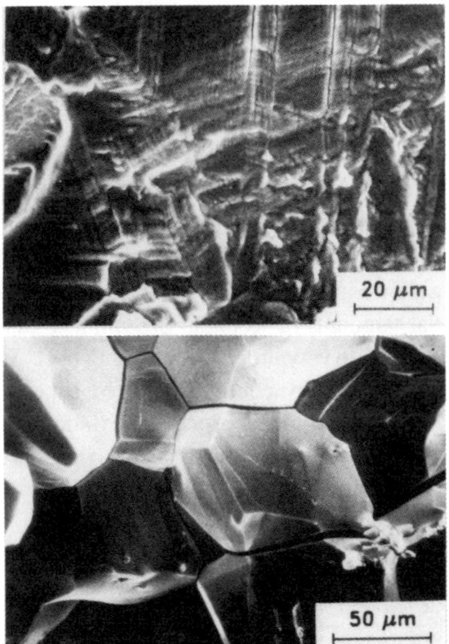

Figure 4 Fatigue cracking fracture mode: (a) transgranular in cycle- dependent regime; (b) intergranular in time-dependent regime.

observed in the cycle-dependent regime, while intergranular cleavage occurs in the time-dependent regime.

One of the most effective methods to improve the crack growth resistance is to heat-treat P/M superalloys above their precipitate solvus. Such a so-called "supersolvus" solutioning can produce a large recrystallized grain structure in contrast to that of the partial-solutioned commercial P/M superalloys. Figure 5 compares the crack growth curves for P/M Rene'95 alloys processed by two different heat treatments. Two testing conditions are used: a 20-cpm sinusoidal fatigue cycle at 400°C simulating the cycle-dependent condition, and a 90-second hold-time imposed on the maximum load of a 20-cpm fatigue cycle at 650°C simulating the time-dependent condition. An order of magnitude improvement in da/dN is observed in the latter condition, although some degree of time-dependence still exists in the supersolvus annealed materials. In the case of cycle-dependent FCP, the improvement is marginal, as expected.

One sacrifice resulting from the supersolvus solutioning is the attainable alloy strength. Large grain structure makes the alloy sensitive to the quench cracking, and a slow cooling rate will generate a substantial amount of cooling precipitates that contribute very little strengthening (11). In addition, many commercial P/M superalloys have a limited processing window for supersolvus solutioning since they are designed to be subsolvus-annealed.

Many new P/M superalloys with excellent crack-growth resistance have been developed in GE internal research programs that were started in the early 1980s. These new alloys combine the maximum advantages of alloy chemistry and metal processing to provide the adequate alloy strength, rupture life, as well as crack growth resistance. One example is shown in Figure 6. The alloy demonstrates a crack-growth behavior without time-dependence at 650°C.

4. DEVELOPMENT OF WELDABLE ALLOYS

Large structural components in turbine engines, such as compressor housings, diffuser cases, turbine and combustor cases, and turbine frames, have been increasingly produced by the investing casting of high-strength superalloys since the 1980s (12). This approach, which involves casting pour weights of up to a thousand pounds level, is an effort to meet new design needs and to reduce engine manufacturing costs. For these engine components, alloy weldability is the vital requirement for fabrication. Repair welding is always necessary to remove the surface defects caused by the residual slag. Welding is also a common procedure to join different parts.

Welding high-strength nickel-base superalloys, which develop their strength through the precipitation-hardening process, has been a challenging fabrication problem in the manufacture of gas turbine engines. A considerable amount of review literature (13-15) is available that deals with the weldability of precipitation-hardened superalloys. In general, most of these alloys are not readily weldable, and welding becomes more difficult as the volume fraction of the precipitation phase is increased. IN 718, which has the most extensive industrial application, partly because of its good welding characteristics, is the only exception among commercially available superalloys. To increase the maximum operating temperature has been one of the most vital design thrusts since the aero gas turbine has existed. IN 718 was developed in the late 1950s (16), and the turbine sections consisting of components made from

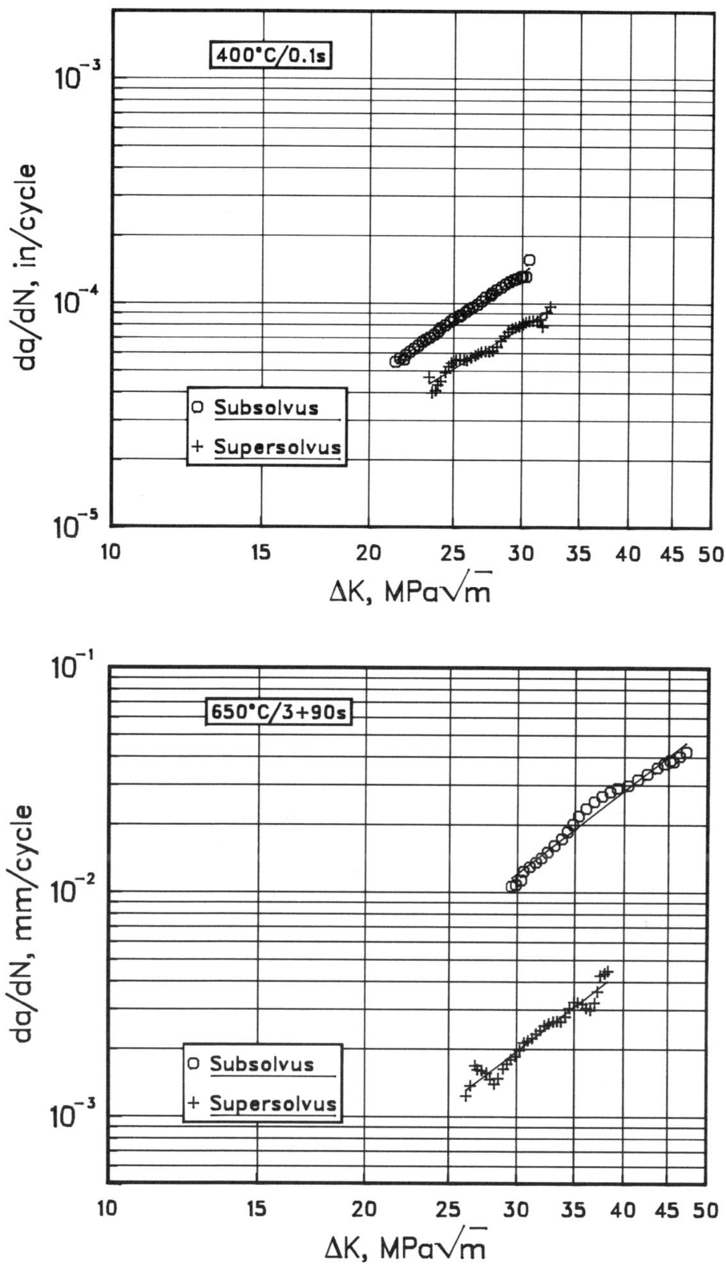

Figure 5 Comparison of fatigue crack propagation for supersolvus- and subsolvus-solutioned P/M Rene'95 alloys under (a) cycle-dependent condition; (b) time-dependent condition.

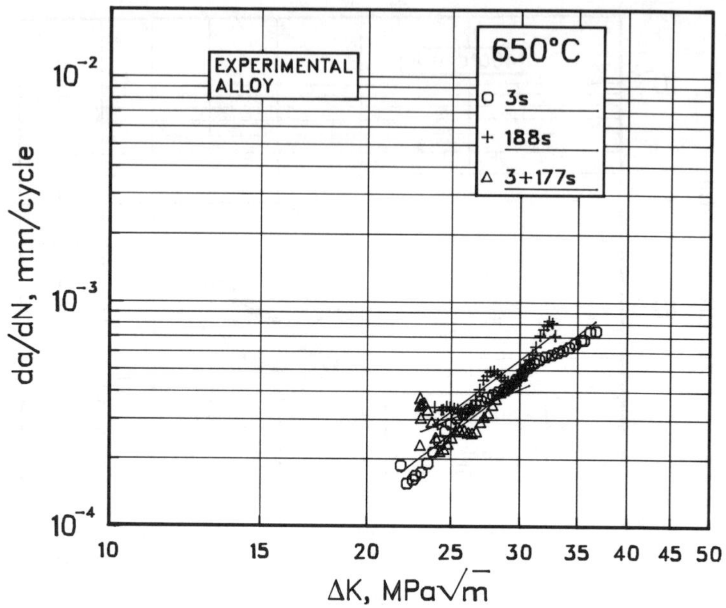

Figure 6 Fatigue crack growth curves of an experimental alloy showing no time dependence at 650°C.

this alloy have already reached the upper limit of its temperature capability in current gas turbine engines. The need for weldable superalloys, which are capable of operating at higher temperatures above IN 718, has been forecast in the designing of the advanced aircraft engines.

The cracking mechanisms that accounted for welding precipitation hardened superalloys directly the kinetics of the precipitation reaction. The alloy could be effectively strengthened as soon as a certain volume of precipitates form in any thermal history. When alloy strength is developed, the critical strain causing cracks becomes sufficiently low. Since the rate of hardening during heat treatment is related in part to the content of precipitation-hardening elements, the tendency for cracking in the γ'-strengthened superalloys is commonly reported to be a function of aluminum and titanium contents (17). The margin line for acceptable weldability is found to be (Al wt.% + 1/2 Ti wt.%) <3.0. This empirical rule implies that all γ'-hardened alloys that have an equivalent weldability, as in IN 718, always show temperatures inferior to IN 718.

The crystal structure of strengthening precipitates is the key difference of IN 718 from other precipitation-hardened superalloys. The addition of 5.1 wt.% Nb in IN 718 makes most coherent precipitates developed during aging treatments to form the DO_{22}, ordered bct structure, though a small percentage of ordered fcc γ' precipitates can still be detected (18). The unique precipitation phase, designated at γ'', inhibits a great lattice misfit along its tetragonal axis (c-axis) with the alloy matrix (19). Large coherency strains associated with

Table 3 - Alloy Series Designed to Study Strengthening Precipitates

Designation	Composition wt.%											C_H*	R**
	Ni	Cr	Co	Mo	Al	Ti	Ta	Nb	C	B	Zr	at%	%
P1	bal.	19	13	4.0	1.5	3.0	--	--	--	.01	0.5	6.71	0.0
P2	bal.	19	13	4.0	1.5	2.0	3.0	--	--	.01	0.5	6.62	14.5
P3	bal.	19	13	4.0	1.0	2.0	6.0	--	--	.01	0.5	6.68	29.6
P4	bal.	19	13	4.0	.5	2.0	6.0	1.5	--	.01	0.5	6.61	44.9
P5	bal.	19	13	4.0	.5	1.0	6.0	3.0	--	.01	0.5	6.38	62.4

* C_H is the total hardening element content (Al + TI + TA + NB) in at%.

** R is the (Nb + Ta) percentage in C_H; i.e., R = (Ta + Nb) /(Al + Ti + Ta + Nb) in at%.

the formation of γ'' precipitates in nickel-base superalloy impose two significant effects with respect to the welding topic studied here. One effect is that γ'' hardening through coherency strengthening becomes so efficient that only a low volume fraction of precipitation phase is required to attain the desirable strength. The second effect is that the nucleation rate of γ'' precipitates is comparably slower because of the additional activation energy from large elastic strain energy. Both effects provide the alloy sufficient advantages from the weldability viewpoint.

Table 3 lists the nominal compositions of an alloy series that was designed to discover the transition of precipitate crystal structure from γ' to γ''. Based on the same alloy matrix and the same total atomic percentage of precipitation-hardening elements (Al+Ti+Ta+Nb $\simeq$ 6.6 at %), the γ'' forming elements Ta and Nb were added to substituting γ'-forming elements Al and Ti with the increment ~15% of total hardening elements.

The transmission electron microscopic (TEM) investigation of the strengthening precipitate phase developed after aging treatments in this alloy series identified both ordered Ni_3X structures, γ', and γ''. The precipitate structure and quantity depend upon the percentage of (Ta+Nb) substitution. Low (Ta+Nb) percentage forms γ', which has a spherical-to-cuboidal morphology as seen in alloy P1, P2 and P3 (Figure 7a). When sufficient (Ta+Nb) replaces (Al+Ti), such as alloy P4 whose (Ta+Nb) percentage is 44.9%, plate-like γ'' precipitates start to appear. Almost equal volume fraction of γ' and γ'' particles are observed in alloy P4. Further increases in (Ta+Nb) percentage above 62% can convert all precipitation phases into γ''. The dark field image taken from a superlattice reflection reveals only γ'' precipitate existing in alloy P5 (Figure 7b).

As mentioned in the prior section, γ' and γ'' are the transition precipitates having the ordered structure determined by the equilibrium phase, which is the function of the (Ta+Nb) percentage in total hardening

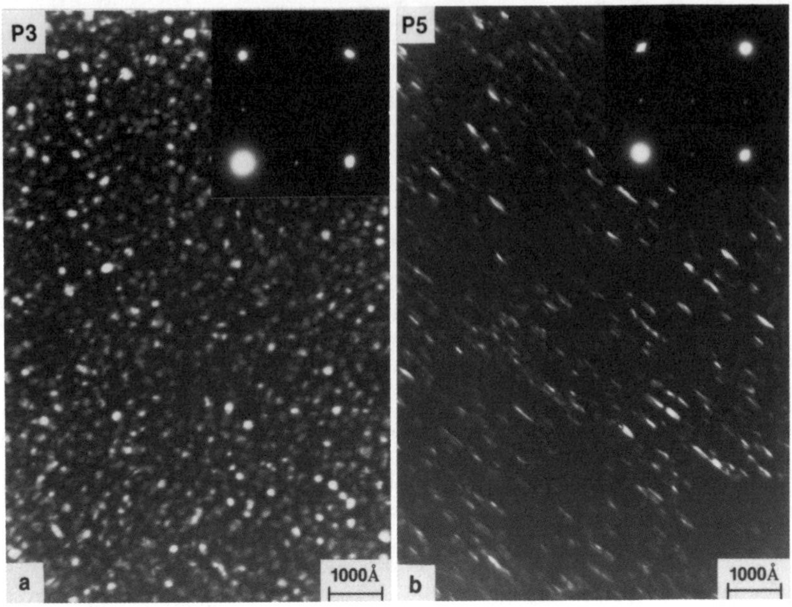

Figure 7 Transmission electron microscopy of strengthening precipitates (a) γ' in alloy P3; (b) γ'' in alloy P5.

element content. Also to be noted is that two transition precipitates, γ' and γ'', coexist in a range of (Ta+Nb) percentage. Prolonged aging of alloy P4 coarsens both γ' and γ'' without any indication that one precipitate will replace or consume another. The idea of a two-phase region in a phase diagram can be applied to alloy P4, and the "transition" phase boundaries can be obtained by analyzing γ' and γ'' precipitate chemistries under analytical STEM. These results are given in Table 4; the complete scheme of microanalysis work is reported elsewhere (20).

The total amount of hardening elements, Al+Ti+Ta+Nb, in the γ' and γ'' precipitates is equal to 23.3. and 22.1 at%, respectively. These values approach the stoichiometric composition 25 at%. The critical (Ta+Nb) percentages for γ' and γ'' precipitates, which may define the "transition" phase boundaries, are 26.6% and 55.2%, respectively. The accuracy of experimental data is expected within 5%, and the results of the present study have an error of no more than 10%.

To ensure the strengthening precipitates being γ'' in nickel-base superalloys, the hardening elements should be controlled so that the (Ta+Nb) percentage with respect to the total hardening element content would be above 62.5%. The precipitation behavior might be influenced by alloying elements other than hardening elements, but the observed effects are secondary.

The single disadvantage of using γ'' strengthening in nickel-base superalloys is the phase stability after prolonged exposure at high temperature. The γ''-precipitation phase is not the equilibrium phase, and it will be replaced, when over-

Table 4 - γ' and γ'' Precipitate Chemistries (at %)

	Ni	Cr	Co	Mo	Al	Ti	Ta	Nb	C_H	R
overall	60.0	20.1	12.9	2.4	1.3	2.5	2.1	1.0	6.9	44.9
γ'	68.9	1.0	6.7	0.2	6.7	10.4	3.8	2.4	23.3	26.6
γ''	64.0	5.2	5.4	3.3	4.5	5.4	6.1	6.1	22.1	55.2

* See Table 3 for the definitions of C_H and R.

aged, by the stable δ phase (DO_a type), which can degrade the alloy properties. Very limited literature is available that discusses how to improve the phase stability of γ''-hardened alloys.

Figure 8 shows the influence of Fe content on the tensile and rupture strengths of the IN 718 alloy system (16). Creep-rupture strength at high temperatures reaches maximum value when no Fe is added. Nevertheless, IN 718 contains 18.5 wt.% Fe to maintain a high level of room temperature tensile properties, especially high yield strength. Improving the phase stability and the temperature capability is feasible through the modification of alloy chemistry.

A five-year GE alloy development program based on this approach involved extensive studies of entire categories of alloying elements, which include solid solution strengtheners, refractory metal elements, precipitation hardening elements, grain boundary refiners, and their combined effects on the high-temperature properties of γ''-hardened superalloys. Based on the above criterion of precipitation reaction, a new alloy, designated as Rene'220C, has been successfully invented (21). The new alloy offers at least a 50°C advantage in temperature capability over 718 alloy, as illustrated by Figure 9. The weldability and the castability were demonstrated through many engine hardware trials (Figure 10). Now Rene'220C has been certified for used in advanced aircraft engines. The first uses will be in a combustion case and in a turbine frame, a

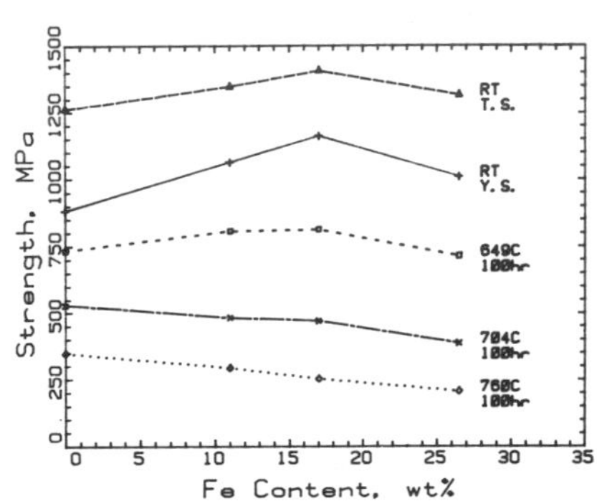

Figure 8 The effect of Fe on the strengths of IN 718 alloy.

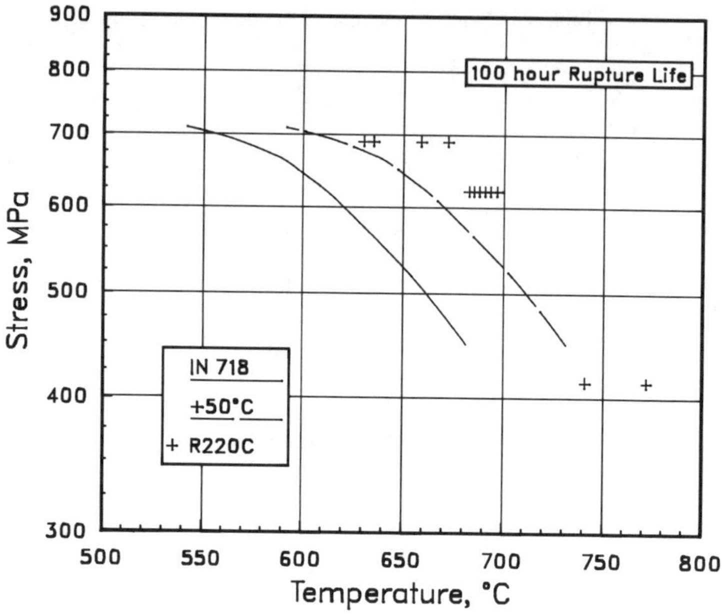

Figure 9 Rupture strength of new casting alloy Rene'220C in comparison with that of IN 718.

major part of the engine's structural skeleton. Until now, the frame had been made from several parts welded together. But the superior castability and the properties of Rene'220 make it possible to produce it as a single casting. This sharply reduces the fabrication cost.

5. CONCLUSION

The success of several GE internal programs on new superalloys fits into a larger overall effort aimed at increasing the operating temperature of aircraft engines. For the long-term, alloys based on more exotic metals or intermetallic compounds, and even beyond, ceramic- or carbon-fiber composite, could become attractive technically and commercially. But until the technical breakthroughs are made, the nickel-base superalloys remain the backbone of turbine engines. Two recent examples, the crack growth resistant alloy and the weldable casting alloy, suggest that significant progress is still available with this well-known class of alloys.

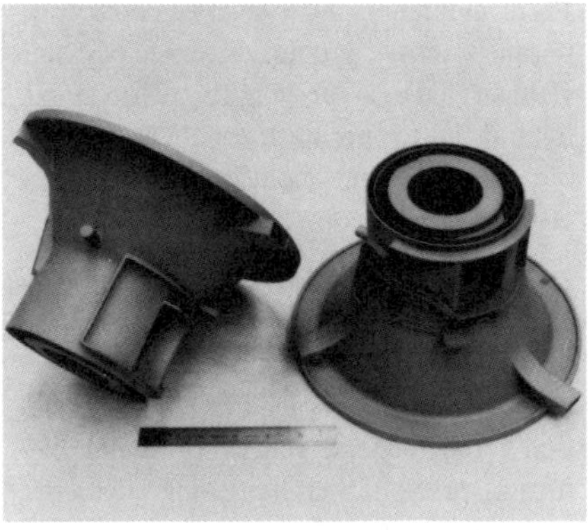

Figure 10 Engine bearing houses made from the investment casting of new alloy Rene'220.

6. ACKNOWLEDGMENT

The author appreciates greatly the encouragement and the helpful discussion from Dr. M.G. Benz and Dr. M.F. Henry, GE-Research and Development Center, Dr. A.H. Nahn, Mr. R. McDaniel, Mr. T.J. Kelly, Mr. R. Carlson, Dr. R.G. Menzies, and Mr. D.D. Krueger, GE-Aircraft Engines. He also acknowledges gratefully the technical assistance of experimental work from his colleagues, E.H. Hearn, B.J. Drummond, E.L. Hall, H. Moran, R.G. Trimberger, and R.O. Auer.

7. REFERENCES

1. Sims, C.T., "A History of Superalloy Metallurgy for Superalloy Metallurgists," *Superalloys 1984,* ed. by M. Gell, et. al., TMS-AIME, 1984, pp. 399-419.

2. Kahn, T., "Recent Developments and Potential of Single Crystal Superalloys for Advanced Turbine Blades," *High Temperature Alloys for Gas Turbines and Other Applications 1986,* ed. by W. Betz, et. al., D. Reidel Publishing Co., 1986, pp. 21-50.

3. Gell, M., Duhl, D.N., Gupta, D.K., and Sheffler, K.D., "Advanced Superalloy Airfoils," *Journal of Metals,* 1987, Vol. 39, No. 7, pp. 11-15.

4. Sims, C.T., Stoloff, N.S., and Hagel, W.C., (ed.) *Superalloys II,* Wiley Interscience, 1987.

5. Donachie, Jr., M.J., (ed.) *Superalloys Source Book,* ASM, 1984.

6. Decker, R.F., "Strengthening Mechanisms in Nickel-Base Superalloys," *Steel-Strengthening Mechanisms,* Climax Molybdenum, 1969, pp. 147-170.

7. Massalski, T.B., (ed.) *Binary Alloy Phase Diagram,* ASM, 1986.

8. Barker, J.F., Krueger, D.D., and Chang. D.R., "Thermomechanical Processing of INCONEL 718 and Its Effect on Properties," *Advanced High-Temperature Alloys,fR ASM, 1986, pp. 125-137.*

9. Howson, T.E., Couts, Jr., W.H., and Coyne, J.E., "High Temperature Deformation Behavior of P/M Rene'95," Superalloy 1984,fR ASM, 1984, pp. 275-284.

10. Chang, K.-M., "Time Dependent Fatigue Crack Propagation in Inconel 718 Superalloys," Mechanical Behaviour of Materials - V Proc. 5th Int. Conf, Beijing, China, June 1987, ed. M.G. Yan et al., Pregamon Press, Oxford, UK, 1139-1147.

11. Bartos, J.L., "Review of Superalloy Powder Metallurgy Processing for Aircraft Gas Turbine Applications," *Micon 78,* ed. H. Abrams et al., ASTM STP 672, 1979, pp. 546-577.

12. Dardi, L.E., Dalal, R.P., and Yaker, C., "Metallurgical Advancements in Investment Casting Technology," *Advanced High-Temperature Alloys,* ed. S.M. Allen et al., ASM, 1986, pp. 25-39.

13. Yeniscavich, W., in *The Superalloys,* ed. C.T. Sim and W.C. Hagel, John Wiley and Sons, New York, 1972, pp. 509-532.

14. Hicks, N.: in *The Development of Gas Turbine Materials,* ed. G.W. Methan, Applied Science, London, 1981, pp. 229-258.

15. Barker, J.F. and McHenry, H.T.: "Alloy Development for High Temperature Strength and Weldability"

presented at ASME Winter Annual Meeting, Philadelphia, PA, Nov. 1963, ASME publication 63-WA-328.

16. Eiselstein, H.L.: U.S. Patent No. 3.046.108, July 1962.

17. Prager, M. and Shira, C.S.: Weld. Res. Counc. Bull. 1968, p. 128.

18. Panlonis, D.F., Oblak, J.M., and Duvall, D.S.: Trans. ASM, 1969, Vol. 62, p. 611.

19. Cozar, R. and Pineau, A.: Met. Trans., 1973, Vol. 4, p. 47.

20. Hall, E.L., Kouh, Y.M., and Chang, K.-M.: Proc. 41th Annu. Meeting EMSA, 1983, p. 248.

21. "The Development of Rene'220C," *R&D Frontiers,* GE-CRD, Winter 1988, pp. 10-12.

DR. KEH-MINN CHANG

Dr. Chang received his BS in Physics from the National Tsing-Hua University, Hsin-Chu, Taiwan in June of 1972, his MS in Physical Metallurgy from the Univ. of Calif. at Berkeley in 1977, and his PhD in Physical Metallurgy at the Univ. of Calif. in Berkeley in 1979. From June 1979 until October of 1980 Dr. Chang was an Assistant Research Engineer at the Univ. of Calif. in Berkeley. He joined General Electric's Research and Development Center in Schenectady, New York on October 1, 1980 where he works as a metallurgist in the Physical Metallurgy Branch of the Metallurgy Laboratory. Since joining General Electric, Dr. Chang's research work has been in the area of high-strength superalloys and ductile intermetallic alloys. Dr. Chang is a member of the American Society of Metals, the Metallurgy Society of AIME, and the Materials Research Society.

2nd International SAMPE Metals Conference
August 2-4, 1988

APPLICATIONS OF HIGH TEMPERATURE/HIGH STIFFNESS ALUMINUM ALLOYS

P. S. Gilman, S. K. Das, D. Raybould,
M. S. Zedalis and J. M. Peltier
Allied-Signal Inc.
Corporate Technology
Morristown, NJ 07960

Abstract

Aluminum-transition metal alloys have been investigated by rapid quenching techniques for more than a decade. Allied-Signal's rapid solidification/powder communition and consolidation technology has developed a series of Al-Fe-V-Si alloys, with superior mechanical properties such as a combination of good room temperature ductility and fracture toughness coupled with excellent elevated temperature strength. The alloys also exhibit high modulii, excellent thermal stability and corrosion resistances. The fabricability of these alloys into extrusions, forgings and sheets has been demonstrated. As a result, a number of applications in gas turbine engines, missiles and airframe structures have emerged. These applications will be discussed and correlated with the excellent property combinations of these Al-Fe-V-Si alloys.

1. INTRODUCTION

Allied-Signal Inc., using its proprietary rapid solidification/ powder communition and consolidation technology has developed a series of quaternary aluminum-iron-vanadium-silicon alloys which combine strength, ductility, fatigue and fracture toughness levels equivalent to present aerospace 2000 and 7000 series alloys with specific strength and stiffness levels competitive with titanium alloys at temperatures up to 370°C (700°F), [1.]. The Al-Fe-V-Si alloys consists of very fine, nearly spherical $Al_{12}(Fe,V)_3Si$ (silicide) dispersoids uniformly distributed throughout the aluminum matrix. The silicides are formed from the decomposition of the rapidly solidified cellular-dendritic, microeutectic or icosahedral structures and the 10-15% transition element content of the alloys forms a very high volume fraction of silicides. The desirable microstructural morphology and high volume fraction of the silicides essentially controls the mechanical properties of these alloys, [2.]. The superior elevated temperature strengths and stability of these alloys are due to the much slower coarsening rates of the silicides compared to the dispersoids found in other high temperature aluminum alloys, [3.]. Since the Al-Fe-V-Si alloys are selectively processed to only contain the fine cubic silicides and be devoid of the embrittling plate or flake like intermetallics good strength and fracture toughness combinations can be obtained, [4.]. This combination of properties is illustrated by plotting the fracture toughness of various planar flow cast Al-Fe-V-Si alloys and atomized Al-Fe-X alloys as a function of their tensile yield strength, Figure 1., [5.]. The rapidly solidified Al-Fe-V-Si alloys are initially destined for a wide range of aerospace applications including aircraft structures, engine components and missile applications where improvements in elevated temperature capability, corrosion resistance and elastic modulus can be utilized.

The two most advanced Al-Fe-V-Si alloys are called FVS0812 and FVS1212. The compositions and selected physical properties of these two alloys are summarized in Table 1., and their tensile properties in Table 2., [6.]. The major difference between the two alloys is simply the volume fraction of silicides. Alloy FVS0812, which contains approximately 27 volume percent silicides, was developed to meet the needs of applications that require light weight in combination with good elevated temperature mechanical properties and fracture toughness. Alloy FVS1212 contains approximately 36 volume percent of silicides and is suited for applications that require light weight along with superior ambient and elevated temperature strength and high specific stiffness.

Table 1. Compositions and selected physical properties of alloys FVS0812 and FVS1212.

	FVS0812	FVS1212
Composition, w/o: (bal. Al)	8.5 Fe–1.3 V–1.7 Si	11.7 Fe–1.15 Fe–2.4 Si
Density, mg/m^3 (lbs/in^3):	2.92 (0.105)	3.02 (0.109)
Coef. of Expansion, /K:	22×10^{-6}	21.6×10^{-6}
Thermal Conductivity, W/cm/K: (400–600 K)	1.05	0.88

Almost half of the research devoted to the aluminum FVS alloys has been directed towards converting these attractive high temperature aluminum alloys into useful product forms by extrusion, rolling and forging. Emphasis has been placed on efficiently forming these alloys to desired products while retaining the alloys' attractive mechanical properties, [7.]. For example alloy FVS0812 has been forged into various aerospace and automotive forms including aircraft wheels, structural members and engine components. Sheet, plate and profiled extrusions have also been made. Alloy FVS1212 has been fabricated into product forms that utilize the alloy's high strength and modulus combinations such as sheet, large extrusions and closed die forgings.

Table 2. Mechanical properties of rapidly solidified aluminum alloys FVS0812 and FVS1212.

Alloy FVS0812

Temp. °C (°F)	TYS MPa (ksi)	UTS MPa (ksi)	El %	Modulus GPa (Msi)
25 (77)	434 (63)	462 (67)	12	88.3 (12.8)
150 (300)	345 (50)	372 (54)	6	81.4 (11.8)
205 (400)	303 (44)	324 (47)	6	77.9 (11.3)
260 (500)	276 (40)	290 (42)	8	74.5 (10.8)
316 (600)	241 (35)	248 (36)	9	71 (10.3)

Alloy FVS1212

Temp. °C (°F)	TYS MPa (ksi)	UTS MPa (ksi)	El %	Modulus GPa (Msi)
24 (75)	531 (77)	557 (81)	7.2	96.6 (14.0)
150 (300)	455 (66)	469 (68)	4.2	86.2 (12.5)
232 (450)	393 (57)	407 (59)	6.0	80.7 (11.7)
316 (650)	297 (43)	303 (44)	6.8	74.5 (10.8)

Specifically, the applications* for the high temperature/high stiffness Al-Fe-V-Si alloys to be discussed include:

1. Aircraft Structural Components
2. Aircraft Wheel Forgings
3. Engine Control Housings for Gas Turbine Engines
4. Missile Fins

2. AIRCRAFT STRUCTURAL COMPONENTS

The ability to save weight in advanced aircraft components that must operate at elevated temperatures is extremely attractive. Consequently, the utilization of FVS aluminum alloy sheet for various aircraft applications such as skins, leading edges and built up structural members is being evaluated. Both alloy FVS0812 and alloy FVS1212 have been converted to sheet and plate of various gages, widths and lengths, Figure 2. Two critical properties for the utilization of these alloys for structural applications are fatigue and tensile creep.

The fatigue crack growth rates of alloy FVS0812 approaches that of alloy 2014-T6, Figure 3. Even though the grain size of alloy FVS0812 is much finer than that of the wrought 2000 series alloy which should facilitate crack growth, by carefully controlling the composition and processing of the alloy to form only the spherical silicides and avoid the coarse

*The described applications, i.e. components, and their property requirements are courtesy of Allied-Signal Aerospace Company including Bendix's Aircraft Brake and Strut Division, and Engine Controls Division.

flake and plate-like aluminum-iron intermetallics the alloy's resistance to excessive fatigue crack growth rates is optimized. Also the high cycle fatigue strength of alloy FVS0812 is at least equivalent to the 2000 series alloy, Figure 4. In this case the less than one micrometer grain size of the rapidly solidified alloy delays fatigue induced crack initiation.

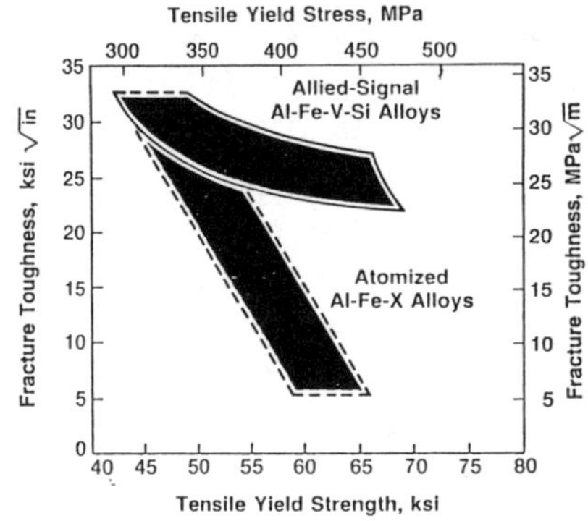

Figure 1. Strength/fracture toughness combinations of rapidly solidified Al-Fe-V-Si alloys.

Critical to the utilization of high temperature aluminum alloys in elevated temperature service is the tensile creep rupture life of the material. A plot of the Larson-Miller parameter as a function of applied stress for aluminum alloys FVS0812 and 2219-T6, clearly shows the superiority of the FVS alloy in regards to elevated temperature creep resistance, Figure 5. As will be discussed later the excellent creep resistance of the rapidly solidified Al-Fe-V-Si alloy is due to the stability of the alloy's microstructure at elevated temperatures (8). Alloy FVS1212 displays similar attractive creep properties (8).

Figure 2. Aluminum alloy FVS0812 sheet.

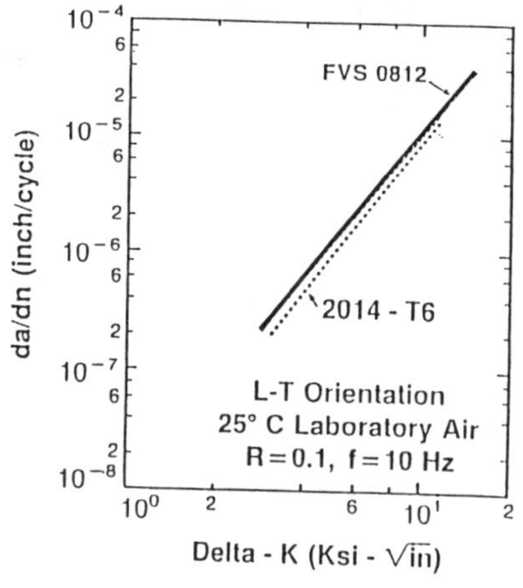

Figure 3. Fatigue crack growth rates of alloy FVS0812.

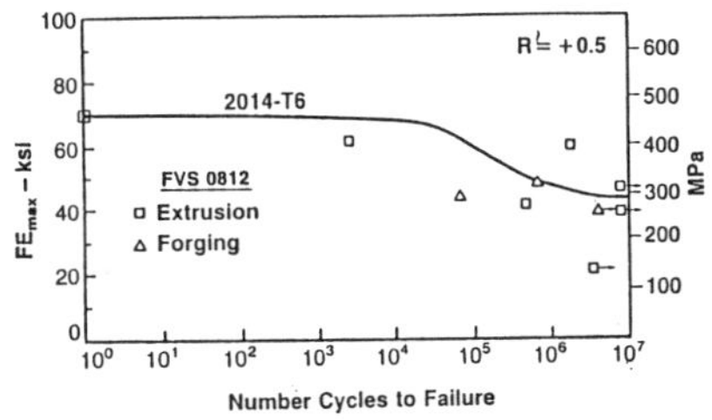

Figure 4. Fatigue strength of alloy FVS0812 extrusions and forgings.

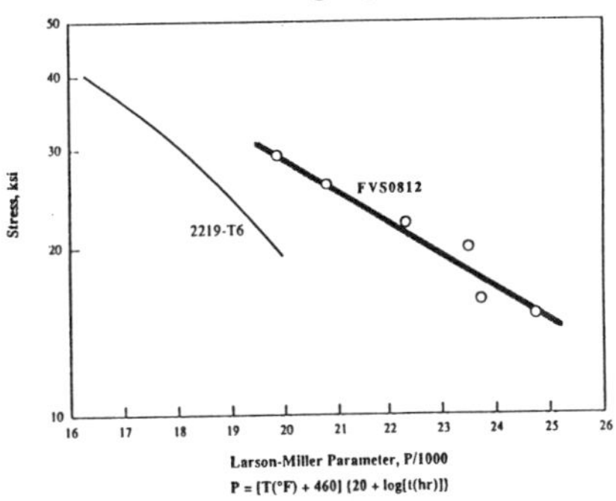

Figure 5. Larson-Miller parameter as a function of stress for aluminum alloy FVS0812.

3. Aircraft Wheel Forgings

Advances in composite aircraft brake design have led to the requirement that advanced aircraft wheel forgings have the ability to operate at temperatures above that of aluminum alloy 2014, the most widely used alloy for aircraft wheels. Also the advanced elevated temperature aluminum alloys must have fracture toughness levels, equal to or greater than 27 MPa $m^{1/2}$ (25 ksi in $^{1/2}$), fatigue crack growth rates, and fatigue strengths at least equivalent to alloy 2014 with enhanced corrosion resistance [5.]. The majority of in service aircraft wheel failures are due to corrosion induced fatigue. With these property requirements in mind including the continuing requirement for light weight and cost effectiveness, alloy FVS0812 has been specifically developed for advanced aircraft wheel applications. Prototype wheels have already been fabricated out of FVS0812 using commercial forging practices.

In addition to being easily forged, alloy FVS0812 has the design requirement of strength/fracture toughness levels equivalent to 2000 series aluminum alloys, Figure 1. At elevated temperatures, FVS 0812 has tensile strengths superior to 2014-T6, Figure 6. These desirable elevated temperature strengths and fracture toughness levels result from the high, 27 volume fracture of fine, spherical and thermally stable silicides. FVS0812 has been found to be extremely corrosion resistant as measured by exposure to a salt fog, Figure 7. The excellent corrosion resistance of alloy FVS0812 is thought to be due to the fine, uniform distribution of the silicides in that the silicides act to increase the adherence of the aluminum oxide scale and act as closely spaced dipoles that negate the chemical polarization of the alloy [4.]. Also, as determined to date, the silicides appear to be chemically inert. The threshold stress for stress corrosion cracking in these alloys is essentially the yield strength of the alloy.

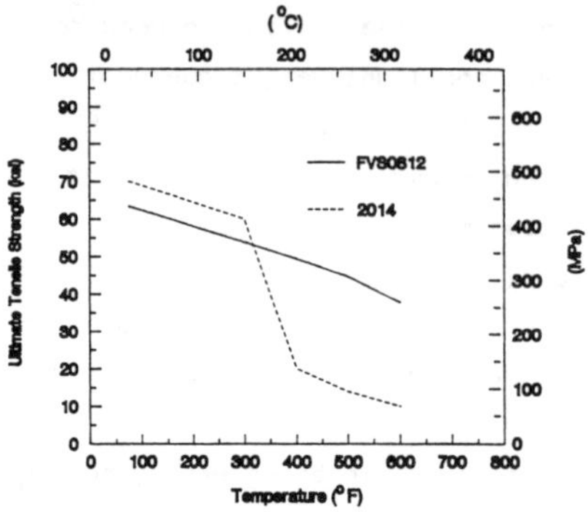

Figure 6. Tensile strengths of aluminum alloys FVS0812 and 2014 as a function of temperature.

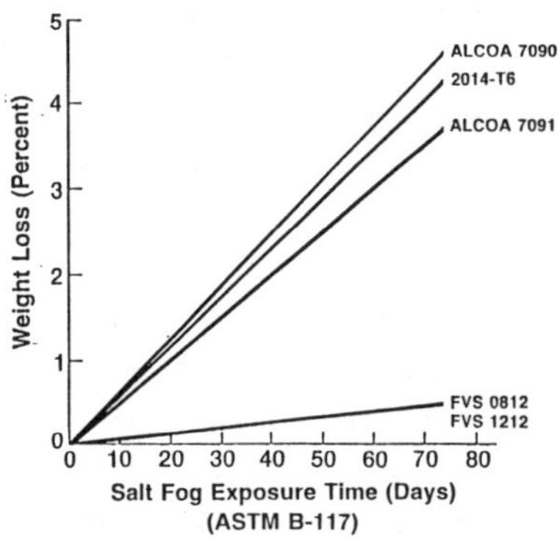

Figure 7. Corrosion performance of alloy FVS0812 in a salt fog (ASTM B-117).

4. ENGINE CONTROL HOUSING - GAS TURBINE ENGINES

The materials requirements for engine control housings and other gas turbine engine static structures include operations at temperatures up to 300°C either in ambient air or the operating fluid. Operating fluid pressures range from 40 to 55 MPa. An increasingly important design criterion is weight savings over titanium, the material at present most widely used. The utilization of high temperature aluminum alloys in engine control housings represents an application that to date has required titanium not because of the extreme high temperature capabilities of titanium alloys but the inability of conventional elevated temperature aluminum alloys to perform in the specified temperature/pressure regimes. Both alloy FVS0812 and alloy FVS 1212 are excellent candidates for engine control housings because of their extreme thermal stability. Alloys FVS0812 and FVS1212 exposed at 425°C for times up to 1000 hours show no degradation in room temperature tensile properties after exposure [2.], Figure 8. When comparing the coarsening rates of various aluminum based intermetallics the $Al_{12}(Fe,V)_3Si$ coarsens three to four orders of magnitude slower than the non silicon containing intermetallics at 425°C, Table 3. Increasing the iron to vanadium ratio from 5, (alloy FVS0812), to 10, (alloy FVS1212) even further decreases the coarsening rate of the silicide dispersoids. Even though the silicides are noncoherent dispersoids having a high interfacial energy, it is felt that their lower than expected coarsen-ing rates are due to the fact that the silicide lattice parameter, a=1.260 to 1.2620 nm, is approximately three times the unit cell edge length for aluminum, 0.4049596 nm. The silicide lattice closely overlays the aluminum unit cell leading to some degree of coherency and slower than predicted coarsening rates [3.]. Also, some components of the complex silicide structure may be coherent with the matrix (9).

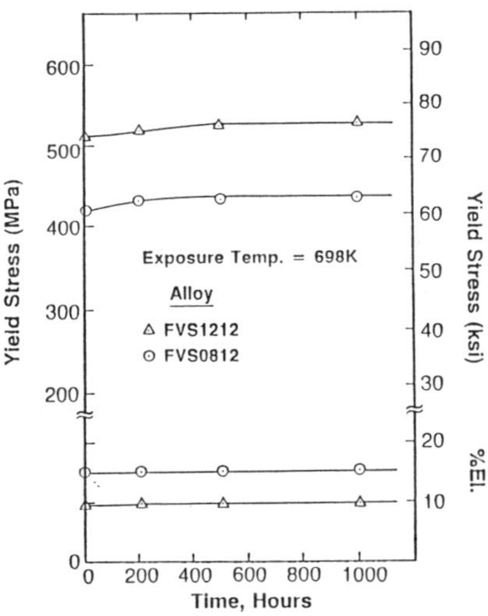

Figure 8. Tensile yield strengths and ductilities of alloys FVS0812 and FVS1212 after 425°C (800°F) exposure.

5. MISSILE FINS

Property requirements for advanced missile fin design include the ability to withstand high temperature excursions, high temperature stability and creep resistance as a function of mach number, transport and flight vibrations, stiffness for aerodynamic designs, and the ability to be fabricated into advanced components. Once the appropriate design requirements are met the material must also give appropriate weight savings and be cost competitive. Alloy FVS1212 has been specifically identified for missile fin applications because of its excellent strengths up to 450°C (850°F) (720°K) and its high modulus. Alloy FVS1212 has already been rolled into sheet over 0.66 m (26 in.) wide and less than 1 mm (40mils) thick., FVS1212 plate has been machined and forged into solid section missile fins and sheet used to construct composite missile wings.

Table 3. Intermetallic coarsening rates at 425°C (800°F) of $Al_{12}(Fe,V)_3Si$ and various Al-Fe intermetallics.

$Al_{12}(Fe,V)_3Si$ Fe: V Ratio: 10 (a = 1.260 nm)	$8.4 \times 10^{-27} m^3h^{-1}$
$Al_{12}(Fe,V)_3Si$ Fe: V Ratio: 5 (a = 1.260 nm)	$2.9 \times 10^{-26} m^3h^{-1}$
Al_3Fe	$2.2 \times 10^{-23} m^3h^{-1}$
$Al_3(Fe,Mo,V)$	$1.7 \times 10^{-23} m^3h^{-1}$
Al_8Fe_4Ce	$4.2 \times 10^{-23} m^3h^{-1}$

Alloy FVS1212 has a Young Modulus of 476 GPa (14 x 10^6 psi) which is approximately 30% greater than the widely used aerospace aluminum alloys, Figure 9. At all temperatures the specific stiffness (modulus/density) of FVS1212 is particulate reinforced aluminum based metal matrix composite systems. Alloys FVS1212 is truly attractive when its excellent specific stiffness is considered along with its excellent strength at temperature, Table 2. Alloy FVS1212 not only has superior elevated temperatured strength/ specific stiffness properties compared to present age hardened elevated temperature aluminum alloys, i.e. 2618, but also when compared to aluminum based metal matrix composites which are expensive and difficult to fabricate and machine. Alloy FVS1212 derives its elevated temperature strength and high modulus from its 36% volume fraction of fine silicide dispersoids. This alloy does not require special machining tools and techniques for fabrication, nor complex heat treatments to achieve optimum strength levels as required for an aluminum based silicon carbide reinforced composite.

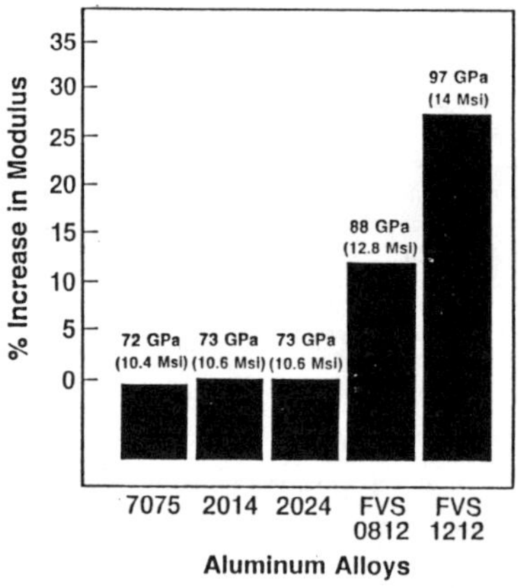

Figure 9. Young's modulus advantage of alloys FVS0812 and FVS1212 compared to other aluminum alloys.

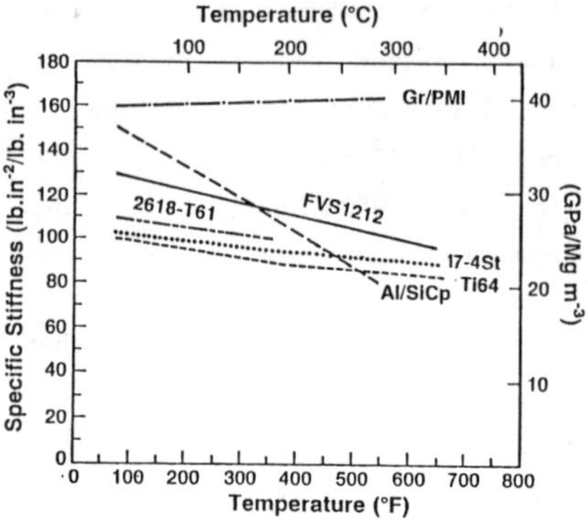

Figure 10. Specific stiffness of alloy FVS1212 and various competing alloys as a function of temperature.

6. PILOT PLANT

Recognizing the potential of the Al-Fe-V-Si FVS alloys and in particular alloys FVS0812 and FVS1212 for light weight high temperature, high stiffness and corrosion resistant applications, Allied-Signal Inc. has independently commissioned and constructed an integrated pilot plant facility that converts raw materials into large vacuum hot pressed billets of these alloys. The constructed pilot plant is located in Morristown, New Jersey and once fully qualified will have a capacity of 110,000 kg/yr. (250,000 lb./yr.) per shift. The pilot plant will produce sufficient quantities of 0.28 m (11 in.) to 0.56 m (22 in.) diameter vacuum hot pressed billets to meet the needs of these and other identified applications. Also, since the pilot plant is a completely integrated facility, accurate product cost determinations can be made. Concurrently, Allied-Signal Inc. is continuously researching the properties and fabricability of the aluminum FVS alloys.

7. SUMMARY

The rapidly solidified Al-Fe-V-Si alloys developed by Allied-Signal Inc. and made by planar flow casting and ribbon comminution are characterized by the formation of high volume fractions of silicides dispersed in the aluminum matrix. These alloys have been shown to possess superior baseline mechanical properties such as a combination of good room temperature ductility and fracture toughness coupled with elevated temperature strength. Also, the FVS alloys have high modulii, excellent thermal stability, creep resistance and corrosion resistance.

Four applications for elevated temperature Al-Fe-V-Si alloys have been discussed including the identification of the material properties that make these particular alloys attractive. Among the important properties for aircraft structural applications are fatigue strength and crack growth resistance, as well as creep resistance. For aircraft wheel forgings the dominate design requirements are fracture toughness and weight savings. The driving material property of elevated temperature aluminum alloys for engine control housings is their weight savings over titanium alloys. Missile fins require advanced aluminum alloys that have comparable properties to existing materials, but show a cost benefit.

8. REFERENCES

(1) D. J. Skinner, R. L. Bye, D. Raybould and A. M. Brown, Scripta Metallurgia, 20, p. 867 (1986).

(2) D. J. Skinner, presented 1988 TMS Annual Meeting, Phoenix, AR January 25-28, 1988.

(3) M. S. Zedalis, D. Raybould, D. J. Skinner and S. K. Das, in Enhanced Properties in Structural Metals via Rapid Solidification, eds. F. H. Froes and S. J. Savage, ASM 1987.

(4) C. M. Adam and S. K. Das, Fourth Conf. on Rapid Solidification Processing: Principles and Technologies, Univ. of CA, Santa Barbara, Dec. 15-18, 1986.

(5) P. S. Gilman, D. J. Skinner, M. Zedalis and D. Raybould, Fourth Conf. on Rapid Solidification Processing: Principles and Technologies, Univ. of CA, Santa Barbara, Dec. 15-18, 1986.

(6) P. S. Gilman and S. K. Das, P/M Aerospace Alloys 87, Lucern, Switzerland, Novermber 1987.

(7) D. Raybould, S. Udvardy, M. Zedalis and S. K. Das, presented 2nd Int. Conf. on Rapidly Solidified Materials, San Diego, CA March 7-10, 1988.

(8) G. M. Pharr, M. S. Zedalis, P. S. Gilman and D. J. Skinner, presented 1988 TMS Annual Meeting, Phoenix, AR January 25-28, 1988.

(9) M. S. Zedalis and D. J. Skinner, presented 1987 TMS Fall Meeting, Cincinnati, OH October 11-15, 1987.

BIOGRAPHIES

P. S. Gilman is presently a senior research associate in the Alloys Department, Corporate Technology, Allied-Signal Inc. with direct responsibility for the high temperature/high stiffness aluminum alloys. He has extensive experience in powder metallurgy, specifically aluminum and nickel base alloys processed by rapid solidification or mechanical alloying, and high temperature mechanical properties. Dr. Gilman was previously with the INCO Research and Development Center and Novamet Aluminum. He has a B.S. in Metallurgy and Materials Science from the University of Pennsylvania and a M.S. and Ph.D. in Materials Science and Engineering from Stanford University.

S. K. Das is the manager of the Alloys Department, Corporate Technology, Allied-Signal Inc. and manages various advanced materials programs including those of rapidly solidified high temperature aluminum alloys, magnesium alloys, aluminum-lithium alloys, titanium alloys, uranium and amorphous materials. Previously he was a scientist at Argonne National Laboratory conducting radiation damage research and at United Technologies Corp. directing the development of rapidly solidified superalloys. Dr. Das has a B.Tech. in Metallurgical Engineering from the Indian Institute of Technology, Kharagpur, and a M.S. and Ph.D. in Materials Science from the University of California, Berkeley.

M. S. Zedalis is a research metallurgist in the Alloys Department, Corporate Technology, Allied-Signal Inc. with direct responsibility for the analytical and mechanical characterization of the rapidly solidified high temperature aluminum alloys and composites. Dr. Zedalis has numerous publications describing the physical metallurgy of these alloys. Dr. Zedalis has a B.S. in Metallurgical Engineering from Lafayette College and a M.S. and Ph.D. in Materials Science from Northwestern University.

J. M. Peltier is a research metallurgist in the Alloys Research Department, Corporate Technology, Allied-Signal Inc. and supervises the mechanical testing laboratory. Dr. Peltier has researched the fracture, fatigue and creep properties of nickel base superalloys and zirconium alloys. Dr. Peltier has a B.S., S.M. and a Sc.D. in Metallurgy from the

Massachusetts Institute of Technology.

D. Raybould is a Research Associate in the Alloys Department, Corporate Technology, Allied-Signal Inc. with responsibility for forming and characterizing a variety of rapidly solidified alloys including high temperature aluminum alloys, magnesium alloys, aluminum-lithium alloys and both soft and hard magnetic materials. His accomplishments include the development of economic techniques to bond amorphous iron ribbon, i.e. Power Core. Previously he worked for Inst. Cerac, S. A. Lausanne, Switzerland where he invented dynamic compaction technology. Dr. Raybould has a B.S. in Metallurgy & Materials Science and a Ph.D. in Metallurgy from Imperial College, London University.

2nd International SAMPE Metals Conference
August 2-4, 1988

RAPID SOLIDIFICATION OF NB-BASE ALLOYS

Gokhale, A.B., Javed, K.R., Abbaschian, G.J.,
Materials Science & Engineering
University of Florida
Gainesville, FL 32611

Lewis, R.E.
Lockheed Missiles & Space Co, Inc.
3251 Hanover Street
Palo Alto, CA 94304

Abstract

New Nb-base alloys are of interest for aerospace structural applications at high temperatures, viz., 800 to 1650°C. Fundamental information regarding the effects of rapid solidification in achieving greatly refined microstructures, extended solid solubility, suppression of embrittling equilibrium phases, and formation of new phases is desired in a number of Nb-X alloys. The microstructures and selected properties of Nb-Si and other Nb-base alloys are presented for materials both rapidly quenched from the equilibrium liquidus and rapidly solidified following deep supercooling. Electromagnetic levitation was used to achieve melting and supercooling in a containerless inert gas environment. A variety of solidification conditions were employed including splatting or drop casting of supercooled samples. The morphology and composition of phases formed are discussed in terms of both solidification history and bulk composition.

1. INTRODUCTION

Niobium-based alloys are of considerable interest for high temperature aerospace structural applications in the range 800 to 1650°C. However, proper alloying and processing are necessary to improve their high temperature oxidation behavior, particularly at temperatures above 1200°C. Factors which are considered to be important for improving the high temperature oxidation resistance include (1) the ability of the alloy to form a protective oxide layer, (2) retardation of oxygen diffusion rate through the oxide film, and (3) lowering the solubility of oxygen in Nb by alloying. Among various alloying additions, Al and Si address all three of the above requirements, while Ti addresses the last requirement. However, alloying with Al or Si can lead to a loss in ductility due to the formation of brittle intermediate phases. Ti additions, on the other hand, improve the strength of the alloys through solution hardening mechanisms. The high temperature behavior of the alloys also depends strongly on their microstructure, compositional homogeneity, and for multiphase alloys, on distribution and shape of the second phase particles. As such, development of a suitable alloy will depend not only on the proper choice of the alloying additions, but also on a proper choice of the processing technique. It is expected that the oxidation resistance and mechanical properties can be improved by rapid solidification processing to obtain refined microstructures with greater chemical and morphological homogeneity.

In this paper, we describe the microstructural and compositional consequences of supercooling of molten alloys under containerless processing conditions, followed by rapid heat extraction during recalescence and solidification. Our research is focused on obtaining a balance between superior high temperature oxidation resistance and high temperature deformation characteristics of binary and ternary Nb-Si, Nb-Al, Nb-Si-Ti, and Nb-Al-Ti alloys. Specific examples of metastable extension in solid solubility, suppression of equilibrium phases, and microstructural refinements will be given for Nb-Si and complex Nb-based alloys.

2. BACKGROUND

Since Nb-based alloys find a variety of uses in high temperature structural applications, their environmental stability at high temperatures is a prime consideration. The recent work of Perkins and coworkers [1] shows that a terminal extended solid solution with b.c.c. structure is formed when substantial amounts of Ti are added to Nb-Al alloys. In addition, a continuous alumina scale can be formed in these alloys at low Al concentrations. These workers also found that additional alloying elements are needed to reduce the solubility and diffusivity of oxygen in the Nb-Ti-Al alloys for optimal coating formation. The two elements most effective for this purpose were found to be Cr and V. Although at temperatures above 1200°C, continuous alumina scale formation was achieved, at lower temperatures (1100°C) growth of Al_2O_3 was found to be much slower.[2] It has been proposed that a substantial increase in Al diffusivity at the lower temperatures can be achieved by grain refinement through rapid solidification of Nb-Ti-V-Cr-Al alloys. Rapid solidification processing may thus be effectively combined with alloy design to yield industrially useful alloys.

One of the most useful and widely recognized effects of RSP is refinement of the microstructure. The refinement may manifest itself in one or more of several forms, e.g., reduction in grain size, fine scale distribution of the second phase, reduction in dendrite arm spacing, decrease in interlamellar spacing, etc.* Microstructural refinement can be achieved by either employing rapid quenching of a superheated melt or solidification from a deeply supercooled melt. In this context, an important distinction exists between the two methods. Because of heat transfer requirements, rapidly solidified superheated materials generally have a thickness limited to very small dimensions

---

* It is of general interest to note that each of these refinements can be quantified in terms of the surface area per unit volume (S_v) in the microstructure. For example, for a singe phase material, the "average grain size" is simply related to the inverse of S_v. Also, in terms of structure-property correlations, use of the total surface area per unit volume in multiphase microstructures (consisting of a sum of the interface and interphase-interface areas) will yield a more general and exact treatment.

(≤ 50 to 100 μm). On the other hand, no such limitation exists for rapid solidification through bulk supercooling, since a substantial portion of the heat of fusion is absorbed by the supercooled liquid itself. These two methods may be usefully combined so that the microstructural coarsening during recalescence of the supercooled liquid can be reduced substantially.

Many properties are enhanced through structural refinement. For example, an increase in the surface area per unit volume (decrease in the average grain size) increases the yield strength by reducing stress concentrations due to dislocation pile-ups. Also, structural refinement has been found to reduce plastic inhomogeneities, resulting in ductility enhancement. Crack growth arrest by limiting the crack length below the critical limit can be achieved through a finer subdivision of the second phase. Examples given in this paper illustrate structural refinement in rapidly solidified Nb-based alloys resulting from using either rapid quenching or bulk supercooling.

It must be noted however, that a finer subdivision of the structure alone does not ensure structural toughening. For example, if the brittle phase in a two-phase (or multiphase) microstructure forms a continuous network, crack propagation through the brittle component can lead to a failure. An obvious way to avoid such failures is to reduce the volume fraction of the brittle component below the 50 volume % limit imposed by geometrical considerations for continuous network formation. The presence of ductile precipitates in the brittle phase can also be used to suppress crack growth.

Effective refinement of the microstructure, particularly grain size, can be achieved through bulk supercooling [2,3]. In addition, the supercooling may cause a drastic change in the solidification morphology: from dendritic to spherical, or to rosette-type microstructure. Another consequence of solidification from an supercooled melt is the formation of phases with a solute content greater than the equilibrium value (terminal extended solid solubility). As a result, the relative amounts of phases which exist in the microstructure can be changed. The exact compositional and morphological consequences, however, depend upon the extent of metastable supersaturation, the recalescence temperature, the type of equilibrium isotherm which may be crossed during recalescence (eutectic, peritectic, etc.), and the growth kinetics of various phases which may form. An example based on our work on Nb-47 at.% Si alloys is given to illustrate the use of metastable extension of solid solubility to obtain greater morphological and chemical homogeneity.

In the foregoing discussion, the nucleation hierarchy and potency of various heterogeneous nucleants in the supercooled liquid has not been considered. These factors can, however, significantly alter the microstructural morphology. For example, during bulk supercooling, one of the potential phases may nucleate preferentially due to various reasons such as clustering in the liquid, relative surface energies, or differences in the degree of effective supercooling for the various phases (stable or metastable). Thus, for relatively simple eutectic systems, a wide range of microstructures can be obtained as discussed elsewhere [4]. In this paper, metastable supersaturation-enhanced eutectoid decomposition of Nb_5Si_3 following its preferential nucleation in bulk supercooled Nb-58 at.% Si alloys is presented. As discussed later, this phenomenon may have some practical benefits.

The formation of metastable crystalline or amorphous phases and suppression of equilibrium phases through RSP has been reported by numerous investigators. As discussed by Turnbull [5], RSP imposes kinetic constraints on the system, thus opening up various metastable states which may otherwise be inaccessible. In practice, the availability of a reliable thermodynamic description of the alloy system under consideration is necessary to determine the various possible metastable forms. This information, together with the constraints imposed by processing, can lead to a reasonable estimate of the expected chemical and morphological state of the microstructure. The energetics of phase suppression for bulk supercooled Nb-20 at.% Si alloys based on our thermodynamic modelling are described together with the possible use of this phenomenon in obtaining microstructures with improved compositional and morphological homogeneity.

3. EXPERIMENTAL PROCEDURE

Arc-melted binary Nb-Si samples of desired composition were prepared from high purity Nb (99.93%) and Si (99.999%). Other complex niobium-based alloys were prepared by Teledyne Wah-Chang, Albany, in the form of multiple-pass, arc-melted buttons weighing about 50 gms. These alloy buttons were broken into smaller pieces suitable for subsequent processing. The details of such processing involving arc melting and electromagnetic levitation practice are reported elsewhere [2,6]. The sample temperature was continuously monitored by a two-color pyrometer for temperatures below 1700°C, while a single-color pyrometer, calibrated by using known equilibrium isotherm temperature as a standard, was used at higher temperatures. The accuracy of temperature measurement was estimated to be ± 10°C. The temperature-time history of each sample was obtained by a strip chart recorder. From the levitated state, the samples were solidified by either dropping onto a stationary copper hearth or splat quenching in a twin-piston device.

Following standard metallographic procedure, microstructures of the processed samples were examined using optical and scanning electron microscopy. Compositional analysis of each phase was carried out in a JEOL JXA-733 electron microprobe using wavelength dispersive spectroscopy with pure Nb and Si as standards.

4. RESULTS AND DISCUSSION

4.1 Metastable Extension of Solid Solubility

As discussed earlier, the segregation profile within supercooled samples depends upon an interplay between the metastable extension of solid solubility curves, the position of the T_0 curve, and the alloy composition. Furthermore, when a eutectic reaction exists, the shape and extent of the zone of coupled eutectic growth must be considered together with the nucleation hierarchy of the constituent phases in the supercooled liquid. Also, the heat of fusion and growth kinetics of the solidifying material, the heat capacity of the liquid and solid, and the external cooling conditions play a major role in determining the maximum recalescence temperature.

A variety of microstructures were obtained for Nb-47 at.% Si samples processed under different conditions, as shown in Figures 1(a) to 1(c). The microstructure of a sample which was superheated 30 K then quenched on a copper substrate (cooling rate approximately 10^2 K/s) is shown in Figure 1(a). This sample, representing "normal" solidification, exhibits primary faceted Nb_5Si_3 dendrites and interdendritic Nb_5Si_3 + $NbSi_2$ eutectic. The eutectic comprises approximately 42% of the total. Figure 1(b) shows the microstructure of a liquid sample that was supercooled 235 K (relative to the liquidus) then quenched on a copper substrate. The volume fraction of eutectic is lower than in 1(a), while the scale of microstructural features is reduced substantially. The microstructure of a sample that was supercooled 350 K then quenched on a copper substrate is shown in Figure 1(c). Note the further refinement of micro- structural features and the reduction in volume fraction of eutectic. Compositional analysis discloses a change in the composition of Nb_5Si_3 from 40 at.% Si for the superheated sample to 44 at.% Si for the sample supercooled 350 K. Based on the variuos factors discussed above and the observed microstructural morphology, the solidification sequence of bulk supercooled Nb-47 at.% Si samples was deduced as described below.

The relevant portion of the Nb-Si equilibrium diagram is shown schematically in Figure 2. The figure includes the metastable extensions and T_o curves for various phases as well as the approximate shape and extent of the zone for coupled eutectic growth (indicated by the shaded region). The shape of the coupled zone is based on our earlier findings [4] that Nb_5Si_3 (γ) is highly facet forming relative to $Nb_3Si(\delta)$. The notation γ^m, $(\gamma\text{-L})^m$, and $(\delta\text{-L})^m$ refers to the metastable extensions of γ, γ-phase liquidus, δ, and δ-phase liquidus, respectively, while $\gamma\text{-}T_o$ and δ-To indicate the T_o curves for γ and δ, respectively.

When a sample of this alloy is supercooled to point "a" (representing a supercooling of 350 K) in a region where the preferred growth form is γ, nucleation of γ takes place with a composition given by point "b" on the curve γ^m.* Upon growth of γ, the temperature and average composition of the liquid shifts, as shown by the path a - c - d. Beyond the point "c" at which the liquid composition crosses the boundary of the coupled zone, nucleation of δ takes place and coupled eutectic forms. The above solidification path is based on our findings that the growth of faceted γ is interfacially controlled [4]. The solidification path for a sample supercooled to a lower extent (e.g. 235 K) is shown by the corresponding letters a'- c' - d'. In both cases, the composition of the solid γ will change along its metastable extension during recalescence. Since the fraction of solid formed during recalescence increases approximately linearly with increased initial supercooling [7,8], the amount of γ formed will be greater for the sample supercooled 350 K compared to that supercooled 235 K. In addition, higher supercooling increases the metastable supersaturation of γ. These two factors cause the volume fraction of the eutectic to decrease with increasing supercooling. This is indeed observed in the microstructure of the two samples presented earlier.

One of the noteworthy consequences of metastable supersaturation is the phenomenon of supersaturation-enhanced solid-state decomposition, an example of which is found in the Nb-Si system. During bulk supercooling of the eutectic 58 at.% Si alloy, it was observed that in some

---

* In these bulk supercooled liquid alloys, either γ or δ may nucleate preferentially depending upon the potency of heterogeneous nucleants present in the liquid as well as thermodynamic criteria such as clustering. However, our microstructural observations suggest that γ "leads" the growth front in these alloys supercooled in the region under discussion.

samples, preferential nucleation of supersaturated γ_2 occured at a supercooling of approximately 60 K. Following a further supercooling of approximately 200 K, during which a continuous growth of γ_2 took place, the samples were quenched on a copper chill plate. These samples exhibited platelets of Nb_5Si_3 containing both single-phase and two-phase regions, Figure 3(a). Compositional electron microprobe analyses using point-by-point line scans indicate that the average compositions of the two-phase and single-phase regions are the same, as illustrated in Figure 3(b). Note that the composition of the single phase region is approximately 44 at.% Si, indicating a metastable supersaturation of approximately 4 at.% Si.

The relevant portion of the equilibrium diagram is shown schematically in Figure 4(a) together with the calculated metastable extensions of various phases. On the figure, γ_1^m, γ_2^m, $(\gamma\text{-L})^m$, and $(\delta\text{-L})^m$ indicate the metastable extensions of γ_1, γ_2, γ_2-phase liquidus, and δ-phase liquidus, respectively, while $\gamma_2\text{-}T_o$, $\delta\text{-}T_o$, and $(\gamma_1\text{-}\gamma_2)\text{-}T_o$ indicate the T_o curves for γ_2-L, δ-L and γ_1-γ_2 equilibria, respectively. The letters a, b, . . . , n indicate the intersection of each of the above extensions with the temperature for which the various Gibbs energy curves in Figure 4(b) are drawn. The observed decomposition of the Nb_5Si_3 platelets can be explained as follows: Upon quenching, supersaturated γ_2, with a composition denoted by γ_2^m, can transform to the energetically more stable $\gamma_1 + \delta$ combination of phases. However, as can be seen from Figure 4(b), the higher the supersaturation, the larger the driving force for decomposition. Although favored energetically, the decomposition process may not be completed, due to the kinetics, as evidenced by the presence of some portions of Nb_5Si_3 plates found to be single phase.

The mechanism and morphology of the solid state decomposition is technologically interesting. One example is the potential for toughening a brittle intermetallic through such "transformation processing" routes.

4.2 Phase Suppression

An example of phase suppression with simultaneous microstructural refinement and enhanced chemical homogeneity is presented below for the Nb-20 at.% Si alloy processed through bulk supercooling. Figure 5 shows the relevant portion of the Nb-Si equilibrium diagram together with metastable extensions of various phase fields based on our thermodynamic modelling of the Nb-Si system. Also, the approximate shape of the zone for coupled eutectic growth morphology is derived based on a consideration of competition between various growth forms [9]. The coupled growth zone is skewed towards Nb_3Si, since this faceted intermediate phase experiences kinetic attachment difficulties during growth.

During "normal freezing" of the 20 at.% Si alloy, Nb_5Si_3 is the first to solidify. This is followed by formation of Nb_3Si through the peritectic reaction, and then the last liquid to solidify is a eutectic between (Nb) and Nb_3Si. The resulting microstructure is shown in Figure 6 for a sample which was superheated 90 K and quenched on a copper chill plate.

When the alloy is supercooled below the eutectic isotherm, two possibilities of solidification sequence exist: either the formation of Nb_3Si is suppressed resulting in a metastable eutectic between (Nb) and Nb_5Si_3, or the formation of primary Nb_5Si_3 is suppressed, resulting in the absence of the peritectic reaction. For the latter, the melting characteristics of Nb_3Si change from peritectic to congruent, while at the same time, the phase is stabilized to lower temperatures as noted by the dotted line. As such, depending on the nucleation sequence, a variety of microstructures will form. As an example, the microstructure of a sample supercooled to 1520°C (380 K below the eutectic or 430 K below the liquidus) and quenched on a copper chill plate, together with its thermal history, are shown in Figure 7. Compositional and microstructural analyses indicate that the structure is almost entirely comprised of a fine coupled eutectic between (Nb) and Nb_3Si with a spacing of approximately 0.5 µm. X-ray analysis also indicates a complete absence of the primary Nb_5Si_3 phase. Based on these results, the zone of coupled eutectic growth must extend to at least 20 at.% Si and 1520°C. However, the microstructure reveals some feathery shaped regions within the eutectic which are single phase, with a composition approximately equal to the initial alloy composition. The fine eutectic microstructure and a region containing the additional phase are shown in Figure 8(a) and 8(b), respectively. The feathershaped regions may be indicative of partitionless formation of a metastable phase, possibly corresponding to Nb_4Si. However, more analysis is necessary to confirm this finding.

Several metastable phases have, in fact, been reported in this composition range under rapid cooling conditions (cooling rate of 10^6 to 10^8 K/s) [10,11,12]. Since rapid quenching can lead to considerable kinetic supercooling for faceted materials [13], this observation is perhaps not surprising. Others [14] have suggested Nb_4Si to be a stable phase in the Nb-Si system.

4.3 Microstructural Refinement

Structural refinement through rapid quenching is illustrated in Figures 9(a) and 9(b) for Nb-62 at.% Si samples quenched from a superheated state. The former shows the microstructure of a sample which was quenched on a copper chill plate with a cooling rate of approximately 10^2 K/s, while the latter is for a sample cooled at a rate of approximately 10^6 K/s (splat quenched). The microstructures show, in both cases, broken dendrites of δ ($NbSi_2$) with interdendritic Nb_5Si_3 + δ eutectic. Upon splat quenching, the microstructural scale is approximately 4 times smaller. These can be compared to microstructural refinement through bulk supercooling as shown in Figures 1(a) to 1(c) for Nb-47 at.% Si alloy. It should be noted that bulk supercooling not only refines the scale of the microstructure, but also refines the grain size, as reported by [8] for Nb-Ti alloys. Several mechanisms have been proposed to explain the change in grain size as a function of supercooling as discussed in detail by Munitz and Abbaschian [6]. These include: (1) selective activation of different nucleants and denucleation process, (2) dynamic nucleation due to cavitation shrinkage, (3) stress induced recrystallization, (4) dendrite remelting and coarsening, and (5) dendrite fragmentation and coarsening.

As noted earlier, microstructural refinement may be used successfully to redistribute the second phase in such a manner as to act as an effective barrier to crack propagation. This is illustrated in Figure 10 which shows the arrest of the thermal-stress induced crack in $NbSi_2$ by interdendritic Nb_5Si_3. However, if the brittle component forms a completely connected network, microstructural refinement alone may not work as an effective toughening mechanism. This is illustrated by a series of micrographs in Figures 11(a) and 11(b), which represent the fracture surface of a Nb-20 at.% Si sample supercooled by 380 K. The approximate volume fractions of (Nb) and Nb_3Si are 25% and 75%, respectively, indicating that the brittle component (Nb_3Si) forms a continuous connected network in the microstructure as shown in

Figure 11(b). The fracture surface indicates classical brittle fracture with almost no evidence of plastic deformation.

4.4 Complex Niobium Alloys

The results given below describe the effect of rapid solidification processing on complex niobium alloys containing Ti, V, Cr, and Al. Examination of all possible binary combinations of these elements revealed that Nb-Ti, Nb-V, Ti-V, Ti-Cr, and Cr-V are isomorphous, while the Nb-Cr system is relatively simple, exhibiting two eutectics and the presence of the intermediate phase Cr_2Nb. The binary systems with Al, on the other hand, are relatively complex, with a total of 22 intermediate phases. Despite the potential microstructural complexities, compositional analyses (described later) indicate that a relatively simple phase identification scheme could be formulated for these alloys. The nomenclature is given in Table 1 with the measured composition of the various phases detected in the microstructures described below. The scheme is based on the Al contents of various phases since Nb, Ti, and V form isomorphous solid solutions. Furthermore, since the maximum solid solubility of Cr in (Nb) is approximately 8 at.%, we have assumed that Cr atoms dissolve substitutionally in the intermediate phases. The designation ξ', η', etc. is used to indicate our belief that these phases may not be distinct from the respective ξ, η type phases, but are merely an indication of the existence of a homogeneity range.

Effects of rapid solidification for the complex niobium alloys are presented in Figures 12 to 14. The microstructure of the as-received arc-melted alloy is shown in Figure 12, which shows coarse platelets of η surrounded by the Al-depleted ξ and ξ' phases. Figure 13(a) shows the microstructural morphology of a sample supercooled 58 K and quenched on a copper-chill plate. The sample contains η' plates in η'' matrix. Another example is given in Figure 13(b) for a sample which was supercooled 290 K and quenched similarly. For this sample, the platelet density was highest at the center, decreased radially, and finally disappeared near the edge. Higher density of platelets is also observed along the grain boundaries and interdendritic regions as shown in Figures 13(c) and 13(a), respectively. Another feature of the microstructure is the directionality of the elongated grains which originated at the contact surface with the copper chill.

Figure 14 shows the microstructure of the above sample after annealing at 955°C for 20 minutes. The microstructure shows a considerable precipitation and coarsening of η' platelets. However, even after annealing, the microstructure is substantially finer than that of the arc melted sample.

5. SUMMARY

Various effects of rapid solidification processing through bulk supercooling of Nb-Si and complex Nb-based alloys have been described. These include metastable extension of solid solubility, phase suppression, structural refinement, and modification of solid state reactions. It was shown that:

(1) In Nb-20 at.% Si alloys, deep supercooling resulted in the suppression of properitectic Nb_5Si_3 and formation of a fine coupled eutectic between (Nb) and Nb_3Si. Also, feather-like single phase regions containing 20 at.% Si were detected, which may correspond to metastable Nb_4Si.

(2) Supercooling of Nb-47 at.% Si alloys was found to lead to a significant microstructural refinement, and a 4 at.% Si extension in the solid solubility of Nb_5Si_3. The volume fraction of Nb_5Si_3 + Nb_3Si eutectic also decreased from 42% to 15% upon 350 K supercooling.

(3) In Nb-58 at.% Si alloys, a partial eutectoid-type decomposition was detected in Nb_5Si_3 plates, which was attributed to supersaturation enhanced eutectoidal decomposition.

(4) In complex Nb-based Nb-Ti-V-Cr-Al alloys, solidification from a supercooled melt was found to suppress the formation of Al-poor A_2B-type phases and a significant reduction in the scale of AB-type precipitates.

6. ACKNOWLEDGEMENT

The authors would like to acknowledge support of this research through the NASA-Vanderbilt Center for Space Processing of Engineering Materials. Special thanks are due to Mark Lipshutz for providing the Nb-20 at.% Si samples, and to Craig Wojcik and John Haygarth of Teledyne Wah Chang, Albany, for supplying the complex alloys.

7. REFERENCES

[1]: Perkins, R.A., Chiang, K.T., and Meier, G.H., Air Force Office of Scientific Research, Contract No. F49620-86-C-0018 (1987).

[2]: Amaya, G.E., Patchett, J.A., and Abbaschian, G.J., Grain Refinement in Casting and Welds (Proc. Conf.), 51-65, Fall Meeting of The Metallurgical Society of AIME, St. Louis, Missouri, October 25-26, 1982, Ed.: Abbaschian, G.J. and David, S.A., Published: 1983.

[3]: Mondolfo, L.F., ibid. [83Ama], pp. 3-50.

[4]: Gokhale, A.B., Sarkar, G., Abbaschian, G.J., Haygarth, J.C., Wojcik, C., and Lewis, R.E., Presented at the TMS Fall Meeting, Cincinnati, Ohio, October 12-14, 1987. To be published in Solidification Processing of Eutectic Alloys.

[5]: Turnbull, D. Met. Trans A, (12A), 695-708 (1981).

[6]: Munitz, A. and Abbaschian, G.J., Undercooled Alloy Phases (Proc. Conf.), 23-48, Annual Meeting of The Metallurgical Society of AIME, New Orleans, LA, March 2-6, 1986., Ed: Collings, E.W. and Koch, C.C.

[7]: Wu, Y., Piccone, T.J., Shiohara, Y., and Flemings, M.C., Rapidly Solidified Metastable Materials, MRS Symposia Proceedings, 37-42, (28), Boston, MA, November, 1983.

[8]: Elder, S.P. and Abbaschian, G.J., to be published in Principles of Solidification and Materials Processing, First Indo-U.S. Workshop, Janaury, 1988.

[9] Kurz, W. and Fisher, D.J., Int. Met. Rev., (5-6), 177-204 (1979).

[10]: Waterstrat, R.M., Haenssler, F., and Muller, J., J. Appl. Phys., 50(7), 4763-4766 (1979).

[11]: Togano, K., Kumakura, H., and Tuchikawa, K., Phys. Lett., 76A(1), 83-85 (1980).

[12]: Wang, W.K. and Spaepen, F., J. Appl. Phys., 58(11), 4477-4479 (1985).

[13]: Cohen, M. and Flemings, M.C., ibid [85Boe], pp. 3-20.

[14]: Panteleimonov, L.A., Seropegin, Yu.D., Guseva, I.A., and Osipova, L.V., Vestn. Mosk. Univ. Ser2, Khim, 20(3), 271-275 (1979).

Table 1: Identification Scheme for Nb-Ti-V-Cr-Al Alloys

Phase	Composition, at.%					Description	Type
	Nb	Ti	V	Cr	Al		
η	23.49	23.23	3.24	1.90	48.14	(NbTiVCr) Al	AB
ξ	28.05	24.43	7.01	5.77	34.74	$(NbTiVCr)_2Al$	A_2B
ξ'	28.79	22.76	7.36	6.33	34.77	$(NbTiVCr)_2Al$	A_2B
η'	25.51	22.89	5.18	3.14	43.28	(NbTiVCr)Al	AB
η"	25.81	23.08	5.22	3.29	42.61	(NbTiVCr)Al	AB

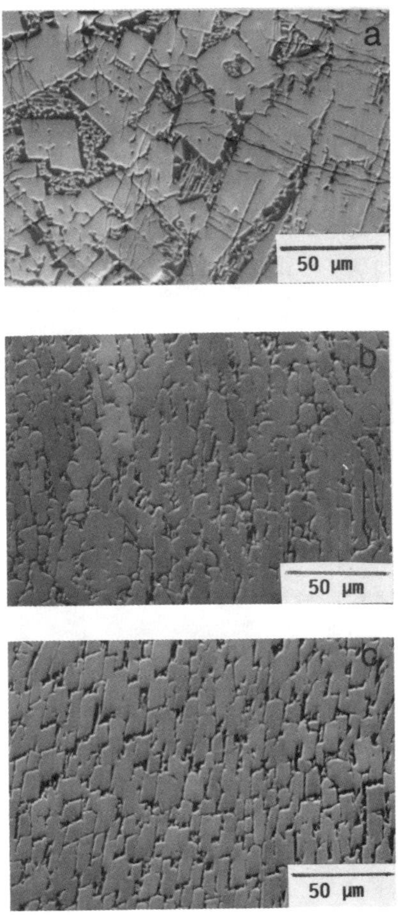

Figure 1 Microstructures of Nb-47 at.% Si samples quenched on a Cu chill plate from (a) 30 K superheat, (b) 235 K supercooling, and (c) 350 K supercooling.

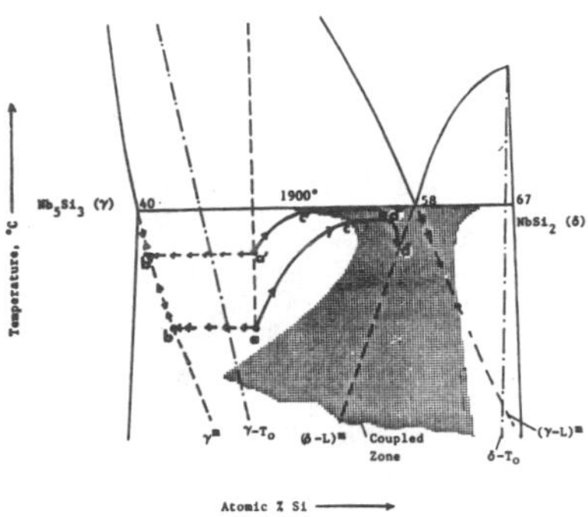

Figure 2 Solidification Sequence in supercooled Nb-47 at.% Si alloy.

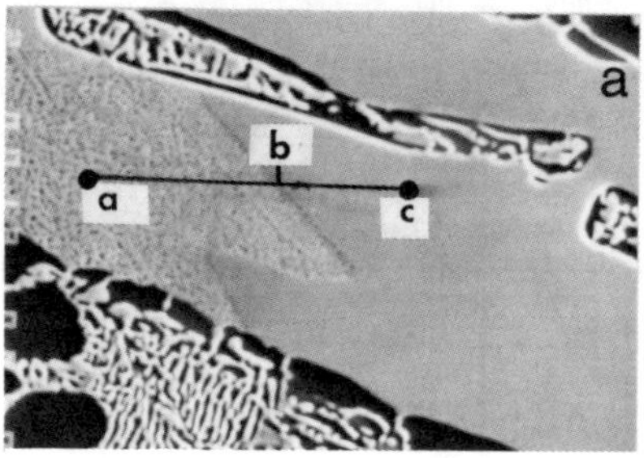

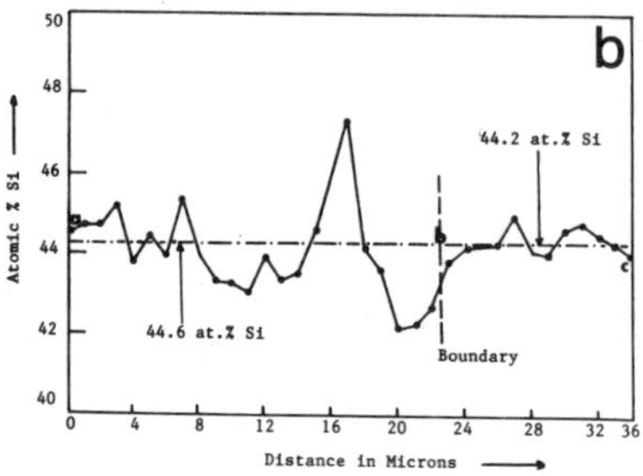

Figure 3 (a) Eutectoidal decomposition in supersaturated Nb_5Si_3 plate. (b) Compositional linescan along line a-c in (a).

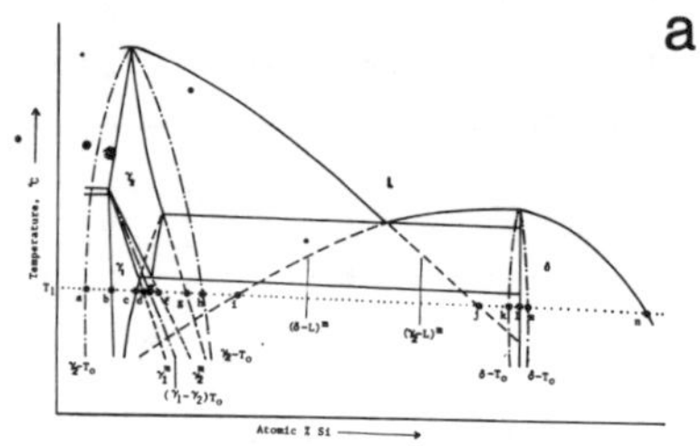

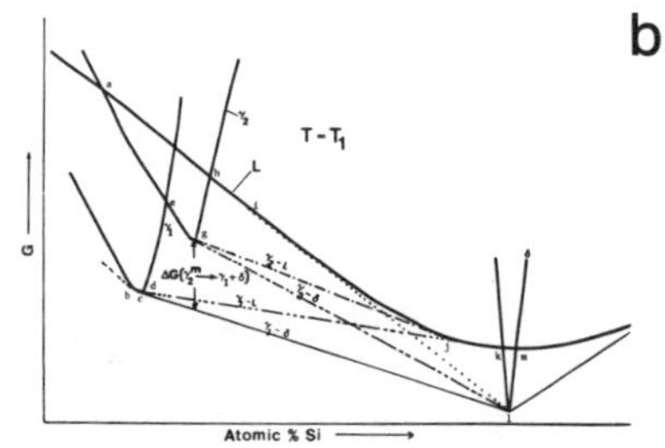

Figure 4 (a) Stable and metastable equilibria in Nb-Si in the range 40 to 70 at.% Si. (b) Gibbs energy of phases shown in (a) at T_1.

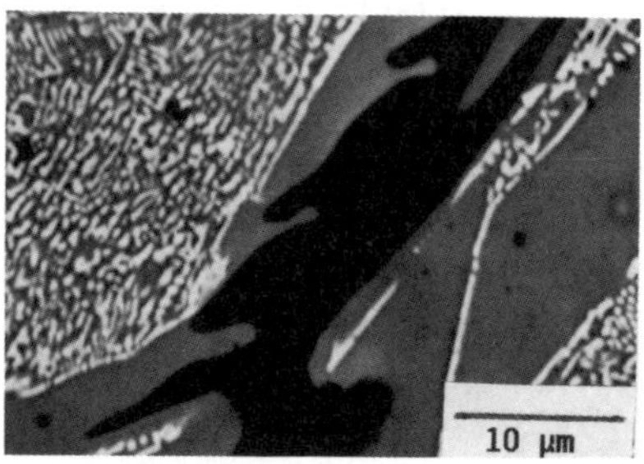

Figure 5 Microstructure of Nb- 20 at.% Si sample solidified from a superheat of 90 K.
Dark phase: properitectic Nb_5Si_3,
Grey phase: Nb_3Si,
Eutectic: (Nb) + Nb_3Si
(Back scattered electron micrograph)

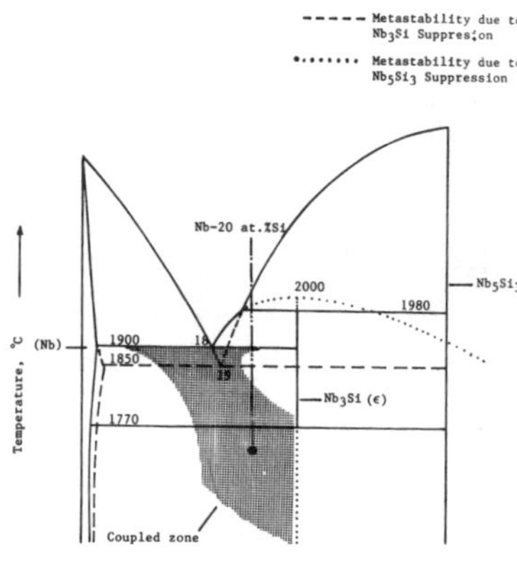

Figure 6 Stable and metastable equilibria in Nb-Si in the range 0 to 40 at.% Si.

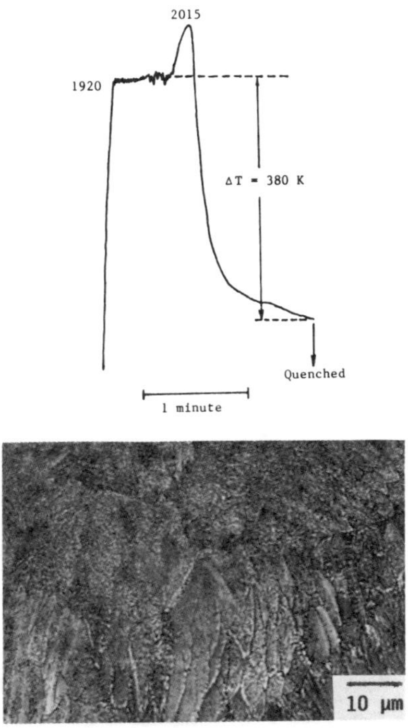

Figure 7 Thermal history and microstructure of an Nb- 20 at.% Si sample supercooled 380 K and quenched on a copper chill plate.

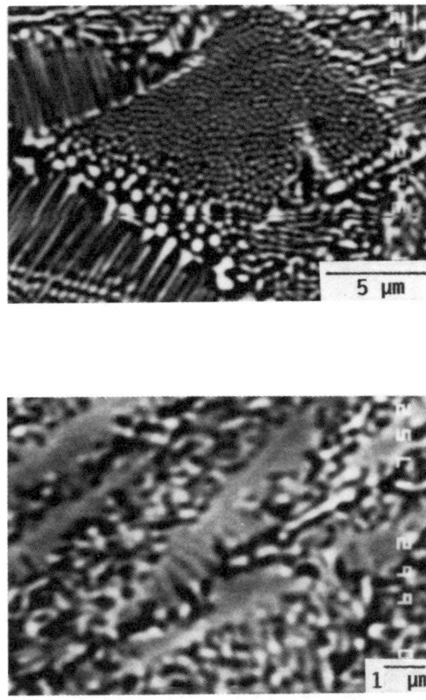

b

Figure 8 High magnification back scattered electron micrographs of sample in Figure 7.
(a) Fine coupled eutectic
(b) Single phase regions containing 20 at.% Si

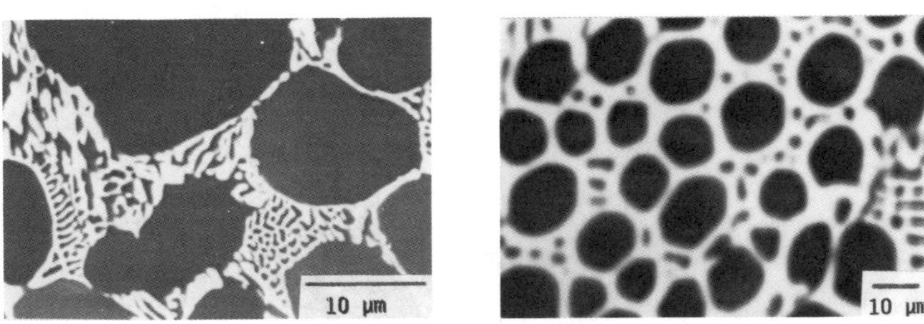

a

b

Figure 9 Microstructures of Nb- 62 at.% Si samples with broken $NbSi_2$ dendrites and interdendritic Nb_5Si_3 + $NbSi_2$ eutectic

(a) Quenched at a cooling rate of approximately 10^2 K/s

(b) Quenched at a cooling rate of approximately 10^6 K/s

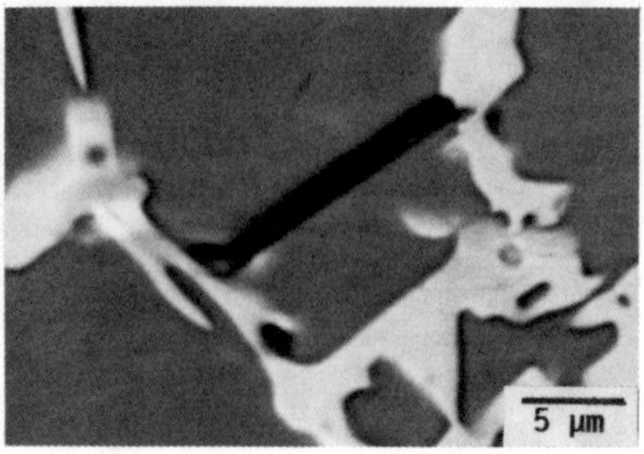

Figure 10 Arrest of thermal crack in $NbSi_2$ by interdendritic Nb_5Si_3.

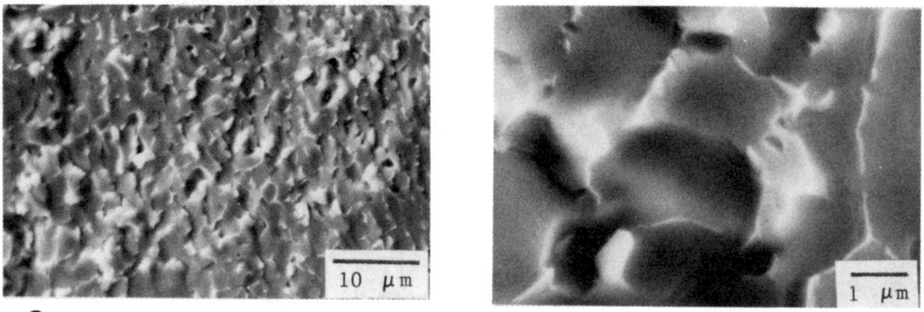

a b

Figure 11 Fracture surface of Nb-20 at.% Si sample solidified from a supercooling of 380 K.
(a) General appearance at low magnification.
(b) Higher magnification shows $NbSi_2$ matrix with (Nb) disributed along cell boundaries.

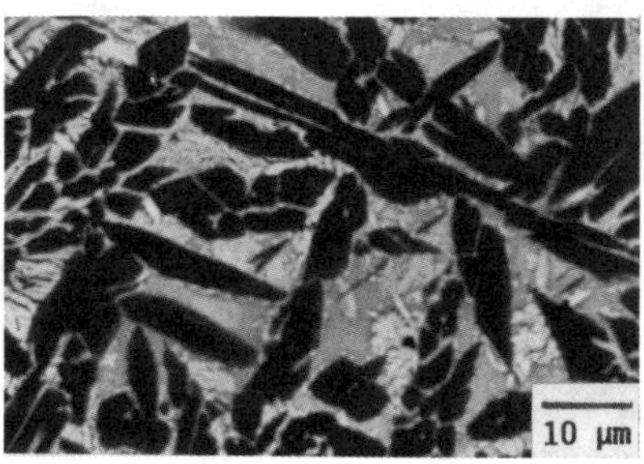

Figure 12 Microstructure of as-received Nb-Ti-V-Cr-Al alloy. (Back scattered electron micrograph)

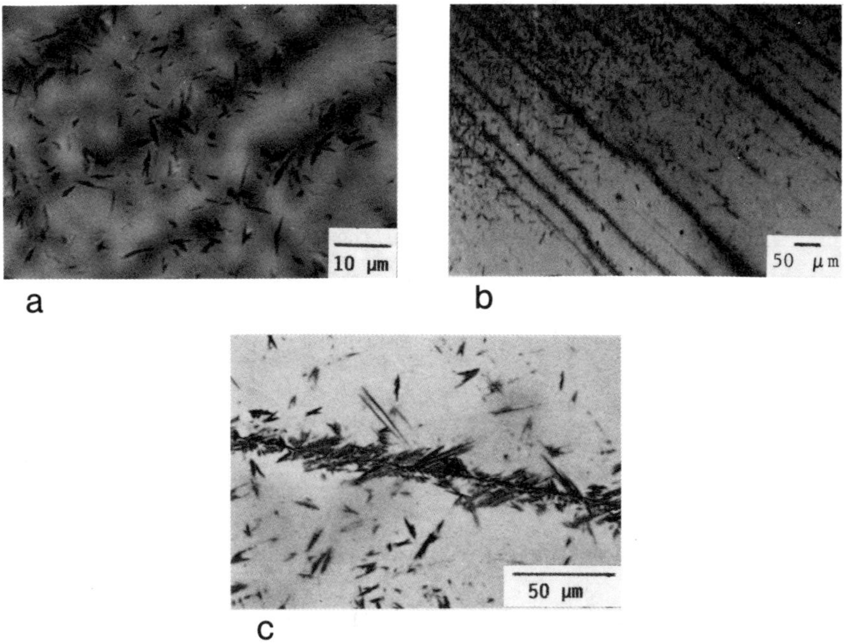

Figure 13 Microstructures of the alloy in Figure 12 quenched on a copper chill plate after processing (back scattered electron micrographs). (a) Supercooled 58 K, (b) Supercooled 290 K, (c) Higher magnification micrograph of (b) showing precipitation along a grain boundary

Figure 14 Microstructure of sample in Figure 13(b) after annealing at 955 °C for 20 minutes. (Back scattered electron micrograph)

2nd International SAMPE Metals Conference
August 2-4, 1988

RAPID SOLIDIFICATION PROCESSING OF COPPER-BASE ALLOY POWDERS

I. E. Anderson
Metals and Ceramics Division
Ames Laboratory, Ames, IA 50011

Abstract

The demands of advanced technology in the defense and energy sectors has promoted much recent interest in research on the rapid solidification processing (RSP) of Cu-base alloys. The results summarized in this report include powder processing research on Cu-Ti, Cu-Ni-Cr-Fe, and Cu-Zr. In Cu-7wt.%Ti, intermetallic phase formation is bypassed in favor of an amorphous phase on cell boundaries in powders finer than about 15μm in diameter. Particles finer than about 5μm exhibit massive (single phase) solidification of the metastable solid solution without any precipitates detectable by TEM. The Cu-16Ni-2Cr-0.7Fe (wt.%) alloy powder was consolidated into a rod form. The TEM revealed a highly refined microstructure with about a 1μm grain size, with a very fine intragranular precipitate, and with only a minor amount of grain boundary Cr-rich phase precipitation. Explosive compaction was used to consolidate Cu-2wt.%Zr powder that was divided into fine (<38μm) and coarse (63-106μm) size fractions. While both samples of this supersaturated alloy composition exhibited age hardening after post-consolidation heat treatments, the effect in the fine powder sample was significantly enhanced. The results indicate the potential for successful RSP fabrication of these types of Cu-base alloys.

1. INTRODUCTION

The challenge of international competitiveness in the industrial and defense sectors has ignited intense interest in research on advanced materials processing. Rapid solidification processing (RSP) techniques have become an important new tool for fabrication of advanced materials and the scope of RSP research has expanded to include several types of copper-base alloy systems. This paper will focus on recent progress in RSP research by a powder metallurgy (P/M) route on three copper alloy systems; Cu-Ti, Cu-Ni-Cr-Fe, and Cu-Zr. Since a previous review (1) discusses earlier results of this research, a summary of subsequent progress is the primary intention of this report.

The P/M approach to RSP has proven to be effective for full scale fabrication of monolithic and composite components. Rapidly solidified particulate can be produced by various methods, including the chopping of melt spun ribbon (2), but gas atomization is often preferred because of its broad applicability. A high pressure gas atomization (HPGA) technique (1,3,4) was used to produce the powders characterized in this study.

Another major issue involved in the processing of rapidly solidified powders is the effective consolidation of such powders containing a metastable microstructure. To accomplish this task many approaches have been devised to minimize microstructural coarsening or decomposition reactions (5). These approaches usually involve high pressure impulse loading and/or high interparticle shear. The Cu-Zr powder was selected for impulse consolidation by an explosive compaction technique (6). This technique was used to permit a test of post-consolidation age hardening of an alloy with an extended solubility level designed for RSP. The consolidation of the Cu-Ni-Cr-Fe powder involved more conventional HIP, hot extrusion, and hot swaging techniques, since the alloy was designed to produce a more stable microstructure. The high interparticle shear of the extrusion and swaging techniques was utilized in this case.

2. EXPERIMENTAL PROCEDURE

Three alloys, Cu-7Ti, Cu-16Ni-2Cr-0.7Fe, and Cu-2Zr (listed in wt.%) were argon-atomized using a high pressure gas atomization technique (1,3,4). As part of the characterization and the size classification, the powder samples were size fractioned using combination of ASTM wire mesh screens, precision electroformed sieves, and Stokes settling techniques. X-ray diffraction (XRD) measurements were performed on sized powder samples using Cu-K_α radiation. Powder samples were mounted in epoxy by a vacuum infiltration method before metallographic preparation of particles in cross section. Microstructural analysis utilized optical microscopy, SEM, and TEM techniques.

The Cu-Ni-Cr-Fe powders were screened to <38μm before outgassing and compaction by HIP at 400 C. The HIP compacted samples were hot extruded at 600 C with a reduction ratio of 9.8:1. Further deformation of the extrusions was by hot rod rolling at about 500 C to achieve an additional reduction ratio of about 4:1. Linear polarization measurements were utilized to examine corrosion resistance in a dearated aqueous solution with 3.5% NaCl.

The Cu-Zr powders were consolidated by an explosive compaction method that has been described in detail elsewhere (6). Two powder size fractions, 63-106µm and <38µm, were consolidated separately in the experiment. Specimens for annealing studies were sectioned from the consolidated material, encapsulated, and heat treated for one hour at temperatures from 300 to 550 C. Hardness was measured with a diamond pyramid indenter at a load of 50g. Resistivity measurements were performed at ambient temperature with a four point probe technique.

3. RESULTS AND DISCUSSION

Since the results of these experiments are presented in detail elsewhere for the Cu-7Ti (7) and the Cu-2Zr (8) systems, this report includes only highlights of the findings. The results for the Cu-16Ni-2Cr-0.7Fe are in a preliminary state but some interesting observations of the rod rolled microstructure are included below.

3.1 Cu-Ti

Particle microstructure was observed to vary as a function of particle size. The cross section of a typical large (44-53µm) particle, shown in Figure 1a, contains a dendritic/cellular microstructure with an apparent intercellular phase. An EDAX analysis of the cells and the intercellular regions, summarized in Figure 1b, reveals compositions of Cu-5at.%Ti and Cu-20at.%Ti, respectively. The crystal structure of this intercellular phase, as examined by selected area diffraction with TEM on a 20-25µm powder sample, was identified as the equilibrium Cu_4Ti (monoclinic) phase.

The fine particles, smaller than at least 10µm, contain regions of both cellular and massive solidification morphologies, when examined with SEM and TEM. However, as Figure 2 illustrates, the cell boundary regions are amorphous, as evidenced by the homogeneous illumination of the entire region in the dark field photograph and by the diffuse ring pattern in the selected area diffraction pattern. Some fine particles observed with TEM, especially those less than 5µm, are completely single phase Cu (fcc) solid solution. This result has been supported by XRD evidence on size fractioned samples.

3.2 Cu-16Ni-2Cr-0.7Fe

Some preliminary observations of the extruded microstructure revealed a fully densified structure with many prior particle boundaries, which suggests poor interparticle bonding.
Measurements of the corrosion resistance also suggested "open" interparticle boundary regions, since the corrosion current reached

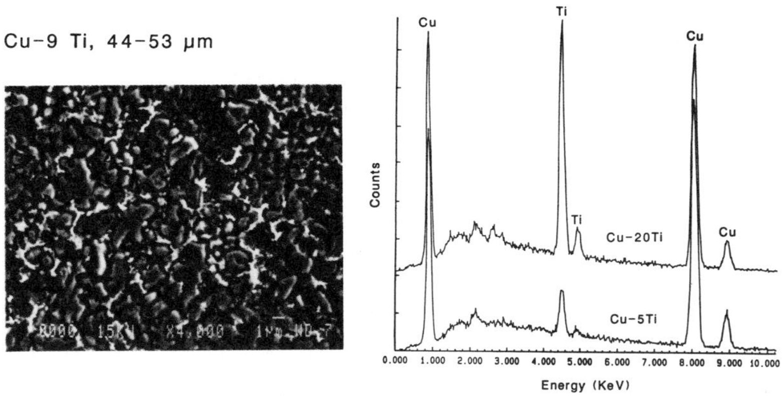

Figure 1: a) SEM micrograph of Cu-7Ti (Cu-9at.%Ti) particle, sectioned and etched, from 44 to 53µm size fraction. b) Compositional analysis (EDAX) of particle cross section to determine Ti content (in at.%) within cells and in cell boundary phase.

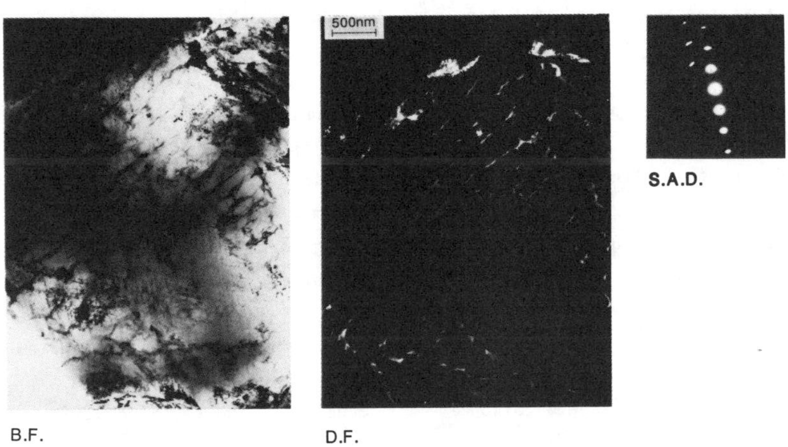

Figure 2: TEM micrographs of Cu-7Ti powder particles <5µm in dia. in bright field and dark field with a selected area diffraction pattern.

an excessively high level. These findings prompted further thermomechanical processing to promote improved interparticle bonding.

Subsequent hot rolling of the extrusions appeared to enhance the integrity of the material. The TEM results shown in Figure 3 indicate the very fine grain size and the lack of apparent prior particle boundaries in this material after the further deformation. An average grain size of about 1μm can be observed in the transverse section given in Figure 3a. Within the grains of Figure 3a, a uniform distribution of dislocations can be noted. This is tentative evidence of a dispersion of extremely fine precipitates associated with the dislocations. Unfortunately, the resolution of the TEM was not sufficient to allow identification of these features. Analysis of the microstructure with TEM did not reveal any coarse grain boundary precipitation; although, SEM and optical microscopy did indicate such precipitates scattered on about 10% of the boundaries. The longitudinal section in Figure 3b shows that some grain elongation has occurred during the thermomechanical processing.

3.3 Cu-2Zr

Optical microscopy of the two shock consolidated samples indicated considerable interparticle flow in the shock propagation direction, as illustrated in Figure 4. The compacts also appeared to be fully dense. The TEM results, as given in Figure 5, indicate local melting and resolidification in the highly deformed interparticle regions.

A summary of the age hardening results for the consolidated samples is presented in Figure 6. The fine particle size sample shows an abrupt increase in hardness on aging at 350 C for one hour. The coarse powder sample has a lower initial hardness and a lower hardness peak at the same annealing temperature. Also included in the figure are data for hot extruded powder samples (9) and melt spun ribbon samples (10) of similar Cu-Zr alloys. The fine particle size sample exhibits a higher initial hardness and a peak hardness value nearly as high as the Cu-1.05Zr melt spun ribbon. The lack of age hardening in the extruded Cu-0.8Zr results demonstrates the detrimental effect of prolonged thermal exposure above the peak aging temperature during consolidation at 650 C (9).

The TEM analysis of consolidated samples in the heat treated condition also revealed differences between the fine and coarse particle size samples. In the peak aged condition the coarse particle sample exhibited a completely recovered microstructure in the dendrite interiors. Few dislocations were found not associated with low angle

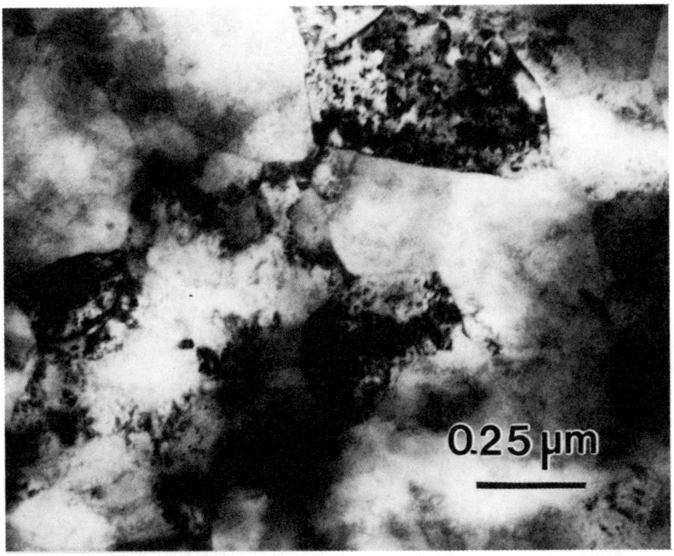

Figure 3: TEM bright field micrograph of rod rolled sample of Cu-16Ni-2Cr-0.7Fe in section transverse to rolling direction.

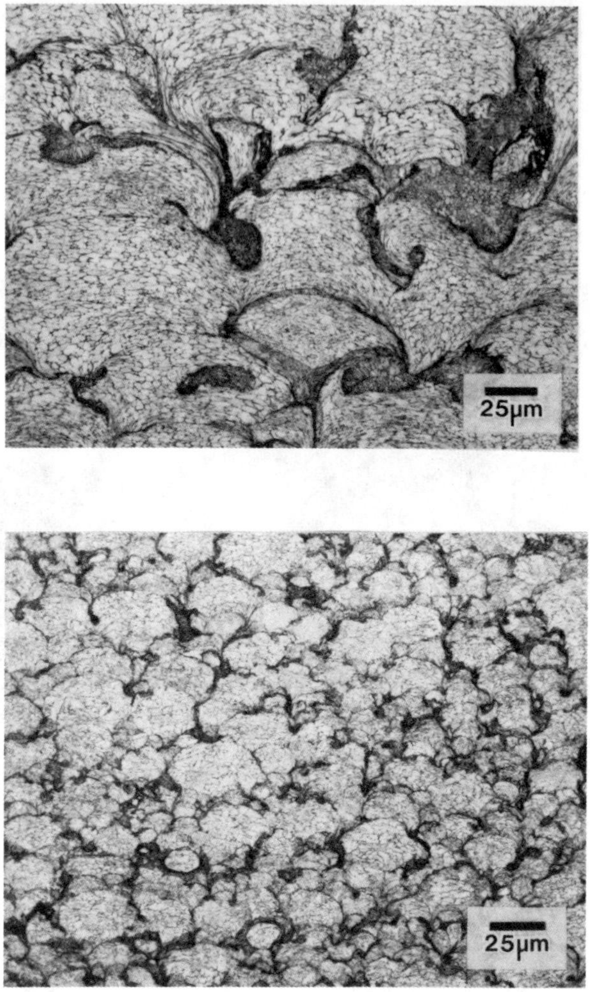

Figure 4: Optical micrographs of dynamically consolidated Cu-2Zr of a) 63 to 106μm and b) <38μm particle size compacts. Shock propogation direction in micrographs is vertically downward in the paper plane.

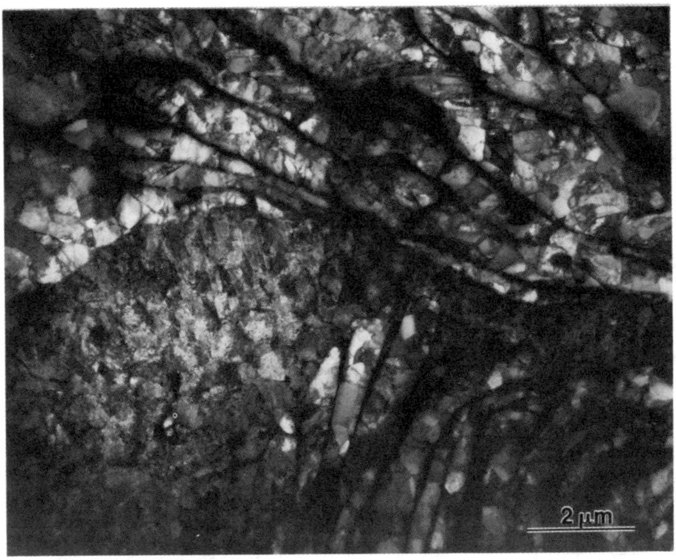

Figure 5: A melted and resolidified region between two dynamically consolidated particles of Cu-2Zr.

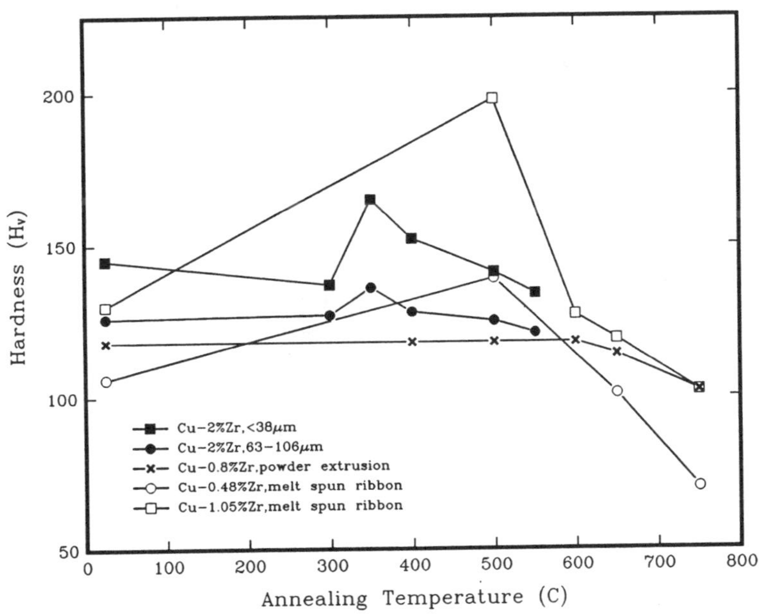

Figure 6: Summary of age hardening response (1 hour anneals) of the dynamically consolidated Cu-2Zr powder samples and other similar rapidly solidified Cu-Zr alloys. The Cu-0.8Zr results (9) are from hot extruded powder samples. The Cu-0.48Zr and Cu-1.05Zr results (10) refer to melt spun ribbon.

boundaries, and there was little evidence of precipitation within the intradendritic regions. The peak aged fine particle sample displayed many dislocations arrayed in loose tangles in the cell interiors and evidence of very fine precipitates on the dislocations. Overaging, at 550 C, was necessary to coarsen these precipitates so that their presence might be confirmed, as shown in Figure 7.

The effect of annealing on conductivity of the shock consolidated samples is given in Figure 8. The fine particle size material exhibits a significant decrease in conductivity on aging at 350 C compared to aging at 300 C. The coarse powder sample has a higher initial conductivity and shows a smaller conductivity decrease at 350 C. After aging at 500 C, the conductivity of the fine particle sample is higher than the coarse sample even though the hardness of the fine particle sample remains higher, as Figure 6 shows.

4. CONCLUSIONS

The results of rapid solidification and powder processing of the copper alloys in this investigation indicates the potential benefit of such processing for improved material performance. In the case of Cu-7Ti alloys, the powder characterization revealed that a fine particle size range, less than about 15μm, is needed to avoid intermetallic phase formation (7). Shock consolidation experiments utilizing fine powder samples of this size range are planned to investigate the strength and age hardening response. For the Cu-16Ni-2Cr-0.7Fe alloy material, the initial microstructural analysis suggests that the series of thermomechanical processing steps have resulted in a useful microstructure. Even though the fine powder billets were exposed to elevated temperatures for prolonged times, the final microstructure was quite highly refined with widely scattered grain boundary precipitates. Corrosion and cavitation erosion resistance measurements are planned for the rolled rod material, as well as a characterization of strength and ductility. The results on the Cu-2Zr material indicate that a lower Zr content, about 0.7 Zr (8), probably can be utilized to gain a similar strengthening response as the Cu-2Zr and to avoid the intercellular precipitation of intermetallics. The enhanced conductivity level of such an alloy modification may be useful for high strength/high conductivity applications.

5. ACKNOWLEGEMENTS

I wish to express appreciation to L. E. Tanner, L. S. Chumbley, and R. N. Wright for valuable technical participation in this investigation. Also, I wish to thank K. M. Rusin, M. Wall, F. Laabs, C. Wu, and A. J. Schwoeble for important analytical and

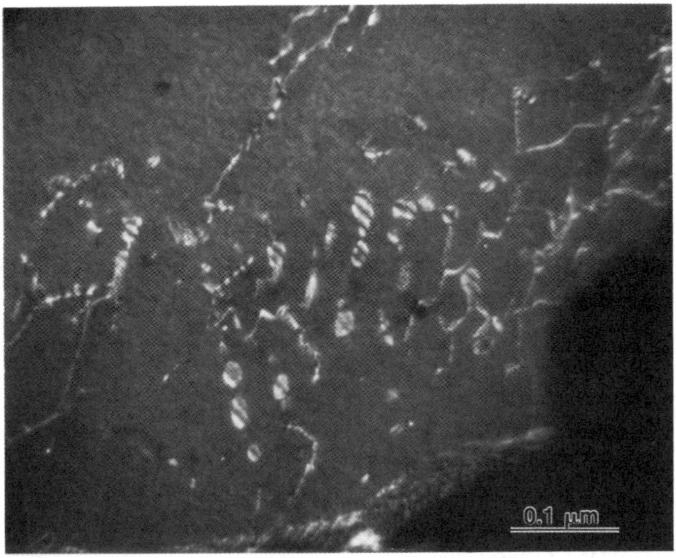

Figure 7: Dark field TEM micrograph of precipitates in overaged fine particle size sample of dynamically consolidated Cu-2Zr.

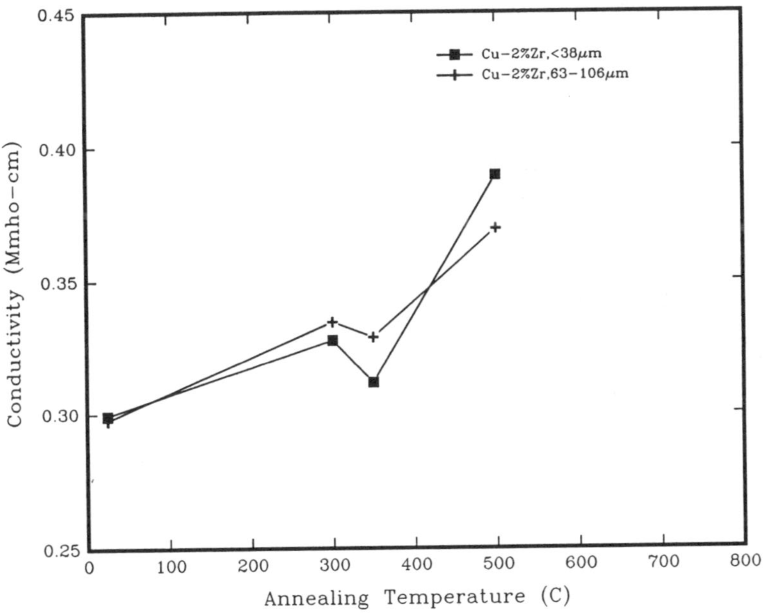

Figure 8: Conductivity of dynamically consolidated Cu-2Zr powder compacts as a function of aging temperature (1 hour anneals).

technical contributions. The support I received at Naval Research Lab from the US DoD and the support I now receive at Ames Lab from the US DoE is greatly appreciated.

6. REFERENCES

1. Anderson, I.E., Rusin, K.M., and Rath, B.B.: in Rapid Solidification Processing; Principles and Technologies, IV, Mehrabian, R., and Parrish, P., eds., Claitor's, Baton Rouge, LA, 1988, p. 114.

2. Adam, C.M. and Lewis, R.E.: in Rapidly Solidified Crystalline Alloys, Das, S.K., Kear, B.H., and Adam, C.M., eds., TMS-AIME, Warrendale, PA, 1985, p. 157.

3. Ayers, J.D., and Anderson, I.E., US Patent 4,619,845, Method for Generating Fine Sprays of Molten Metal for Spray Coating and Powder Making, October 1986.

4. Anderson, I.E., and Figliola, R.S.: in Modern Developments in Powder Metallurgy, vol. 18,19, or 20, MPIF, Princeton, NJ, 1988, (accepted for publication).

5. Cline, C.R., and Raybould, D., p.111 of ref. 2.

6. Berry, R.A., and Williamson, R.L.: in Metallurgical Applications of Shock-Wave and High-Strain-Rate Phenomena, Murr, L.E., Staudhammer, K.P., and Meyers, M.A., eds., Marcel Dekker, New York, NY, 1986, p. 167.

7. Anderson, I.E., and Tanner, L.E., Materials Science and Eng., (submitted in collection of papers from "Micrometallurgy '88", sponsored by TMS-AIME, March, 1988).

8. Wright, R.N., and Anderson, I.E., Materials Science and Eng., (submitted for publication).

9. Sarin, K.V., and Grant, N.J., Metall. Trans., 3, 875 (1972).

10. Arnberg, L., Backmark, U., Backstrom, N., and Lange, J., Materials Science and Eng., 83, 115 (1986).

7. BIOGRAPHY

Dr. Iver E. Anderson graduated from Michigan Technological University in 1975 with a B.S., from the University of Wisconsin-Madison with a M.S. in 1977 and a Ph.D. in 1982; all degrees in Metallurgical Engineering. Dr. Anderson was at the US Naval Research Laboratory from 1982 until 1987. Dr. Anderson has been employed from 1987 to the present at the Ames Laboratory (US DoE), Iowa State University. He is an active member of ASM-MSD, APMI, and TMS-AIME and serves on several TMS and ASM committees. He has delivered numerous presentations and his publications include 12 papers and one patent. Dr. Anderson's research interests include rapid solidification, atomization, and powder processing.

2nd International SAMPE Metals Conference
August 2-4, 1988

ELECTRON BEAM COLD HEARTH MELTING (EBCHM) OF SPACE AGE METALS

Howard R. Harker and Charles H. Entrekin
Axel Johnson Metals, Inc.
P.O. Box 529
Exton, PA 19341

Abstract

Electron beam cold hearth melting (EBCHM) is being used on a production basis to process titanium and titanium alloys for aerospace applications. Development of the process for nickel base superalloys is underway. The process and associated equipment is discussed. Various products and their production techniques are illustrated. The quality assurance program and equipment for aerospace applications are reviewed.

1. INTRODUCTION

The desire to achieve improved performance from titanium and nickel base alloys for critical aerospace applications has caused material scientists to investigate methods of producing "clean (defect free) metals." These evaluations indicate that EBCHM can provide refined alloys with improved properties such as low cycle fatigue.[1] EBCHM is a production process with full scale equipment commercially available to fulfill this unique melting and refining requirement.

2. HISTORICAL

Electron beam melting dates back to the early 1900's when Pirani built a laboratory electron beam furnace to melt tantalum. Electron tube developement in the 1930's led to electron gun and high vacuum technology. Atomic physics requirements in the 1940's led to large high vacuum systems. Electron beam melting then evolved commercially in the 1950's as a drip melting process to provide Tantalum and Niobium for rocket and nuclear materials applications.[2]

Electron beam cold hearth melting originated in 1963 at Temescal Metallurgical Corp. in Berkeley, California to solve problems the firm had experienced in applying the drip melt electron beam melting process to several new applications. These problems included:

- Segregation of alloying elements when melting nickel base superalloy ingots from prime raw materials.
- Oxide inclusions in nickel base superalloy ingots.
- Inadequate removal of impurities when drip melting aluminothermic reduced Niobium.

The concept of separating the melting, refining and casting by the use of a horizontal water cooled copper trough or hearth solved these problems. This concept has been further developed today to provide the basis for EBCHM "clean metal" processing technology for producing "inclusion free" metals for critical aerospace applications.[3]

FIGURE 2
Recent EBCHM (Two Skimmer Dams)
Axel Johnson Metals, Inc. 1987

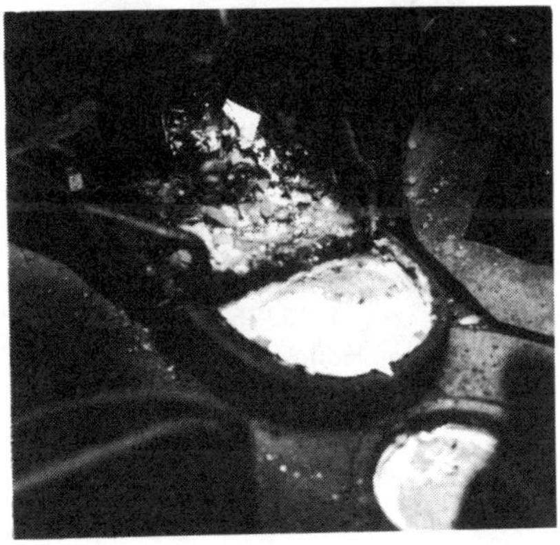

FIGURE 1
First EBCHM (One Skimmer Dam)
Temescal Metallurgical Corp.
1963

3. THE PROCESS

As shown in Fig. 3, the process involves introducing the metal into a vacuum chamber and melting it by electron bombardment in a horizontal water cooled copper hearth. The molten metal then flows by gravity along the hearth allowing time for gravimetric separation of oxide, nitride or high density metallic inclusions, mixing of alloying elements and evaporation of volatile impurities. The molten metal is contained by a skull or liner of solidified metal which is formed next to the water cooled copper hearth. Various devices and techniques are used to aid the hearth in performing the desired refining reactions (water cooled copper skimmer dams, high intensity beam patterns, specific hearth geometries and various sensors.)

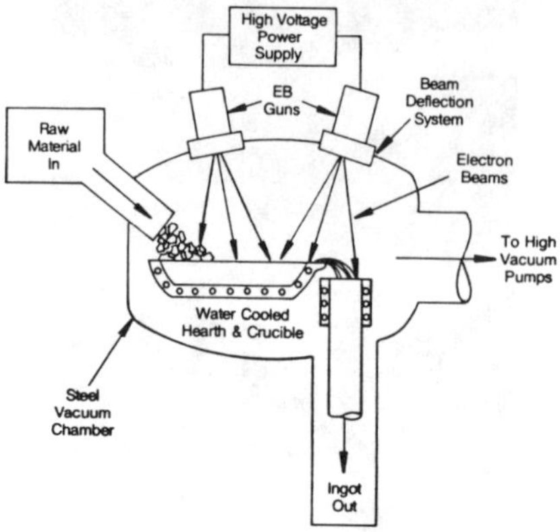

FIGURE 3
EBCHM Schematic

The refined molten metal then flows by gravity to a casting station where it is solidified into an ingot in a water cooled copper collar-type mold in a continuous casting process. This mold or molds (multi-strand casting is possible) can be a variety of shapes and sizes.[4]

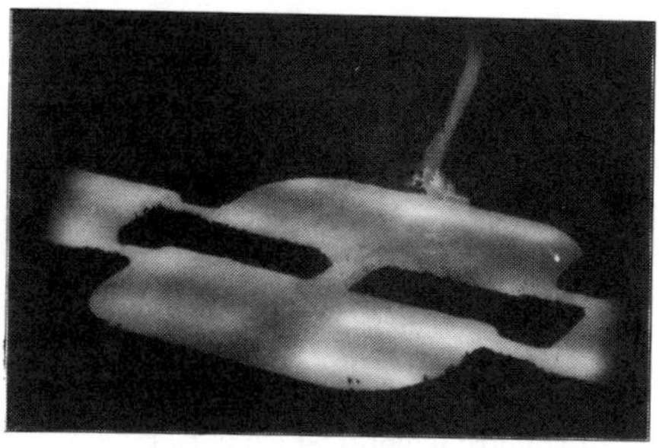

FIGURE 4
Four Strand Casting
of Titanium Alloys

The molten metal is sampled and analyzed during casting.

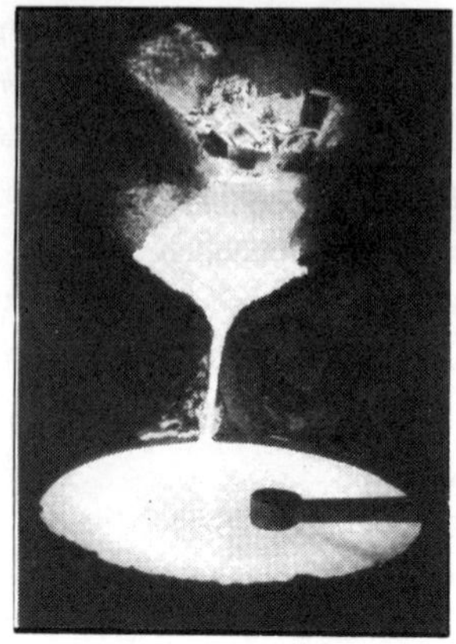

FIGURE 5
Dip Sample Taking

The flexibility of energy management during casting allows the casting of hollow ingots using a water cooled copper mandrel.

FIGURE 6
Casting Hollow Titanium Ingot

The solidification process can be very closely controlled which offers the opportunity for control of grain size and segregation during solidification.

4. THE EQUIPMENT

The EBCHM process is typically performed in a large water cooled steel vacuum chamber with a high vacuum pumping system capable of maintaining the pressure in the range of 13.33 to

4000×10^{-3} PA ($0.1 - 30 \times 10^{-3}$ Torr) during melting. Material handling is done using vacuum locks to preserve the melt chamber vacuum environment during loading, unloading, sampling and gun service operations.

The Axel Johnson Metals, Inc. EBCHM furnace is typical of a large megawatt class EBCHM furnace used primarily for titanium melting. This furnace has a 3.66M (12 ft.) diam. melting chamber, 4 - 600 KW electron beam guns, 4 - .89M (35 in.) diam. oil vapor diffusion pumps, 5 roots type mechanical vacuum pumps and 5 - 8490 L/M (300 Ft^3/Min.) rotary piston mechanical vacuum pumps. This furnace is capable of melting solid or particulate titanium scrap at rates up to 1362 kg/hr. (3000 lb./hr.) into ingots or slabs up to 6810 kg (15,000 lbs.)

5. PRODUCT

The EBCHM process has found a large commercial application in recycling titanium scrap.[5] Approximately 4.09 million kg/yr. (9 million lbs./yr.) are being presently processed in this manner. Possibly another 0.45 million kg/yr. (1 million lbs./yr.) of nickelbase superalloy ingots are being melted by EBCHM.

Mill Ready Products:

- Pure titanium ingot 635 mm diam. x 3.68 m long (25 in. x 145 in.)
- Pure titanium slab 305 mm x 1.1 m x 3.68 m long (12 in. x 44 in x 145 in.)
- Alloy titanium slab 305 mm x 1.1 m x 3.68 m long (12 in. x 44 in. x 145 in.)
- Pure titanium ingot 432 mm diam. x 3.68 m long (17 in. x 145 in.)
- Pure titanium hollow ingot 305 mm O.D. x 90 mm I.D. x 3.68 m (12 in. x 3.5 in. x 145 in.)

FIGURE 7
EBCHM Titanium Slab

For Remelt:

- In-spec titanium alloy electrodes 635 mm x 3.68 m long (25 in. x 145 in.)
- Depleted alloy electrode 280 mm x 480 mm x 3.68 m long (11 in. x 19 in. x 145 in.)
- Depleted alloy multi-strand electrode 203 mm x 406 mm, 203 mm x 203 mm x 3.68 m (8 in. x 16 in.) (8 in. x 8 in.) (x 145")
- In-spec nickel base superalloy electrode 203 mm diam. to 635 mm diam. x 3.68 m long. (8 in. - 25 in. x 145 in.)

FIGURE 8
Four Strand EBCHM
Titanium Ingot

The in-spec titanium electrodes and nickel base superalloy products should be of particular interest to aerospace materials scientists because these products take advantage of the EBCHM process to produce "clean metal" for critical part applications. In the case of certain superalloy investment casting scrap, EBCHM is necessary to allow this scrap to be reverted.(6)

6. QUALITY ASSURANCE

Quality assurance and quality control in EBCHM are particularly important in light of an apparent dichotomy of purpose with regard to raw materials. A large portion (roughly 60 to 100%) of the titanium input to industrial-scale EBCHM furnaces is in the form of scrap. It is the purpose of the EBCHM process to refine and consolidate this raw material into premium grade products for critical applications.

Quality control for the process is accomplished by close adherence to established standard manufacturing procedures and frequent in-process inspection operations. These standard manufacturing procedures are negotiated with end users and are "frozen" through the use of rigid written vendor agreements.

A key concept in quality control is to fully characterize the chemistry of the incoming raw materials so that there are "no surprises" in either the performance or the products of the EBCHM furnace. This is accomplished by selective purchasing from established vendors; by rigorous sorting, sizing, and cleaning; and by complete chemical analyses of all materials before they are released for melting. No raw material can be released for melting before a statistically significant number of samples (usually 3 to 6) are analyzed. Additionally, all solid scrap which is of large enough size (usually over 500 mm^2 (1 in^2) is individually inspected and sorted using a portable optical emission spectrometer. Smaller solids and turnings are blended and statistically sampled.

The programmable controls of the EBCHM furnace allow for the precise and repeatable distribution of beam energy over the surface of the molten metal. This is particularly advantageous in creating zones of superheat for refining purposes, for creating unique flow patterns in the hearth, and for generating a sweeping motion with a beam to keep floating inclusions (in superalloys) away from the pouring area.

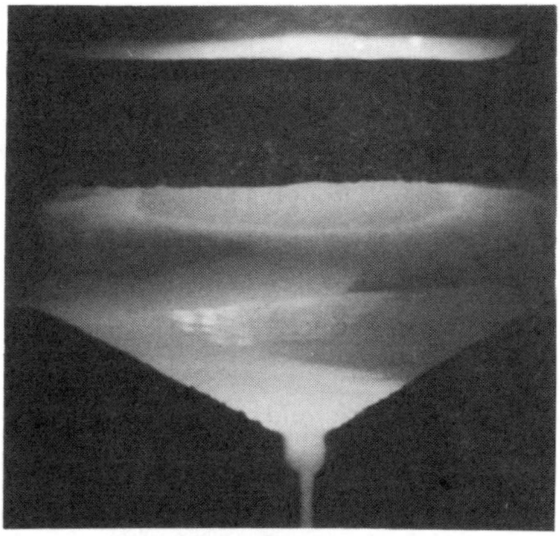

FIGURE 9
Ni-Base Alloy Flowing Under Skimmer Dam

Chemistry control of the finished products is achieved both by controlling the input stock and by controlling the melt rate and beam power during melting. Certain elements with high vapor pressures such as aluminum and chromium preferentially evaporate in a vacuum melting system such as EBCHM. The raw materials for alloys containing these elements are selectively enriched and then "evaporated down" to the appropriate specifications during melting. Rapid chemical feedback and rigid melt rate control allow uniform chemistries to be achieved.

FUTURE

The need for automated process control will require the use of a variety of sensors such as:

- Ingot pool height
- Molten metal temperature and profile
- Furnace residual gas analysis
- Molten metal chemistry
- Input material feed rate
- Floating inclusion count

These sensors are all available or are being developed. Computerized process operations and data recording will soon be possible leading to intelligent material processing to optimize product quality and efficiency.

Alloys will be formulated to take advantage of the "clean metal" refining capability of EBCHM (defect free materials.)

EBCHM will be used more frequently with other melting processes such as VIM, VAR, or VADER.

The ability to cast to near net shape and to cast multiple strands will be further utilized.

The ability to closely control solidification will lead to a commercial "fine grained" ingot casting process.

CONCLUSION

The EBCHM process is being utilized as an efficient titanium scrap recycle technique, a method to revert certain superalloy investment casting scrap and a means to produce "super clean" alloys for critical aerospace parts. These capabilities along with the ability to cast to near net shape and closely control ingot solidification insure a growing role for EBCHM in aerospace materials processing. The use of "clean metals" in aerospace applications is on the rapidly increasing slope portion of a typical "S" shaped experience curve.

(1) Menzies, R.G. and Adasczik, C.B.: "Clean Melting of Ni-Base Alloys", Proc. 1986 Vacuum Metallurgy Conf., Iron & Steel Society, Inc., Warrendale, PA 15086, p 65.

(2) Smith, H.R., Hunt, C.d'A, and Hanks, C.W.: "Electron Bombardment Melting", Journal of Metals, Vol II, No. 2, Feb. 1959, pp 112-117.

(3) Tien, J.K. and Jarrett, R.N.: "Current Status of Electron Beam Melting and Refining of Superalloys", Proc. Conf. Electron Beam Melting and Refining - State of the Art 1984, Bakish, R., Ed., (Bakish Materials Corp., Englewood, NJ 1984) pp 321-329

(4) Knecht, II, J.A., and Entrekin, C.H.: "Electron Beam Casting of Shapes and Hollows", Proc. Conf. Electron Beam Melting and Refining State of the Art 1987, Bakish, R., Ed., (Bakish Materials Corp., Englewood, NJ 1987) pp 46 - 59

(5) Harker, H.R.: "Electron Beam Melting of Titanium Scrap" Proc. Conf. Electron Beam Melting and Refining - State of the Art 1983, Bakish, R., Ed., (Bakish Materials Corp., Englewood, NJ 1983) pp 187-190

(6) Lowndes, J.C.: Superalloy Refining Technique Lowers Rejections, Boost Performance. Aviation Week and Space Technology, Aug. 19, 1985, pp 103 - 108

AUTHOR BIOGRAPHIES

Howard R. Harker is Vice President of Manufacturing and Engineering for Axel Johnson Metals, Inc. in Lionville, PA. He graduated from the Univ. of California at Berkeley with a Bachelors Degree in Mechanical Engineering and is a Registered Metallurgical Engineer in the State of Calif. He has been employed in the field of Electron Beam, High Vacuum metallurgy for the past twenty-six years. Previous employers were Airco-Temescal and Viking Metallurgical Corp.

Charles H. Entrekin is Vice President - Technology for Axel Johnson Metals. His primary responsibilities include quality control, research, and process development. He received a B.S. in Metallurgy and Materials Science from Lehigh University, an M.S. in Materials Engineering from Drexel University, and an M.B.A. from the University of Delaware. Prior to joining Axel Johnson Metals, Inc. in 1983, he was employed by Lukens Steel Company as a research engineer.

2nd International SAMPE Metals Conference
August 2-4, 1988

MELTING TECHNIQUES FOR TITANIUM

A. Mitchell and D.W. Tripp
Dept. of Metals & Materials Engineering
The University of British Columbia
309-6350 Stores Road
Vancouver, B.C., V6T 1W5

Abstract

The present melting methods for titanium are reviewed and the short-comings identified. It appears that the VAR systems suffer from inherent disadvantages in the control of the thermal field for both ingot making and for casting. The use of induction heating methods overcomes some of these disadvantages, but cannot be used with refractory crucibles for high-quality applications. It is suggested that a combination of heating and stirring techniques will be the most advantageous for casting purposes, whilst the hearth melting systems have proven application in ingot making.

1. INTRODUCTION

Titanium melting technology has been in general industrial use for many years and the problems involved are reasonably well-defined. The same, unfortunately is not true for the potential solutions and until recently the techniques were relatively static in development. As a result, titanium casting remains a costly and difficult procedure with very few practicioners in the market. Titanium ingot-making is somewhat less fraught with difficulties, but also has remained until recently a static technology in spite of some obvious drawbacks in both quality and yield, together with a substantial cost.

The purpose of the present survey is to highlight the key problems in the technology, and to suggest potential solutions which may be applied.

1.1 Containment Problems

Liquid titanium has quite justifyably been nick-named "the universal solvent", since it will dissolve almost all common refractory containers very rapidly. Silica, alumina, MgO, refractory metals, carbides and nitrides are all highly soluble in liquid titanium. This feature has two consequences for titanium melting:

- it is difficult to ensure alloy homogenization, particularly when the alloy is made from elementary components.
- it is difficult to produce a large volulme of liquid titanium at a uniform superheat.

The most usual container for liquid titanium is water-cooled copper, although NaK cooling is also used. With suitable attention to the thermal design of crucible, either of these systems is satisfactory as a container, but neither solves the above two problems. In consequence we are compelled to make melts of alloy ingots to ensure homogeneity: even this technique does not serve to homogenize some systems, notably alloys rich in Nb or Mo. We also must use a high level of automation in the casting processes to ensure that the liquid casting temperature falls into the desired range.

Two alternative possibilities have recently become feasible on an industrial scale. Both employ induction heating as the power source, one with a water-cooled crucible[1]; one with a CaO crucible[2].

The inductoslag process[3] has been used as a laboratory tool for reactive metal melting for some time, but has only recenly become useful on an industrial scale. The charge is heated by a conventional induction coil which is placed outside a water-cooled copper crucible constructed so as to permit adequate field penetration for charge heating. The crucible is made of electrically-isolated segments with the copper wall thickness minimized. The original version of the process used calcium fluoride as an electrically-insulating frozen layer between the charge and the crucible, but subsequent design improvements have shown this to be unnecessary. This latter point is important as the fluoride slag not only makes high quality casting difficult, but also produces fluoride inclusions in the product through the slight fluorine solubility in liquid titanium. (It is probable that this latter phenomenon also rules out electroslag remelting for the production of high-quality titanium). The charge size, crucible design and power characteristics must be matched carefully, but when this is done the process has an efficiency close to other similar processes such as plasma or electron-beam heating. The major advantages of the process for casting lie in its ability to give a well-mixed volume at a reasonably-uniform temperature, and to hold that volume in a steady-state without the need to cast immediately on forming the liquid volume.

The use of conventional induction melting with a refractory crucible has the same advantages as the inductoslag technique, together with the potential of higher efficiency. However, of the refractory container materials available, only CaO shows a significant resistance to liquid titanium. Calcium is essentially insoluble in titanium and so if the melt is carried out under a pressure of argon sufficient to prevent calcium evaporation, the titanium liquid will not pick-up oxygen. However, as with inductoslag casting, if the casting process is carried out under argon there is a potential risk of argon-filled porosity which will not densify after HIP treatment. At this point, the results of CaO crucible casting are not sufficiently detailed to permit an evaluation, but the process has been used to manufacture titanium cast parts.

Both of the above processes can be used to reclaim and consolidate titanium scrap to form an ingot by either casting or by a withdrawal technique. In this respect they would resemble techniques using arc or plasma heating already in commercial use[4]. In common, however, with these latter processes, they have a potential direct route from charge to ingot for defect-forming inclusion material such as HDI or LDI, and are therefore not regarded as equivalent to the hearth melting processes in quaranteeing freedom from this type of defect in the ingot product.

1.2 Temperature Regime

Temperature distributions in the processes are of concern for both ingot making and for casting. We must have adequate superheat for the solution of both alloy additions and unwanted material such as LDI's, but in the ingot making processes we must also control the superheat in alloy systems to maintain adequate structure control in the ingot. Large superheat necessarily means long solidification times which in turn lead to the macro-segregation responsible for β-fleck defects. Also, a large pool volume in an ingot making process will lead to difficulties in controlling solidification during hot-topping. The formation of segregation at the last stages of solidification during hot-topping has been cited as the cause of α-II defects[5]. The growth of a shrinkage cavity during hot-topping of a VAR ingot is shown in Fig. 1. Where such a cavity exists, most nuclear and aerospace specifications require its removal before forging; a costly procedure which is difficult to make reproducible.

The VAR process has a heat balance which contains two pertinent features:

- the response to added power is a direct increase in the melting rate, but only a small increase in the superheated pool volume.
- the electrode acts as a radiation shield for the ingot pool surface and makes it impossible to obtain a shallow pool (and high solidification rates) at any but very slow melting rates.

These two features severely limit the potential of VAR in titanium alloy production. In respect of providing liquid metal for casting, VAR has the additional disadvantage that once melting has ceased, it is not possible to provide thermal energy to the pool. The timing of the melting and casting sequence is therefore very critical.

The alternative melting techniques using cold-crucibles and a surface heat source i.e. non-consumable arc, plasma or EB, offer the advantages of controllable superheat and also of less-critical timing, but all of them have the common disadvantage of being surface-heated. Titanium is not a good thermal conductor, and unless the liquid pool is well-stirred, a surface-heated process will result in very high surface temperatures, but only a small pool volume at significant superheat, as illustrated in Fig. 2. The required stirring can be input by a mass flow of metal, but once the melting/feeding sequence has stopped, the static thermal balance re-asserts itself and the pool bottom starts to freeze. The timing of casting, whilst less critical than in VAR, nonetheless remains important.

1.3 Ingot Structures

It is fortunate that our present range of titanium alloys freezes as simple solid-solution β-alloys. In consequence, there are few problems of microsegregation and most ingot defects result from either macro-segregation or from extraneous second-phase particles. The control of macrosegregation through the manipulation of VAR melting variables has distinct limits, because of the high melting points of the alloys. If we adopt the simple approach of reducing the solidification time by lowering the melting rate, even with other parameters such as geometry and magnetic stirring suitably optimized, we are still limited by the yield requirement to have an acceptable ingot surface. In the case of the more complex alloys (e.g. Ti-17), this requirement means a choice between yield loss due to poor surface or yield loss due to β-fleck rejections.

The solution appears to lie in hearth-melting, using either EB or plasma as the heat-source. The advantages of hearth melting in respect of extraneous particles have been well-described in the literature[6], but less emphasis has been placed on an equally important aspect, namely temperature control. Since the feed metal to the ingot does not originate from an electrode held above the ingot pool, we have accomplished two important new changes to the ingot's thermal field as distinct from that in VAR:

- the melting rate and metal feed superheat have been un-linked.
- the ingot thermal field is now controlled by its top-surface free radiation.

These two features combine to give a much shallower pool in hearth melting than in VAR at the same melting rate, without a great loss in electrical efficiency. (The large VAR pool causes a great heat loss to the crucible cooling system the removal of which in the hearth melting shallow pool compensates for the additional radiative losses in hearth melting). It is probable that the segregation-prone alloys will be much easier to make successfully in hearth melting than is the case in VAR. The problems introduced by hearth melting, such as evaporation in the EB system[7], or gas contamination and temperature control in plasma[8], appear to have been successfully resolved in large-scale practical systems and at least the EB version may be considered to be a normal industrial technique.

1.4 Casting Techniques

The future for casting techniques appears to lie in the successful resolution of temperature homogenization in the water-cooled crucible process. Whilst the refractory crucible techniques are certainly of interest in lower-quality systems, or in systems for melting alumnides, they do not as yet offer the potential for oxygen control required in high-quality castings. The induction cold-crucible technique has been shown to have good application in respect of temperature control, but as yet has not been scaled-up to the size required for many of our larger castings. There are almost certainly difficult problems in this regard concerning the matching of crucible and coil design to the induction power-supply characteristics.

However, the body-force required to attain thermal homogenization in a large volume of liquid titanium is rather small. It could be obtained by the application of either inductive or electromagnetic stirring coils to a cold-crucible in which the melting power was applied by other means. This combination of the heat generation simplicity of the established systems with the temperature homogenization capability of the induction process would satisfy the prime requirements of casting for melting, i.e. capability of using loose feed, of holding a good superheat distribution for long periods and of generating the required accuracy of composition control.

2. CONCLUSIONS

1. The VAR techniques for both melting and casting have now matured to demonstrable process limits.
2. The shortcomings in our present ingot-making techniques can be solved by hearth-melting.
3. The most promising route for large titanium alloy castings is a combination of induction and non-consumable electrode techniques such as EB and plasma.
4. The refractory crucible/ induction techniques offer some potential advantages over induction cold-crucible for small, low-quality castings, but not for high quality products.

3. REFERENCES

1. Chronister, D.J., Scott, S.W., Stickle, D.R., Eylon, D. and Froes, F.H., Journal of Metals, Vol. 38, 9, Sept., 1986, pp. 51-64.
2. Desawa, T. and Ototani, T., Tetsu-to-Hagane, Vol.73 (1987), No. 14 (October), pp.1691-1697.
3. Clites, P.G. and Beall, R.A., Proceedings of the Fifth International Conference on Electroslag and Special Melting Technology, G.K. Bhat, editor, Pittsburg, PA, 1975, pp.477-496.
4. Stocks, S. and Hiatt, D., Proceedings of the 1986 International Conference on Titanium Products and Applications, Vol. II, Titanium Development Association, Dayton, Ohio, pp. 918-927.
5. Froes, F.H. and Bomberger, H.B. Journal of Metals, Vol. 37, No. 6, July 1985, pp. 27-37.
6. Entrekin, C.H. and Clarkson, D.S., Metals Progress, Vol. 129, 7, June 1986, pp. 35-39.
7. Mitchell, A. and Takagi, K., Electron Beam Remelting and Refining - State of Art 1984, 1984, Bakish Materials Corporation, Englewood, NJ, pp. 88-99.
8. Schlienger, M.P. and Eschenbach, R.C., Proceedings of the 1986 International Conference on Titanium Products and Applications, Vol. II, Titanium Development Association, pp. 394-403.

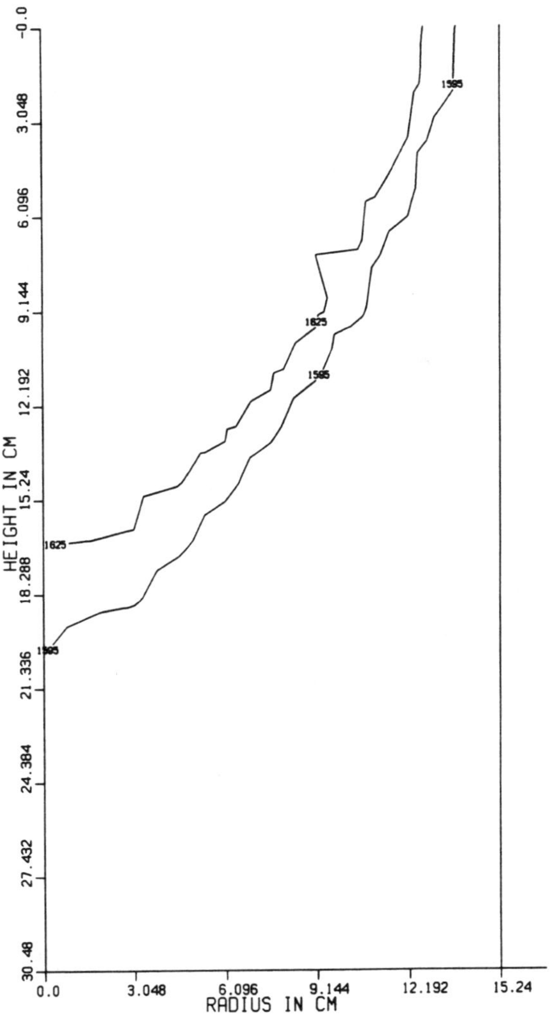

<u>FIGURE 1</u>

POOL PROFILES FOR AN ELECTRON BEAM REMELTED TITANIUM 6 ALUMINUM 4 VANADIUM INGOT. TOTAL BEAM POWER INPUT = 115 KW. MELT RATE = 189 KG / HR (7 LB / MIN). INGOT DIAMETER = 30.5 CM (12")

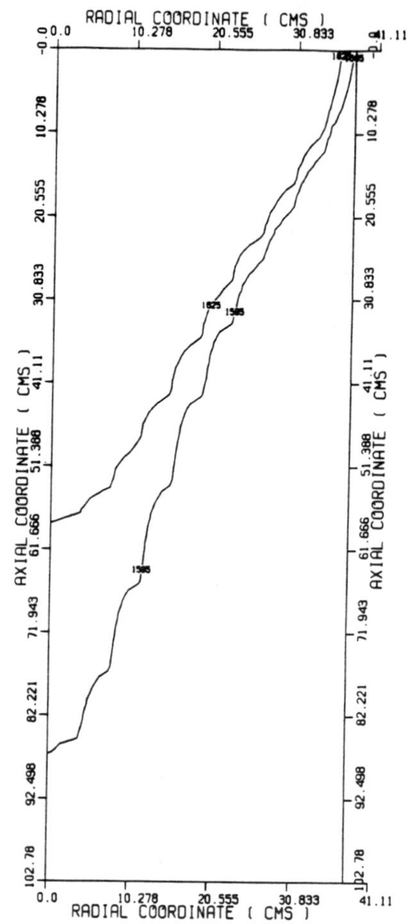

FIGURE 2A

POOL PROFILES FOR TITANIUM 6 ALUMINUM 4 VANADIUM VAR INGOT 10 MINUTES AFTER THE START OF THE HOTTOP SEQUENCE.

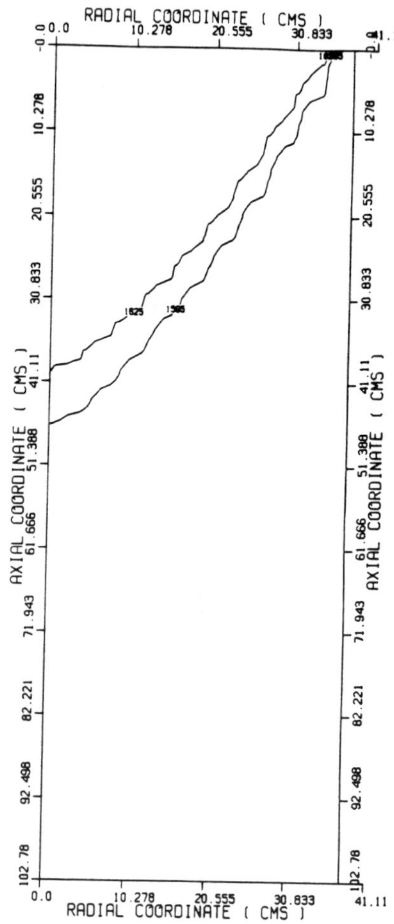

FIGURE 2B

POOL PROFILES FOR TITANIUM 6 ALUMINUM 4 VANADIUM VAR INGOT 60 MINUTES AFTER THE START OF THE HOTTOP SEQUENCE.

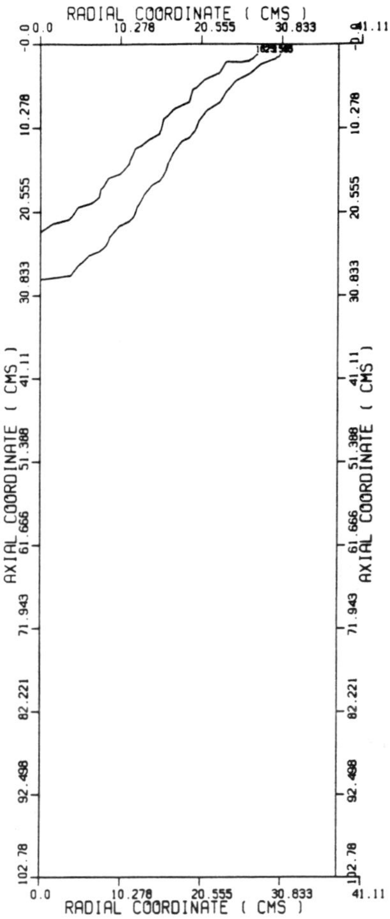

FIGURE 2C

POOL PROFILES FOR TITANIUM 6 ALUMINUM 4 VANADIUM VAR INGOT 100 MINUTES AFTER THE START OF THE HOTTOP SEQUENCE.

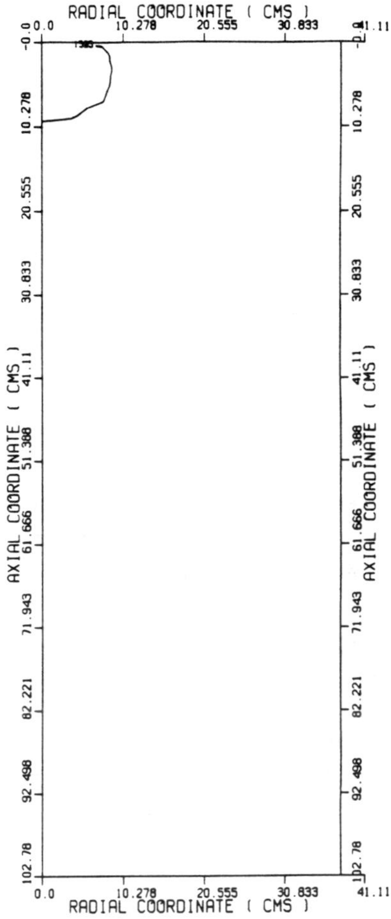

FIGURE 2D

POOL PROFILES FOR TITANIUM 6 ALUMINUM 4 VANADIUM VAR INGOT 150 MINUTES AFTER THE START OF THE HOTTOP SEQUENCE.

2nd International SAMPE Metals Conference
August 2-4, 1988

ADVANCED PRODUCTION OF SUPERALLOY AND TITANIUM INGOTS BY COLD HEARTH MELTING

H. Stumpp, H. Pannen, H. Ranke, G. Sick, W. Dietrich
Leybold AG, West-Germany

ABSTRACT

Conventional remelting processes like VAR and ESR are limited in their development to a more advanced process control. Significant control parameters cannot be measured in these processes. The interdependence of various process parameter limits the ability to further optimization. Electron Beam- and Plasma Cold Hearth Melting offer excellent access for various sensors and on line product quality control. Under these aspects, the state of the technology of a modern industrial size Electron Beam Cold Hearth furnace and Plasma Furnace will be presented and discussed.

1. INTRODUCTION

The advanced applications of high performance metals and alloys like titanium and superalloys in the aircraft industry are demanding a continiuous improvement in properties like cleanliness, chemical homogeneity, and structure of semifinished products but also for a search of new control essential parameters for the manufacturing processes.

Conventional remelting processes like Vacuum Arc Remelting (VAR) and Electron Slag Remelting (ESR) are limited in their ability to be adapted to the demands of the future. It is impossible to measure significant control parameters. These processes suffer from the strong interdependence of various process parameters as e.g. melt rate, refining efficiency, and ingot structure.

Cold Hearth Melting Processes - up to now only sparsely applied for aircraft alloys - are offering a number of promising new aspects to fulfil the requirements for advanced ingot manufacture [1]. The two main types of heat sources available today for cold hearth melting are Electron-Beam Guns and Plasma Torches - preferably Low Pressure Plasma Torches. The essential characteristic of Cold Hearth Processes is the local separation of melting, refining, and solidification. This concept allows an independent access to the refining zone in the hearth and to the pool in the withdrawal crucible.

2. FEEDSTOCK AND REACTIONS IN A COLD HEARTH PROCESS

In principle a variety of very different material can be fed into a watercooled copper hearth. It is obvious that cold hearth melting technology offers a high degree of flexibility concerning size and preparation of the feedstock. The Cold Hearth furnaces which melt in production today profit only to a limited degree from this flexibility. The numerous possible reactions in the cold hearth add still more flexibility to the process.

Fig. 1 summarizes the different kinds of feedstock and possible reactions in a cold hearth process. There is no other melting or remelting process known, which offers a similar degree of process-flexibility.

In addition, there is a free choice of product form (hollow, slab ingot, multiingot, disc, pellets, powder, sprayed near net shapes) and solidification structure, providing a heat source that is flexible enough to achieve the required power densities in the product forming stage of the process.

Fig. 2 shows different solidification structures and a selection of products of cold hearth melting processes.

Compared to conventional remelting processes like VAR and ESR, the interrelation between heat input, melt rate, degree of reaction (e.g. refining) and solidification in a cold hearth process is on a substantially lower level. This offers more freedom for metalurgical optimization at each stage of the process.

This statement is anything but new, however, today's process automation and equipment reliability offer the chance for the realization of complex hearth melting processes in industrial scale for a variety of materials. Process automation requires access of intelligent sensors to control parameters, fast signal processing and sophisticated software. In this aspect, the cold hearth melting process offers first of all an excellent access to the reaction stage (hearth) and the product forming stage as apposed to conventional remelting processes.

3. MEASUREMENT OF PROCESS DATA AND AUTOMATION

Important process data in a complex remelting process are:

1) melt rate
2) temperature distribution in the melt and product
3) melt composition
4) degree of removal of non soluble inclusions
5) degree of melt degassing
6) shape and location of liquid-solid boundary in the continuously cast product

In conventional remelting processes like VAR and ESR only the measurement of the melt rate via load cells is possible. In VAR melting, degassing of the melt can be monitored additionally. The time dependence of the slag composition (oxygen potential) in the ESR-process enhances difficulties to monitor and control the chemical composition of the product. Slag technology has to be developed for each alloy. Process specific contamination or defects like white spots in VAR can occur. The parameters for hot-topping in ESR and VAR are arrived at empirically. Contrary to that a <u>Cold Hearth Process</u> offers a wide range of process control and automation:

1) The melt rate can be measured and controlled by ingot withdrawal and feeding rate.

2) The temperature distribution on the melt surface can be measured with automated two colour pyrometric devices or thermal imaging.

3) The melt composition (or better the concentration of some important elements) will be monitored with insitu energy dispersive x-ray analysis (EDX) in EB and with emission spectroscopy in plasma melting.

4) A satisfactory degree of moval of high density inclusion has been demonstrated. (W_xC_y in Ti-remelting)

5) Continuous control of refining efficiency for low density inclusions is possible, due to the effect that solid inclusions usually are differing in their light emissivity from the base melt.

6) Degassing of the melt can be monitored by residual gas analysis via mass spectrometer.

7) Hot topping of the ingot can be automated to achieve highest yield.

8) A high degree of process safety and reproducibility can be achieved in combination of EB position detection and microprocessor controlled Electron Beam deflection [2],[3].

All quality affecting parameters can be measured and recorded with digital process data acquisition systems and are then available for extensive statistical quality control (SQC).

4. STATE OF TECHNOLOGY OF INDUSTRIAL SIZE EB COLD HEARTH MELTING FURNACES

4.1 Electron Beam Furnace for 13 t ingots

As an example a large Cold Hearth Melting Furnace, recently installed at Nippon Mining Corporation, Japan will briefly described.

4.1.1. Basic Design Criteria

The furnace with 1200 kW available beam power is basically designed for Cold Hearth Melting. The main task is the production of very clean specialty steels, nickel- and cobalt base alloys - in the long run - super clean superalloys. The feeding electrodes will be cast in a 5 ton VIM furnace.

The second main application of this furnace will be the production of ingots of pure titanium starting from sponge and the recycling of titanium scrap contaminated with high density WC-tool tips.

The products of this Cold Hearth Melting Furnace are slabs with cross sections up to 470 x 1500 mm or round ingots between 400 mm and 800 mm diameter. The maximum ingot length is 3000 mm.

Besides these main applications the furnace is equipped with all necessary facilities for the production of directly drip-melted ingots of refractory and reactive metals and alloys. Horizontal and vertical feeding is possible for this process. The vertical feeder is limited to a diameter of 350 mm of the remelt electrode.

4.1.2. Furnace Layout

Fig. 3 shows a perspective view of the furnace. Not shown are the electrical control facilities. The round meltchamber (diameter 3.6 m) is arranged 8 m above ground level of the building and is supported by a heavy steel structure. The main operating floor is situated on the same level as the melt chamber. An intermediate floor serves for access to furnace components installed below the bottom of the chamber such as vacuum pumps, chamber bottom valve, water connections etc.

The ingot withdrawal unit can handle ingot weights of up to 13 tons by means of a telescopic hydraulic cylinder.

4.1.3 Electron-Beam Guns and Power Supplies

Four electron-beam guns KSR 600/40 and one auxiliary gun K150/40 are installed. Each of the KSR 600/40 guns can provide a maximum beam power of 600 kW at an acceleration voltage of 40 kV. The guns are equipped with double differential vacuum pumping by one 360 l/sec and one 1000 l/sec turbomolecular pump and allow an operating pressure of up to 0.1 Torr in the meltchamber.

4.1.4 Arrangement for Continuous Flow Melting

The watercooled copper hearth is 350 mm wide. The length of the liquid metal pool can be adjusted according to the requirements of the actual process by adjusting the rear wall of the hearth. The maximum length is 1200 mm. A watercooled mechanical barrier, which is hydraulically adjustable in its height, can be utilized for holding back of floated low density inclusions.

EB-Cold Hearth processes are critical concerning the selective evaporation of certain alloy elements. For this reason, it is important to avoid any process interruptions. Therefore, a horizontal feeder was designed which allows a fully continuous feeding to the melt position. The feeding arrangement is shown in Fig. 4. Billets of up to 300 mm diameter and up to 2000 mm length can be fed via a lockvalve system. The interruption free feeding is achieved with an external and internal drive system. Near the rear end of the hearth the small auxiliary eb-gun takes care of the join welding of the individual billets. Thus a falling of billet end pieces into the hearth with the resulting splashing and stirring is completely avoided. Provisions are made for the later installation of a feeding system for loose scrap.

Fig. 5 shows the EB Cold Hearth Melting Furnace during installation. To achieve optimum product quality but also to facilitate operation the withdrawal unit can be controlled by an automated pool level control system. Independent of the melt rate the pool level controller keeps the level of the pool in the withdrawal crucible constant. During the usual operation, the eb-guns are controlled by means of the LEYBOLD computer beam power distribution system BD 564. This system is laid out for establishing complex power distributions as they are required in Cold hearth melting. It is equipped with a disk memory for storage and reinsertion of programs at any time.

5. PLASMA HEARTH MELTING

Plasma Cold Hearth Melting has been generally recognized to be a promising recycling process for titanium alloy scrap. The attractive features of plasma melting include good control of alloy composition and the ability to melt smoothly volatile elements containing raw materials such as titanium sponge. There are also some drawbacks like reduced power efficiency relative to other established melting processes and the necessity to recycle and clean the plasma gas, but these do not outweigh the advantages of the process. Removal of high density inclusions (HDIs) by gravity separation has already been demonstrated and interest has now focused on the extent to which hearth melting, either EB or plasma is effective in eliminating the so called low density or hard alpha inclusions (LDIs).

The behaviour of individual TiN particles during plasma melting has been studied in a large scale plasma melting furnace at LEYBOLD AG, West-Germany. Results were presented at the Ti-Conference in Cannes, France, 1988 [4].

5.1 Plasma Melting Furnace at LEYBOLD, Hanau, West-Germany

This furnace has been previously described [5]. Unique features are programmable movement of a 1.5 MW rated vortex stabilised, modified plasma torch for melting with argon and/or helium in the pressure rang between 150 and 1000 mbar. The furnace power supply is rated at 650 kW. A front view of the furnace is shown in Fig. 6.

It has been demonstrated that titanium alloys can be plasma melted without contamination or loss of alloying elements [4]. Loose, not compacted Ti-sponge and scrap and alloy feed material can be melted into a homogeneous product after plasma melting of raw, loose material. which is shown in Fig. 7

5.2 Concept of a High Productivity Plasma Hearth Melting Furnace

The experience and special know-how gained with LH's industrial size plasma melting and casting furnace points the way to the design of production Plasma Cold Hearth Melting Furnaces. The furnace will be capable of melting any standard titanium ingot size from scrap or sponge with an output in the range of several million pounds per year in 3 shift operation. Fig. 8 shows such a Plasma Cold Hearth Furnace Concept. The furnace has three automatically controlled plasma torches. The power rating of the first torch is 1200 kW. It controls the output of the furnace and yields a melt rate of more than 1200 lbs/h. The second hearth torch is specifically designed to supply heat into front part of the hearth without creating turbulence. The third torch serves for controlled solidification in the mold.

The watercooled hearth is critical for the ingot quality, it is divided into two zones separated by a gas barrier. Melting of the feed material takes place in the first zone whilst the second allows smooth flow of liquid metal to the ingot mold. Only molten metal can flow into the second zone and no unmelted material can short circuit the system and pass into the calm flow region or into the mold. One torch is provided for each hearth zone and for the crucible so that torch design and operation parameters can be optimized for each location.

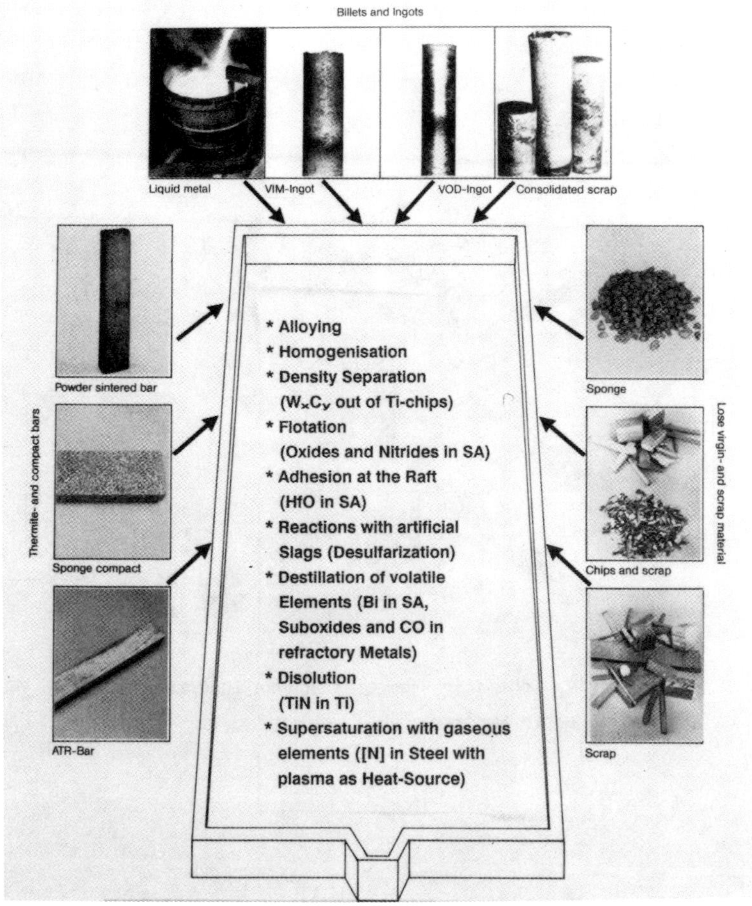

Fig.1 Variety of Feedstock and Possible Reactions in a Cold Hearth Melting Process

Fig. 2 Variety of Products and their Solidification Structure in EB Cold Hearth Melting

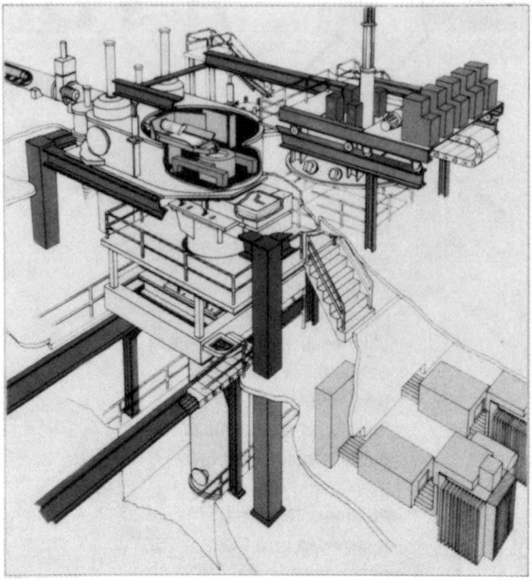

Fig. 3 Perspective Schematic View of a Modern Industrial Size EB-Cold Hearth Furnace

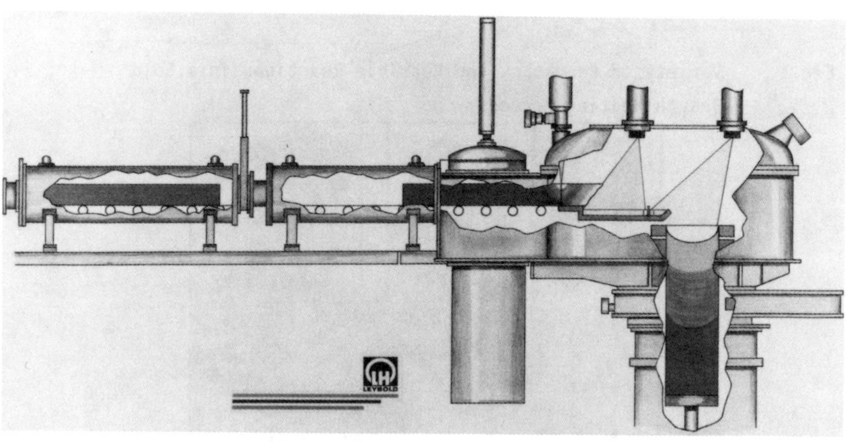

Fig. 4 Furnace Arrangement for Uninterrupted Bar Feeding

Fig. 5 EB Hearth Melting Furnace during Installation

Fig. 6 Industrial Size Plasma Melting and Casting Furnace (650 kW) at LEYBOLD AG, Hanau

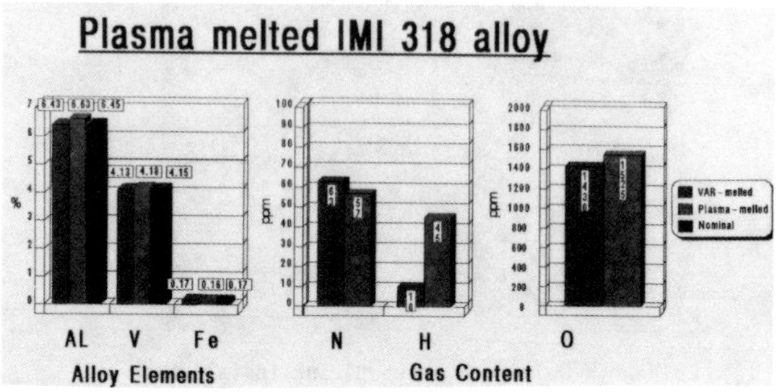

Fig. 7 Chemical Analysis of Ti6Al4V after Plasma Melting of raw, loose Material

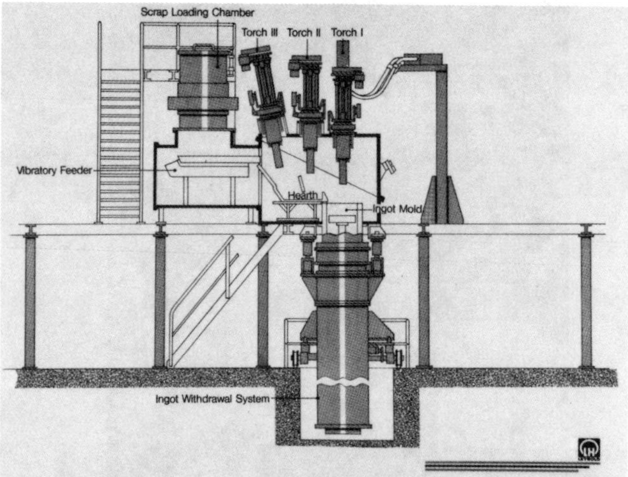

Fig. 8 Plasma Cold Hearth Melting Furnace Concept

REFERENCES

[1] Harker, H., "The Present Status of Electron Beam Melting Technology" in ibid, p. 3 - 7 in Proceedings of the Conference, Electron Beam Melting and Refining, State of the Art, 1986, Bakish, R., ed., Englewood, New Jersey. p. 98 - 126

[2] Ranke, H.; Bauer, V.; Heimerl, J.; "Process and Equipment Developments in Four Different High Power EB-Melting and EB-PVD Coating Pilot Installations at LEYBOLD-HERAEUS" in ibid p. 98 - 126

[3] Thomas, F.W.; Bittenbrünn, H.; Ranke, H.; Thorn, G.; "Advancements in Automation of EB-Melting Processes" to be published in Proceedings of the Ninth International Vacuum Metallurgy Conference on Special Melting, San Diego, CA, April 11 - 15, 1988

[4] Pannen, H.; Sick, G.; Wainhouse, D. M.; "High Power Plasma Melting of Titanium" to be published in Proceedings of the Sixth World Conference on Titanium, Cannes, France, June 6 - 8, 1988

[5] Sick, G., "Large Scale Plasma Melting and Remelting Tests" in Proceedings of the 1986 Vacuum Metallurgy Conference on Specialty Metals Melting and Processing, Pittsburgh, PA., Iron and Steel Soc., 1986, p. 179 - 186

BIOGRAPHY

Dr. Hermann Stumpp, born 12-19-48, holds a Doctor degree in physics of the University of Tübingen, West Germany. He actually is Head of department of vacuum melting and casting technology at LEYBOLD AG, West Germany.

Hartmut Pannen, born 11-09-58 holds a M.A. degree in chemical engineering of the Technical University of Clausthal, West Germany, and a M.A. degree in metallurgy of the Pennsylvania, State University. He actually is project manager for plasma melting at LEYBOLD AG, Hanau, West Germany.

Dr. Horst Ranke, born 12-17-48 holds a Doctor degree in physics of the University of Tübingen, West Germany. He actually is manager of R&D for EB-technology at LEYBOLD AG, Hanau, West Germany.

Dr. Georg Sick, born 03-16-58 holds a Doctor degree in metallurgy of the Technical University of Clausthal, West Germany. He actually is head of department of vacuum melting and casting technology at LEYBOLD AG, Hanau, West Germany.

Dr. Dr. Walter Dietrich, born 02-28-26 holds a Doctor degree in physics. He entered W.C. HERAEUS, Hanau, West Germany in 1961. He actually is division president of the division of vacuum metallurgy at LEYBOLD AG, Hanau, West Germany. He received a Doctor of honors in 1986 for his contribution in the field of electron-beam technology.

2nd International SAMPE Metals Conference
August 2-4, 1988

DEVELOPMENTS IN ELECTRON BEAM MELTING OF AEROSPACE MATERIALS

D. M. Ferris, L. C. Hainz
CONSARC, P. O. Box 1839, Albany, OR 97321 U.S.A.

Abstract

CONSARC has dedicated itself to the development of industrial electron beam melting equipment with the installation of a research, development, and manufacturing facility in Albany, Oregon. This facility has been utilized for the development of the first U.S. manufactured high powered E.B. guns. Current developments include equipping the R & D furnace with ingot casting crucibles and chamber, material feeding devices, water-cooled hearth and other components for the purpose of evaluating and developing melting processes for production of aerospace materials. This furnace is dedicated to the development of E.B. guns along with continuous development of power supplies, beam guidance systems, programmable logic controller and microprocessor based control systems, numerous types of control and feedback systems for quality control, and for the future evaluation of melting processes and equipment. The E.B. gun development will be discussed, as will many of the integral systems necessary for their operation.

1. INTRODUCTION

In September, 1986, CONSARC acquired North Hill Enterprises' staff and electron beam technology which included developed power supplies and beam guidance systems. Less than two months later, at the 1986 Electron Beam Melting Conference in Reno, Nevada, we announced that CONSARC would be developing and manufacturing electron beam guns to complement the other systems. We believed that we could design a Pierce type gun that would provide reliable service, longer life of consumable parts and be very easy to maintain. This belief came directly from the fact that we produced two very successful guns in 1979 that are still in production today.

Prior to acquisition by CONSARC, NHE imported E.B. guns to the United States from the Institute

Manfred von Ardenne in East Germany even though NHE had previously developed and manufactured 75 kilowatt E.B. guns. As new furnaces were built requiring larger guns, it was a convenience to procure the von Ardenne guns. NHE was small and the required development work for power supplies, beam guidance systems, controls and mechanical equipment precluded providing the attention necessary for the development and testing of larger guns.

An evaluation of the market by CONSARC led to the decision to develop and manufacture E.B. guns for the following reasons:

1. No domestic manufacturer of high powered E.B. guns in the United States
2. Restricted availability of E.B. guns from other suppliers for world markets
3. An internal need to procure or manufacture E.B. guns at lower cost in order to remain competitive within the furnace market

With the other systems developed and operating, CONSARC began E.B. gun development, initially with a model size for approximately 100 kilowatts.

2. FACILITIES

With the commitment to proceed with development of an E.B. gun line, facilities at CONSARC WEST (in Albany, Oregon) were evaluated. The staff of CONSARC WEST was being increased with the addition of mechanical and electrical engineers for research and design efforts. Relocation into a larger facility was deemed necessary. An 18,000 square foot building located nearby accommodated our needs for manufacturing and installation of the R & D furnace. Existing offices in the new facility were remodeled and expanded. The furnace installation and remodeling proceeded concurrently and relocation was completed during May.

In addition to the furnace installation, facility improvements were made for support of R & D test melting. These included the installation of a cooling tower and closed loop water system for the furnace. Expanded electrical service for the testing of high power E.B. guns up to one megawatt was also installed.

The engineering, manufacturing and operations of CONSARC WEST are linked directly with CONSARC in Rancocas, New Jersey, via an on-line Hewlett-Packard 3000 computer with Manufacturing Resources Planning (MRP) software. This system provides continuous engineering and manufacturing control, including labor and materials, for planning and scheduling. Engineering improvements include the use of computer aided drafting (CAD) for electrical systems. The addition of mechanical software and CAD work stations is currently being implemented.

The E.B. R & D furnace was installed with adequate equipment to facilitate erection and testing of the E.B. guns. The furnace chamber is a double-walled horizontal cylinder 7 feet in diameter by 7 feet long, erected on an 8 foot high working platform. An internal water-cooled hearth and copper crucible mold for casting ingots is utilized to evaluate E.B. guns and melting processes. Porting for two E.B. guns is located on the top with one port dedicated for a development gun and the second port for the testing of all E.B. guns manufactured at CONSARC WEST. The chamber vacuum is provided through porting for a 35 inch diffusion pump and an 18" oil jet booster pump. Additional porting is included for horizontal bar feeding, scrap or loose material feeding, and ingot withdrawal. Viewing of the furnace is through four viewports.

With this furnace we are able to test our gun and control system designs under all operating conditions. With the material handling equipment in place, we can test our control systems under actual production conditions or develop melting process improvements. Future development plans include installation of EDAX equipment for continuous on-line monitoring of alloy chemistry.

3. PRODUCT DEVELOPMENT

CONSARC's approach to all aspects of the E.B. melting furnace is the same. We look for significant improvements in performance, reliability, and flexibility in the equipment and systems we design. The level of maintenance and operating skills required to keep systems in operation must be kept to a minimum. Complexities of operation are simplified with programmed control systems allowing the operator to concentrate on the process.

Electron Beam Gun

Our design provides a Pierce type E.B. gun that is constructed in modular units. The most significant of these units are the cathode holder, the beam forming head, and the beam forming head support column. These components fit into place using precision machined and aligned surfaces. Nothing is provided that requires operator judgement, skill, or experience to adjust. The assembly fits only one way, thus eliminating the need for skilled technicians to perform adjustments when re-assembling the gun after the replacement of consumable parts.

The challenge presented to us in this design has been to maintain the proper alignment over the extreme temperature ranges achieved during the E.B. gun operation. The design had to be simple and sturdy with close tolerances to guarantee alignment of the gun during assembly, yet still allow for thermal expansion under operating conditions. Our tests have proven that we have achieved this.

Based on our development experience, E.B. gun frame sizes have been defined.

Gun Frame Size	Power Range
A	10-60 Kw
B	50-200 Kw
C	150-450 Kw
D	400-1000 Kw

The appropriate gun frame size is matched with a power supply to meet its voltage and power output requirements for each customer application. The wide power range for each gun frame simplifies engineering and manufacturing, with corresponding cost savings.

The development gun (frame size B) has been operated over a range of 10 to 40 kilovolts and produced outputs from 10 to 280 kilowatts. Within its intended operating range, the gun will produce a high quality beam at virtually any combination of beam current and beam voltage. A newly developed control system automatically varies focus lens current according to variations of the beam voltage through innovative state-of-the-art PLC software. This allows the operator to concentrate on the melting process, not the dynamic responses of the gun.

Power Supply Regulator

During the testing of the development gun, we began to realize the degree of interdependence between a gun and its power supply. An E.B. gun requires more than the proper voltages to be applied to run with a stable and reliable output. The filament, exciter, high voltage, and focus supplies must be interconnected in a regulator system that can separate each parameter from outside influence. The regulator system must also automatically maintain the proper relationship of these parameters according to the desired operating conditions.

Beam focus and quality is the result of a delicate balance of the mechanical alignment, electrostatic fields, and the magnetic fields within the E.B. gun. If the gun is to operate with stability over a wide range of output, it requires precision control within the power supply system.

Typically, E.B. power supplies are designed to operate at fixed high voltage potential with variable filament control, or with variable high voltage control with fixed filament current. CONSARC has developed a unique power supply regulator capable of independent control of the beam current and beam voltage. Within the limits of the system, the gun is operable with any combination of beam voltage and current while still maintaining beam focus and quality. This significantly increases the flexibility of the E.B. gun output. Specifically, it is no longer necessary to make any mechanical adjustments of spacings to achieve desired outputs.

Beam Guidance Controller

The new CONSARC beam guidance controller is a stand-alone device that is capable of driving the desired beam sweep patterns with minimum interface from the PLC or the operator.

The PLC merely defines the size, location, speed, and shape of the beam pattern desired. The controller does the rest by driving the deflection coils to accomplish the desired result. The controller is designed to compensate for projection distortion resulting from the beam pattern (derived in polar coordinates) striking a planar surface at oblique angles.

We have made the task of defining the beam sweep patterns for any melting process very simple for the operator. A CRT allows the operator to fill in blanks to define each step. For example, a single entry for pattern, speed, gain, offset, and time will instruct the beam to sweep the perimeter of the hearth. In this way, most complex sweep patterns can be accomplished with a few entries.

To tell the system to proceed from pre-heat to drip melting to hot-topping, the operator simply indicates to the system the order in which the steps are to be performed. The CONSARC automatic control system will do the rest.

The beam sweep patterns can be edited on-line as the melt process is in operation. This allows the operator to make minor corrections to the system as necessary, or actually enabling a "teach" mode if conditions permit.

4. SUMMARY

CONSARC's new research facility is dedicated to the development of electron beam technology. Goals for development of systems focus on eliminating or reducing the significance of variables in the operation of the furnace, thereby making control of the process easier for the operator.

Highlights of E.B. gun development include modularized assembly, matched gun frame size with power supply, and automatic beam focussing to compensate for voltage variation. Improvements to power supply outputs utilize a regulator that independently controls beam current and/or voltages. An upgraded beam guidance controller automatically compensates for projection distortion. The integration of these improved systems provides control of the E.B. power output and scanning of the beam for optimum control of power distribution to the work piece despite any complexities of geometry.

CONSARC's plans for the future encompass the further automation of variable control in E.B. furnace operation. This will provide integrated systems that will meet more

exacting specifications for melting
and evaporation applications.

Figure 1: Consarc West E.B. Research Furnace

Figure 2: Consarc West E.B. Research Furnace Chamber Vacuum System

Figure 3: Consarc E.B. Development Gun, Frame Size B

Figure 4: Quantity (2) Frame Size B E.B. Guns Mounted on the Research Furnace

Figure 5: Operator's Control Console and PLC Controls for Research Furnace

2nd International SAMPE Metals Conference
August 2-4, 1988

ALUMINUM-LITHIUM ALLOYS FOR FIGHTER AIRCRAFT APPLICATIONS

V. M. Vasey-Glandon, E. A. Walker-Heckman and K. K. Sankaran
McDonnell Aircraft Company
McDonnell Douglas Corporation
P. O. Box 516, St. Louis, MO 63166

Abstract

The improved properties of aluminum-lithium (Al-Li) alloys and the potential for weight savings have resulted in their consideration by McDonnell Aircraft Company for use in both new aircraft and redesigns of existing aircraft. These efforts include the successful fabrication from 8090 plate and the subsequent flight testing of the upper aft inboard wing skins of an F-15 aircraft. Several potential applications have also been identified for sheet forms of Al-Li alloys, and demonstration parts have been fabricated for installation and evaluation on carrier based F/A-18 aircraft. However, it is evident from this work that significant differences exist between the response of Al-Li alloys and conventional aluminum alloys to forming and processing operations commonly used in the manufacture of aircraft parts. An additional issue is the need to segregate Al-Li scrap material from conventional aluminum alloy scrap. In this paper, we discuss these issues with particular emphasis on our experience in manufacturing Al-Li aircraft parts. Results of our investigation of the formability (conventional and superplastic), chemical processing, mechanical properties and corrosion behavior of aluminum-lithium alloys are also discussed.

1. INTRODUCTION

Aluminum-lithium (Al-Li) alloys possess about 8 to 10% lower density, 10 to 15% higher modulus, and equivalent to higher strength levels compared with present, high-strength 2XXX and 7XXX series aluminum alloys. Studies have shown that use of Al-Li alloys will result in significant weight savings, as high as 15 to 20% for specific components, over current aluminum aircraft structure. In addition, as shown in Figure 1,

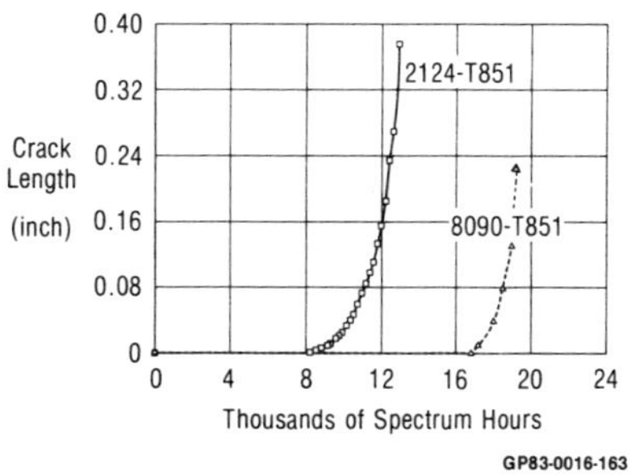

Figure 1. Spectrum Fatigue Crack Growth Curves for Open Hole Unflawed Specimens Subjected to Wing Spectrum Loading

Al-Li alloys offer about a 2- to 3-fold increase in fatigue crack growth lives which results in a potential increase of damage tolerance. The alloys also retain their strength at temperatures up to 350°F which enhances their potential use in high-performance aircraft structures. Al-Li alloys are superplastically formable, and further cost and weight reductions are possible through designs that reduce the number of parts and fasteners.

Al-Li alloy development has been directed at achieving selected combinations of property targets. Alloys have been developed for applications requiring varying combinations of strength, toughness, durability and damage tolerance. Alloys 2090, 2091 and 8090 can be considered production-ready, although not all mill product forms of all of these alloys are yet available.

2. EARLY AL-LI APPLICATION STUDIES

The promising early results of the development of Al-Li alloys resulted in their consideration in 1985 for the three upper and two lower wing skins of the F-15E aircraft. Use of alloy 8090 for these applications was shown to result in a potential weight reduction of 83 lb per aircraft. However, Al-Li alloys could not be incorporated into the F-15E design because production lots of plate were not available in time. Subsequently, Al-Li alloys were selected to replace the 2124-T851 upper aft inboard wing skins of the F-15 Short Takeoff and Landing Maneuvering Technology Demonstrator (S/MTD) aircraft.

Following the evaluation of 1.8 inch thick 2090-T8E41, 8090-T851 and 8091-T851 plate, alloy 8090 was chosen for the F-15 S/MTD application. The wing skins (Figure 2) were manufactured using the same sequence of operations as the production 2124 wing skins. The 8090 plate was machined at Cleveland Pneumatic Corporation, Tullahoma, TN, peen formed to moldline configuration at Metals Improvement Company, Pacoima, CA, fluorescent penetrant inspected at

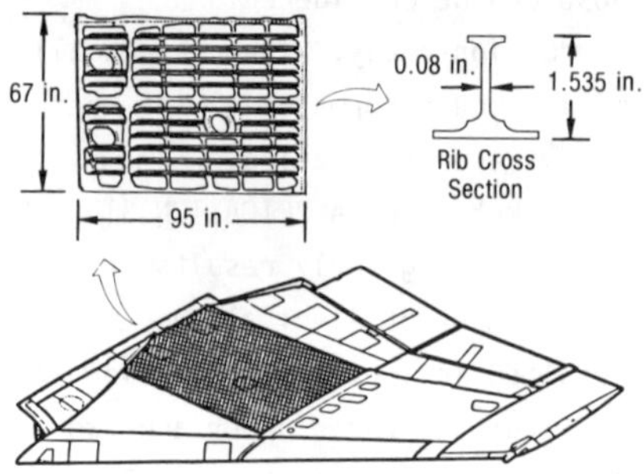

Figure 2. F-15 Upper Inboard Aft Wing Skin

CED-TEK Corporation, Los Angeles, CA, and finally sulfuric acid anodized/dichromate sealed at Anadite Corporation, Los Angeles, CA. The wing skins were then drilled, painted, and installed on the S/MTD aircraft at MCAIR, and successfully flight tested in September 1986 (Figure 3). The fabrication of the 8090 F-15 S/MTD

Figure 3. First Flight of Aluminum-Lithium Wing Skins on the F-15 S/MTD Aircraft

wing skins revealed no notable differences in the response of this alloy to machining, peen forming, inspection and finishing operations normally performed on 2124 plate. However, to fulfill a special scrap handling requirement, the skin was both rough and finish machined on one mill and care was taken to prevent the mixing of Al-Li machining chips with conventional alloy chips. Purging of coolant was not necessary, and no adverse effects from lithium contamination of the coolant were noted on subsequent parts. While an additional cost of $2000 was incurred to segregate scrap for the two wing skins, the impact of this cost can be expected to be lowered for production parts by the development and institution of scrap handling procedures for Al-Li alloys. In a parallel machining study, face milling and peripheral milling of 1.8 inch thick 8090-T851 plate was compared with 7075-T7351 plate. The force, horsepower and tool insert wear during these operations were monitored and the results for 8090 were equivalent to or lower than those for 7075-T7351. No special tooling or setup requirements were identified.

From the above discussion it appears that fabrication of Al-Li parts from plate products (which are not normally subjected to forming operations) should not pose any special problems, once consistent properties are achieved in production lots and scrap handling procedures are established. However, as discussed subsequently, the unique characteristics of Al-Li alloys do necessitate special consideration in the fabrication of sheet metal parts which undergo considerable forming.

3.0 APPLICATION OF AL-LI ALLOYS FOR FORMED SHEET STRUCTURES

The potential for manufacturing high-performance, low-cost net-shape structures is the major driver for the consideration of formed aluminum alloy sheet in aircraft structures. The importance of formed sheet structures in the manufacture of fighter aircraft can be illustrated by the thousands of such components presently in production at MCAIR in the F-15, F/A-18 and AV-8B aircraft. Al-Li alloys have the added advantage of superplastic formability which can potentially be used to replace difficult operations such as drop-hammer and deep-draw forming. However, as mentioned earlier, application of Al-Li alloys in formed sheet structures requires special considerations which are discussed below.

Because of the availability of 2090 sheet, much of the work at MCAIR has been conducted using this alloy.

3.1 Anisotropy of Mechanical Properties

The anisotropy of the mechanical properties of 2090, shown in Figure 4, is a significant

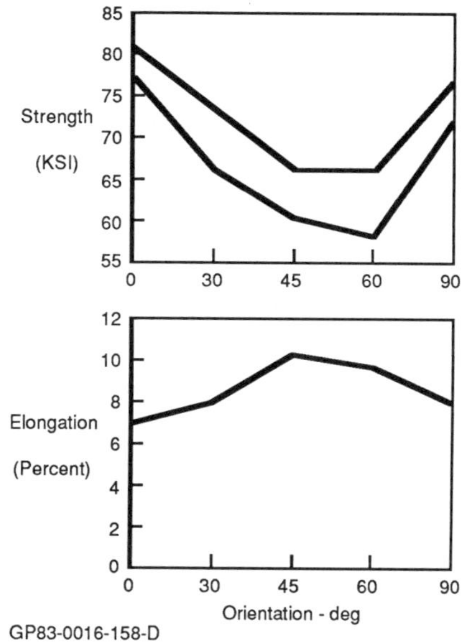

Figure 4. Anisotropy of the Tensile Properties of 0.063 - inch Thick 2090-T8E41
(0° Corresponds to the Rolling Direction)

challenge in the design of sheet metal parts. The difference of 20 ksi in strength between the maximum value in the longitudinal (L) rolling direction and the minimum value 60° to the L direction may make it difficult to substitute 2090-T8E41 for 7075-T76 if the design is based on minimum directional properties, or it will require strict attention to orientation during manufacturing. In conventional aluminum alloys, the properties are the lowest in the long transverse (LT) direction, and designs are developed using the minimum LT properties without requiring special orientation of the part, thereby enabling maximum sheet utilization by computer generated nesting plans in production.

3.2 <u>Formability of Al-Li Alloys in Various Tempers</u>

Manufacturing of high performance fighter aircraft parts requires maintaining the tightest practical forming limits to allow for both less weight and better load transfer in the structure. These goals are typically achieved for brake and hydroformed parts by forming in the "O" (annealed) or the "AQ" (recently solution heat treated and quenched) temper. Forming of Al-Li represents a departure from this approach in that to attain the high strength T8 properties in formed parts, the alloys must be formed in the T3 temper prior to aging. The ductility and formability are lower in the T3 temper, and therefore require more generous forming limits than are typically used. As shown in Figure 5, the formability of 2090 sheet is nearly equal to that of 2024 and 7075 in equivalent tempers. Even with its anisotropy in formability, 2090 still compares favorably with conventional aluminum alloys, especially in the AQ and W tempers. It should be noted that the minimum bend radii presented for 2024 and 7075 are nominal values used in the design of parts without any regard to the part orientation. Figure 6 shows examples of small 2090-W parts with varying details such as multiple bends, joggles, and curvature which were hydroformed on tooling designed for forming 2024-O and 7075-O. No problems were encountered during forming of 2090-W, and acceptable surface finish was achieved. However, these W formed parts can only be aged to a T6 strength level with a reduction of about 10 ksi in strength compared with T8 properties.

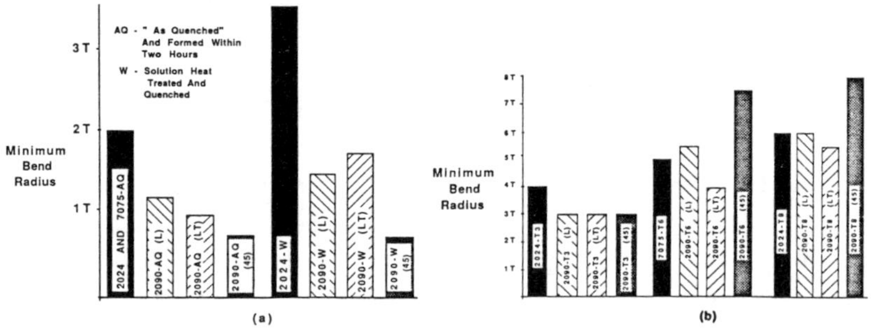

Figure 5 - Minimum Bend Radii as a Function of Temper and Orientation: (a) "AQ" and "W" Tempers and (b) "T3", "T6" and "T8" Tempers

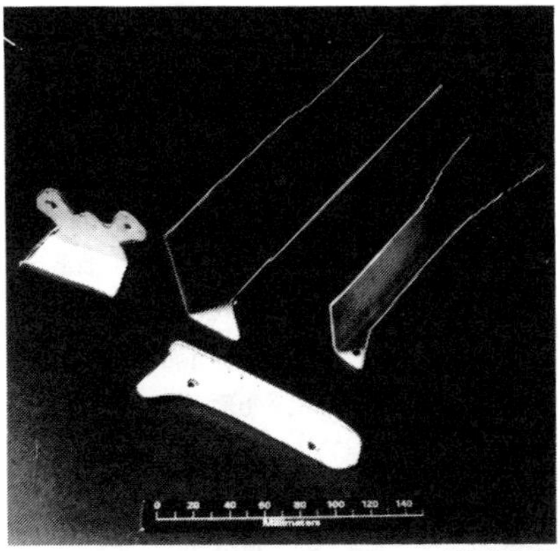

GP73-0454-11-R
JG-3

Figure 6. Examples of Hydroformed 2090-W Parts

While the anisotropy and formability characteristics of Al-Li alloys need special consideration in part design, they will nevertheless find application in a large number of parts. As an example, the F/A-18 Air 45/Detector Access Cover (Figure 7), normally fabricated from roll formed 7075-T6 alclad sheet, was fabricated without

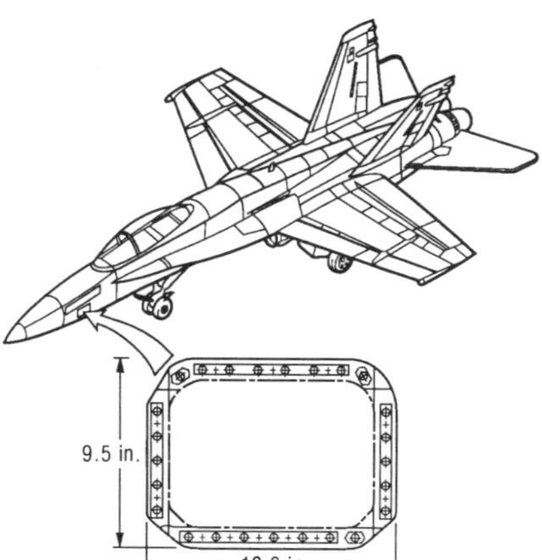

GP73-0454-16-A
JG-3

Figure 7. Schematic of F/A-18 Air 45/Detector Access Cover

difficulty from 0.10 inch thick 2090-T8E41 alclad sheet supplied by ALCOA. The cover was roll formed, and chemically milled to reduce the sheet thickness and produce the required pocket. The chemical milling was accomplished using a proprietary MCAIR solution developed for 2090 alloy. The surface roughness achieved on 18 parts varied from 30 to 90 RHR, well within the required 150 RHR maximum. The fabrication was completed by drilling and countersinking holes, and applying an exterior moldline finish (Figure 8). Plans are to install

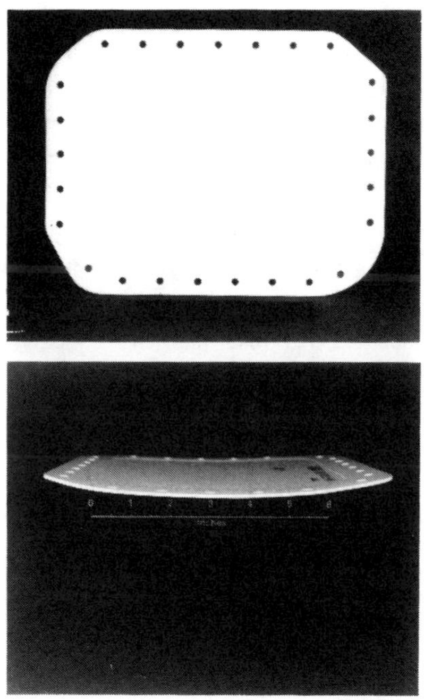

GP73-0454-26-R
JG-3

Figure 8. F/A-18 Air 45/Detector Access Cover Fabricated From Alclad 0.100-Inch Thick 2090-T8E41 Sheet

the covers on carrier based F/A-18 aircraft and evaluate their in-service performance.

3.3 Post-superplastic Forming Properties

As in 7475 and 2004, superplastic forming (SPF) results in cavitation and a corresponding reduction in the strength of Al-Li alloys. The cavitation can be suppressed by the application of a suitable back pressure (Reference 1). Of more serious concern with Al-Li alloys is the inability to achieve the T8 properties in superplastically formed parts. In this respect, 8090 is an attractive candidate for superplastic forming, since its strength in the T6 temper is only slightly lower than in the T8 temper (ultimate strength minimum values of 69 ksi for 8090-T8 vs. 67 ksi for 8090-T6) (Reference 2). As was discussed earlier, the corresponding decrease is higher for 2090. An added advantage of 8090 is its lower quench sensitivity which eliminates the need for solution heat treatment and water quenching prior to aging, thereby avoiding potential distortions in thin and complex formed parts. Rather, the alloy can be air cooled from the forming temperature and aged to achieve near -T6 properties (Reference 3).

Work is in progress to identify and demonstrate applications for superplastic Al-Li alloys. MCAIR is presently considering the F-15 Gun Fairings that make up the leading edge forward extension of the wing surface. The SPF structure will combine into one component the presently used two 2024-T8 pieces and a casting which are mechanically fastened in the assembly. The use of superplastically formed 8090 structure will also reduce assembly time and improve structural efficiency.

3.4 Chemical Processing of Al-Li Alloys

Techniques for cleaning, anodizing, and conversion coating of conventional alloys appear to be generally applicable to Al-Li alloys. MCAIR has found that 8090 can be chemical milled using the bath (sodium hydroxide etchant with triethanolamine and sodium sulfide) designed for conventional aluminum alloys (Figure 9). The etch rate for 8090 is about 20% higher and therefore it may not be possible to mix 8090 with conventional aluminum alloys during processing.

Alloy	Etch Rate (10^{-3} in./min)	Undercut Ratio	Setback	Surface Finish: Before Chem-Mill (RHR)	Surface Finish: After Chem-Mill (RHR)
2024	2.6	1.0	0.4	20	35
8090	3.3	1.3	0.3	140	55
2090	2.9	0.9	0.5	40 – 50	190 – 235

GP63-0603-13-R
S-4

Figure 9. Chemical Milling Response

The chemical milling of 2090 with conventional solutions results in an unacceptable surface finish. A modified chemical milling solution has since been developed and has demonstrated successful chemical milling of 2090 sheet and extrusion (Figure 10) including

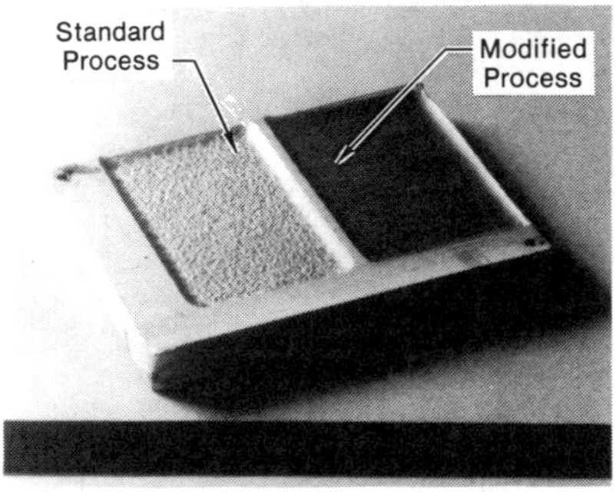

Figure 10. Chemical Milled Surface of 2090-T8E41 Extrusion

the F/A-18 access covers discussed earlier. The modified process uses the same chemical tank operating temperature and life as the standard process and does not require extreme process control. The etch rate of 2090 with the modified solution is 2.9×10^{-3} in/min. Compared to 2024, the undercut ratio for 2090 is slighlty less, and the setback is slightly higher. The modified chemical milling solution is not satisfactory for conventional aluminum alloys, and consequently, a dedicated tank will be required to chemical mill 2090 on a production basis.

3.5 <u>Corrosion Resistance of Al-Li Alloys</u>

The corrosion resistance of Al-Li alloys has not been characterized thoroughly. Limited studies at MCAIR have shown that the corrosion resistance of these alloys is typically comparable to those of conventional aluminum alloys. Limited testing at MCAIR on pre-production alloys has shown that Al-Li alloys are more resistant to corrosion than conventional aluminum alloys in finish system tests. Results of 32 day SO_2-salt fog exposure of 8090-T851, 2090-T8E41 and 2124-T851 plate and 7075-T76 sheet are presented in Figures 11 and 12. Both conversion coated and sulfuric acid anodized/dichromate sealed surfaces were tested; the corrosion resistance of the Al-Li alloys was superior to that of 2124 and 7075. Accelerated laboratory tests such as the EXCO (ASTM G34) and MASTMASSIS (ASTM G85, A2) tests were developed to determine the susceptibility to exfoliation corrosion of conventional 2XXX and 7XXX series alloys. However, for Al-Li alloys, the results of these accelerated tests have not correlated well with those of atmospheric exposure tests in sea coast, industrial and carrier environments (References 4 & 5). The evaluation of the in-service performance of the F/A-18 access

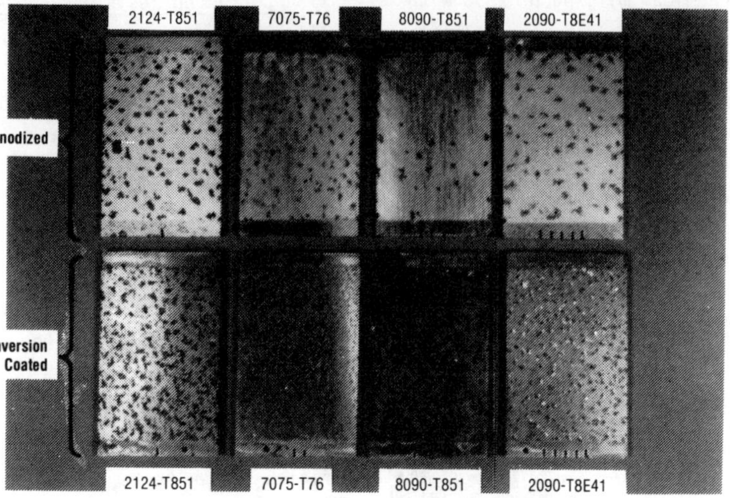

Figure 11. Appearance of the Surface of Specimens Exposed to SO_2 - Salt Fog for 32 Days

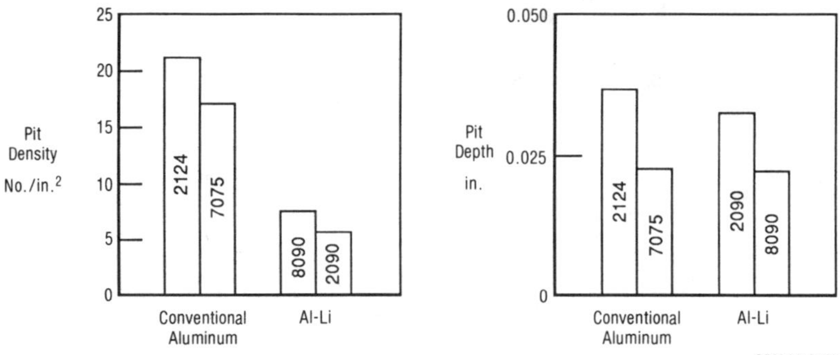

Figure 12. Pitting Characteristics of the Alloys Exposed to SO_2-Salt Fog for 32 Days

covers in a carrier environment will provide further information on the extent of correlation between the accelerated and field service tests. It is clear from experience gained thus far that extensive additional testing is required on production material to determine the influence of various environments, surface conditions resulting from processing, and surface finishes on the corrosion behavior of these alloys.

3.6 Handling of Al-Li Scrap

It is very difficult and expensive, if not impossible, to segregate Al-Li scrap from conventional aluminum alloy scrap in a large production shop handling large amounts of aluminum alloys. Scrap segregation is thus a problem, but is required to avoid the presence of lithium in recycled aluminum products for the following reasons:

i. Lithium reacts with and damages the conventional refractory materials in the melting furnaces used for recycling scrap.

ii. The presence of lithium in molten aluminum potentially increases the severity of explosion when in contact with water.

iii. The mechanical properties of conventional aluminum alloys are degraded by small amounts of lithium.

iv. No information is presently available to assure that aluminum alloys containing lithium are safe for use in contact with food products.

According to Wilson et al. (Reference 6), the most economical short term solution to Al-Li scrap handling is to remove the lithium followed by blending the bulk scrap with Al-Si foundry alloys. However, to ensure the long-term cost effectiveness of the use of Al-Li alloys, work is necessary to quantify the amount of Al-Li scrap that can be mixed with conventional aluminum alloy scrap without requiring special handling and to refine the techniques which allow for the reclamation of lithium.

4.0 SUMMARY

Aluminum-Lithium alloys have demonstrated a potential for improving the structural efficiency of fighter aircraft structures. Their applicability is enhanced by their superplastic forming potential and also the amenability to processing by using techniques essentially similar to those used for conventional aluminum alloys. Their use is expected to increase as the differences between the characteristics of Al-Li alloys and conventional alloys are better understood, and parts are designed, fabricated and used accordingly.

5. REFERENCES

1. Ridley, N., "Cavitation and Superplasticity", AGARD Lecture Series No. 154, September 1987, pp. 4-1 - 4-14.

2. Alcan International Limited Publication "Expected Guaranteed Minimum Property Values for 8090 and 8091", January 1987.

3. Partridge, P. G., McDarmaid, D. S., Bottomley, I., and Common, D., "The Mechanical Properties of Superplastically Formed Titanium and Aluminum Alloys", AGARD Lecture Series No. 154, September 1987, pp. 6-1 - 6-23.

4. Lane, P. L., Gray, J. A., and Smith, C. J. E., "Comparison of Corrosion Behavior of Lithium-Containing Aluminum Alloys and Conventional Aerospace Alloys", Aluminum-Lithium Alloys III, Institute of Metals, London, 1986, pp. 273 - 281.

5. Scutti, J. J., Bresnahan, K. M., Bohlen, J. W., and Fetchko, P. A., "Stress Corrosion and Exfoliation Corrosion Resistance of Near-Commercial Aluminum-Lithium Alloys", TMS-AIME Fall Meeting, Orlando, FL, October 1986.

6. Wilson, W. R., Worth, J., Short, E. P., and Pygall, C. F., "Recycling of Aluminum-Lithium Process Scrap", Journal De Physique, September 1987, pp. C3-75 - C3-83.

6.0 BIOGRAPHY

Virginia Vasey-Glandon has been an engineer at McDonnell Douglas for three years, the last year in the Material & Process Development Department responsible for metallic materials R & D. She has been principally responsible for the engineering activities necessary to implement aluminum-lithium alloys into production. Virginia has a B. S. degree in Metallurgy from Iowa State University.

Elizabeth Walker-Heckman has been a senior engineer in the Material & Process Development Department for five years, during which she has been a project metallurgist on the AV-8B Harrier, and a staff metallurgist responsible for metallic materials R & D. Currently she is the project metallurgist for the Advanced Tactical Fighter program. She has a B. S. degree in Metallurgical Engineering from the University of Minnesota.

K. K. Sankaran has been with McDonnell Douglas Corporation for ten years of which he has spent the last two as a section chief responsible for metallic materials R & D at McDonnell Aircraft Company. He has a Ph. D. degree from the Massachusetts Institute of Technology.

2nd International SAMPE Metals Conference
August 2-4, 1988

LOW DENSITY AL-MG-LI ALLOY FOR AEROSPACE FORGINGS

by: R. D. Schelleng, A.I. Kemppinen, and J. H. Weber
Inco Alloys International
P. O. Box 1958
Huntington, WV 25720

Abstract

The development and manufacture of a new lightweight Al-Mg-Li alloy to replace 7075-T73 in forgings is described. The density of the new alloy is 8% less than that of 7075-T73 and the elastic modulus is 15% greater while other properties are equivalent. Attractive transverse properties and SCC resistance are obtained. The alloy is produced by mechanical alloying elemental and master alloy powders followed by vacuum hot pressing into billets. The VHP billets are then extruded and forged. A new production plant has been built and put into operation replacing pilot facilities previously employed to make the alloy. Improvements in the alloy composition and manufacturing practices have been established. Thermomechanical processing routines, through which the properties of the alloy are controlled, have been defined and put into practice.

1. INTRODUCTION

The aerospace industry's need for aluminum alloys with reduced density has focused a substantial effort by the aluminum industry on alloys containing lithium. Several I/M alloys have been developed and are now available as sheet, plate and extrusions. Because the currently available I/M alloys must be cold worked after solution heat treating and prior to aging to develop optimum properties, manufacture of forgings with adequate properties from these alloys has been difficult.

Many forgings are required in airframes for multi-axially stressed components of various sorts. Such components typically comprise about 10% of the weight of an airframe. The transverse mechanical properties and resistance to stress corrosion cracking are often critical for these parts. To maximize cost effectiveness these parts should be produced as precision forgings which require minimal amounts of machining. The development of an alloy to meet this combination of properties and characteristics is the subject of this paper.

The invention of the mechanical alloying process by Inco researchers (1,2) and the subsequent application of the process to aluminum systems (3,4) has resulted in several new alloys with unique combinations of properties. The alloys capitalize on the ability of mechanical alloying (MA), a powder metallurgy, high energy milling process, to produce materials with dispersions of ultra fine particles which strengthen the alloys by stabilizing a submicron grain size. Strengthening of alloys by this method rather than through age hardening precipitates, as in conventional aluminum alloys, gives great flexibility in choices of alloying additions. Solid solution alloys can be strengthened to levels achievable heretofore only in precipitation hardenable alloys.

A MA light weight Al-Mg-Li alloy designed for aerospace applications, designated IncoMAP[(R)] alloy AL-905XL, has been developed (5,6). Trial lots of this material were produced in pilot scale facilities in 1983-86 by Novamet, an Inco business unit, and the alloy is now being commercialized by Inco Mechanically Alloyed Products, IncoMAP, a business unit of Inco Alloys International. The strong interest shown by potential users of this and other MA alloys prompted the construction of a full-scale production plant for these novel alloys,(7). This plant, located in Pittsboro, NC, includes new powder production units to increase capacity to about 400,000 lb/yr and newly designed equipment to ensure powder cleanliness. The plant began operating in late 1987 and tightly controlled manufacturing routines have been put in place to ensure consistency and uniformity of the products manufactured. Forging stock up to 250 mm (10 in.) diameter is produced.

This paper describes IncoMAP alloy AL-905XL and recent experience in the manufacture of the alloy and conversion into forgings.

2. INCOMAP ALLOY AL-905XL DESIGN

The design goal for IncoMAP alloy AL-905XL was to develop a lightweight alloy to replace alloy 7075-T73. Substantially equivalent or improved mechanical properties were desired, i.e., minimum values for tensile strength: 455 MPa (66 ksi), yield strength:

[(R)]Registered trademark of the Inco Family of Companies

386 MPa (56 ksi), and elongation: 7% in forgings (8). Three composition guidelines were selected: (1) Use only alloying ingredients which reduce density; (2) Limit alloy levels to fully soluble amounts; and (3) Strengthen via MA generated dispersions. Considerations of practicality limited the choice of alloying ingredients to magnesium and lithium. Limiting the alloy levels to fully soluble amounts was decided on mainly because of the potential reduction of ductility due to the presence of shearable lithium-rich precipitates and/or precipitate-free zones (9). Elimination of the need for quenching and aging heat treatments also was desired to reduce production costs, to avoid warpage and costly straightening of parts, and to minimize residual stresses and dimensional instability during machining.

From an initial compositional survey, an alloy containing 4% magnesium and 1.5% lithium was selected for detailed evaluation and manufacturing trials. Pilot scale lots of this material, produced prior to the construction of the new plant, provided material for extrusion and forging trials and subsequent evaluation. Testing of many lots of this material showed properties in the desired ranges but also identified areas for improvement. The properties developed by these materials are summarized in Table 1. A density reduction of 8% compared with 7075-T73 was achieved with equivalent strength values. Elastic modulus is increased by 15%. Stress corrosion cracking resistance is as good or better than that of 7075-T73.

Although substantially complete solubility of the alloying ingredients was indicated in this initial work, it was found that simulated prolonged warm ambient exposure triggered microstructural instability and caused an increase in strength and loss of ductility. In addition, inclusions of nonmetallics and iron-base particles from wear of milling equipment signaled a need for improved manufacturing practices.

3. CHEMICAL COMPOSITION

Recent work resulted in a slight downward adjustment of the Li content of IncoMAP alloy AL-905XL. The nominal composition of the alloy now is:

Mg	Li	C	O
4.0	1.3	1.2	0.5

This new lithium level for the alloy was selected based upon two-week aging tests at 100°C (212°F) of several alloys with varying lithium and magnesium contents. The alloys were tested as 16 x 64mm (5/8 x 2-1/2 in.) bar re-extruded from 152 mm (6 in.) diameter bar initially extruded from 280 mm (11 in.) diameter vacuum hot pressed billets. The compositions tested are given in Table 2, and strength changes are plotted in Figure 1. These data, summarized in Table 2, show that reduction of lithium content to 1.3% prevented any significant strength increase after aging. Similarly no evidence of significant loss of ductility after aging was observed. A very small decrease in strength due to the slightly lower lithium level is not significant. Stress corrosion cracking tests of this new composition are underway. Earlier tests at 1.5% Li showed the threshold stress for short transverse SCC to be 345 MPa (50 ksi).

4. INCLUSIONS

A major concern with any powder metallurgy material is control of inclusions. Depending upon their size, inclusions may constitute flaws which trigger premature failure. MA processing must guard against the presence of inclusions originating not only from exogenous sources but also from wear of the ball mill and milling media. Several systems have been put in place to prevent occurrence of both exogenous and indigenous inclusions in MA products. These are:

1. Stringent control of cleanliness of raw materials.
2. Enclosed handling during all powder transfers.
3. Magnetic separation of MA powder.
4. Inert Atmosphere protection.

Three different test methods were used to validate the effectiveness of these controls. Samples of MA powder after processing through the magnetic separator were digested in caustic and no measurable magnetic residues were observed. In contrast, the magnetic material removed contained particles of iron. Examination of microstructures after consolidation disclosed only the presence of very fine iron aluminide particles approximately 1-2 microns in diameter, Figure 2. Rotating beam fatigue tests were employed to seek out otherwise undetected inclusions. Ten such tests were run at a stress of 43 ksi. The average life for these tests was 370,000 cycles. For reference, alloy 7075 at the same strength level shows a life of about 50,000 cycles (8).

5. THERMO-MECHANICAL PROCESSING

The properties of IncoMAP alloy AL-905XL are controlled by adjustment of the thermo-mechanical processing conditions imposed during the several manufacturing steps. Hot working within prescribed limits is required. No subsequent heat treatment is necessary. Because the material is a solid solution alloy, control of grain size is the major factor. The grain size is inherently very small, about 1/2 micron, and is stabilized by a dispersion of aluminum oxides and carbides introduced by the MA process. These dispersoids are extremely fine, about 30-40 nm in diameter. Variations of the conditions for vacuum hot pressing the MA powder, extrusion of the VHP billet into forging stock and in the conditions for making forgings control the properties of the final part.

The data in Table 3 obtained on pilot plant material, show the effects of extrusion temperature and forging temperature on the properties of the final forging. Lower fabrication temperatures give higher strengths. Data on material made in the new plant in Figure 3 show similar trends. The long transverse yield strength is about 7% lower than the longitudinal. Complete tensile data on these forgings are presented in Table 4.

6. CONCLUSIONS

1. IncoMAP alloy AL-905XL, a lightweight Al-Mg-Li mechanically alloyed replacement for 7075-T73, has been developed and put into production in a new plant designed and built specifically for MA light metals.

2. Forgings of the new alloy are 8% lighter and 15% stiffer than 7075-T73 but otherwise equivalent in properties. Excellent short transverse toughness and stress corrosion cracking resistance is obtained.

3. The strength of the alloy is controlled by varying the billet making, extrusion and forging conditions. No final heat treatment is required.

4. The alloy is stable under extended aging at 100°C, with no loss of ductility or increase in hardness.

5. Protective atmospheres, enclosed powder transport vessels and magnetic separation of processed powder ensure cleanliness and optimum properties of the material.

7. REFERENCES

1. Benjamin, J. S., "Dispersion Strengthened Superalloys by Mechanical Alloying", Met. Trans., 1 p 2943 (1970).

2. Gilman, P.S. and Benjamin, J.S., "Mechanical Alloying", Ann. Rev. Mater. Sci., 13, 279 (1983).

3. Benjamin, J.S. and Bomford, M.J., "Dispersion Strengthened Aluminum Made by Mechanical Alloying", Met. Trans. 8A, 1301 (1977).

4. Weber, J.H. and Schelleng, R.D., "Mechanical Alloying of Dispersion Strengthened Aluminum: A Retrospective Review", to be published in Dispersion Strengthened Aluminum Alloys, TMS-AIME Proceedings, Y. W. Kim and W. M. Griffith eds. 1988.

5. Donachie, S.J. and Gilman, P.S., "The Microstructure and Properties of Al-Mg-Li Alloys Prepared By Mechanical Alloying", Aluminum-Lithium Alloys II, ed, T.H. Sanders, Jr. and E.A. Starke, Jr. (1984) p. 507, The Metallurgical Soc. of AIME.

6. Schelleng, R.D., Gilman, P.S., Donachie, S.J., "Aluminum Magnesium Lithium Forging Alloy Made by Mechanical Alloying", Overcoming Materials Boundaries 17, SAMPE (1985) p. 106.

7. Schelleng, R.D., "Production of Mechanically Alloyed Aluminum", to be published, Metal Powder Report (1988).

8. Military Standarization Handbook 5C, Department of Defense, (1976) p. 3-273.

9. Sanders, T.H., Jr. "Al-Li-X Alloys - An Overview" Aluminum - Lithium Alloys, ed. Sanders, T.H. Jr. and Starke, E.A., Jr., TMS AIME Warrendale, PA (1981) p. 63.

Table 1. Measured Properties of IncoMAP AL-905XL Forgings

Density

2.58 Mg/m^3 (0.093 lb/in^3)

Mechanical Properties

		Longitudinal	Short Transverse
Ultimate Tensile Strength	MPa (ksi)	520 (75)	480 (70)
Yield Strength 0.2%	MPa (ksi)	450 (65)	420 (60)
Elongation	%	9	6
Fracture Toughness K_{ic}	MPa (Ksi $\sqrt{in}$)	30 (27)	30 (27)
Elastic Modulus	GPa (Msi)	80 (11.6)	

Corrosion

General 0.003 mm/yr (0.12 mils/yr)
Stress Corrosion Cracking Threshold ST 345 MPa (50 ksi)

(Alternate immersion in 3.5% NaCl ASTM G44-75 and G38-75, 90 Day Exposure)

Thermal Expansion Coefficient

°C	93	149	204	260	316	371	427	482
	21.5	22.1	22.8	23.2	23.2	23.7	24.3	25.1

Melting Point

593°C (1100°F)

Table 2. Effect of Chemical Composition of MA Al-Li-Mg Alloys on Strength After Aging 334 Hr at 100°C

Heat Number	% Mg	% Li	% C	% O	Change in Yield Strength MPa F*	W**
85V099	3.93	1.49	1.15	0.46	96	159
85V100	3.95	1.47	1.12	0.43	122	149
85V102	3.46	1.45	1.09	0.41	62	141
85V103	3.00	1.48	1.07	0.54	33	88
85V104	3.45	1.27	1.08	0.46	21	3
85V105	3.01	1.26	1.28	0.47	3	2
86V051	3.78	1.92	0.80	0.33	97	200
86V070	4.04	1.28	1.18	0.38	-7	7
86V071	4.02	1.08	1.17	0.38	-3	3

*F - Initially as -extruded
**W - Initially solution annealed

Table 3. Effect of Fabrication Temperature on Strength of IncoMAP alloy AL-905XL

Operation	Fabrication Temp. °C (°F)			
Extrusion:	260 (500)		370 (700)	
Blocker Forge:	260 (500)	357 (675)	260 (500)	357 (675)
Finish Forge:	320 (610)	320 (610)	320 (610)	320 (610)
Properties				
Yield Strength - MPa (Ksi)	462 (67.0)	452 (65.6)	411 (59.6)	412 (59.8)
Tensile Strength, MPa (Ksi)	525 (76.2)	510 (74.0)	502 (72.8)	483 (70.0)
Elongation, %	13.0	14	12	14

Table 4. Tensile Test Data on IncoMAP alloy AL-905XL Forgings From New Production Plant

Forging Temp. °C (°F)	Longitudinal			Long Transverse		
	UTS MPa (ksi)	YS MPa (ksi)	Elong. %	UTS MPa (ksi)	YS MPa (ksi)	Elong. %
343 (650)	506 (73.4)	438 (63.5)	11.1	488 (70.8)	405 (58.7)	10.2
288 (550)	512 (74.3)	450 (65.3)	10.2	500 (72.5)	421 (61.1)	9.5
232 (450)	520 (75.4)	462 (67.0)	7.5	510 (74.0)	438 (63.5)	4.5

Note: Forgings side flattened from 76 mm (3 in) round bar stock to 25 mm (1 in) thick. Average of duplicate tests.

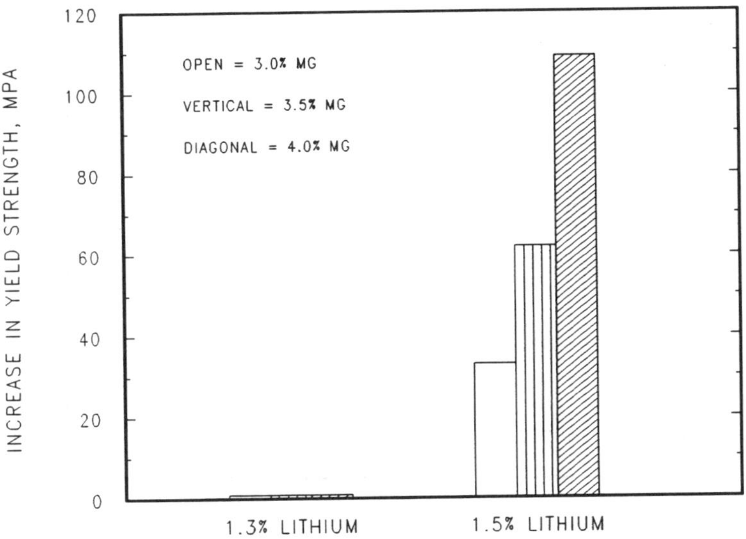

Figure 1. Effect of Mg and Li contents of IncoMAp alloy AL-905XL on increase in strength after simulated long-time, warm ambient exposure, 100°C (212°F)/334 hr.

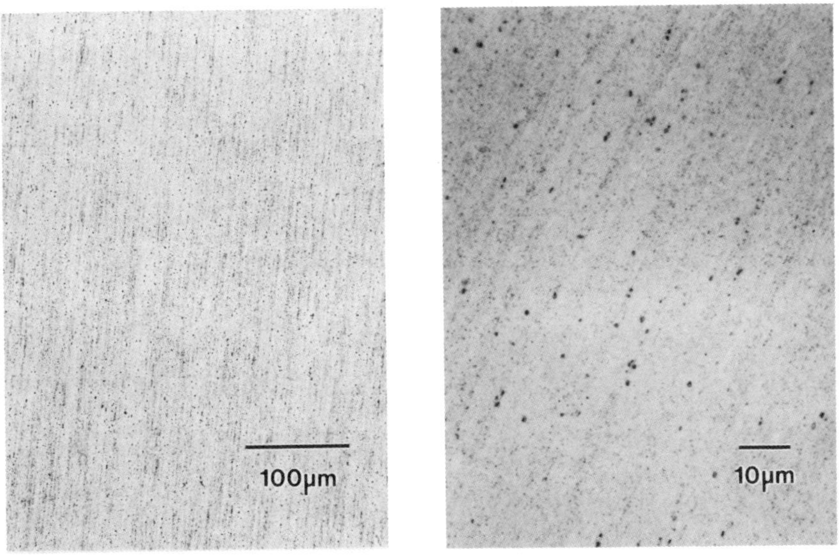

Figure 2. Microstructure of IncoMAP alloy AL-905XL showing presence of very small inclusions.

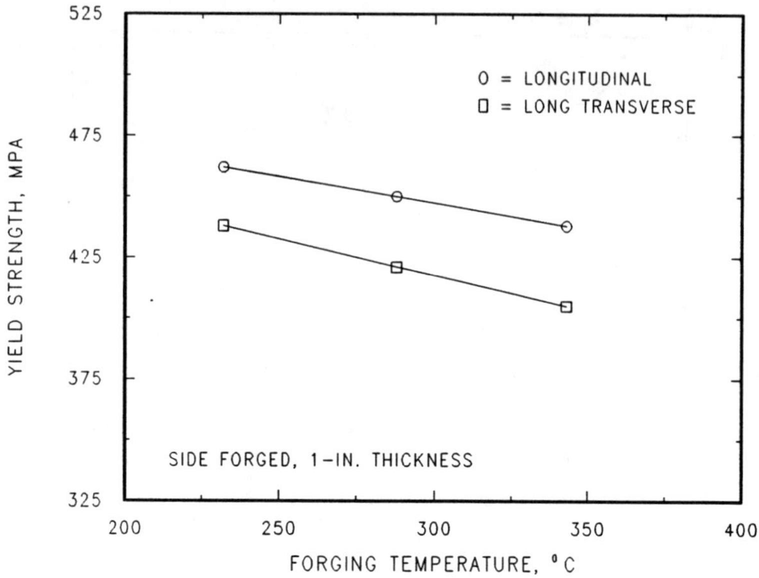

Figure 3. Yield strength of IncoMAP alloy AL-905XL forgings.

R. D. Schelleng is Manager Aluminum Sales and Marketing - IncoMAP Light Metals. His background includes supervision of mechanically alloyed aluminum research and market development studies for commercialization of these new materials. He holds BS and MS degrees in metallurgical engineering from the University of Michigan and is the author of a large number of publications and patents. He is a member of ASM International, TMS-AIME and APMI.

A. I. Kemppinen has extensive experience in the fabrication of aluminum alloys with prior experience at Reynolds Metals, Consolidated Aluminum Corp. and Arco Chemical Co. He now is responsible for extrusion, rolling and forging of mechanically alloyed aluminum at IncoMAP Light Metals. He holds BS and MS degrees in metallurgical engineering from Michigan Technological University and has published numerous papers and patents. He is a member of ASM International and TMS-AIME.

J. H. Weber is Development Manager - IncoMAP Light Alloys and is responsible for marketing, process and product development as well as quality assurance and control functions for the IncoMAP Light Alloys Plant in Pittsboro, NC. He has been involved in the mechanical alloying development since its inception in the Inco R and D Center holding positions in research, research and engineering management. He holds BS, MS and PhD degrees in Metallurgy and Materials Science from Lehigh University and is the author of many papers and patents. He is a member of ASM International, TMS-AIME, ASTM and APMI.

2nd International SAMPE Metals Conference
August 2-4, 1988

EXPERIMENTAL HIGH MODULUS ELEVATED TEMPERATURE Al-Ti BASED ALLOYS BY MECHANICAL ALLOYING

P.K.Mirchandani and R.C.Benn
Inco Alloys International, Inc.
Huntington, WV 25720
USA

Abstract

It is generally recognized that the presence of hard, high melting point phases can contribute to the modulus and elevated temperature strength of aluminum alloys. One approach to incorporating such phases involves the in-situ formation of high melting point transition metal-based intermetallic dispersoids within aluminum alloys. Mechanical alloying (MA) has been used in this study to fabricate Al-Ti alloys containing large volume fractions (up to ~ 35 v/o) of submicron sized Al_3Ti dispersoids. Test results have shown that such alloys exhibit very attractive combinations of elevated temperature strength, elastic modulus and ductility. Further, such alloys have been found to exhibit excellent microstructural stability after long-term (500 hr) exposure at elevated temperatures up to 510°C (950°F).

1. INTRODUCTION

The increasing demand for higher performance aircraft has stimulated the development of advanced aluminum alloys, with a view to identifying materials which could replace heavier titanium alloys for skin and structural applications in advanced aircraft, and for structural/drive-train applications in the power plants of such aircraft [1,2]. In order to receive consideration as replacements for titanium alloys in such applications, advanced aluminum alloys would be required to exhibit a significantly superior combination of elevated temperature strength, stiffness (or elastic modulus) and thermal stability compared to their conventional counterparts. In addition, advanced aluminum alloys would be required to exhibit reasonable ductility.

High modulus alloy development has hitherto centered around the incorporation of large volume fractions of ceramic (usually SiC) particles or whiskers within conventional aluminum alloy matrices (metal-matrix composites) [3-5]. Currently available high modulus composites based on conventional aluminum alloys do not, however, exhibit useful strength at elevated temperatures (for example, in the 260-425°C (500-800°F) range). Also, such composites are usually characterized by low levels of ductility. On the other hand, many of the currently available rapid solidification (RS) processed elevated temperature aluminum alloys (for example, Al-Fe-Ce, Al-Fe-Mo etc.) do not exhibit a high elastic modulus (for example, in the preferred 95-110 GPa (~ 14.0-16.0 x 10^6 psi) range). It has been difficult, therefore, to obtain aluminum alloys which simultaneously exhibit good elevated temperature strength retention, high stiffness and good ductility.

This paper describes some aspects of the processing, structure, and properties of a series of experimental Al-Ti based alloys prepared using mechanical alloying (MA) techniques. Such alloys are currently under development at Inco Alloys International, Inc. MA Al-Ti alloys, having Ti concentrations ranging from 7.5 up to 15.0 w/o, have been found to exhibit attractive combinations of room and elevated temperature strength, elastic modulus and ductility. Further, such alloys are very stable after exposure to elevated temperatures for extended periods of time. Also, such alloys have been found to exhibit significantly lower coefficients of thermal expansion compared to conventional aluminum alloys.

2. MA Al-Ti ALLOYS

The mechanical alloying (MA) process has been described in detail elsewhere [6-11]. Briefly, MA is a specialized technique for processing metal powders prior to consolidation into full-density wrought alloys. The process involves the dry, high energy milling of metal powder blends to produce particles with unique microstructures. The nature of the process permits the solid-state incorporation of very fine dispersoids (metallic, intermetallic, ceramic etc.) within powder metal particles. During MA processing of Al-Ti alloys an extremely fine dispersion of Ti is produced within Al particles during the milling operation. Further, during elevated temperature exposure of such "composite" particles, the aluminum rapidly reacts with the dispersed titanium to form an ultra-fine dual-phase $Al-Al_3Ti$ structure. Equilibrium is rapidly established since reaction kinetics are enhanced by the presence of the defect structure generated during milling, and also because diffusion distances are extremely small. Since the formation of most of the intermetallic phase does not occur via a precipitation mechanism in MA Al-Ti alloys, and since processing occurs entirely in the solid state, the scale of microstructure is not dictated by phase diagram considerations such as freezing ranges or solubility limits. Instead, the scale of microstructure is controlled primarily by the scale of Ti dispersion within the aluminum particles. The typical morphology of MA Al-Ti powders is as shown in the scanning electron micrograph in Figure 1.

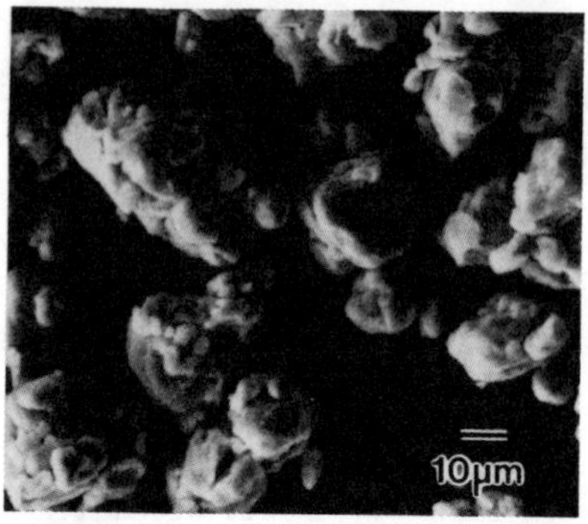

Figure 1. Typical morphology of MA Al-Ti powders.

3. EXPERIMENTAL PROCEDURE

MA Al-Ti alloys were prepared with titanium concentrations ranging from 7.5 to 15.0 w/o using the mechanical alloying technique. In addition, a baseline alloy containing no Ti, was also prepared for comparison. Table I provides a listing of the chemistries of the alloys prepared. The C and O in these alloys arises from the organic process control agent added during mechanical alloying. Following MA, the powders were canned, hot compacted, and extruded. Information on the densities of the alloys in the as-extruded condition is provided in Table I. Selected alloys were also prepared in the form of 1.5 mm (0.06 inch) thick sheet. Sheet materials were produced by hot rolling of the alloy extrusions.

Table I: Chemical compositions and densities of experimental MA Al-Ti alloys.

Alloy	Composition (wt.%)				Density
	Al	Ti	C	O	(Mg/m^3)
E1	bal.	--	1.6	0.8	2.69
E2	bal.	7.3	1.5	0.8	2.82
E3	bal.	8.6	1.9	0.8	2.84
E4	bal.	10.0	1.7	0.8	2.85
E5	bal.	11.6	1.9	0.7	2.89
E6	bal.	12.5	1.5	0.8	2.90
E7	bal.	15.0	1.8	0.9	2.94

Microstructural evaluations were carried out using optical microscopy and transmission electron microscopy (TEM). Dynamic modulus measurements on the alloys were made following the procedure described in reference 12. Room and elevated temperature tensile tests were performed on the alloys according to the methods prescribed in ASTM E8-87 and E21-79 respectively. Hardness measurements were made on selected alloys, before and after exposure to elevated temperatures for various times, to assess their thermal stability. The coefficients of thermal expansion of selected alloys were determined according to the method described in reference 13.

4. RESULTS AND DISCUSSION

4.1. Microstructures

MA Al-Ti alloys appear almost featureless when observed by optical microscopy (see photomicrograph in Figure 2 which is representative of all of the Al-Ti alloys prepared). A few small (2-3 µm) Ti-rich particles are in evidence. Details of microstructure become more apparent when the alloys are observed by TEM (see photomicrograph in Figure 3 which again is representative of all of the alloys). The structure consists of ultrafine (~100-300 nm in size) grains of Al randomly mixed with Al_3Ti grains which vary in size from 20-250 nm. In addition, very small (~30-40 nm) Al_2O_3 dispersoids, as well as cylindrically shaped (20 nm in diameter, 30-100 nm in length) Al_4C_3 dispersoids, are present mainly along grain boundaries. The grain sizes in these alloys are significantly smaller than those generally observed in ingot metallurgy or even RS processed Al alloys.

Assuming that all of the Ti present in the alloys is in the form of Al_3Ti, the volume fraction of Al_3Ti present can be calculated as being 2.5 times the weight fraction of Ti present in the alloys.

Figure 2. Typical microstructure of MA Al-Ti alloys as observed by optical microscopy (Alloy E3). Keller's etch.

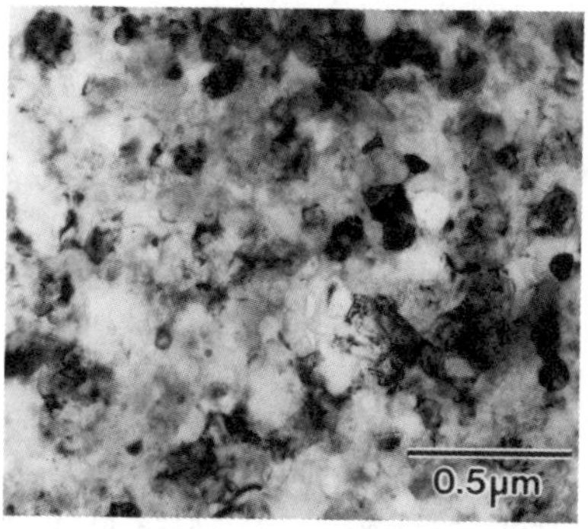

Figure 3. Typical microstructure of MA Al-Ti alloys as observed by TEM (Alloy E3).

4.2. Elastic Properties

The modulus of MA Al-Ti alloys increases almost linearly with increasing Ti content (see Figure 4). Results indicate that a modulus increase of approximately 2.45 GPa (~ 0.35 x 10^6 psi) is obtained for every 1 w/o Ti present in the alloy. Results show, therefore, that Ti additions are very effective in increasing the stiffness of Al alloys. For example, alloy E5 (Al-11.6Ti) exhibits a modulus about 50 % higher than the modulus of conventional aluminum alloys (i.e., ~70 GPa (~ 10 x 10^6 psi)). Indeed, the modulus of alloy E5 (~105 GPa (~ 15 x 10^6 psi)) is almost as high as that of most Ti alloys (~112 GPa (~ 16 x 10^6 psi)). Thus, on a density compensated basis, alloy E5 exhibits a modulus significantly higher than that of Ti alloys (~ 36.2 GPa.m^3/Mg for alloy E5 versus ~ 24.9 GPa.m^3/Mg for Ti alloys). The modulus of alloy E5 is compared with those of Ti-6Al-4V and a conventional Al alloy (2014) in Figure 5.

Modulus improvements in MA Al-Ti alloys are undoubtedly due to the Al_3Ti intermetallic phase present in these alloys. Assuming a linear (i.e., "rule of mixtures") increase in modulus with Al_3Ti volume fraction, the elastic modulus of the Al_3Ti phase can be calculated as being approximately 175 GPa (~ 25.4 x 10^6 psi).

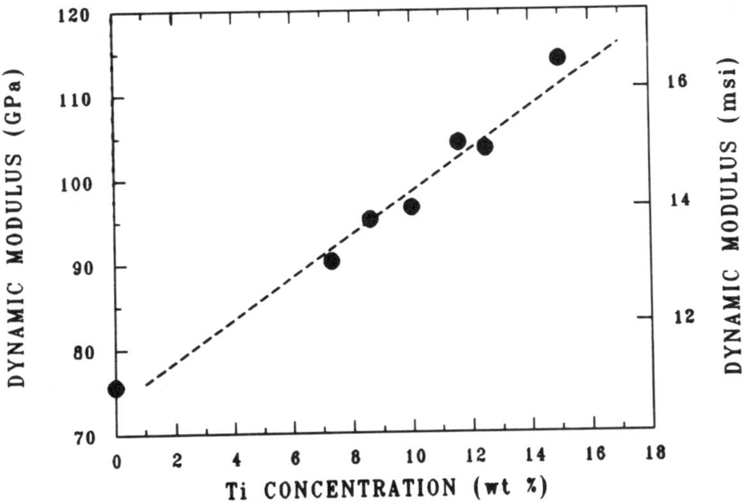

Figure 4. Influence of Ti concentration on the dynamic modulus of MA Al-Ti alloys.

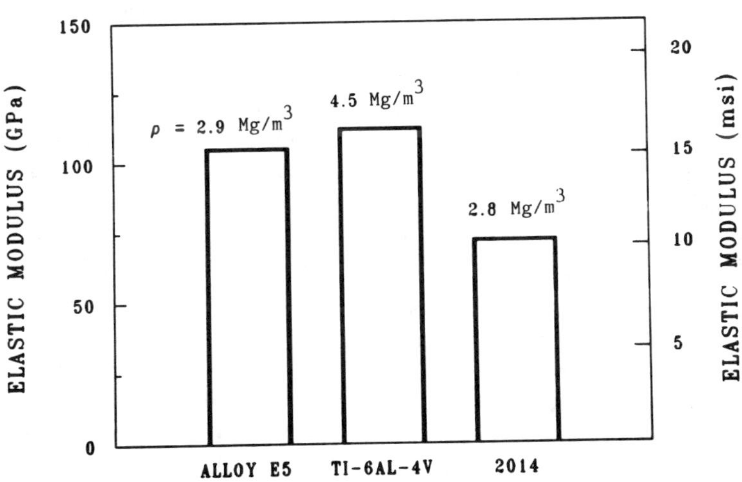

Figure 5. Comparison between the modulii of alloy E5, Ti-6Al-4V, and a typical conventional aluminum alloy (2014).

4.3. Tensile Properties

Besides contributing to the elastic modulus, Ti additions are also effective in increasing the tensile strength of the alloys over a wide temperature range. Figure 6 compares the as-extruded tensile strengths of alloys E1 (no Ti), E3 (8.6 Ti), and E5 (11.6 Ti) over a temperature range extending from room temperature up to 427°C (800°F). As may be observed from Figure 6, alloys containing higher levels of Ti exhibit higher strength levels at any temperature, within the Ti concentration and temperature ranges considered.

The tensile strength of alloy E5 is compared with that of cast and wrought aluminum alloy 2014-T6 as well as that of an RS processed Al-8Fe-4Ce alloy in Figure 7. The MA Al-Ti alloy is substantially stronger than 2014-T6 at temperatures exceeding about 200°C (392°F). Also, the MA Al-Ti alloy exhibits superior strength retention at temperatures greater than about 300°C (572°F), as compared to the Al-8Fe-4Ce alloy [2,14].

Figure 8 shows the tensile elongation to fracture exhibited by alloys E1, E3, and E5 over the same temperature range as in Figures 6 and 7. In general, MA Al-Ti alloys exhibit good ductility at all test temperatures. In particular, the room temperature ductilities of these alloys are very attractive, and appear to be significantly higher than the ductility levels typically exhibited by Al-SiC composites at equivalent modulus levels. An interesting feature of MA Al-Ti alloys is that, unlike MA Al (no Ti) alloys, which exhibit a monotonic decrease in ductility with increasing temperature, MA Al-Ti alloys exhibit an initial decrease in ductility with increasing temperature up to some intermediate temperature, following which the ductility increases with increasing temperature.

The good ductility of these alloys is related, in all likelihood, to the effectiveness of the MA process in producing a uniform dispersion of Ti within the Al particles (and hence, of the Al_3Ti phase in the final alloy). Further, since the Al_3Ti grains are likely to be well bonded to the Al grains (having been formed in-situ rather than "artificially" incorporated as in the case of composites), void nucleation (by interface decohesion) is inhibited until the alloy has experienced

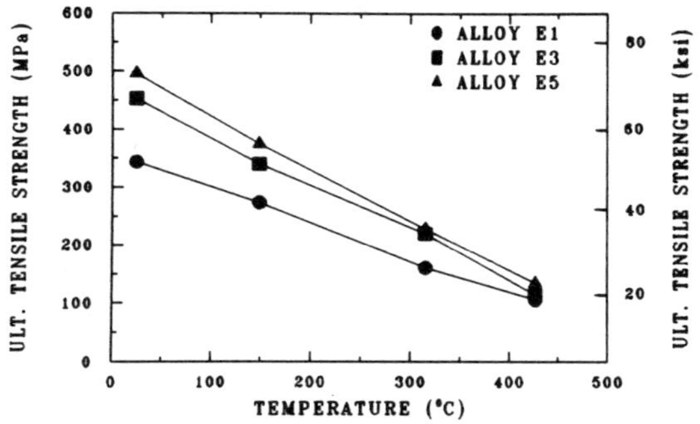

Figure 6. Variation in the tensile strength of MA Al-Ti alloys with temperature.

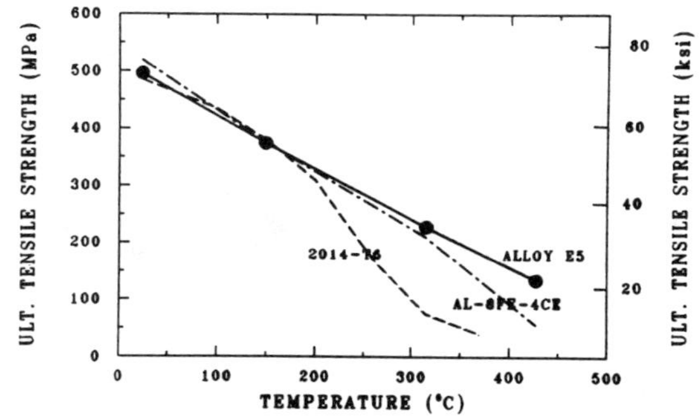

Figure 7. Comparison between the tensile strengths of alloy E5, cast and wrought 2014-T6, and Al-8Fe-4Ce over a range of temperatures.

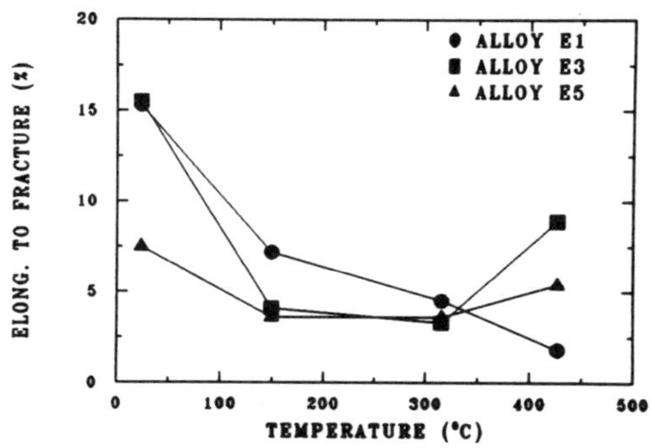

Figure 8. Variation in the elongation to fracture of MA Al-Ti alloys with temperature.

substantial macroscopic plastic strain. For these reasons, MA Al-Ti alloys exhibit good ductility even at relatively high volume fractions (up to 30-40 v/o) of the hard and relatively non-deformable Al_3Ti phase. Thus, a very attractive feature of these alloys is that substantial improvements in the modulus and strength can be obtained with only a marginal compromise in ductility.

4.4. Thermal Stability

The microstructure of MA Al-Ti alloys changes very little when the alloys are exposed to elevated temperatures. For example, the microstructure of alloy E3 is compared before and after extended elevated temperature exposure in Figure 9. As may be observed, the scale of microstructure remains virtually the same, i.e., the grain size within the alloy does not increase significantly after the elevated temperature exposure. It is most likely that the Al_2O_3, Al_4C_3, and to some extent, the Al_3Ti dispersoids, are responsible for the resistance to grain growth exhibited by these alloys. The dispersoids are thus responsible not only for the fine grain sizes in the as-fabricated condition, but also for the thermal stability of these alloys.

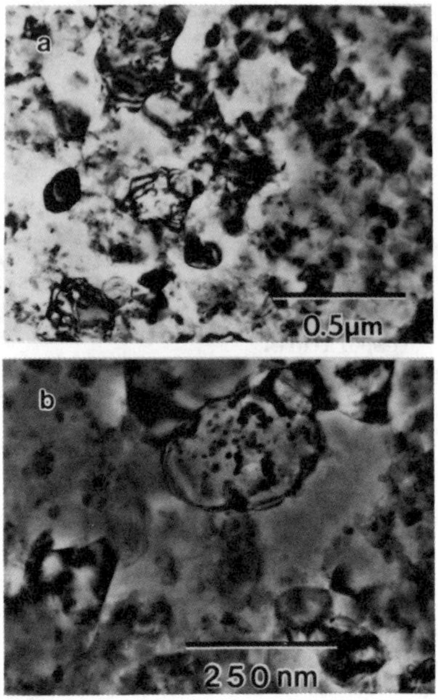

Figure 9. Comparison between the microstructure of alloy E3 (a) before, and (b) after elevated temperature exposure (100 hr @ 510°C (950°F).

Figure 10 summarizes the results of hardness measurements made on selected alloys before and after exposure to elevated temperatures for various periods of time. As may be observed, MA Al-Ti alloys exhibit a slight initial increase in hardness following which the hardness remains virtually stable even after long-term (500 hr) exposure at very high temperatures up to 510°C (950°F). The observation that the alloys are able to maintain their hardness is a reflection on the microstructural stability of these alloys. MA Al-Ti alloys are thus characterized by excellent thermal stability and could be expected to retain their properties over extended service lives.

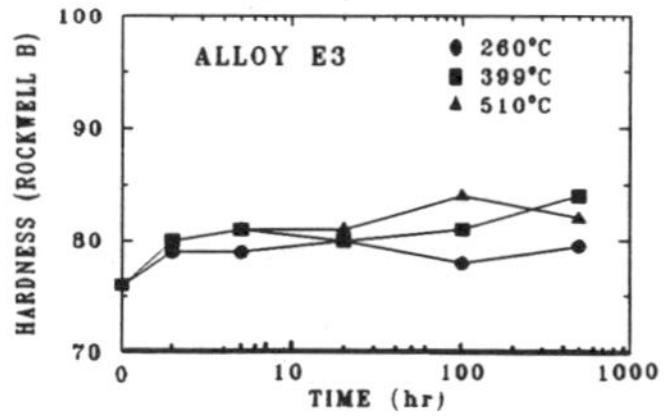

a

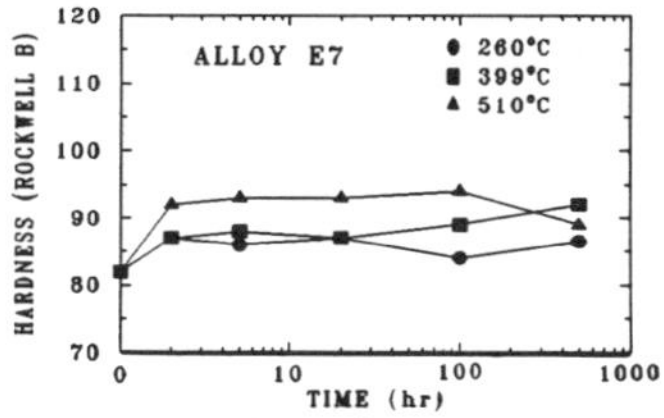

b

Figure 10. Variation in the hardness of MA Al-Ti alloys after exposure to elevated temperatures for various periods of time (a) Alloy E3, and (b) Alloy E7.

4.5. Secondary Fabrication

The excellent thermal stability combined with the good elevated temperature ductility suggests that fabrication of MA Al-Ti alloys can be carried out at high temperatures without serious degradation in room temperature strength and modulus. To demonstrate this concept, 10 mm (0.4 inch) thick extrusions of alloys E4 and E6 were hot rolled to 1.5 mm (0.06 inch) thick sheet at approximately 482°C (900°F). Figure 11 provides an illustration of the sheet material so prepared, showing practically no edge cracking.

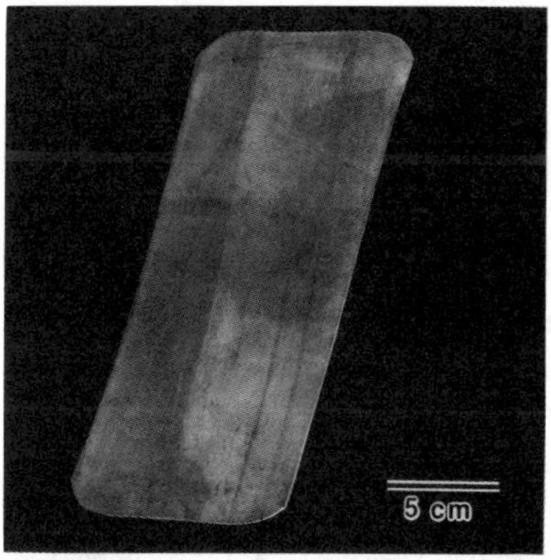

Figure 11. Sheet material produced by hot rolling of extruded alloy (Alloy E4).

The room temperature properties of the above alloys in the as-extruded condition as well as in sheet form are compared in Table II. As may be observed, the tensile and elastic properties remain virtually the same even after the rather severe high temperature TMP treatment. Results of the rolling study suggest that it should be possible to fabricate MA Al-Ti alloys at high temperatures into a variety of useful shapes without any loss in mechanical properties. In this respect, it appears that MA Al-Ti alloys offer a significant advantage over many RS processed high temperature Al alloys which cannot be worked at very high temperatures without incurring significant microstructural coarsening, and hence, degradation in strength.

4.6. Thermal Expansion

MA Al-Ti alloys exhibit significantly lower coefficients of thermal expansion compared to conventional aluminum alloys (see Figure 12). Therefore, MA Al-Ti alloys are likely to be more compatible with high temperature Ni-, Fe- and Ti-based alloys compared to conventional Al alloys, particularly in structures requiring a combination of materials for optimum performance.

5. SUMMARY AND CONCLUSIONS

1. Mechanical alloying is a very useful technique for preparing Al-Ti based alloys with ultrafine microstructures.

2. MA Al-Ti alloys exhibit a superior combination of elevated temperature strength and elastic modulus compared to conventional as well as many RS processed elevated temperature Al alloys.

3. The excellent thermal stability combined with good high temperature ductility suggests that high temperature fabrication of MA Al-Ti alloys can be carried out without significant degradation in strength and modulus.

4. MA Al-Ti alloys thus offer potential for selection as structural materials in some advanced gas turbine engine and airframe applications for which property requirements would normally dictate selection of titanium alloys.

6. ACKNOWLEDGEMENTS

The authors gratefully acknowledge the technical assistance of W.E. Mattson, M.D. Wyrick, K.A. Heck and F.J. Veltry. The authors are also grateful to Inco

Table II. Comparison between the room temperature mechanical properties of MA Al-Ti alloys in the as-extruded and hot-rolled conditions.

Alloy	Condition	E GPa(msi)	UTS MPa(ksi)	0.2% YS MPa(ksi)	Elong. (%)
E4	as-extruded	96.6(14.0)	451(65.4)	367(53.2)	16.0
E4	hot-rolled	95.9(13.9)	487(70.2)	434(62.9)	11.0
E6	as-extruded	103.6(15.0)	448(64.9)	371(53.3)	10.0
E6	hot-rolled	104.0(15.1)	495(71.7)	445(64.5)	8.4

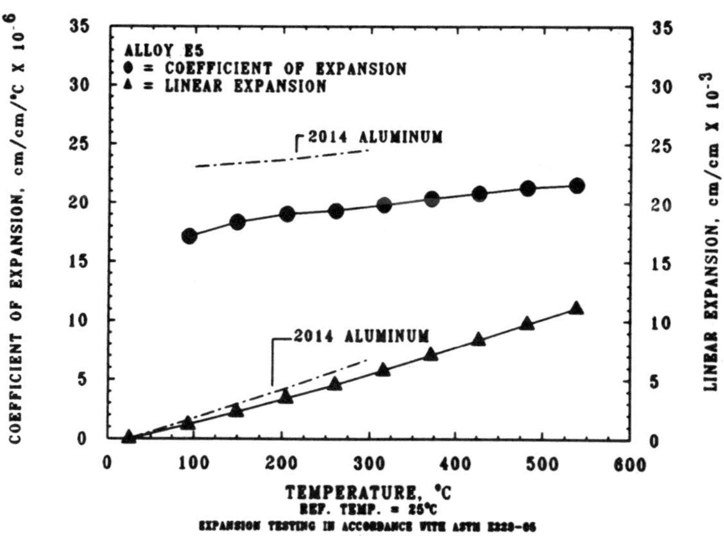

Figure 12. Variation in the coefficient of thermal expansion of alloy E5 with temperature. Expansion data for a conventional aluminum alloy (2014) also shown for comparison.

Alloys International, Inc. for permission to publish these results.

7. REFERENCES

1. Griffith, W.M., Sanders, R.E., and Hildeman, G.J., in High Strength Powder Metallurgy Aluminum Alloys, eds., Koczak, M.J., and Hildeman, G.J., TMS-AIME, Warrendale, PA, 1982, pp 209-224.

2. Lagenbeck, S.L., et. al., Final Report AFWAL TR-86-4027, May 1986.

3. Divecha, A.P., Fishman, S.G., and Karmarkar, S.D., Journal of Metals, September 1981, p 12.

4. McDanels, L., Met. Trans., 16A, June 1985, p 1105.

5. Rack, H.J., Cook, J.L., and Baruch, T.R., in Progress in the Science and Engineering of Composites, Japan Society for Composite Materials, Tokyo, 1982, vol. 2, pp 1465-72.

6. Benjamin, J.S., Met. Trans., 1A, 1970, pp 2943-2951.

7. Gilman, P.S., and Benjamin, J.S., Annual Rev. Mater. Sci., 1983, pp 279-300.

8. Benjamin, J.S., Scientific American, vol. 234, 5, 1976, pp 40-48.

9. Benjamin, J.S., and Bomford, M.J., Met.Trans., 8A, 1977, pp 1301-1305.

10. U.S. Patent # 3,740,210 (1973).

11. Benjamin, J.S., and Schelleng, R.D., Met. Trans., 12A, 1981, pp 1827-1832.

12. Spinner, S., and Tefft, W.E., ASTM proceedings no. 61, pp 1221-1232.

13. ASTM standard E228-86.

14. Yaney, D.L., Ph.D Thesis, 1986, Department of Materials Science and Engineering, Stanford University, CA.

8. BIOGRAPHIES

Prakash K. Mirchandani is Research Metallurgist, Mechanical Alloying R & D Group, Inco Alloys International, Inc. Having joined the company in 1987, his responsibilities include development of advanced MA aluminum-based alloys. He earned a Ph.D degree (1987) and a M.S degree (1983), both in metallurgical engineering, from

Michigan Technological University. He also holds a M.S degree in mechanical engineering from the State University of New York, as well as a B.Tech degree from the Indian Institute of Technology. He is a member of ASM, TMS-AIME, and APMI.

Raymond C. Benn, B.Sc., C.Eng., M.I.M, graduated from the University of Surrey in England with First Class Honors and the "Hayward Award" in metallurgy in 1970. He is currently Senior Metallurgist and leader of the Mechanical Alloying R & D Group, Inco Alloys International, Inc. He is responsible for initiation and execution of R & D on a range of MA materials including heat-resistant (incl. refractory) and light metal alloys. He has held various research and management positions within the R & D functions of Inco Alloys International in the U.S.A. and U.K. He is a member of ASM, TMS-AIME, Institute of Metals, and a chartered engineer of C.E.I.(U.K.).

2nd International SAMPE Metals Conference
August 2-4, 1988

HIGHLY EFFECTIVE HIP PROCESS FOR INTRICATELY-SHAPED PARTS FROM POWDER ALLOYS DIFFICULT TO DEFORM

Vsevolod A. Domoratsky,
Anatol F. Sanin,
Vladimir K. Nikolayev,
Lioudmila L. Scheglova
Powder Metallurgy Scientific Labôratory,
Dniepropetrovsk State University, USSR

Abstract

HIP methods of powder materials of various purposes have been widely applied in industry during the last two decades. It causes the necessity for perfection of traditional technological pressing variants and for new variants development.

A short review of achievements in the sphere of powders HIPing with the help of rigid metal containers is presented in the paper. Major disadvantages of this method are pointed out. Deformation processes of rigid containers are analysed. The questions of a free-filled powder compressibility under high temperature and pressure are studied. The speed of compaction front spread depends upon strength and density of the material as well as upon temperatures distribution along the container cross-section. The technology of obtaining preforms of intricate shape without using metal containers or other rigid ones is discussed. The preforms are made by way of a bar sintering which is molded before in the process of combined cold isostatic pressing (CIP). The methods of powder particles consolidation are dealt with in connection with prospective development of the studies.

1. INTRODUCTION

While producing powder goods of more than overall dimensions with mass nearing several tens of kilogramms a usual technology of "pressing-sintering" can not be applied because of a probable considerable deformation of a preform shape. The process of hot isostatic pressing (HIP) guarantees an almost ideal shrinkage unoformity while powder materials compacting, permits to obtain finished articles of practically any configuration. Gas atomized powders with particles of spherical shape are usually used in this conditions.But as they are notable for their bad compressibility and moldability, then, to obtain a transportable preparatory form, the powder in the state of free-filling is placed into rigid metal containers, which are being subjected to a uniform external pressure under high temperatures after a vacuum processing and making air-tight. And that is why the finished shape of an article is determined wholly by the character of a jacket form changing, which in this case plays the part of a tool. On the whole the containers design determined the stability of form changing under HIP, and their geometry - the powder material utilization factor.

To prevent the variations of density and thermal diffusivity, a powder vibro-packing is provided. The latter, in case the metal containers were made air-tight beforehand, leads to the formation of voids in the most dangerous cross-sections (at the welds of an upper lid). Besides, the surface layers are somewhat loosened. While heating such containers with powder in the state of filling within the range of temperatures from 1273 up to 1503 K (1100-1230°C) and while applying isostatic pressure some temporary microdamages of the containers air-tightness, which might occur, can be the cause of interparticle porosity and of oxides, nitrides and other compounds formation on the particles surface. Because of all that up to the present time the questions of filling and developing of containers were the most important while organizing a production unit with HIP.

2. DISADVANTAGES OF TECHNOLOGY AND CONTAINERS DESIGN

Certain types of non-regulated form changing, the deplantation of a preform cross section (distortion), variations in the sizes of preforms even of a simple configuration after HIP, caused by the variations in the filling density, by the non-uniform wall-thickness of a container by the non-uniform deformation picture in a plate and in a weld zone are in proportion with sizes and can achieve 5%. It demands the increase of allowances for mechanical treatments, which even for comparatively small articles of 250 mm (10 inch), achieve 8 mm (1/3 inch) and introducing additional straightening after HIP on hydraulic presses. Metal utilization coefficient on the parts of a complex shape does not exceed 0,5, which is a high indicator, but results in practically irretrievable losses of 50% of an expensive powder.

Great attention is to be paid not only to the location but also to the type of a welded joint, to the increase of transition radii among various elements because of the complex loading character of the welded joint. For example, investigations showed that the safety margin of a T-form weld while HIPing exceeds 4 times the safety margin of a corner welded joint. To reduce the powder material expence and the further mechanical treatment volume, the steel insets are used, obtained by casting and mechanical treatment. It demands a rational location of forming cavities, as their unforesaid transition under HIP causes the appearance of "steps" on the inner surfaces.

Demands for a high production culture, a great amount of metal used for containers and metal insets (metal utilization factor even at the advanced sheet forming of the container elements and the precision casting of the metal insets does not exceed 0,38-0,42), a great amount of mechanical treatment and expenses for die equipment lead to the fact that the containers price achieves 24-27% of the general cost of a production unit for making preforms under HIP. Another disadvantage is that after HIP a container is grinded off or pickled off by an acid together with the insets, and this either considerably cheapens waste products or results in the loss of a container metal.

2.1 Providing air-tightness

And even in spite of large expenses, there are cases when containers turn out to be non-air-tight. At that, the appearance of a discontinuity in the weld metal in the process of a pressure increase which disappears due to a diffusive defect healing, while holding for a long term under high temperatures, is especially dangerous. A dynamic pressure differential, emerging after HIP, leads to a container destruction and to a hitting by fragments of heaters, gas-containers inner elements and attendants. Thus, to provide safety for the attendants, it is necessary to to make expenses for special premises and special discharging sets.

3. PROCESSES OF CONTAINERS DEFORMATION

The difficulties which arise while studying the processes going under HIP of the powders free filled, lie in the necessity of taking into account simulteneous deformation of the compressible (powder) and the non-compressible (jacket) materials. These difficulties also depend strongly on the plastic flow character, on the jacket parameters (thickness, linear sizes) and on the location of welds difficult to deform. Thus, practical scientists making suppositions about the uniformity of distribution of deformation speed, have to deal simulteneously with the deformation of compressible and non-compressible plastic bodies, with casted metal of welds and sheets.

Theoretical and experimental studying of different calculation models of containers designing, proceed from the assumption that a jacket is made from an ideal plastic material, that powder is a compressible compacted plastic material, the density of which is assumed as the parameters of reinforcement. It is also taken into account that at the boundary "powder-container", influenced by the uniform external load, the conditions of the transports equality are true, which fits the hypothesis about the complete sintering of the material at the boundary.

While modeling the process of the volume modification of the containers of intricate shape using electronic computers, linear dependance is usually taken into consideration between stresses and speeds of deformation. Also the coefficients determining this connection, are taken as prescribed functions of a powder material density. Some of the decisions are obtained by a variational method, which considers functional minimization being the upper evaluation of the ultimate load, which allows to determine from the prescribed class of kinematically possible speeds, the speed carrying the minimum for the upper evaluation of the ultimate load.

3.1 Distortions of containers configuration

A non-regulated change of form, analogous to the loss of stability, causes a wave-like article shape, which points to the possibility, under certain parameters of a jacket and a powder, of the specific plastic flow.

Levelling of densities goes only at the final stagr of HIP, when the external element achieves the ultimate density and stops to deform. At that, inner elements continue to be deformed only owing to settling. This is rather important for the articles with the variations of cross-sections and while determining scale factor. That is why it is necessary to pay particular attention to experimental and theoretical substantiation of the coefficients of the form changing (shrinkage) of air-tight containers with powder while HIPing, to the elaboration of the special methodics of the autonomous form changing of various-volume containers and to the large amount of containers sketching work.

In the numerous works done, in spite of the substantiation of the interrelation between the material ultimate strength and the shrinkage coefficient both in diametrical and axial directions, experimental data on the relationship of the shrinkage coefficient and the wall thickness of a sheet are practically absent. A non-uniform filling with material and various degrees of deformation lead to the situation, when shrinkage coefficients in diametrical direction are substantially higher than the coefficients of axial shrinkage.

3.2 Compaction front spread

The cases are possible, when in consequence of the heightened hardness of a container isolated elements under HIP, the guided deformation of a powder material takes place. It has been noticed, in particular, that the deformation of disk-like elements fits into a transition law, under which boundary elements begin to deform first, and then - inner elements.

To understand this phenomenon it is necessary to take into account a considerable non-uniformity of the temperature pattern at certain HIP stages. At that, effects are possible, connected with the abrupt modification of certain structural material parameters in the narrow temperature range (yield point of a solid phase toughness, thermal diffusivity). It causes such situation that, when a jacket achieves the state of an all-round compression, the elements densities will be different. Owing to this, the compaction of a powder body has a non-uniform character, which is displayed by the presence of substantial density and temperature gradients along the cross-section. It is possible to speak about a specific "densification wave" moving from the surface into the inner part of the body. It is evident that spread speed of such a "wave" not only influences the article final structure and properties, but also defines technological parameters of HIP-process. Solving of a one-dimentional problem made it possible to obtain the expression for defining the compaction front movement speed. In this expression both temperature-strength parameters of the process and physical properties of a material under pressing are taken into consideration. The aforesaid expression describes well the mechanism of the HIP-process which is being daily watched by experimentors. Firstly, to make the "densification wave" emerge and spread it is necessary to have a temperature gradient in a preform, the pressure of compacting must exceed yeild stress. At that front movement speed and, consequently, compacting speed increases with the decrease of yeild stress. Secondly, the speed is increasing when levelling of the temperatures along a preform cross-section. It can be achieved both while decreasing of the heating speed up to operating temperature and while increasing thermal conduction due to increase of a preform density in its initial state or while HIPing. However, a small heating speed causes the decrease of the process productivity. At the same time increasing of the initial material density leads to the lessening of compaction front motion speed. Thus, to make HIP-process more efficient, it is necessary to use preforms with optimal initial density.

In front of such wave (or compaction front) the material is not warmed-up and, consequently, is not compacted. As soon as the narrow layer, adjoining the front of compacting, of the material is warmed-up, compacting and heating become highly intensified. Behind the front, due to a much more high heat conduction, material temperature and density are actually uniform.

4. SPECIFIC FEATURES OF HIP TECHNOLOGY OF A FREE-FILLED POWDER

Preforms forming frompowders with spherical particles to provide a dense, transportable preparatory briquette is a complex process, as such powders have a smooth surface and a spherical shape of particles. That is why, while using conventional methods, static or hydrostatic (CIP) and vibro pressing, it is impossible to provide the results demanded. Besides, to provide the uniform warming of a capsule up to the necessary temperature before the final packing, tens of hours may pass.

The decrease of power expenses is achieved due to powders pre-packing by CIP, which favours the increase of particles contact areas, the reduction of a preform volume and the increase of its specific mass before HIP. As it is shown in the works of some scientists, as well as by the authors of the present paper, the rise of density from 60 (free-filled)÷ 68 (vibropacked) up to 74% reduces the heating time from 24 to 5-6 hours. A further rise of density is restricted due to the following reasons.

Firstly, the rise of density might be achieved by particles plastic deformation, but it will demand considerable efforts (pressure) while CIPing, because in the process of static fixing of particles within a compact volume pressing energy will be consumed only by aggregate and fragments crushing or by a plastic deformation. At that, energy dissipation degree is determined by granulometric composition of powders, particles strength and initial density. That is why on the non-plasticized spherical powders, due to high friction, the compacting goes mainly on account of an insignificant particles deformation and the compacting up to 74% goes quickly, but the strengthening becomes slower.

Secondly, it is a broad range of particle sizes, used for HIP (from 25 up to 800 μm, or 0,001 ÷ 0,032 inch), while the rise of the coefficient of a mold filling by spherical particles demands the maintaining of a two-fraction powder composition in the particles sizes Dmax=2,5Dmin or the crushing of atomized powders in an inert atmosphere to fasion the particles into irregular shape. This facilitates the increase in the number of contacts between the particles and the decrease of CIP forces to get a high bars density.

As the calculations, made by the authors, showed, the shrinkage values of the containers with a free-filled powder, containers metal deformations and ceramics deformations achieve so significant values, that it is difficult to guarantee their strength at high temperatures and pressures, especially when using metal containers of intricate shape with a large extent of welds. Besides, the increase of compacts density brings the shrinkage to minimal values. Though it is rather difficult to place a briquette, formed beforehand, into the container of an intricate shape.

The authors have carried out the researches, directed towards the increase of density, of technological strength and of economic indices of HIP, towards the decrease of the process duration and of the deformation degree while HIPing, in particular, towards the realization of HIP-process without "rigid" metal and ceramic jackets.

5. OBTAINING DENSE ARTICLES OF AN INTRICATE SHAPE WITHOUT RIGID CONTAINERS

At present one of the variants of powder materials CIPing became widely spread, that is, pressing in elastic jackets, which allows to obtain articles with a uniform density at a considerably higher precision of sizes.

The aim was to study the possibility of pressing of intricately shaped articles with a variable cross-section using elastic medium and combined loading and to study the process of vacuum sintering. The powder in an elastic jacket is subjacted to a combined high pressure both in diametrical and axial directions.

At that, maximum axial force values might constitute 1,7 of maximum diametrical force value. Densification goes due to re-packing of particles, which make alternating travels both in axial and diametrical directions. It favours the elimination of an "arch" effect and prevents the generation of the voids devoid of powder.

As a result a dense compacting of particles in a briquette with a through porosity is achieved, constituting only

3-4% for the powders with particles of irregular shape and 5-6% for the powders with particles of spherical shape. High technological strength of the briquette against compressing and special elastic equipment allows to extract the compressed bar from a high-pressure vessel and to transport it into a vacuum furnace for sintering.

The results of the studying of vacuum sintering made it possible to state that a gas-emission process from the pore space ends after holding in the range of temperatures from 1223 up to 1373 K (950-1100°C) with density rising only for 3 ÷ 4%. After that the preforms can be subjected to a mechanical treatment to give them sizes and geometry, nearing the article shape after HIP, taking into consideration the shrinkage. The bar, after a regeneration annealing, can be successfully processed with a cutting tool made from an instrumental steel.

The process of final sintering can go in a HIP unit, which reduces the power losses considerably, as HIP goes without any intermediate cooling or heating. A linear shrinkage at constant sintering conditions depends on the original density of a preform, that is why the distribution of a relative density along the preforms height before and after sintering is interesting. The difference of density along the preforms height before sintering constitutes only 4%, which may be explained by the densification of powder on the preform end faces not only towards a radial, but also towards an axial direction under the influence of the high pressure of an axial compression. After sintering a density levelling is going (the difference of density along the preform height is not more than 1%).

To rise the bar density after the combined CIP to provide its mechanical finishing before vacuum heating up to 1223-1373 K (950-1100°C) and to reduce power losses for the re-heating after mechanical treatment, the experiments were carried out on welding the particles by a powerfül momentary electric current pulse, stored by a capacity accumulator. The contact "necks" between particles are formed along as well as across, already while passing the first half-cycle of the electric current pulse.

The reasearches, which were made, allow to increase the

metal consumption coefficient up to 0,92-0,95 at the absense of expenses for containers production.

6. SUMMARY

The analyses of the methods of the preforms obtaining and their preparation for HIP shows that further intensification of researches in the sphere of the process theory and the quickest industrial realisation of the positive results received, will serve the production of new materials and new articles with high operation characteristics, which will satisfy the demands of the modern machinery and will become the economic key-factor for the further market development by the HIP-process.

BIOGRAPHIES

1. <u>Vsevolod A. Domoratsky</u> was born in 1946. He graduated from the Dniepropetrovsk State University in 1974. He has carried out scientific researches connected with powder metallurgy (PM) beginning from 1970. At present he is the Head of PM Scientific Laboratory of the Dniepropetrovsk State University (DSU). V. Domoratsky is the author of 42 scientific publications and of 12 inventions.

2. <u>Anatol F. Sanin</u> was born in 1956. He graduated from the Charkov State University in 1979, specializing in metallophysics. He has worked over PM problems since 1981 and defended the candidate thesis on PM HSS in 1987. At pres sent he is a scientific worker of PM Scientific Laboratory of the DSU.

3. <u>Vladimir K. Nikolayev</u> was born in 1945. He graduated from the DSU in 1971. He has dealt with PM problems since 1975. At present V. Nikolayev is a scientific worker of PM Scientific Laboratory of the DSU. He is the author of 31 scientific publications and of 10 inventions.

4. <u>Lioudmila L. Scheglova</u> was born in 1943. She graduated from the Dniepropetrovsk Metallurgy Institute in 1967, specializing in metals thermal treatment. She has worked over PM problems since 1970. At present she is a scientific worker of PM Scientific Laboratory of the DSU. L. Scheglova is the author of 12 scientific publications and of 1 invention.

2nd International SAMPE Metals Conference
August 2-4, 1988

LOCALIZED CORROSION OF GRAPHITE FIBER/6061-T6 ALUMINUM ALLOY METAL-MATRIX COMPOSITES IN AERATED AND DEAERATED SODIUM SULFATE SOLUTIONS

L.H. Hihara and R.M. Latanision
The H.H. Uhlig Corrosion Laboratory
Massachusetts Institute of Technology

ABSTRACT

The localized corrosion of a graphite fiber/6061-T6 aluminum alloy metal-matrix composite (MMC) plate was investigated in 0.5 M Na_2SO_4 solutions. The microstructure, corrosion morphology, and electrochemical behavior were studied. Cl-containing compounds were found in the microstructure of the composite, and their origin was traced to the liquid metal infiltration process. Potentiodynamic polorization data and corrosion morphology indicated that pitting occurred in the regions containing the Cl-containing compounds (e.g., precursor wire-wire diffusion bonds). Thus, these results implied that the inherent corrosion resistance of graphite fiber/aluminum alloy MMCs should be improved by eliminating the Cl-containing compounds from the microstructure.

1. INTRODUCTION

The relatively poor corrosion resistance of graphite/aluminum (G/Al) metal-matrix composites (MMCs) in comparison to their monolithic matrix alloys is well known (1,2,3,4). Although results in the literature are not unambiguous, corrosion of G/Al MMCs is generally attributed to galvanic corrosion between the graphite fibers and the aluminum matrix (1,2,3), and to inferior corrosion resistance of the precursor wire-wire and wire-foil diffusion bonds (4). However, it appears that detailed studies relating corrosion morphology to microstructure have not been previously published. In this study we have acquired compelling evidence correlating localized corrosion of wire-wire diffusion bonds and general pitting of the matrix to Cl-containing compounds in the microstructure introduced through the liquid-metal infiltration process.

A graphite fiber/6061-T6 aluminum alloy (G/6061-T6 Al) MMC was the subject of this study. Examination of the MMC in a sodium sulfate solution (with only trace amounts of chlorides) was essential to observe the effects of the Cl-containing compounds in the matrix of the composite. Substantial corrosion of the diffusion bonds and some general pitting of the matrix occurred during anodic polarization of cross-sectional areas of G/6061-T6 Al MMC plates in the sodium sulfate solution. Localized corrosion of the 6061-T6 Al matrix was not expected because monolithic 6061-T6 Al passivated under similar conditions, which is characteristic of aluminum (5). The results indicated that chlorides from the

Cl-containing compounds promoted pitting of the matrix, and it is well established that halides (e.g., chlorides) promote pitting in aluminum [(6)]. Correlation between the microstructure, corrosion morphology, and electrochemical data provided substantial evidence for the attribution of localized corrosion to Cl-containing compounds in G/6061-T6 Al MMCs.

1.1 Source Of Cl-Containing Compounds

The source of Cl-containing compounds was traced to the liquid-metal infiltration process where graphite fibers (in a tow containing 2000 fibers) are coated with a Ti-B compound in an argon atmosphere containing $TiCl_4$ (g), BCl_3 (g), and Zn (v) to enhance the wettability of the fibers in liquid 6061 Al. The coated fibers are immediately made into precursor wires by infiltrating the tows in molten 6061 Al [(7)]. Cl-containing compounds are presumably integrated into the precursors' skin while the precursors are extracted from the 6061 Al melt which is covered by the argon atmosphere containing the $TiCl_4$ (g), BCl_3 (g), and Zn (v) [(8)]. For example, Amateau [(9)] reported that $TiCl_2$, $TiCl_3$, and $ZnCl_2$ are some of the thermodynamically favorable products of the deposition process. The precursors are then washed in detergent and rinsed in acetone. When plates are made, layers of precursor wires are bound with a consumable acrylic binder prior to diffusion bonding at 538 $\pm$ 11°C (1000 $\pm$ 20°F) and 13.8 MPa (2000 psi) for approximately 1200 seconds [(8)].

2. MATERIALS

The G/6061-T6 Al MMC precursors were produced by Material Concept, Inc. (666 N. Hague Ave., Columbus, OH 43204), and the six-ply (six layers of precursor wires) plates were consolidated by DWA Composite Specialties, Inc. (21119 Superior St., Chatsworth, CA 91311). The composites contained approximately 50 vol.% of high-modulus graphite fibers (Thornel P100S: pitch-based with an elastic modulus equal to 690 GPa (100×10^6 psi)) which were unidirectional and continuous. Graphite electrodes were also made with the Thornel P100S fibers. Monolithic 6061 Al was supplied in the form of 1.27 cm (0.50 inch) diameter wrought rod.

Solutions were prepared with 18 MΩ•cm water and reagent grade Na_2SO_4 (< 0.0002% Cl) and NaCl.

3. PROCEDURE

Optical microscopy and scanning electron microscopy were used for microstructural and corrosion morphological characterization. Chemical analysis was performed by energy-dispersive spectrometer (EDS) analysis and Auger electron spectroscopy (AES). The scanning electron microscope with the EDS was operated at 20 KeV. Microscopy specimens were polished to 0.05 microns (2.0×10^{-6} inch) with gamma alumina powder.

The G/6061-T6 Al MMCs and monolithic 6061 Al were heat treated to the T6 condition.

Precursor wire and six-ply plate G/6061-T6 Al MMCs and monolithic 6061-T6 Al were made into planar electrodes. The surface aluminum foils from the six-ply plate were removed. The specimens were coated with an epoxy paint (AMERCOAT 90 RESIN, Ameron) prior to mounting in an epoxy resin (EPON 828 RESIN, Miller-Stephenson Chemical Co., Inc.). Graphite electrodes were made by infiltrating 15 tows (2000 fibers/tow) with the epoxy resin. An average fiber diameter of 10 microns (3.9×10^{-4} inch) was used to calculate the surface area of the graphite electrode. In the MMC and graphite electrodes, the graphite fibers were oriented perpendicularly to the electrode surface. The specimens were polished to 0.05 micron (2.0×10^{-6} inch) finish with gamma alumina powder and rinsed with 18 MΩ•cm water about 300 seconds before immersion in the test solution.

Precursor wire electrodes that exposed only the 6061-T6 Al skin (figure 1) were made by sealing the tip of the wire with the epoxy paint. This ensured that the cross section of the precursor was not initially exposed to the solution. These electrodes will be referred to as precursor wire 6061-T6 Al skin electrodes. The electrodes were rinsed in methanol and then in 18 MΩ•cm water about 300 seconds prior to immersion in the test solution.

The electrochemical examinations were conducted in 0.5 M Na_2SO_4 (~ pH 6.4) and 3.15 wt% NaCl solutions, which will be referred to as the sodium sulfate and sodium chloride solutions, respectively for convenience. The solutions were maintained at 30 ± 0.2°C (86 ± 0.4°F), and deaerated with purified hydrogen or aerated with air (19.5 - 23.5 % oxygen balanced nitrogen).

Potentiodynamic polarization examinations were conducted with a Model 173 Princeton Applied Research (PAR) Potentiostat/Galvanostat and a Model 376 PAR Logarithmic Current Converter. The accuracy of the logarithmic current converter was measured to be better than about 5% while measuring currents in the nA range.

Prior to polarization, the corrosion potential of the specimens were allowed to stabilize for the times indicated in the polarization diagram figures (RESULTS section). Multiple polarization diagrams were generated for each specimen, but only the average current densities and sample standard deviation bars were plotted as a function of potential. The scan rate was 0.1 mV/S. All electrochemical potentials reported in this paper are with respect to the saturated calomel electrode. Also, reversible potentials were calculated assuming a pH 6.4 solution and 101.4 KPa (14.7 psi (1 atmosphere)) gas pressure.

In situ corrosion observations were made using an optical microscope equipped with a video camera and recorder. A specially designed cell held the specimen horizontally beneath the objective lens of the microscope. There were several millimeters of solution above the specimen surface.

4. RESULTS

4.1 Microstructural Analysis

A Cl-containing compound was found in a large fraction of the wire-wire diffusion bonds, as shown in figure 2. A typical EDS spectrum is shown in figure 3 showing a well defined chlorine peak. Although the oxidation state of chlorine can not be determined by EDS analysis, the chlorine in the Cl-containing compounds is likely to be in the form of chlorides. Chlorine was not detected in the matrix regions next to the diffusion bonds, but an argon peak may have concealed a small chlorine signal. Precursor wire skins (figure 1) were also analyzed, and chlorine was detected, as shown in the EDS spectrum in figure 4.

Fiber-matrix interfaces were also examined. The elemental map (figure 5) produced by scanning AES confirmed that Ti was concentrated in the region of the interface where one fiber was pulled out of the matrix (figure 6). A similar map was produced at an energy of approximately 180 eV, but could not be exclusively attributed to B or Cl since their Auger energies are similar, 179 and 181 eV, respectively. EDS analysis of interfaces such as in figure 7 produced spectrums which were typical of figure 8. Titanium but not chlorine was detected. Because the X rays are generated from a range on the order of microns, the spectrum contained information from the interface as well as the adjacent matrix. Chlorine was also undetectable in the matrix regions away from the fibers. Occasionally, neighboring fibers were bridged by a compound rich in Ti, Cl, and Mg, as shown in the SEM micrograph in figure 9 and the EDS spectrum in figure 10. Titanium and chlorine were also detected in the fiber-matrix interfaces of these bridged fibers.

In summary, chlorine was detected only in diffusion-bond regions and in regions of bridged fibers.

4.2 Corrosion Morphology

The wire-wire diffusion bonds of the G/6061-T6 Al six-ply plate corroded preferentially in the sodium sulfate solutions during anodic polarization. The surface of one such corroded electrode was wafered and mechanically broken. The wafer parted predominantly along corroded diffusion bonds (figure 11) which contained chlorine-containing corrosion products which were detected by EDS analysis. General pitting of the matrix was also observed, and shown in figure 12. Vigorous hydrogen evolution from corroding diffusion bonds of a specimen that was at open circuit potential, but previously anodically polarized at 2.1 and 2.5 V for a total of 4800 seconds in the aerated sodium sulfate solution was observed at about 800x magnification and recorded on video tape in real time.

Gas evolution from a diffusion bond of a freshly polished surface at open circuit potential was also observed in the aerated sodium sulfate solution and in aerated distilled water (figure 13).

The planar G/6061-T6 Al precursor wire electrodes were also anodically polarized in the deaerated sodium sulfate solution. As shown in figure 14, pits were concentrated near the perimeter of the these precursor electrodes. When the precursor wire 6061-T6 Al skin electrodes were anodically polarized in the deaerated sodium sulfate solution, pits penetrated the skin (figure 15). Sectioning and polishing these precursors revealed a 6061-T6 Al shell that encased a severely corroded precursor interior where corrosion propagated towards the core (figure 16). A detailed micrograph of the 6061-T6 Al shell shows where pits had initially bored through the skin (figure 17). In figures 16 and 17, the large cavities (black) were the result of mechanical damage during polishing.

4.3 Electrochemical Data

The anodic polarization diagrams of the graphite fiber and monolithic 6061-T6 Al electrodes will be presented first, followed by the G/6061-T6 Al MMC electrodes, and finally by the cathodic polarization diagrams of the graphite fiber electrodes. In each of the polarization diagram figures, the following additional information is given: 1) the average stabilized open circuit potential, 2) the sample standard deviation (s) of the open circuit potential, and 3) the time the samples were allowed to stabilize at open circuit.

In the anodic polarization diagram of the graphite fibers in deaerated sodium sulfate solution (figure 18), the Tafel slope from 3×10^{-5} to 10^{-3} A/cm^2 was 0.26 V/decade, and at current densities greater than 10^{-3} A/cm^2, it was 0.16 V/decade. The anodic polarization diagram of the monolithic 6061-T6 Al in deaerated sodium sulfate solution (figure 19) showed a sharp increase and decrease in current density near the open circuit potential followed by a passive region. At a potential greater than 0.15 V, the passive current density increased slightly and was potential dependent, but did not exceed about 10^{-5} A/cm^2. In figure 20, the monolithic 6061-T6 Al in deaerated sodium chloride solution had a pitting potential of approximately -0.73 V; however, the sharp increase and decrease in current density near the open circuit potential was not observed.

Anodic polarization diagrams of the planar G/6061-T6 Al MMC electrodes in the forms of six-ply plate and precursor wire in deaerated sodium sulfate solution are shown in figures 21 and 22, respectively. Both diagrams were similar and had a pitting potential of approximately -0.5 V. The anodic polarization diagram of the planar six-ply G/6061-T6 Al MMC plate in aerated sodium sulfate solution was similar to that in the deaerated solution, but with the exception that the open circuit potential was closer to the pitting potential (figure 23). In figure 24, the precursor wire 6061-T6 Al skin electrode had a pitting potential of about -0.6 V in the deaerated sodium sulfate solution. The decrease in current at potentials more noble than 0.0 V was a result of the shrinkage of the precursor-skin electrode area (core perimeter) due to corrosion (figure 16).

Cathodic polarization diagrams of the graphite electrodes in deaerated and aerated sodium sulfate solutions are shown in figures 25 and 26, respectively. The polarization diagram of the graphite electrode in the deaerated solution had three distinct regions of apparent Tafel behavior. The slope in the first region was -1.35 V/decade followed by -0.04 V/decade and finally -0.15 V/decade. In the aerated solution, the Tafel region (-0.10 V/decade) was followed by diffusion limited oxygen reduction.

5. DISCUSSION

The relationships between microstructure, corrosion morphology, and electrochemical data will provide new insights into the corrosion mechanisms of G/6061-T6 Al MMCs. The anodic polarization diagrams of the graphite fibers and the monolithic 6061-T6 Al in the deaerated sodium sulfate solutions will be analyzed first. From these diagrams, a theoretical polarization curve of the composite (based on the mixed electrode theory) will be generated and compared to polarization curves of the actual composites. The characteristics of the actual anodic polarization diagrams of the G/6061-T6 Al composites will be related to the corrosion morphology and microstructure. Finally, the corrosion of G/6061-T6 Al MMCs in aerated solutions, and cathodic polarization of the graphite fibers will be discussed.

The anodic currents from the graphite fiber electrode can mainly be attributed to CO_2 formation, based on work of Kokhanov and Milova [(10)] who found that during the anodic polarization of graphite, 91% of the current is contributed from CO_2 formation and the balance from CO and O_2 formation in 50°C, pH 1 to 6.7 buffered phosphate solutions at a total anodic current of 3.7 mA/cm^2. The Tafel slope, 0.16 V/decade (from 1.5 to 3.7 mA/cm^2), of the graphite fibers was in relatively good agreement with the value published by Kokhanov and Milova in the same current range [(11)], 0.2 V/decade (in the same phosphate solutions described above), of a graphite anode of unspecified structure and quality. Also, Tafel slopes ranging from 0.15 to 0.25 V/decade have been reported in the literature [(11 - 15)]. Ngo Dai Viet, et al. [(16)] reported that the composition of the anodic gases in acid to neutral solutions is only slightly affected by the current density. Also worth noting is that the open circuit potential of the graphite fiber electrode (about -0.2 V) was approximately 0.4 V more noble than that of the reversible hydrogen electrode, and in better agreement to the reversible CO/C electrode (-0.1 V [(17)]), as shown in figure 18. The monolithic 6061-T6 Al passivated in the deaerated sodium sulfate solution (figure 19). The increase in anodic current density of 6061-T6 Al at potentials more noble than 0.15 V can be attributed to thickening of the oxide film by anodization [(5)].

In figure 27, the theoretical polarization diagram of the MMC is compared to the actual diagram of the G/6061-T6 Al MMC six-ply plate. The immensely larger currents of the actual MMC was attributed to pitting of the 6061-T6 Al matrix predominantly in the diffusion-bond regions, as shown in the SEM micrographs (figures 11 and 12). Attention should also be drawn to the pitting potential exhibited in the actual G/6061-T6 Al MMC polarization diagram, which is characteristic of aluminum in chloride-containing solutions during anodic polarization.

The anodic polarization diagram of the precursor wire 6061-T6 Al skin electrode in the deaerated sodium sulfate solution is compared to the anodic polarization diagram of monolithic 6061-T6 Al in the deaerated sodium chloride solution in figure 28. Both diagrams showed typical pitting potentials even though the precursor wire 6061-T6 Al skin electrode was in the sodium sulfate solution. Examination of the precursor skin electrode verified that the mode of corrosion was pitting (figures 15, 16, and 17). Presumably, chlorides from the Cl-containing compounds caused the skin to pit, having the same effect that the chloride ions in the sodium chloride solution had on the monolithic 6061-T6 Al. The differences in the pitting potential (~ 0.1 V) of the two diagrams was not unexpected because the pitting potential is dependent on chloride-ion concentration [(6)].

The anodic polarization diagram of the planar G/6061-T6 Al MMC precursor wire electrodes (figure 22) in the deaerated sodium sulfate solution were similar to those of the planar six-ply plate electrodes, as shown in figure 21. In the planar precursor wire electrodes, extensive pitting occurred near the perimeter of the precursors (figure 14) which indicated that corrosion of diffusion-bond regions in plates is caused by pitting in regions of Cl-containing compounds rather than being a manifestation of the diffusion-bonding process.

The polarization diagrams of the G/6061-T6 Al MMC six-ply plate in deaerated (figure 21) and aerated (figure 23) sodium sulfate solutions showed that the open circuit potential of the MMC was nearer to the pitting potential in the aerated solution. The cathodic polarization diagrams of the graphite fibers in aerated and deaerated sodium sulfate solutions (figures 25 and 26) were in accordance with the former observations; namely, that the graphite fibers were effective cathodes only in the aerated solution. These observations are good indications that the G/6061-T6 Al MMC is susceptible to localized corrosion in aerated, but not deaerated sodium sulfate solutions in the open circuit condition. In figure 13, the evolution of gas at open circuit potential in aerated distilled water from the diffusion-bond region of a freshly polished G/6061-T6 Al MMC plate was evidence of localized corrosion in the diffusion-bond region. The gas was probably hydrogen which is not unusual under these circumstances because the local potential at regions of localized corrosion can be more reducing than the bulk potential due to ohmic drop in the solution. For example, we have also observed hydrogen evolution from pits in 6061-T6 Al under anodic polarization. However, the possibility of CH_4 due to the chemical decomposition of Al_4C_3 should not be overlooked, but gas evolution due to the oxidation or reduction of graphite is unlikely since gas evolution predominantly occurs at diffusion-bond regions rather than at the ubiquitous graphite fibers.

Before ending this discussion, some important points regarding the cathodic polarization of the graphite fibers will be made. The open circuit potential of the graphite fibers in the aerated solution was about 0.66 V less noble to the reversible oxygen potential, 0.61 V. The discrepancy between the open circuit and the reversible potential is not unusual, and others have reported that the open circuit potential of carbon or graphite electrodes is unstable, difficult to reproduce and usually about 0.1 - 0.3 V less noble than the reversible oxygen potential [18,19,20]. The Tafel slope of the graphite fibers (-0.10 V/decade) was also in the range of published values, -0.09 to -0.16 V/decade in acid and alkaline solutions in the high polarization regime [21 - 24]. In regards to the deaerated solution, as mentioned previously, the open circuit potential of the graphite fibers was in better agreement to the reversible CO/C than it was to the H^+/H_2 electrode. Also, the cathodic reaction operating from the open circuit potential to about -1.15 V where the slope was -1.35 V/decade could be the reduction of CO or diffusion limited oxygen reduction (assuming trace oxygen contamination) based on the relative positions of the reversible O_2/H_2O, CO/C, CO_2/C, H^+/H_2, and C/CH_4 potentials (figure 25); this, however, is only speculation. The dominating cathodic reaction at potentials more reducing than -1.15 V was assumed to be the hydrogen evolution reaction where two distinct Tafel slopes were observed (-0.04 and -0.15 V/decade), which are in the realm of theoretical values assuming various hydrogen evolution mechanisms [25]. Tafel slopes reported in the literature range from -0.110 to -0.210 V/decade for glassy carbon and various graphites in acid electrolytes [26,27,28].

6. CONCLUSION

The good correlation between microstructure, corrosion morphology, and electrochemical data suggested that localized corrosion in G/6061-T6 Al MMCs in chloride-free solutions was primarily caused by pitting of the 6061-T6 Al matrix in regions of Cl-containing compounds. Although the chemical or electrochemical dissolution of the Cl-containing compounds may have contributed to localized corrosion, we believe that the complementary results between the electrochemical data and pitting corrosion morphology provided sufficient evidence to disregard the hypothesis of dissolution of the compounds to be the primary mechanism for localized corrosion. More likely, the dissolution of these compounds created an aggressive chloride medium that initiated and accentuated pitting in the 6061-T6 Al matrix.

The practical implication of this research is that the inherent corrosion resistance of G/Al MMCs should be improved by coating the graphite fibers with suitable Cl-free compounds.

7. ACKNOWLEDGEMENTS

We are grateful to the Shell Foundation for support of this research. We are also grateful to Dr. W.C. Harrigan, Jr. for useful discussions regarding the processing of the G/6061-T6 Al MMC.

8. REFERENCES

1) Czyrklis, W.F., Corrosion 88, NACE, paper # 196, 1985.
2) Vassilaros, M.G.; et al., in Mechanical Behavior of Metal-Matrix Composites (Hack, J.E.; Amateau, M.F., Eds.), The Metallurgical Society of the AIME, 1983, p.335.
3) Dull, D.L.; Harrigan, Jr., W.C.; Amateau, M.F., Aerospace Corporation, AD - A011 761, 1975.
4) Pfeifer, W.H., in Hybrid and Select Metal-Matrix Composites (Renton, W.J., Ed.), American Institute of Aeronautics and Astronautics, 1977, p.231.
5) Moshier, W.C.; Davis, G.D.; Ahearn, J.S., Corrosion Science, vol. 27, no. 8, 785 (1987).
6) Ovari, F.; Tomcsanyi, L.; Turmezey, T., Electrochimica Acta, vol. 33, no. 3, 323 (1988).
7) Harrigan, Jr., W.C.; Flowers, R.H., in Failure Modes in Composites IV (Cornie, J.A.; Crossman, F.W., Eds.), The Metallurgical Society of the AIME, 1977, p.319.
8) Harrigan, Jr., W.C., private communication, 1988.
9) Amateau, M.F., p.166 in ref. 4.
10) Kokhanov, G.N.; Milova, N.G., Elektrokhimiya, 5, 93 (1969).
11) Kokhanov. G.N.; Milova, N.G., Elektrokhimiya, 6, 73 (1970).
12) Mulina, F.I.; Krishtalik, L.I.; Kolotukhin, A.T., Zh. Prikl. Khim, 38, 2808 (1965).
13) Krishtalik, L.I.; Rotenberg, Z.A., Elektrokhimiya, 2, 351 (1966).
14) Constantinescu, M.; Neata, M.B., Rev. Chim. (Bucharest), 23, 410 (1972).
15) Ngo Dai Viet; Kokoulina, D.V.; Krishtalik, L.I., Elektrokhimiya, 8, 221 (1972).
16) Ngo Dai Viet; Kokoulina, D.V.; Krishtalik, L.I., Elektrokhimiya, 8, 384 (1972).
17) Van Muylder, J.; Pourbaix, M., in Atlas of Electrochemical Equilibria in Aqueous Solutions (Pourbaix), NACE, 1974, p.449.
18) Thiele, H., Z. Anorg. Allg. Chem., 206, 407 (1932), 207 340 (1932).
19) Thiele, H., Z. Elektrochem., 40, 26 (1934).
20) Kordesch, K.V., in Handbook of Fuel Cell Technology (Berger, C., Ed.), Prentice-Hall, Englewood Cliffs, New Jersey, 1968, p.359.
21) Tarasevich, M.R.; et al., Elektrokhimiya, 4, 432 (1968).
22) Tarasevich, M.R.; Sabirov, F.Z.; Burshtein, R.Kh., Elektrokhimiya, 7, 404 (1971).
23) Yeager, E.; Krouse, P.; Rao, K.V., Electrochim. Acta, 9, 1057 (1964).
24) Marcos, I.; Yeager, E., Electrochim. Acta, 15, 953 (1970).
25) Bockris, J.O'M., Modern Electrochemistry, vol. 2, Plenum/Rosetta, 1977, p.1242.
26) Drossback, P.; Schulz, J., Electrochim. Acta, 9, 1391 (1964).
27) Despic, A.R.; et al., Croat. Chem. Acta, 44, 79 (1972).
28) Brennan, M.P.J.; Brown, O.R., J. Appl. Electrochem., 2, 43 (1972).

9. FIGURES

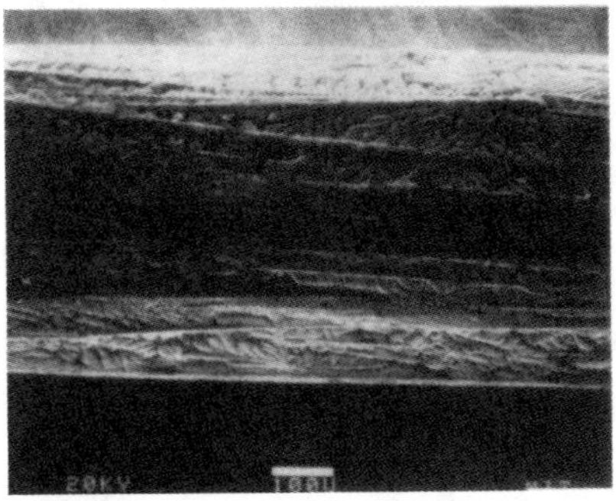

Figure 1. SEM micrograph of precursor wire 6061-T6 Al skin electrode showing the 6061-T6 Al skin and subsurface fibers.

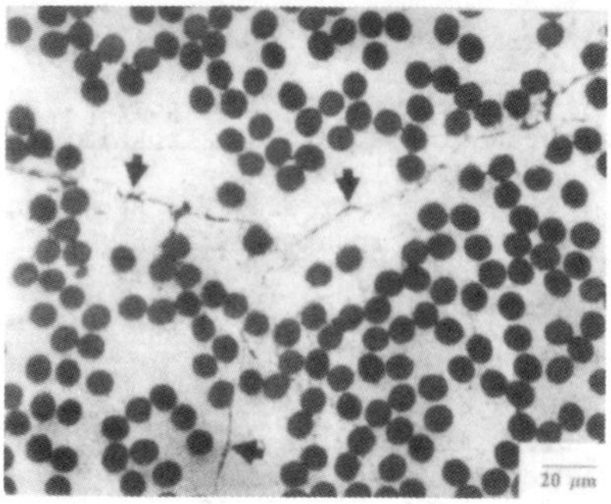

Figure 2. Cross section of G/6061-T6 Al MMC plate showing the Cl-containing diffusion-bond compound (indicated by arrows) at the intersection of three precursor wires. Fibers - black. Matrix - gray.

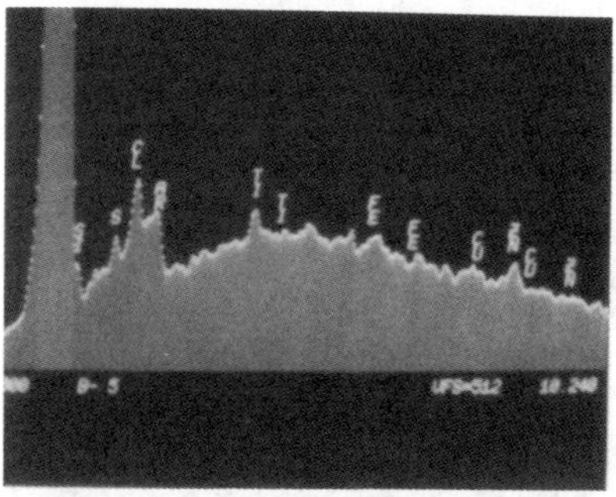

Figure 3. EDS spectrum of the diffusion-bond compound. Notice the Cl, Ti, and Zn signals.

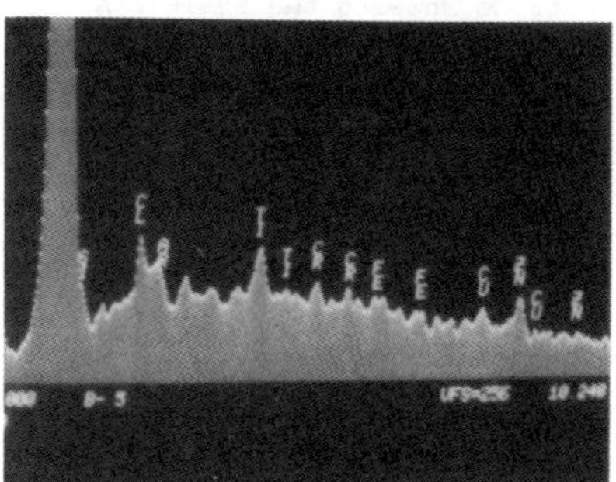

Figure 4. EDS spectrum of the precursor wire 6061-T6 Al skin (the electron beam was focussed on a region not containing subsurface fibers). Also notice the Cl, Ti, and Zn signals.

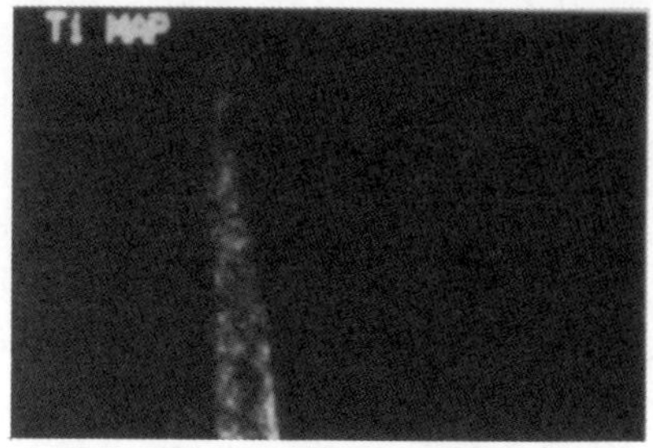

Figure 5. Titanium elemental map produced by scanning AES.

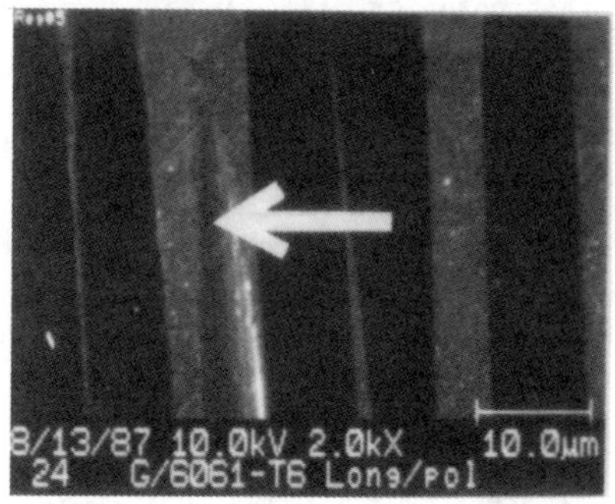

Figure 6. SEM micrograph corresponding to figure 5. The graphite fibers (black) are parallel to the surface. The arrow points to the trough where a fiber was pulled out of the matrix.

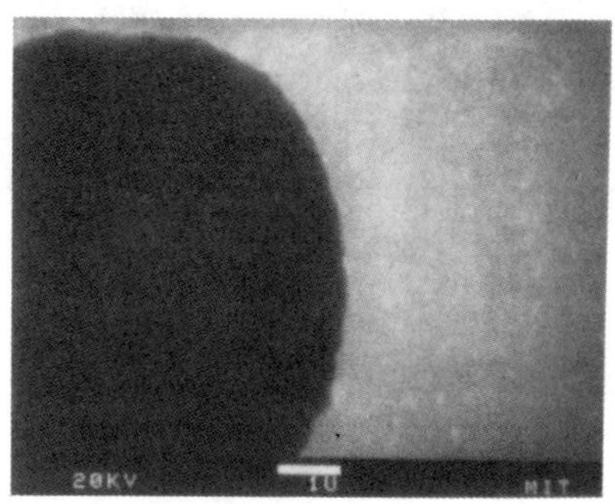

Figure 7. SEM micrograph of fiber(black)-matrix(gray) interface.

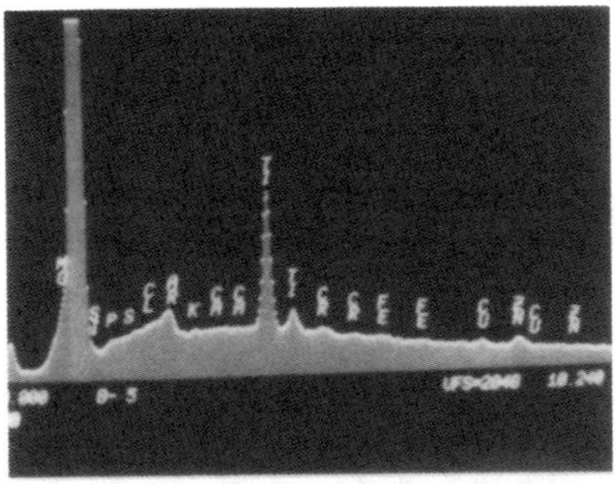

Figure 8. EDS spectrum corresponding to figure 7. Notice the Ti signal, and absence of Cl.

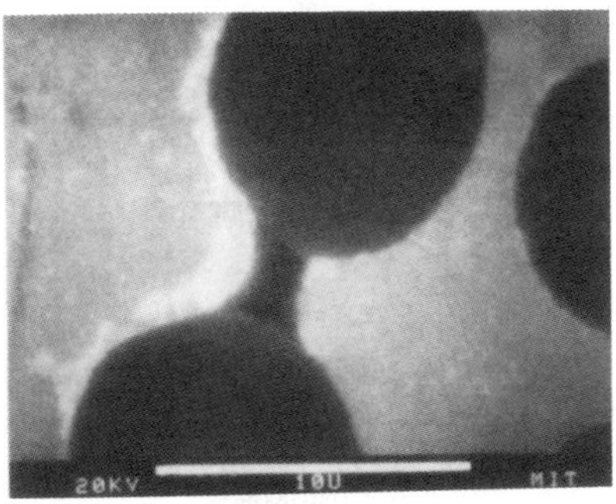

Figure 9. SEM micrograph of two graphite fibers bridged by a Ti-Cl-Mg rich compound.

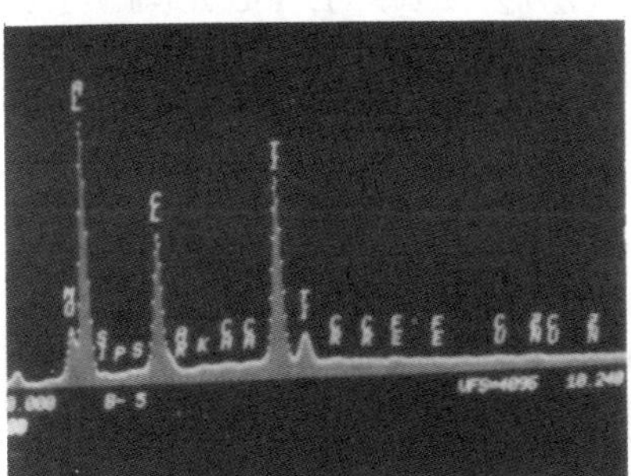

Figure 10. EDS spectrum of bridge in figure 9. Notice the strong Ti, Cl, and Mg signals.

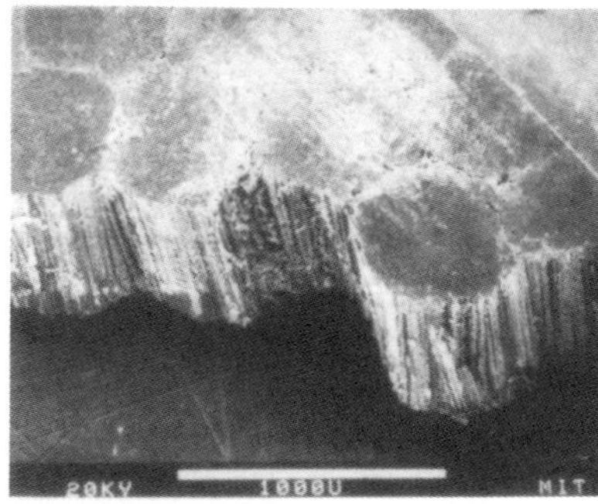

Figure 11. SEM micrograph of G/6061-T6 Al MMC plate anodically polarized in deaerated 0.5 M Na_2SO_4. Corrosion occurred predominantly at wire-wire diffusion bonds.

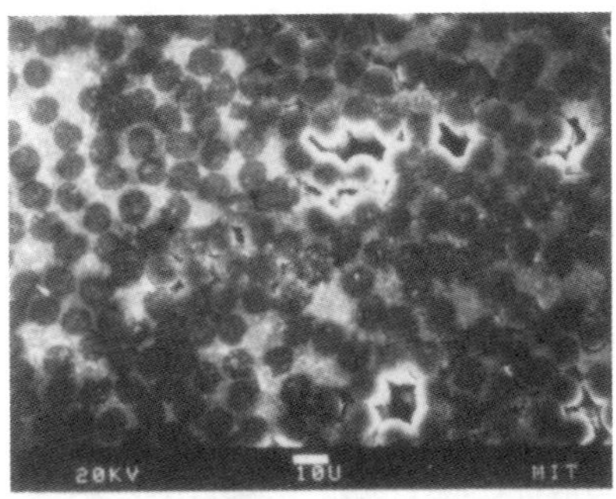

Figure 12. SEM micrograph showing some localized corrosion in a precursor interior caused by anodic polarization in deaerated 0.5 M Na_2SO_4.

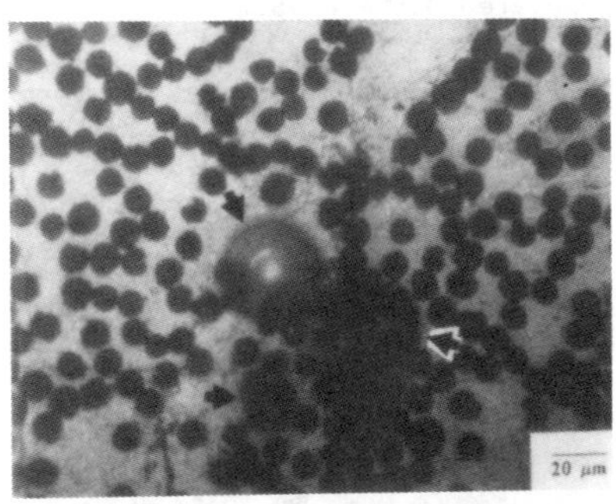

Figure 13. Gas bubbles (indicated with arrows) evolving from a diffusion-bond region of a G/6061-T6 Al MMC plate in aerated distilled water. Fibers - black. Matrix - gray.

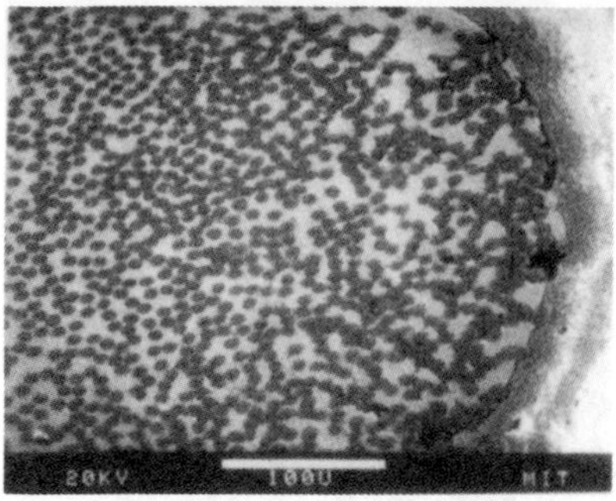

Figure 14. Substantial localized corrosion concentrated near the perimeter of a planar precursor wire electrode.

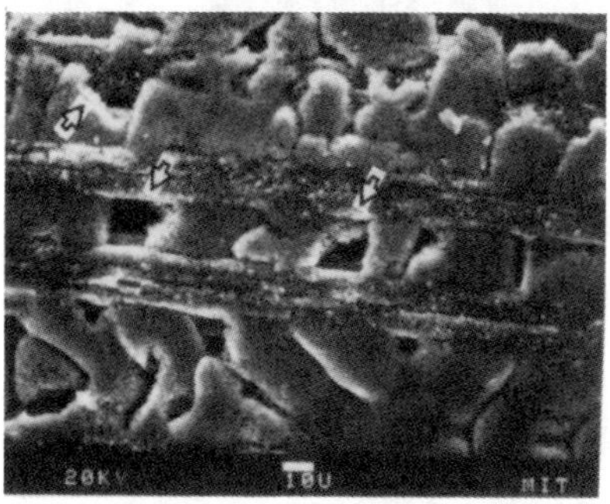

Figure 15. Pits (indicated with arrows) in the skin of a precursor wire 6061-T6 Al skin electrode that was anodically polarized in deaerated 0.5 M Na_2SO_4.

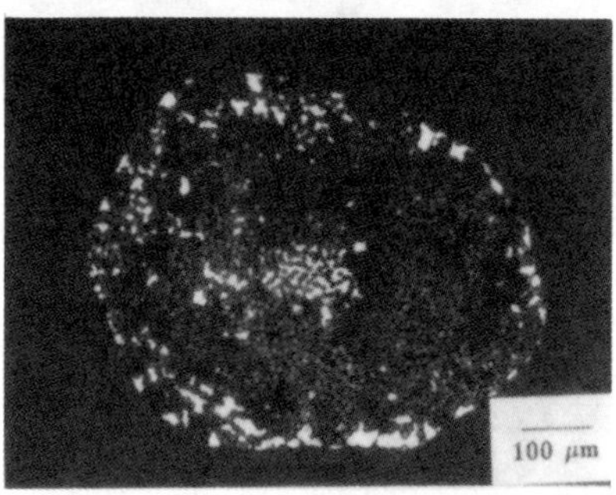

Figure 16. Cross section of the corroded precursor wire 6061-T6 Al skin electrode (figure 15) showing the 6061-T6 Al shell.

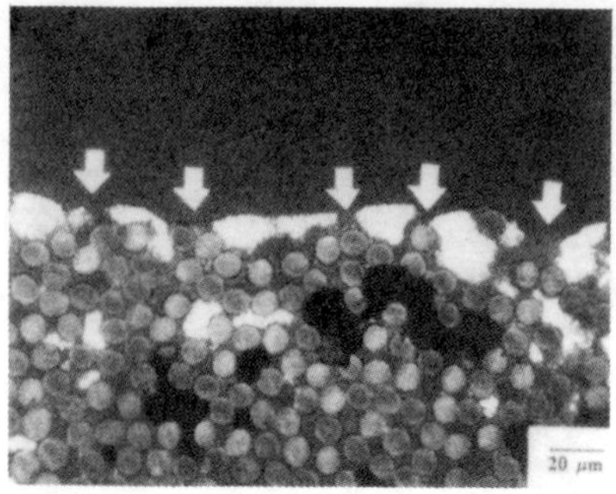

Figure 17. Enlarged view of 6061-T6 Al shell showing where pits (indicated with arrows) initially bored through the 6061-T6 Al skin.

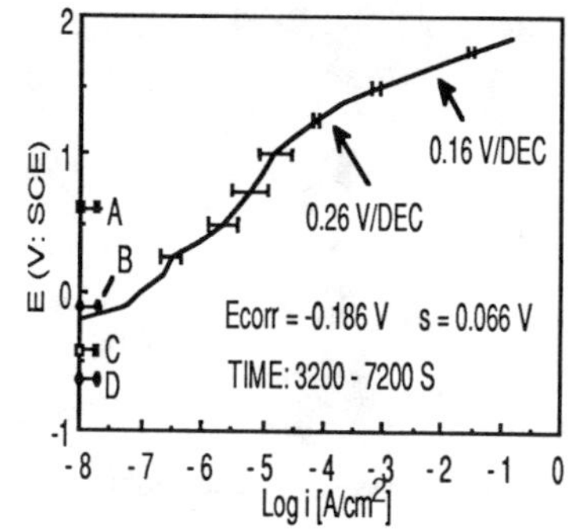

Figure 18. Anodic polarization diagram of the graphite fiber electrode in deaerated 0.5 M Na_2SO_4. Reversible potentials (17): A:O_2/H_2O(0.609 V), B:CO/C(-0.101 V), C:CO_2/C(-0.412 V), D:H^+/H_2(-0.619 V)

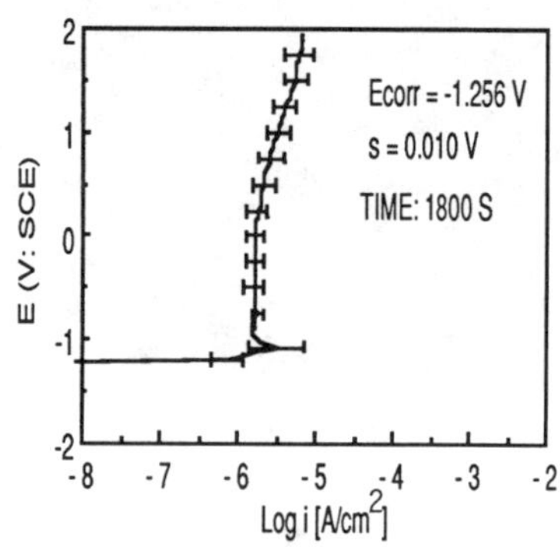

Figure 19. Anodic polarization diagram of monolithic 6061-T6 Al in deaerated 0.5 M Na_2SO_4.

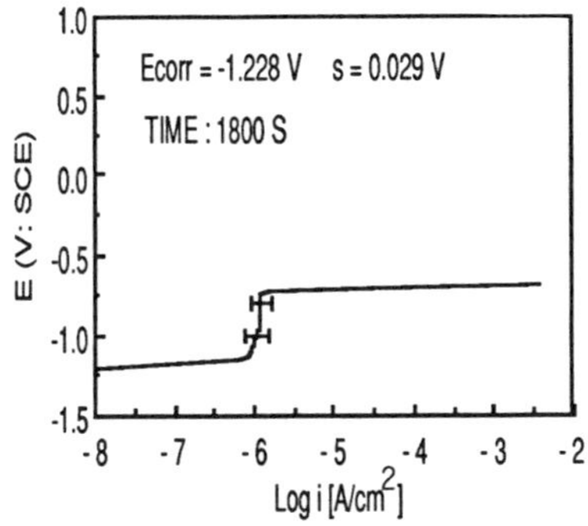

Figure 20. Anodic polarization diagram of monolithic 6061-T6 Al in deaerated 3.15 % NaCl.

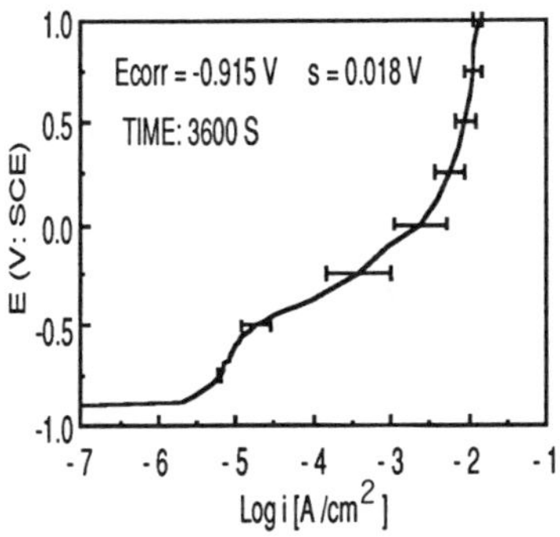

Figure 21. Anodic polarization diagram of the planar six-ply plate G/6061-T6 Al MMC electrode in deaerated 0.5 M Na_2SO_4.

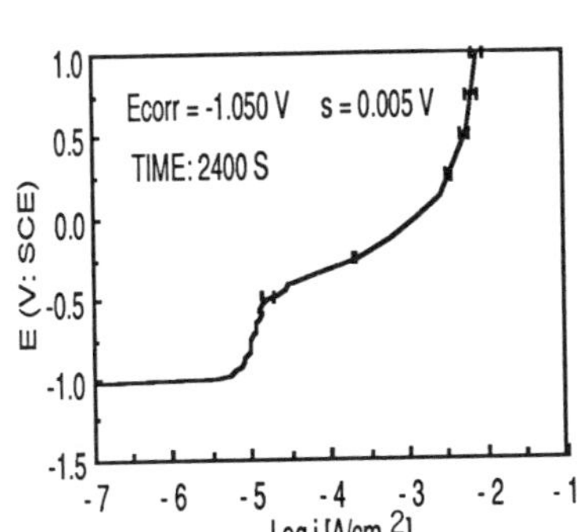

Figure 22. Anodic polarization diagram of the planar precursor wire G/6061-T6 Al MMC electrode in deaerated 0.5 M Na_2SO_4.

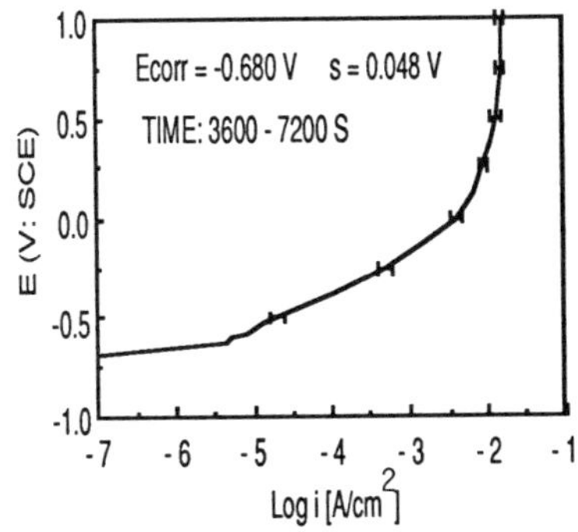

Figure 23. Anodic polarization diagram of the planar six-ply plate G/6061-T6 Al MMC electrode in aerated 0.5 M Na_2SO_4.

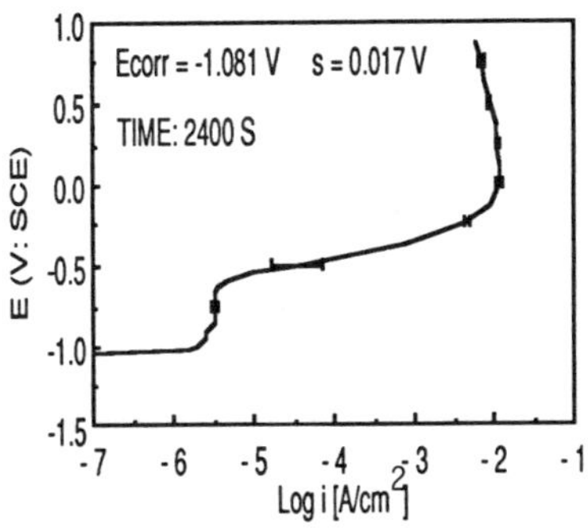

Figure 24. Anodic polarization diagram of the precursor wire 6061-T6 Al skin electrode in deaerated 0.5 M Na_2SO_4.

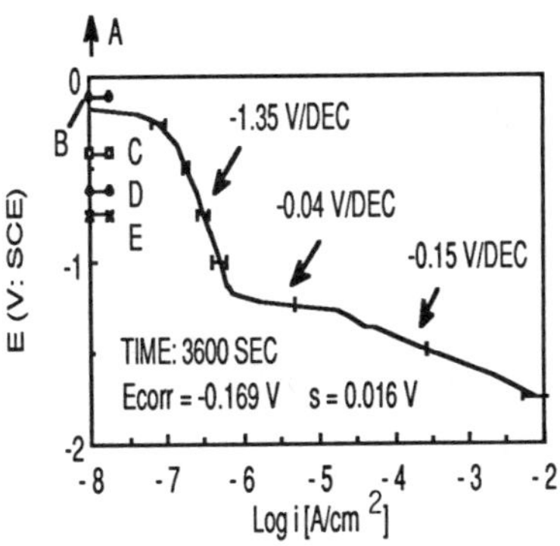

Figure 25. Cathodic polarization diagram of the graphite fibers in deaerated 0.5 M Na_2SO_4. Reversible potentials: A,B,C,D same as in figure 18. E:C/CH_4(-0.751V) [17].

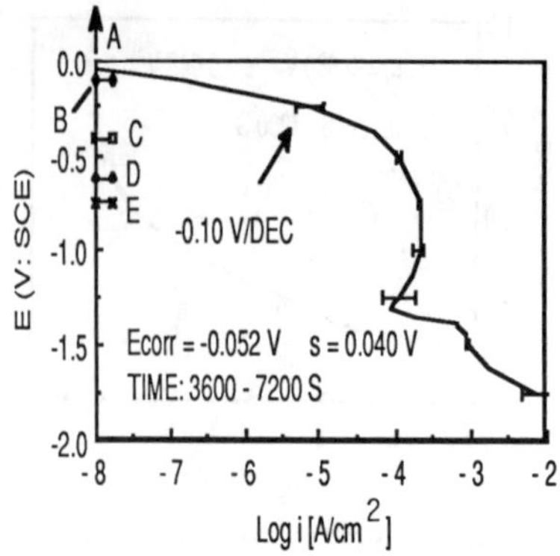

Figure 26. Cathodic polarization diagram of the graphite fibers in aerated 0.5 M Na_2SO_4. Reversible potentials: A,B,C,D,E same as figure 25.

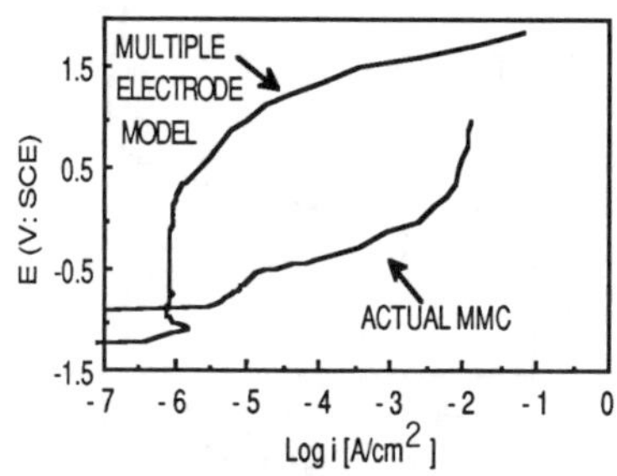

Figure 27. Comparison of the theoretical anodic polarization diagram of a graphite(50%)/6061-T6 Al(50%) MMC to the actual polarization diagram of the planar six-ply plate G/6061-T6 Al MMC electrode in deaerated 0.5 M Na_2SO_4.

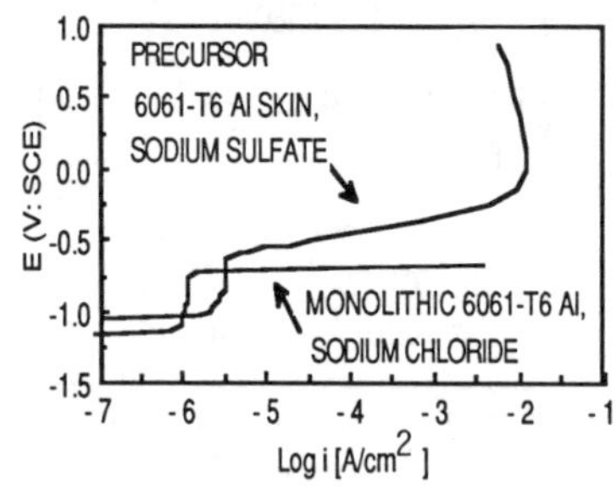

Figure 28. Comparison of the anodic polarization diagram of the precursor wire 6061-T6 Al skin electrode in deaerated 0.5 M Na_2SO_4 to the monolithic 6061-T6 Al in deaerated 3.15% NaCl.

10. ABOUT THE AUTHORS

10.1 Lloyd H. Hihara

Lloyd H. Hihara is a doctoral student at M.I.T. in the Department of Materials Science and Engineering. He received his B.S. from the University of Hawaii at Manoa and his S.M. from M.I.T. in Mechanical Engineering. He is a member of Sigma Xi, The National Association of Corrosion Engineers, The Electrochemical Society, and The Metallurgical Society of AIME, ASM International.

10.2 Ronald M. Latanision

Ronald M. Latanision, Shell Distinguished Professor of Materials Science, is the Director of the Materials Processing Center at M.I.T., and a member of the National Academy of Engineering.

He received his B.S. in Metallurgy from Penn State and his Ph.D. from Ohio State. He also received a Humboldt Foundation Senior Scientist Award at the Max-Planck-Institut in Dusseldorf, F.R.G.. He has served as Science Advisor to the U.S. House of Representatives Committee on Science and Technology -- and currently chairs the Advisory Committee on the Massachusetts Materials Center, a statewide initiative, linking government, industry, and academia.

Professor Latanision's interests are focussed in materials processing and the corrosion of metals and other materials in aqueous environments. He has authored or co-authored more than 100 scientific publications and is co-editor of Advances in the Mechanics and Physics of Surfaces.

2nd International SAMPE Metals Conference
August 2-4, 1988

CORROSION IN THE EARLY STAGES OF CONDENSATION

J. H. Payer and S. K. Chawla
Department of Materials Science and Engineering
Case Western Reserve University
Cleveland, Ohio 44106

Abstract

The severity of corrosion and rate of attack is often greatest for a metal during the early stages of condensation. The thin film of aqueous electrolyte on the metal surface is most aggressive. The traditional rationale for this observation is that the mass transfer effects, i.e. concentration polarization, do not slow the corrosion reactions in thin films of electrolyte. Recent observations indicate that additional factors can contribute to the accelerated attack. Specifically, the formation of clusters of water on the metal surface during the early stages of condensation result in heterogeneous coverages and concentration of corrosive species from the atmosphere.

The factors leading to the formation of micro-cells of corrosion are presented. The processes underway and the reactions controlling the rate of corrosion under the conditions of micro-cell corrosion are identified.

The need for a better understanding of these processes is required in a variety of applications: microelectronic devices, electronic equipment, and structural components exposed to either alternate wetting and drying or to conditions that result in condensation on the metal. Trace levels of corrosive constituents, e.g. chloride and sulfur species, in the gas phase concentrate in the electrolyte and result in accelerated attack of corrosion resistant alloys.

1. INTRODUCTION

Aqueous corrosion requires an electolyte to support the flow of current from anodes to cathodes in the operating electrochemical cells on the metal surface. The electolyte can be a bulk liquid phase, as in the case of a metal immersed in seawater, or small amounts of moisture distributed on the metal surface. The severity of corrosion and rate of attack are often greatest for a metal during the early stages of condensation. The thin film of aqueous electrolyte on the metal surface is most aggressive. This paper addresses the corrosion processes during the early stages of condensation.

The need for a better

understanding of these processes is required to increase durability and reliability in a variety of applications where corrosion shortens the useful life: microelectronic devices, electronic equipment, and structural components exposed to either alternate wetting and drying or to conditions that result in condensation on the metal. The factors leading to the formation and operation of micro-cells of corrosion are of principle interest.

2. OBSERVATIONS OF CORROSION PHENOMENA

2.1 Atmospheric Corrosion

Significant corrosion does not occur until an aqueous phase is formed on the metal surface. Vernon (1) observed a critical relative humidity between 60 and 75% above which corrosion rate increases dramatically. The corrosion rate of steel as a function of relative humidity is shown in Figure 1. No condensation occurs below the critical relative humidity. Once an aqueous phase is formed on the metal surface corrosion is observed. Vernon also demonstrated the critical relative humidity to be a function of SO_2 in the gas phase.

The concept of a critical relative humidity is well established. Furthermore, it is clear that this value is a function of gas composition (2), particles on the metal surface (3) and surface irregularities (4). Factors that promote the formation of an aqueous phase, e.g. hygroscopic particles, reduce the critical relative humidity.

Corrosion rates are greatest over an alternate wetting and drying cycle during the very first moments of wetting and the very last moments of drying (5). This phenomena is shown schematically in Figure 2. While the surface is dry, no corrosion occurs. When moisture first forms the corrosion rate is relatively high , and the corrosion rate decreases on further wetting of the surface. As the metal dries, a spike in the corrosion rate is again observed prior to complete drying. High corrosion rates are observed before a film of moisture can be observed on the metal surface with the naked eye (6). Thin films of moisture result from the alternate wetting and drying of metal surfaces.

2.2 Corrosion under Condensing Conditions

Severe corrosion can be observed for metals subjected to condensing conditions. As the temperature decreases below the dewpoint for a gas containing water vapor, moisture will condense on metal surfaces. This results in thin films of moisture and conditions similar to those observed for alternate wetting and drying. These conditions can prevail in many types of service, e.g. cooling of high temperature gases from engines, power generators or process equipment. Similarly, furnaces and heating equipment can be subjected to these conditions.

The severity of these conditions was demonstrated in a study of the corrosion in condensing heat exchangers of residential and commercial furnaces and boilers(7). A series of stainless steels and nickel alloys were exposed to a laboratory test rig that simulated the conditions in a condensing heat exchanger. A schematic diagram of the test system is shown in Figure 3. Hot gases from a residential furnace were passed through the five zones of the heat exchanger. Corrosive gases were added to the flue gases ahead of the test section. The furnace was cycled on and off to expose the specimens to alternate wet/dry cycles.

The temperatures and flow rates were adjusted to control the condensing conditions in each of the zones. Condensation in zone 1 was relatively low at 0.07 ml/min, and the amount of condensate increased in zones 2 to 5, i.e 0.3, 0.9, 0.9 and 1.3 ml/min, respectively. Specimens exposed near the entry of the heat exchanger were continuously dry. Subsequent rows in zone 1 were exposed to thin films of condensate, and the tubes in the later zones were exposed to greater amounts of condensate which could be seen to run down the sides of the tubes.

The specimens in zone 1 experienced severe attack. Several austenitic stainless steels and Alloy 20 (20 Cr-29Ni-2Mo) exhibited severe, localized corrosion. Pits penetrated the tube walls of several specimens. Corrosion was much less severe in the later zones. Bulk condensate samples from zone 1 contained approximately 150 ppm chloride and were pH 2.7. The observed corrosion behavior was indicative of considerably greater chloride concentrations and much more acidic solutions. The chloride concentration at the exit end of the heat exchanger (zone 5) was less than 20 ppm. The lower concentration was due to greater quantities of condensate in the later zones, and the removal (scrubbing) of chloride from the flue gas as it passed through the prior zones.

3. FACTORS AFFECTING CORROSION IN THE EARLY STAGES

3.1 Moisture Film Thickness

The traditional explanation of higher corrosion rates in the early stages of wetting is that mass transfer effects (concentration polarization) do not slow the corrosion process in thin films of moisture. The variation of atmospheric corrosion rate with thickness of the moisture film on a metal surface is shown schematically in Figure 4 after Tomashov ([4]). Region I corresponds to the initial stages of moisture adsorption where the adlayer cannot yet be considered to behave as an electrolyte. In region II, as a result of the thickening of the adsorbed layer, significant ionic conductivity is developed in the electrolyte film leading to high corrosion rates. Further thickening of the film causes lowering of the corrosion rate as in region III due to the impairment of oxygen diffusion. A very slight reduction in corrosion rate is observed in region IV because thereafter the thickness of the diffusion layer hardly changes.

3.2 Water Clusters

The important role of water clustering on metal surfaces as a factor in the initial stages of atmospheric corrosion has not been widely recognized. Clustering of water molecules at sub monolayer coverage leads to the accumulation of significant local surface volumes of water where the properties of adsorbed water approach that of bulk water. This inhomogeneous adsorption of water from the atmosphere presents the condition for initiation of microcell corrosion.

Water does not adsorb as a uniform film on the surface but rather as water clusters. Klier , Shen, and Zettlemoyer ([8]) have demonstrated the formation of water clusters on partially hydrophobed silica via infrared spectroscopy. They concluded that silica surfaces at monolayer water coverage had equivalent amounts of clustered and surface adsorbed water. Yan, Meilink, Warren, and Wynblatt ([9]) have studied water adsorption on alpha-alumina substrates using thermogravimetric analysis and surface electrical conductivity measurements. They describe a discontinuous or "patchy" nature of water adsorption on alumina.

Rice et. al. ([10,11]) related the morphology of corrosion observed on thin foils of metal to the presence of water clusters on the surface. Their calculations based on adsorption isotherm measurements and assuming a circular cap configuration of the clusters with a radius of 50 nm gave a cap height of 17.5 nm. Determination of the heat of adsorption for monolayer coverage of water on a cobalt surface from adsorption isotherms resulted in values very similar to the latent heat of vaporization of water. This is consistent with the formation of clusters rather than a uniform film of water.

Our view of the stages of adsorption of water is shown schematically in Figure 5. The first water forms an oxyhydroxide layer of the metal. Higher relative humidity results in the formation of water clusters, and still higher relative humidities result in a continuous moisture film.

3.3 Composition of the Aqueous Environment

Moisture condenses on the metal surface and provides the aqueous electrolyte for the electrochemical corrosion cell. Water clusters or a thin layer of moisture on the surface can effectively scavenge atmospheric pollutants thereby attaining high ionic concentrations and exhibiting increased conductivity. Under such conditions, local microcorrosion cells can be set up on the metal surface. The condensate provides an ionic conduction path between anodic and cathodic regions within the microcorrosion cell.

The chemistry of the electrolyte and its relationship with the nature and concentration of atmospheric gases is critically important for the corrosion behavior of metals. The corrosivity of the microcell environment is a controlling factor. Gaseous species which would be physisorbed on metal surfaces only in insignificant amounts are effectively captured by the condensate. Thus, a microcell environment quite different from the atmospheric environment is developed locally on the metal surface.

A theoretical development of condensate cluster chemistry has been attempted by several researchers. The model of Rice et. al. ([10]) calculates the concentration of species dissolved in the gaseous phase through the use of ionization constants, solubility of gases, and the partial pressures of gases in the atmosphere. A similar approach is used by Byrne and Miller [12].

Graedel [13] used Henry's law constants for computing the concentration of dissolved gases from their concentration in the atmosphere and their molecular density in an adsorbed water monolayer. Mikhailovskii et. al. [14] proposed a Langmuir type equation for determining the quantity of an adsorbed gas from its concentration in air.

All such models implicitly use thermodynamic equilibrium criteria of equating the chemical potentials of gaseous species in the vapor phase and in the aqueous phase. This tacitly assumes that the chemical equilibrium between the vapor phase and the condensate is attained within a time period shorter than the period of any adsorption-desorption cycles occurring on the metal surface. The changes in ionic concentration as a result of corrosion reactions and desorption have not been accounted for by the above models.

4. THE ROLE OF MICROCELL CORROSION

The emphasis of current research in our laboratory is on the development of a model for microcell corrosion. The processes contributing to this have been identified. Our view of the overall process is shown schematically in Figure 6. Water clusters form and soluble gases build up in the aqueous environment. The oxyhydroxide layer on the copper dissolves by either a concentration cell mechanism or by the bulk electrolyte moving into a region of instability for the layer. Local anodic and cathodic areas result and the electrochemical corrosion cell becomes operational. As this process continues, corrosion products are generated, reducible species are consumed, and metal dissolves. The electrolyte composition is affected by each of the above.

Multiple surface analysis techniques are used to investigate the process. The initial work examines the behavior of copper and silver foils in moist environments containing corrosive gases. Preliminary results for copper specimens are presented below.

Thin films of copper were exposed to a humid, corrosive environment and analyzed using Scannning Electron Microscopy (SEM), Auger Electron Spectroscopy (AES) and Transmission Electron Microscopy (TEM) in order to study the surface morphology, surface composition and microstructure during the early stages of corrosion. These films were prepared by vacuum-deposition of copper at 10^{-7} torr on TEM-grids with a carbon substrate. During deposition, foil thickness was controlled using a quartz crystal monitor and foils of 600 A° thickness were prepared.

One such copper foil was exposed to an atmosphere containing sulfur dioxide and 79% relative humidity at room temperature. This corrosive atmosphere was generated in a closed chamber by placing saturated sodium chloride solution at the bottom of the chamber and adding sodium hydrosulfite to the solution. The foil was placed on a glass slide in the environment above the solution within the chamber for 90 minutes. An unexposed control foil and the exposed foil were then analyzed.

4.1 SEM Study of Surface Morphology

The surface features of the foils were examined using a JEOL JSM 35CF Scanning Electron Microscope. Whereas the unexposed foil showed a clean, smooth surface at high magnification (30000 X), the exposed foil revealed a heterogeneous corrosion morphology. A uniform distribution of small black and white spots (~100 nm diameter) as shown in the micrograph in Figure 7 were found. Upon tilting the foil, it was observed that the white regions were protrusions from the surface and the black spots were pits.

4.2 AES Study of Surface Chemistry

Surface composition of the foils was studied using a Perkin-Elmer Phi 660 Scanning Auger Multiprobe. A preliminary Auger survey revealed the presence of copper and oxygen in the surface of the unexposed foil. Depth profiling was conducted using Argon sputtering at a rate of 2 A°/sec. The depth profile indicated the presence of an oxide layer of about 20 A° thickness.

The preliminary Auger survey of the exposed foil revealed the presence of copper, oxygen and sulfur in the surface layer. The Cu, O and S contents in relation to various topological features at 10000 X were mapped using three line-scans. The results of the Cu and O line scans are illustrated in Figures 8 and 9, respectively. The Cu line-scans indicated a relatively higher Cu content in the areas corresponding to the white outgrowths and a slight depletion near the dark pits in comparison with the average composition. Oxygen line-scans also showed an enrichment at the outgrowths but no significant change around the pits. Since the size and spacing of the topological corrosion features was of the same order as the smallest Auger probe diameter, the line scans do not indicate sharp changes of composition, but rather smooth transitions due to overlap of the boundaries by the beam.

The nominal surface composition corresponded to approximately Cu-23at% O - 9at% S. On depth profiling through a protrusion, the Cu and O signals remained unchanged to a depth of about 120 A° while the S signal decreased gradually with depth. Beyond a depth of 120 A°, the O signal showed a steady attenuation and the Cu signal showed a corresponding increase which indicated the existence of an interface at a depth of 120 A° under the protrusion. A depth profile through an adjoining flat area of the surface was similar to that through a protrusion except that the changes in O and Cu signals were more gradual in this case and indicated the presence of a uniform layer of corrosion products greater than 250 A° thick.

4.3 TEM Study of Microstructure

The structure of the unexposed and exposed foil

was examined in a Philips 400T Transmission Electron Microscope. The unexposed foil had an equiaxed, fine grained structure with a grain size of about 200 nm. An electron diffraction pattern taken from this foil showed rings corresponding to polycrystalline copper. Faint rings of Cuprite (Cu_2O) were also observed which indicated the presence of an air-formed oxide film on the foil.

The exposed Cu foil showed the orignal microstructure but now with a uniform dispersion of small black and white spots. The bright field image of the corroded foil is shown in Figure 10 at 10000X and in Figure 11 at 46000X. The size and distribution of the dark and white spots was analogous to that observed in the SEM image which suggested that the spots corresponded to similar features on the foil. Only in this case the contrast between pits and protrusions was reversed so that the former appeared white and the latter dark. The contrast reversal was expected as a result of the difference in the attenuation of the transmitted electron beam caused by these topological features.

Further work will use the above techniques to characterize the chemistry, structure and morphology of corrosion products and deposits. The results will form the basis for a model of microcell corrosion in the early stages of condensation.

5. CONCLUSIONS

Condensing conditions can result in thin layers of moisture on metals and severely corrosive exposures. Water in the early stages of condensation is likely to be distributed in the form of tiny clusters on the metal surface. The early stages of corrosion involve microcell corrosion processes which control the rate and distribution of corrosion.

Service applications where it is critical to control microcell corrosion processes include the use of corrosion resistant alloys in water containing vapor phase, alternate wet/dry cycles, and where condensation occurs on metal surfaces. These processes are also central to the prediction and control of indoor atmospheric corrosion of microelectronic, computer and communication devices. The application of multiple surface analysis techniques provides an effective means to investigate the microcell corrosion processes. Increased understanding of these corrosion processes will lead to improved corrosion prevention designs and strategies.

6. REFERENCES

1. Vernon, W.H.J. Trans. Far. Soc. 27, 255 (1931).

2. Mansfeld F., Kenkel J.V. Corros. Sci. 16, 111 (1976).

3. Mansfeld F., Kenkel J.V. Corrosion 33 no.1, 13 (1977).

4. "Tomashov N.D., Theory of Corrosion and Protection of Metals, The Macmillan Co., 1966."

5. Woelker G. Z. Meteorologie 13 no.30 (1972).

6. "Mansfeld F., New Approaches to Atmospheric Corrosion Research using Electrochemical Techniques" in "Corrosion Processes".

7. Cudnik, R.A. et al ASHRAE Trans. 91 pt. 2 (1985).

8. Klier K., Shen J.H., Zettlemoyer A.C. J. Phys. Chem. 77 no.11, 1458 (1973).

9. Yan B.D., Meilink S.L., Warren G.W., Wynblatt P. Proc. 36 Conf. on Electronic Components, IEEE, 1986,pp 95-99.

10. Rice D.W., Phipps P.B.P., Tremoureoux R. J. Electrochem. Soc. 126 no. 9, 721 (1979).

11. Rice D.W., Phipps P.B.P., Tremoureoux R. J. Electrochem. Soc. 127 no. 3, 563 (1980).

12. "Byrne S.C., Miller A.C., Atm. Corr. of Metals, ASTM STP 767, eds. Dean S.W., Rhea E.C., ASTM, 1982, pp 359-373.

13. Graedel T.E. Corros. Sci., 27 no.7, 721 (1987).

14. Mikhailovskii Yu. N., Strekalov P.V., Agfanov V.V. Zashch. Met. 16 no. 4, 396 (1980).

8. BIOGRAPHIES

Joe H. Payer is Professor of Materials Science at Case Western Reserve University. Dr. Payer applies corrosion science and technology to understand and solve corrosion problems. Current interests include corrosion in thin films, adhesion and durability of polymer/metal interfaces, development of corrosion sensors, reliability of microelectronic and electronic devices, localized corrosion processes, and the breakdown and repassivation of protective films. He received his B.S. and PhD in Metallurgical Engineering at Ohio State University. He is Past President of the National Association of Corrosion Engineers, is a Fellow of the American Society of Metals, and received the ASTM Sam Tour Award for distinguished contribution to the development of testing methods.

Sandeep K. Chawla is a graduate student in Materials Science and Engineering at Case Western Reserve University. He received his MS degree in that Department in May 1987 and is now a PhD candidate. Mr. Chawla received his B.Tech. in Metallurgical Engineering at the Institute of Technology, Banaras Hindu University in Varanasi, India. He received the University Medal in 1982 and the Vidya Bharati Award from the Indian Institute of Metals for securing the highest marks in Metallurgical Engineering in 1982. In 1988, he was a student poster award winner at the Annual Conference of the National Association of Corrosion Engineers.

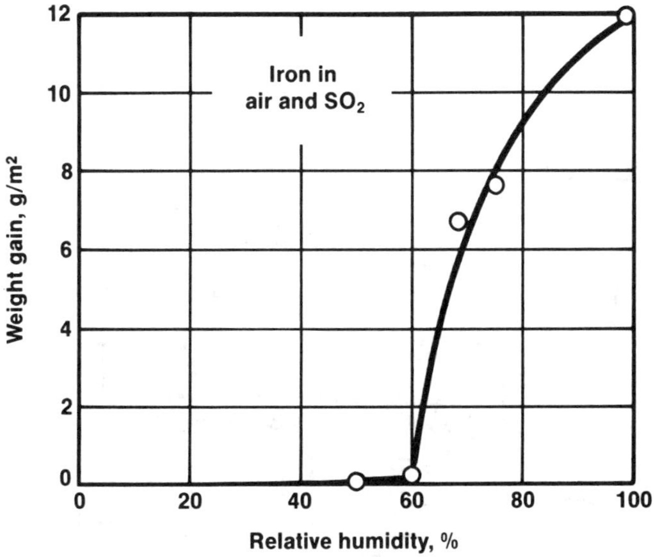

Figure 1 : Corrosion rate of steel as a function of relative humidity; after Vernon (1).

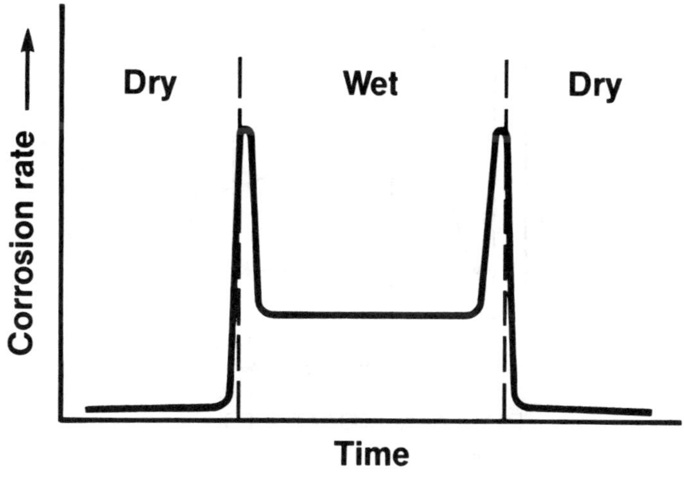

Figure 2 : Variation of corrosion rate as a function of time during a wetting-drying cycle.

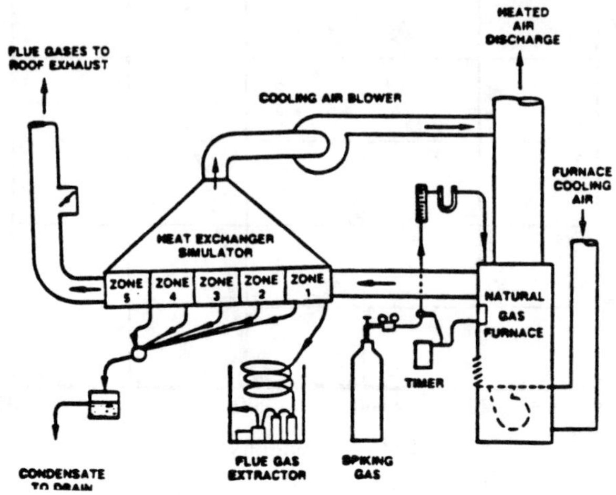

Figure 3 : Schematic of a condensing heat exchanger simulator; after Cudnik et al (7).

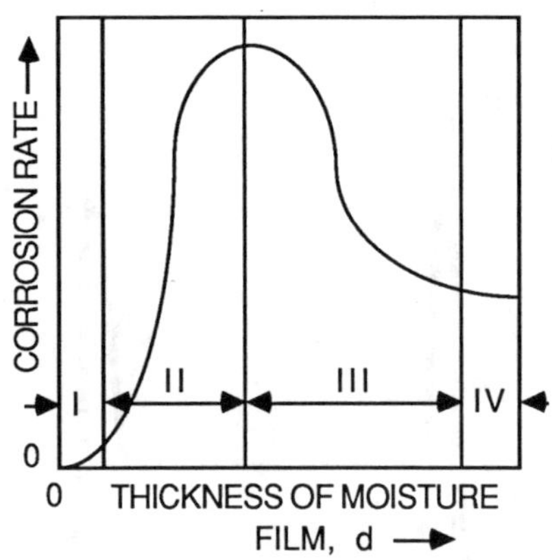

FILM THICKNESS IN THE FOUR REGIMES :

I : d = 10 - 100 A
II : d = 100 A - 1 micron
III : d = 1micron - 1 mm
IV : d > 1 mm

Figure 4 : The variation of atmospheric corrosion rate with thickness of moisture film on the metal surface; after Tomashov (4).

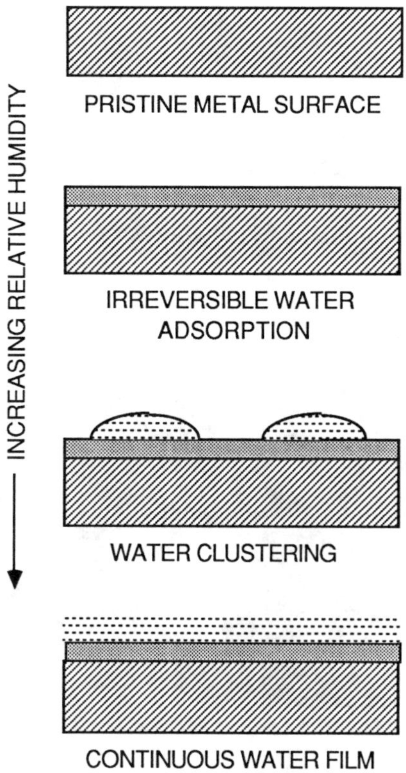

Figure 5 : Stages of water adsorption on a metal surface.

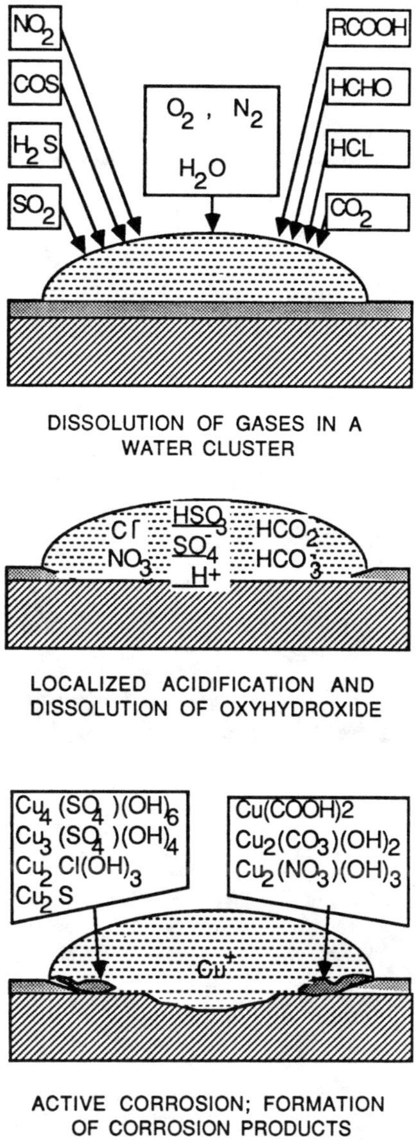

Figure 6 : Overall view of corrosion under water clusters; after Graedel (13).

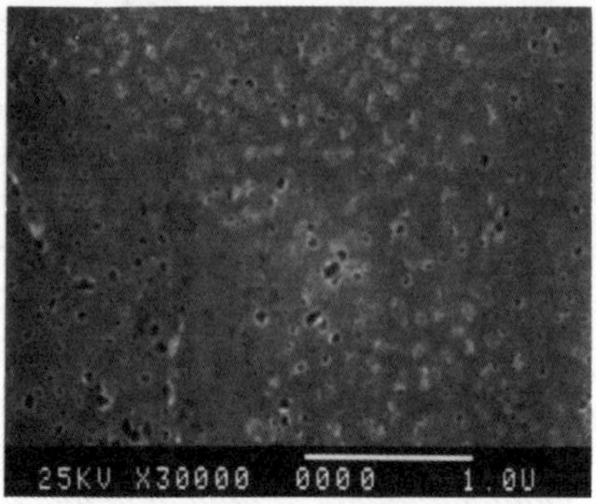

Figure 7 : Scanning electron micrograph of the copper foil exposed to 79% RH and SO_2.

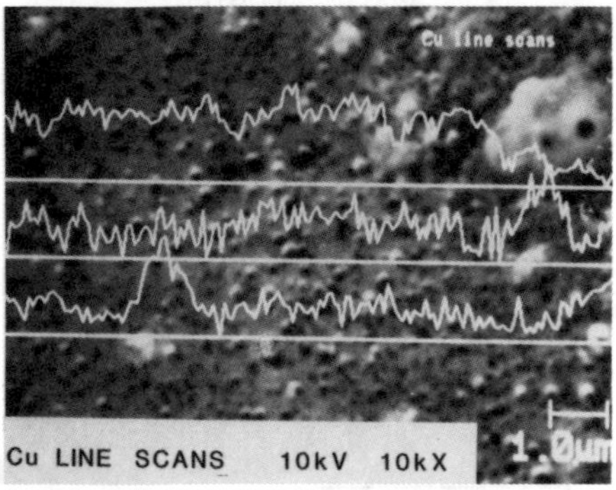

Figure 8 : Auger mapping (line scan) of the copper content of topological features on the exposed foil.

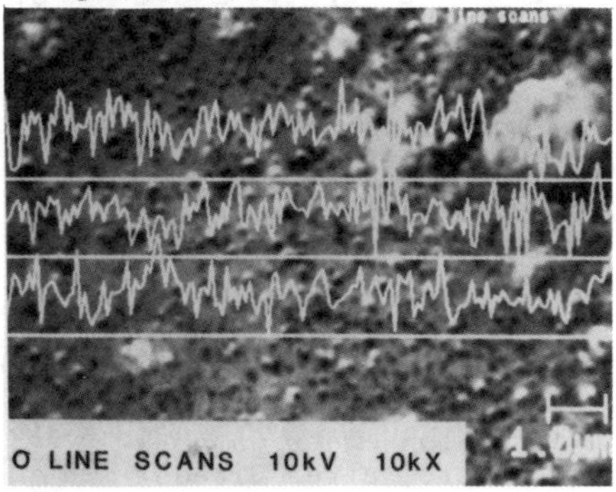

Figure 9 : Auger mapping (line scans) of the oxygen content of topological features on the exposed foil.

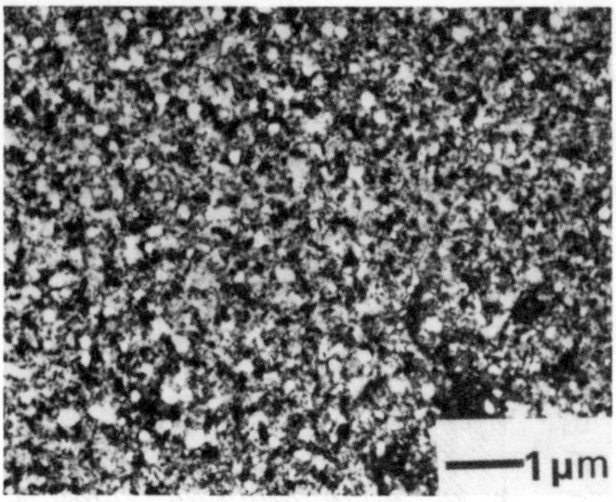

Figure 10 : Transmission electron micrograph of the copper foil exposed to 79% RH and SO_2; X10000.

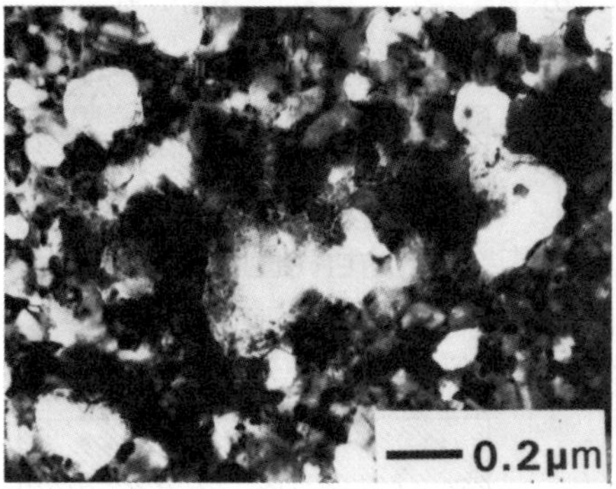

Figure 11 : Transmission electron micrograph of the copper foil exposed to 79% RH and SO_2; X46000.

2nd International SAMPE Metals Conference
August 2-4, 1988

OXIDATION OF ION-IMPLANTED NICKEL-BASE ALLOYS

V. Srinivasan
Universal Energy Systems, Inc.
4401 Dayton-Xenia Road
Dayton, OH 45432

Abstract

Isothermal and cyclic oxidation tests were performed on two Ni-base alloys before and after surface modification by ion-implantation. The species implanted directly or introduced by ion-beam mixing were Y, Ce and Pt. The scales and the substrates were characterized by scanning electron microscope, energy dispersive analysis by X-rays and Rutherford backscattering spectroscopy. The Y-implantation produced significant increase in scale adherence under cyclic oxidation conditions. Other effects of implantation and the experimental details are discussed in this paper.

1. INTRODUCTION

Protective α-alumina or chromia scales that form on high temperature alloys and coatings are poorly adherent, and more often they spall and crack because of their negligible ductility and thermal mismatch and growth stresses. Minor additions of oxygen active elements, dispersoids and certain other elements improve significantly the scale adherence and integrity (1). With negligible solid solubility some elemental additions are likely to lead to matrix and grain boundary segregations and constraints on fabricability. As these elemental additions are required only in the near surface regions, ion beam techniques provide the flexibility of selectively enriching the surface with beneficial elements. By such selective surface modifications the oxidation behavior of the substrate may be altered temporarily till the implanted regions are consumed or with enduring influence. Both types of behavior modifications have been reported (2,3). Y-implantation improved the long term oxidiation resistance of Fe-20Cr-25Ni/Nb stabilized stainless steel (2). However, Bennett, Houlton, and Dearnaley reported no improvement in the oxidation resistance of Y- implanted Fe-15Cr-4Al. The authors attributed this observations to the loss of Y-implants in the subsurface regions along the spalled alumina scales (3). Therefore, any generalization with regard to the effect of ion-implanted species on the oxidation behavior of alloys requires further study. Investigation was carried out to understand the oxidation response of ion-implanted nickel-base alloys, and the results of this investigation are presented in this paper.

2. EXPERIMENTAL MATERIALS AND METHOD

IN939 and single crystal RenéN4 were the experimental alloys. Their nominal chemical compositions in wt.% are given in Table 1.

Coupons of approximate size 1cmx1cmx0.1cm were cut from the heat-treated blocks, and a hole of diameter about 0.1cm was drilled near one of the edges of the coupon for suspension from the hang-down of the Cahn balance. They were metallographically polished upto 600 grit. Coupons for kinetics study were implanted both sides with Y, Pt or Ce at ion energies 2 Mev, 1 Mev, or 180 Kev. The fluences were between 10^{14} and 10^{16} ions/sq.cm.

Isothermal oxidation tests were performed in Cahn microbalance at 950° or 1000°C with a continuous flow of air or oxygen at 1 atm. Weight gain was recorded continuously in a strip chart recorder. Cyclic tests were done in a custom-made unit. The sample was periodically introduced and withdrawn from the furnace. A typical cyclic period consisted of about 45 min. exposure to the test temperature and 15 min. in laboratory atmosphere. Samples were removed at regular intervals during cyclic oxidation test for weight change measurements.

After isothermal or cyclic exposure the samples were examined in scanning electron microscope provided with energy dispersive analysis by X-rays (EDAX).

Table 1.
Nominal Compositions in wt%

Alloy	Cr	Mo	Al	Ti	W	Co	Ta	Cb	Zr	Cu	C	Ni
IN939	22.5	0.05	1.96	3.69	2.06	19.1	1.4	1.0	0.14	0.06	0.06	Bal.
RenéN4	9.25	1.5	3.7	4.5	6.0	7.0	4.0	0.5	--	--	--	Bal.

3. RESULTS AND DISCUSSION

Typical weight gain data at 950°C for unimplanted and implanted IN939 are shown in Figure 1.

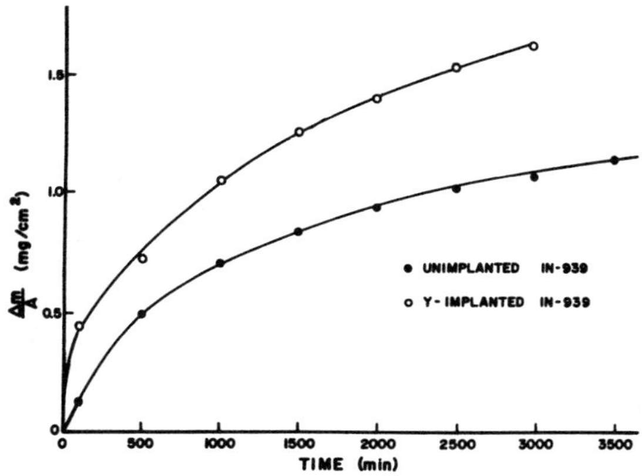

Figure 1. Typical weight gain data for unimplanted and Y-implanted IN939 alloy during the isothermal oxidation test at 950°C in flowing air at 1 atm.

Here the implanted species was yttrium. The implanted alloy gained more weight at a given time than the unimplanted one. Similar larger weight gains were observed in the cases of Ce- and Pt- implanted IN939. Weight gains by the unimplanted and Y- and Pt- implanted RenéN4 at a given time were not meaningfully different, and appear to fall within the experimental scatter. Faster oxidation kinetics observed in the implanted IN939 differs from that of Pivin et al. on implanted Ni-33Cr and Ni-20Cr binary alloys (4). Y and Ce implantation reduced the oxidation kinetics in Ni-33Cr and Ni-20Cr as their bulk additions to Ni-Cr did (4,5). At percent the reasons for the faster kinetics in the implanted IN939 are not clear, and further work is necessary to understand the mechanism of oxidation of implanted IN939.

Figure 2 shows weight change data obtained from the cyclic tests on unimplanted and implanted RenéN4 single crystals. Weight gains in the case of Y-implanted RenéN4 went through a maximum initially and leveled off after about 200 cycles (hours) at a positive value. On the other hand the unimplanted RenéN4 gained more weight initially but spallation later on took over resulting in weight loss.

Both alloys, IN939 and RenéN4 single crystals form multilayered oxides scales. Figures 3a - 3d show scanning electron micrographs of scales observed in

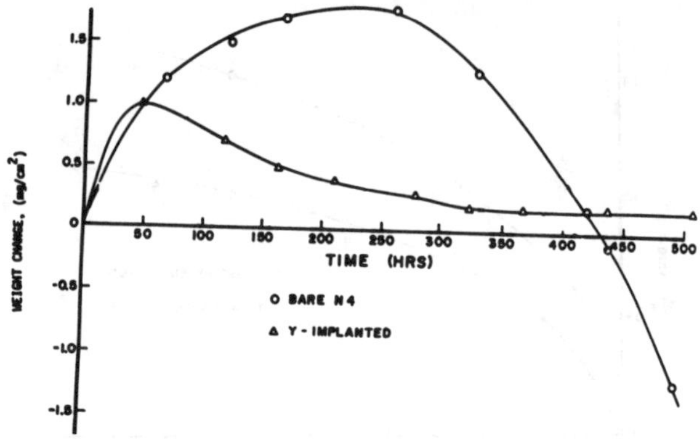

Figure 2. Weight change data for unimplanted and Y-implanted RenéN4 single crystal during the cyclic oxidation test between 1000°C and room temperature; 45 min. at 1000°C and 15 min. in the laboratory air cyclic period, 1 hr.

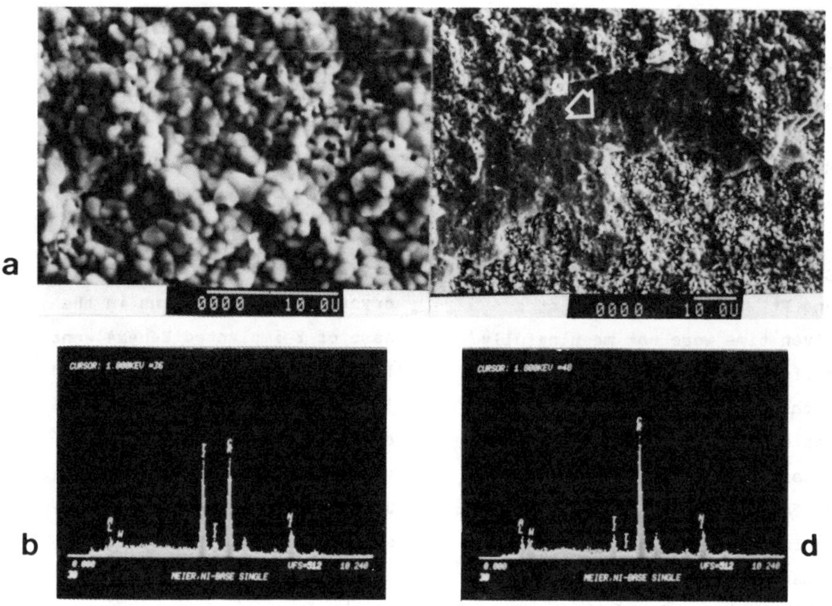

Figures 3a - 3d. Scanning electron micrographs and EDAX spectra respectively of continuous scale (a,b) and the spalled area (c,d) observed on unimplanted RenéN4 oxidized at 1000°C for 150 hrs. in flowing air at 1 atm.

RenéN4. While chromia was the major constituents at the scale/matrix interface in IN939, alumina with refractory elements such as Ta was observed at the scale/matrix interface in RenéN4. In the scale implants could not be detected with EDAX. The interface at the scale/matrix was smoother in isothermally oxidized Y-implanted RenéN4 than that in unimplanted RenéN4, Figures 4a and 4b. Similar smooth interface was observed in cyclically oxidized coupons also. This smooth interface is likely to be a microstructural evidence for less or insignificant scale spallation during cyclic oxidation as deduced from the kinetics data.

CONCLUSION

Unimplanted and implanted IN939 and RenéN4 formed multilayered oxide scales under isothermal and cyclic oxidation conditions. Implantation did not reduce the weight gain in IN939 at a given time during isothermal oxidation. Scale adherence and hence the cyclic oxidation resistance was considerably improved in Y-implanted RenéN4 as revealed by the kinetic data and the smooth scale/matrix interface.

ACKNOWLEDGEMENT

The author thanks Howmet Corporation and General Electric Company for experimental alloys, Mr. A.W. McCormick for performing implantation, Mr. A. Smith for experimental assistance and Prof.

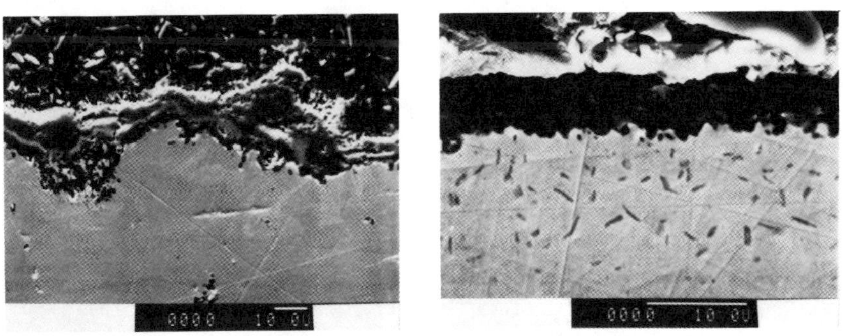

a b

Figures 4a - 4b. Scanning electron micrographs of cross-sections of unimplanted (a) and Y-implanted (b) RenéN4 oxidized at 1000°C for 150 hrs. in flowing air.

G.H. Meier for valuable discussion. DoE and NSF supported the work under their SBIR programs, contract no. DE-AC01-83ER80044 and DMR 8460741.

REFERENCES

1. Whittle, D.P., and Stringer, J., Phil. Trans, Roy, Soc. London A295, 309(1980).
2. Bennet, M.J., Dearnaley, G., Houlton, M.R.G., Hawes, R.W.M., Goode, P.D., and Wilkins, M.A., Corrosion Science, 20, 73(1980).
3. Bennet, M.J., Houlton, M.R., and Dearnaley, G., Corrosion Science 20, 69(1980).
4. Pivin, J.C., Roques-Carmes, C., Chawmont, J., and Bernas, H., Corrosion Science 20 947(1980).
5. Ecer, G.M., and Meier, G.H., Oxidation of Metals, 13, 159(1979).

2nd International SAMPE Metals Conference
August 2-4, 1988

STRESS INDUCED HOT CORROSION OF RENE' 80

S. J. Balsone[1] and M. Khobaib[2]

[1]Air Force Wright Aeronautical Laboratories,
Materials Laboratory, AFWAL/MLLN, Wright-Patterson AFB, Ohio 45433-6533

[2]University of Dayton Research Institute, Dayton, Ohio 45469

Abstract

Stress has been known to play a significant role in environmental degradation of materials. The influence of stress on environmental degradation was studied by conducting sustained-load creep tests on smooth round bar tensile specimens of a representative Ni-base superalloy, Rene' 80. The test temperature used was 900°C, and experiments were conducted in laboratory air and a salt mixture of 90% Na_2SO_4 and 10% NaCl. A few selected tests were conducted under sustained load with periodic superimposed cyclic loading to investigate the effect of stress cycling on oxide cracking. All tests were conducted to failure, and specimens were sectioned below the necked region of the failure. The cross section was then metallographically analyzed under SEM.

1. INTRODUCTION

The hot corrosion resistance and elevated temperature mechanical behavior are the two most important criteria for selecting materials for certain aircraft turbine engine structural components. As a common practice, environmental testing such as laboratory hot corrosion tests, burner rig tests, oxidation tests, and mechanical property testing such as creep, fatigue, and thermal mechanical fatigue tests are conducted and evaluated independently. Information from such investigations are used for design purposes without considering the synergistic effects of stress and environment in degrading material properties.

In the last decade, experimental investigations have indicated that unstressed high temperature oxidation and hot corrosion tests may not be adequate in predicting the corrosion behavior of a material in actual service. Floreen and Kane [1] reported that the environments which markedly increase fatigue crack growth rates in a nickel-base superalloy did not produce significant corrosion attack or degradation of material in unstressed specimens. Even in some of the very aggressive sulfur-containing environments which are known to accelerate cracking, unstressed samples showed very little general corrosion damage or visible surface attack.

Whitlow [2] observed a creep threshold stress level above which the rupture life of a nickel-base superalloy diminished significantly in the presence of an environment of molten salt. The synergistic effect of stress and environment was more dramatically demonstrated by Allen and Whitlow [3] who noted that superalloys in combustion turbine environments could experience high-frequency fatigue failure if design values of vibratory and maximum steady stress were based on mechanical property data obtained in air. At high mean stresses (in the presence of combustion gases), fatigue failures occurred in orders of magnitude fewer cycles than for corresponding loading conditions in air. In related studies [4,5], the presence of molten salt resulted in more than one order of magnitude reduction in the low-cycle fatigue life of a superalloy. Toshiba, et. al, [5] showed that for salt coated specimens, stress rupture life was drastically reduced. These results are in agreement with recent observations [6,7] on the influence of environment on the creep behavior of nickel-base superalloys. These investigations clearly demonstrate the importance of understanding the synergistic effect of environment and stress on material degradation and the potential problems that may be encountered in designing structural components from laboratory hot corrosion/oxidation data obtained from unstressed samples.

This investigation examines the role stress plays in the hot corrosion/oxidation process and deals with the microstructural changes as a result of environment-mechanical property interactions.

2. MATERIAL

Smooth round bar creep specimens as shown in Fig. 1 were machined from cast bar stock of Rene' 80. The composition of this alloy is given in Table 1. Table 2 describes the standard heat treatment given to the material for optimum stress-rupture, creep, and fatigue properties.

A continuous and smooth thin coating of a mixture of 90% Na_2SO_4 + 10% NaCl on the specimen was used as the hot salt environment.

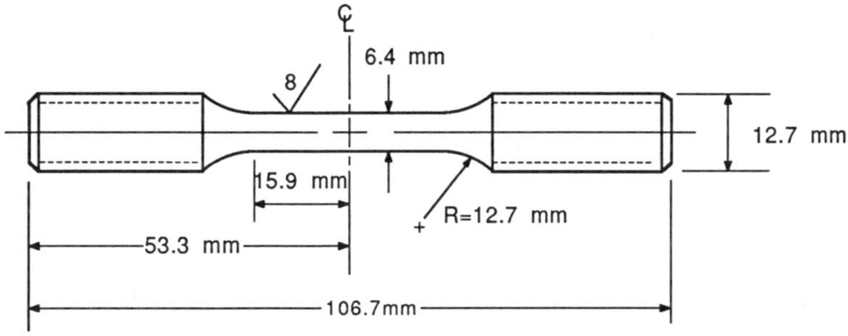

Fig. 1. Round bar tensile creep specimen.

TABLE 1. COMPOSITION OF RENE' 80 (WT %)

Ni	Cr	Co	Mo	Ti	Al	C	Other
Bal	14.8	9.8	3.8	5.4	3.2	0.17	7.4W

TABLE 2. HEAT TREATMENT FOR RENE' 80

(a) 1204°C(2200°F)/2 Hrs/Vac/FC
(b) 1093°C(2000°F)/4 Hrs/Vac/FC
(c) 1052°C(1925°F)/4 Hrs/Vac/FC
(d) 843°C(1550°F)/16 Hrs/Vac/FC

3. EXPERIMENTAL

The effect of stress on hot corrosion behavior was investigated by studying the corrosion attack due to a coating of 90% Na_2SO_4 + 10% NaCl on stressed samples, while the oxidation studies were conducted in laboratory air. Sustained-load creep tests provide a simple condition where the desired environment is easily provided along with selected levels of stress. The effect of cyclic stress was also studied on a limited scale.

All sustained- and cyclic-load tests were performed on Arcweld creep frames equipped with a tubular resistance furnace. The specimen temperature was controlled through a thermocouple mounted adjacent to the midpoint of the gage length of the specimen. Additional thermocouples were used to monitor the temperature distribution over the gage length of the specimen. All tests were conducted at 900°C (1652°F), and a temperature control of nearly ± 2°C was obtained along the gage length of the specimen. The creep strain was determined from the specimen elongation which was recorded continuously through a specially designed extensometer-LVDT (linear variable differential transducer) [6]. The unstressed samples were tested in a box furnace with a temperature control of ± 2°C.

The desired hot corrosion environment was provided by depositing a thin layer of 90% Na_2SO_4 + 10% NaCl (by weight) on mechanically polished and cleaned specimens. A continuous coating of approximately 5.0 mg/cm^2 of salt throughout the gage section was obtained in each case by a repeated spraying and heating technique [6].

Sustained-load creep tests were conducted at stress levels of 206.7 MPa (30 ksi), 241.2 MPa (35 ksi), 275.6 MPa (40 ksi), and 310.1 MPa (45 ksi). Most of the tests were run to specimen failure. Some tests were interrupted prior to failure after specific times to compare damage at a fixed time interval. In the case of unstressed samples, all the specimens were exposed for a specific time. In addition, sustained-load tests with a periodic superimposed 0.25 Hz (4 second period) unloading/reloading cycle were conducted at a stress level of 275.6 MPa to investigate the effect of cyclic stress. The time interval between these 0.25 Hz fatigue cycles ranged from zero (pure fatigue with mean load) to 900 seconds. After all tests, the specimens were sectioned at the midpoint of the gage section or directly below the necked region of the failure to determine the depth of corrosion attack. The environmental degradation was analyzed using a scanning electron microscope (SEM) and X-ray energy spectroscopy (XES).

Depth of the environmental attack was measured as shown in Fig. 2. The amount of sound metal along each of twelve diameters, d_i, 15° apart in the cross-section, was measured. The average depth of penetration (DOP) was calculated as:

$$DOP = \frac{1}{2}\left[d_o - \frac{\sum_{i=1}^{12} d_i}{12} \right] \quad (1)$$

where d_i represents the individual measurements, and d_o is the initial specimen diameter.

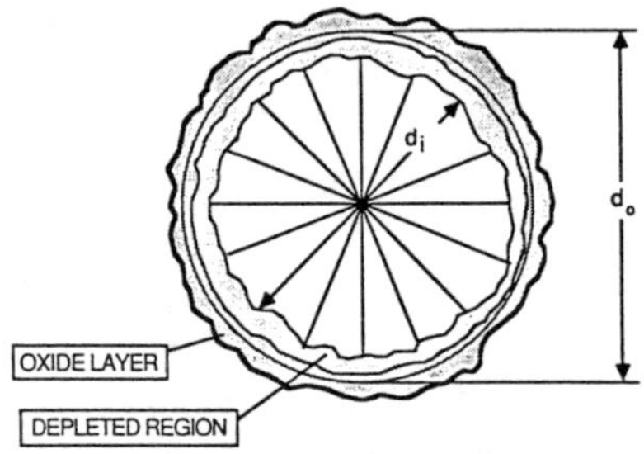

Fig. 2 - Metallographic technique for quantifying the depth of environmental attack.

RESULTS AND DISCUSSION

Hot Corrosion/Oxidation Tests

The results of the hot corrosion tests conducted on 3 mm diameter pins coated with 90% Na_2SO_4 + 10% NaCl and oxidation experiments in laboratory air at 900°C are given in Table 3. It is clear that the hot corrosion produces a higher degree of environmental degradation due to the combined action of S and O_2. Fig. 3(a,b) shows the micrographs taken from such samples. For both environments, the corrosion attack can be divided into three zones as

TABLE 3. DEPTH OF CORROSION ATTACK ON SPECIMENS TESTED UNDER NO STRESS

Time (Hrs.)	Depth of Penetration (μm) Laboratory Air	90% Na_2SO_4/10% NaCl
100	13.67	16.65
200	19.33	23.55
300	23.68	28.84
400	27.34	33.30
500	30.56	37.23

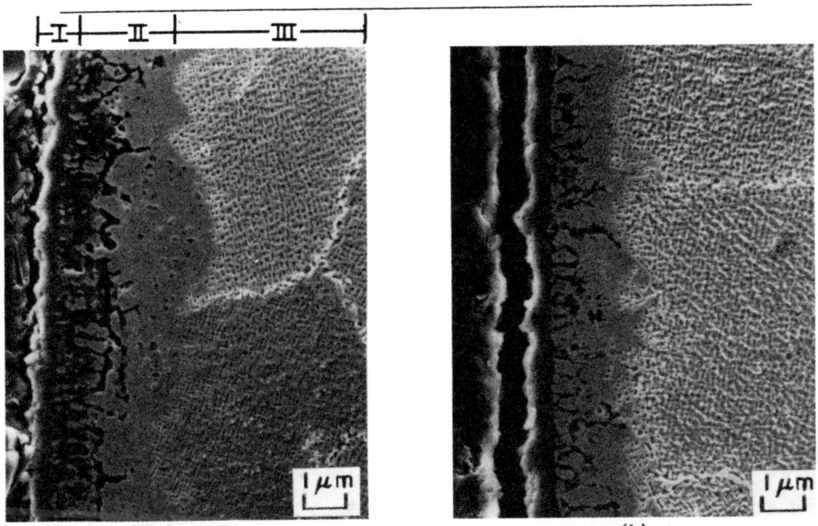

(a) (b)

Fig. 3 - Scanning electron micrographs (1000X) showing hot corrosion attack coated with (a) 90% Na_2SO_4 + 10% NaCl, and (b) no salt for 72 hours.

shown in Fig. 3. Zone I is a tight, adherent oxide layer. There was no evidence of this oxide layer being stripped away from the surface of the specimens even during sectioning for metallographic analysis. Zone II is directly beneath this oxide layer and is a region of metal depleted of some of its alloying elements. Zone II also contains sulfide particles for the salt coated specimen. Zone III is the base metal which was not affected by corrosion attack. This zone contains a fine distribution of the precipitation strengthening gamma-prime phase. Note that for the uncoated specimens, there is little oxide penetration into the depleted zone of metal. However, in the salt-coated pin, there was localized penetration of oxide into the depleted zone. The depleted zone was slightly smaller in the uncoated pins than in the salt-coated pins, and as expected, there were no sulfide particles present in Zone III of the uncoated pins. There was no evidence of cracking or mechanical damage to the surface oxide layer and no subsurface cracking along oxide-metal interfaces.

Hot Corrosion/Oxidation Under Sustained Load

The effect of 90% Na_2SO_4 + 10% NaCl on the sustained-load creep behavior of Rene' 80 is shown in Fig. 4. The laboratory air data are also plotted for comparison. The reduction in

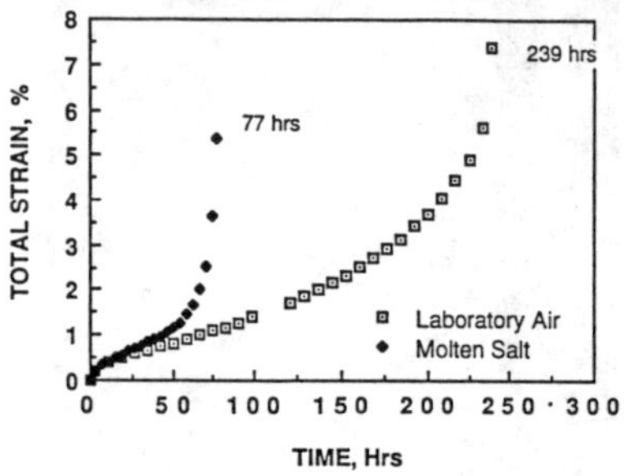

Fig. 4 - Creep behavior of Rene' 80 at 275.6 MPa and 900°C.

rupture life due to the presence of molten salt is quite clear. A reduction in strain to failure is also observed, indicating some degree of embrittlement of the material by molten salt. In the initial stages, the creep behavior is noticed to be similar in both environments until the more aggressive environmental attack of the molten salt causes a more abrupt onset of tertiary creep. The secondary stage creep under the influence of molten salt is not well defined in many cases, but after detailed analysis, the steady-state creep rate was found to be greater in the molten salt environment [6].

The rupture lives in molten salt environment were reduced by almost a factor or two as compared to those obtained in laboratory air. The rupture lives are plotted in Fig. 5 as a function of stress level and are found to obey a simple power-law relationship

$$T_{RUP} = A\sigma^{-P} \quad (2)$$

where A is a constant and P is the power-law exponent.

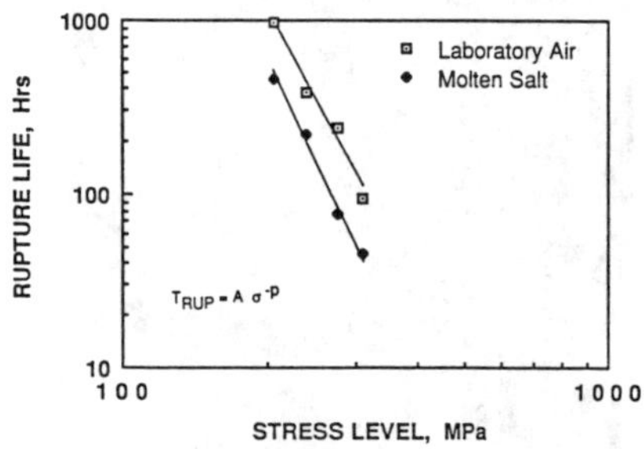

Fig. 5 - Stress rupture life of Rene' 80 at 900°C.

Depth of Environmental Attack

All the specimens were analyzed through a detailed metallographic examination after each test. Table 3 gives an indication of the depth of attack for specimens tested in laboratory air under no stress while Fig. 3 shows the details of the corrosion affected zone.

The severity of the attack is seen to increase when the specimen is tested under sustained load. Fig. 6 shows the depth of penetration of the environmental attack by laboratory air as a function of time at various stress levels. Fig. 7 is a similar plot

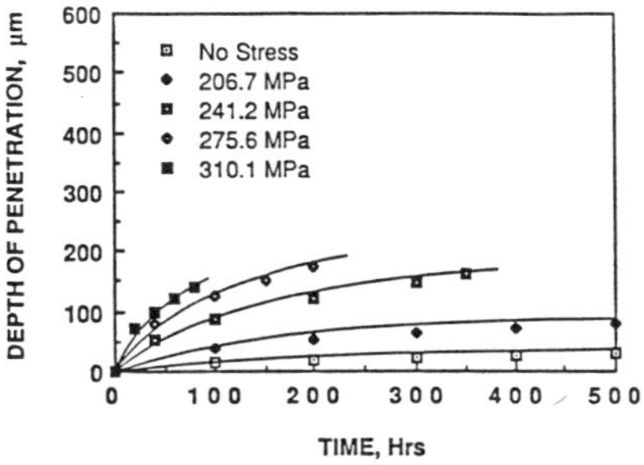

Fig. 6 - Depth of environmental attack by oxidation at 900°C.

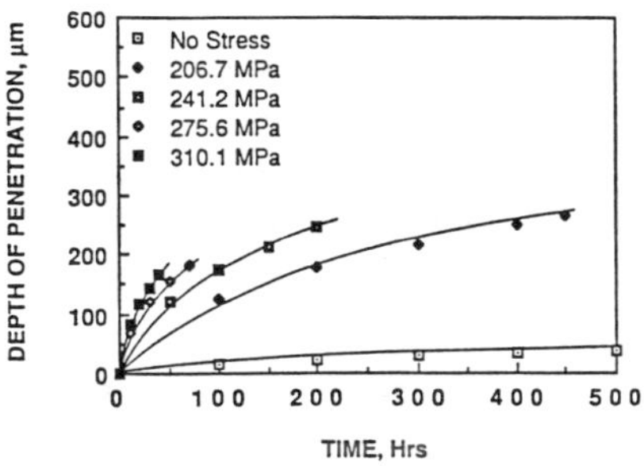

Fig. 7 - Depth of environmental attack by hot corrosion at 900°C.

for tests conducted on specimens coated with 90% Na_2SO_4 + 10% NaCl. The depth of penetration values were obtained through metallographic analysis of specimens tested in sustained-load creep for fixed times shorter than the rupture life at the particular stress level. The results clearly show that environmental attack is significantly greater under an applied stress than in the unstressed condition. Again, the depth of environmental attack at any fixed time is found to increase as the stress level increases. Note also that this effect increases with longer exposure times. The depth of penetration (DOP) data in both Figs. 6 and 7 has been found to follow a simple parabolic relationship [8]

$$DOP = \sqrt{Dt} \qquad (3)$$

where D is a parabolic rate constant and t is time.

The environmental degradation is further accelerated when the specimen is coated with 90% Na_2SO_4 + 10% NaCl. The hot corrosion attack by the molten salt film is illustrated in Fig. 8a. In this case, the outermost oxide layer is much thicker than that formed in the laboratory air, as shown in Fig. 8b, and it penetrates into the alloy in a "finger-like" fashion. Above all,

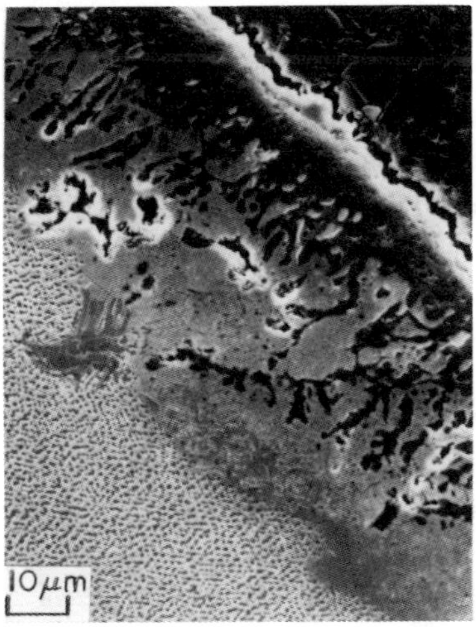

Fig. 8a - Typical stress assisted environmental degradation of Rene' 80 at 900°C - (a) hot corrosion with 90% Na_2SO_4 + 10% NaCl.

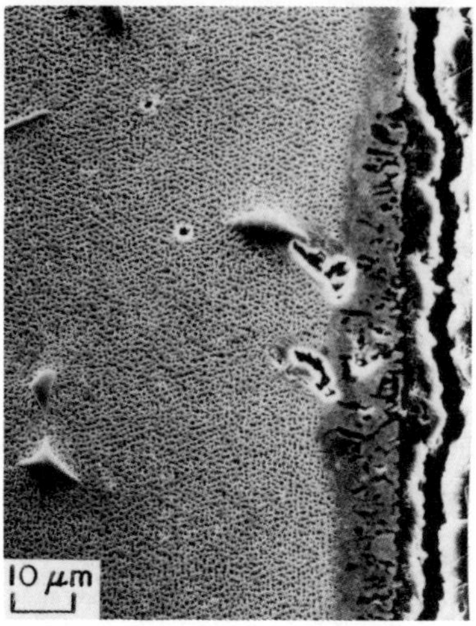

Fig. 8b - Typical stress assisted environmental degradation of Rene' 80 at 900°C - (b) no salt.

the extent of the alloy depleted zone and the depth of environmental attack are much greater for specimens tested under sustained load in molten salt environment. The depth of environmental attack remained nearly constant in the stressed specimen until a critical stress level was reached. A synergistic effect of stress was noticed above this critical stress, and the environmental degradation increased significantly with stress level as shown in Fig. 9. However, there was only a little difference in the microstructural details of unstressed and stressed specimens. In the case of molten salt environment, the surface layer in either case was found to be rich in chromium and contained significantly more titanium and aluminum than in the oxide layer formed during tests in laboratory air. The relatively

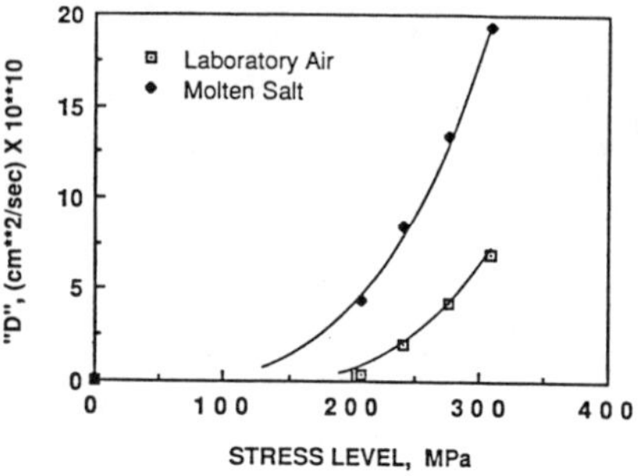

Fig. 9 - Parabolic rate constant, D, for environmental attack at 900°C.

thick surface oxide layer formed in the molten salt environment was very porous and non-adherent. In many cases, its exact thickness was difficult to determine since some of the oxide was often stripped away during metallographic preparation. The depth of environmental attack was relatively uniform around the circumference of the specimen, and only a few instances of "spikes" caused by environmental attack down grain boundaries were observed as reported by other investigators [1,4,5].

These results clearly suggest that the effect of stress level on the magnitude of the environmental attack is more pronounced in the molten salt film environment, and obviously the depths of environmental attack are greater in the molten salt environment than in laboratory air at any fixed exposure time.

Effect of Cyclic Stress

The effect of cyclic stress on the magnitude of environmental attack was investigated by conducting sustained-load tests with a superimposed periodic unloading/ reloading fatigue cycle. Fig. 10 shows the waveform used in the experiment consisting of the sawtooth fatigue cycle and a sustained-load hold-time. The

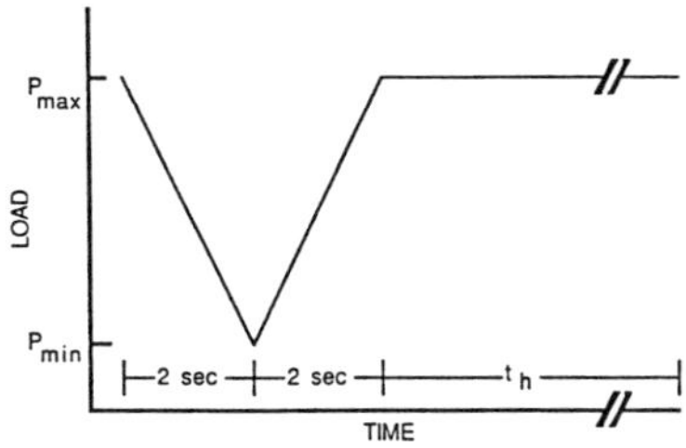

Fig. 10 - Waveform of periodic fatigue cycle; $P_{min} = 0.15\ P_{max}$.

effect of cycling on rupture life in both environments is shown in Table 4. The superimposed cyclic

Examination of the failed specimens after tests in the molten salt environment showed extensive cracking and spallation on the specimen surface. In this case, the surface oxide layer was very fragile and non-adherent and crumbled even during careful removal of the specimen from the test frame. This cyclic-stress-assisted environmental degradation led to a greater depth of penetration and an associated reduction in rupture life. Fig. 11 summarizes the effect of unloading/reloading on the depth of environmental attack. Note that the depth of penetration in laboratory air remained essentially constant as the

TABLE 4. STRESS RUPTURE LIFE OF RENE' 80 AT 900°C AND 275.6 MPa WITH SUPERIMPOSED PERIODIC UNLOADING CYCLE, HRS.

	Environment	
Hold Time, Sec.	Laboratory Air	90% Na_2SO_4/10% NaCl
0	294	[Not Tested]
90	289	28
900	281	37
No Cycling	239	77

stress was found to have little effect for tests conducted in laboratory air. The metallographic analysis of the failed specimen showed a continuous adherent surface oxide layer with no visible mechanical damage, similar to that observed for sustained-load specimens. However, the cyclic stress had a noticable effect on the rupture lives of specimens coated with 90% Na_2SO_4+10% NaCl.

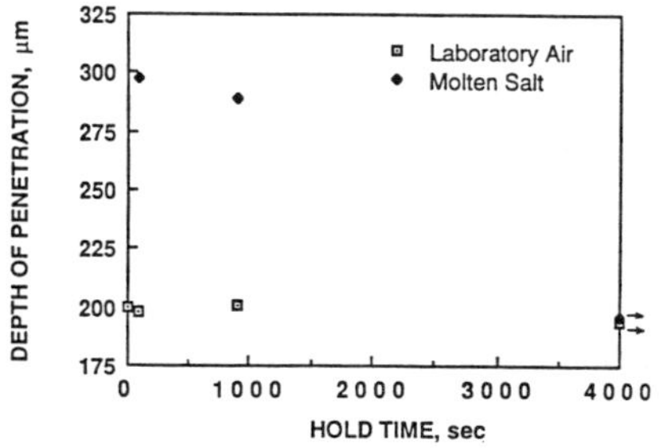

Fig. 11 - Effect of periodic fatigue cycle on the depth of environmental attack at 900°C.

sustained-load hold-time was reduced. However, in the molten salt environment, reducing the sustained-load hold-time (increasing the number of fatigue cycles) caused an increase in the depth of penetration of the environmental attack. Apparently, the synergistic effect of cyclic stress is more pronounced with aggressive environment.

SUMMARY AND CONCLUDING REMARKS

1. The depth of penetration of environmental attack in both laboratory air and molten salt film was found to be signficantly greater under an applied stress than in the unstressed condition. The depth of penetration was observed to increase as the stress level increased from 207 MPa to 310 MPa.

2. In the molten salt film environment, superimposed fatigue cycling led to signficant reduction in rupture lives. The spallation of the surface oxide layer was extensive, and the depths of hot corrosion attack were greater than in the case without the periodic fatigue cycle. However, in laboratory air, a periodic unloading/reloading cycle superimposed on the sustained-load creep had no effect on the depths of environmental attack or rupture life. In that case, the surface oxide layer remained adherent with no visible cracking.

3. The depth of environmental attack by the 90% Na_2SO_4 + 10% NaCl molten salt was significantly greater than the depth of environmental attack in laboratory air in both stressed and unstressed conditions. Rupture lives of Rene' 80 specimens tested in molten salt environment were nearly half as compared to the lives obtained in laboratory air.

ACKNOWLEDGEMENTS

This work was conducted at the Metals Behavior Branch, Materials Laboratory, under in-house project 2302P101. The support of the U.S. Air Force is gratefully acknowledged. Special thanks are extended to the Aircraft Engine Group, General Electric Company, Evendale, Ohio, for supplying the test material for this study.

REFERENCES

1. Floreen, S. and Kane, R. H., "The Sulfidation Attack of a Nickel-Base Alloy at Intermediate Temperatures," Met. Trans., 15A, January 1984, pp. 5-10.

2. Whitlow, G. A., et al., "Intermediate Temperature, Low-Cycle Fatigue Behavior of Coated and Uncoated Nickel-Base Superalloys in Air and Corrosive Sulfate Environments," Journal of Eng. Mats. and Tech., 106:1, January 1984, pp. 43-49.

3. Allen, J. M. and Whitlow, G. A., "Observations on the Interaction of High Mean Stress and Type II Hot Corrosion on the Fatigue Behavior of a Nickel-Base Superalloy," Journal of Eng. for Gas Turbines and Power, 107:1, January 1985, pp. 220-224.

4. Whitlow, G. A., et al., "The Effects of a Liquid Sulfate/Chloride Environment on Superalloy Stress Rupture Properties at 1300°F (704°C)," Met. Trans., 15A, January 1984, pp. 23-28.

5. Yoshiba, M., et al., "Effect of Hot Corrosion on the Creep Rupture Properties of a Nickel-Base Superalloy," Journal of Iron and Steel Inst. of Japan, 68:1, January 1982, pp. 120-129.

6. Balsone, S. J., "The Effect of Stress and Hot Corrosion on Nickel-Base Superalloys," M.S. Thesis, Air Force Institute of Technology, Wright-Patterson AFB, Ohio, 1985.

7. Jianting, G., et al., "The Effect of Hot Corrosion on Creep and Fracture Behavior of Cast Nickel-Base Superalloy IN738LC," High Temperature Alloys for Gas Turbines, Brunetaud et al., Eds., D. Reidel Publishing Company, Dordrecht, Holland, 1982, pp. 805-819.

8. Balsone, S. J., Nicholas, T., and Khobaib, M., "The Effects of Stress and Stress History on Environmental Attack in Rene' 80," Proceedings of the International Symposium on Environmentally Assisted Cracking: Science and Engineering, November 1987.

Stephen J. Balsone received his BS in Metallurgical Engineering and Material Science from Carnegie Mellon, Pittsburgh, PA, in 1982 and his MS in Aeronautical Engineering from the Air Force Institute of Technology, Wright-Patterson AFB, OH, in 1985. He is currently a structural materials engineer, Air Force Wright Aeronautical Laboratories, Materials Laboratory, Metals Behavior Branch, at Wright-Patterson AFB. His research activities include the study of environmental effects on mechanical properties of gas turbine engine structural materials.

M. Khobaib received his Ph.D. in Materials Science from the University of Connecticut. He is currently associated with the University of Dayton Research Institute and for over ten years, is doing research in areas related to the environmental assisted failure of aerospace materials.

2nd International SAMPE Metals Conference
August 2-4, 1988

CORROSION/OXIDATION BEHAVIOR OF SURFACE MODIFIED NICKEL

M. Khobaib[1], R. S. Bhattacharya[2], C. N. Raffoul[2], and A. K. Rai[2]
[1]University of Dayton Research Institute, Dayton, Ohio 45469
[2]Universal Energy Systems, Inc., Dayton, Ohio 45432

Abstract

The effect of varying amounts of Cr and Ti additions in improving the environmental resistance of Ni was studied by investigating the corrosion resistance of thin layers of Ni-Cr, Ni-Ti-Cr, $Ni_{48}Ti_{34}Cr_{17}$ $Ni_{42}Ti_{30}Cr_{28}$, etc. rf-sputtering was used to fabricate a thin layer of such metallic coatings. Alternate thin layers of Ni, Cr, and Ti with different thicknesses were prepared by electron beam evaporation, to obtain the desired composition. The aqueous corrosion behavior of these coatings was studied by conducting potentiodynamic polarization tests in 1 N HNO_3 and 0.1 N NaCl solutions while the elevated temperature oxidation behavior of the modified surface layers was investigated by exposing the specimens at 750°C for 125 hours. Detailed studies of oxide penetration were conducted by analyzing the cross section of the specimen tested under SEM.

INTRODUCTION

Nickel is known to be resistant to many corrosives. Its resistance to environmental degradation is significantly improved by proper alloy addition. The large family of corrosion resistant nickel-base alloys owe their resistance to corrosion attack due to its ability to form solid solutions with several elements which enhance its resistance to corrosion. For example, chromium addition provides higher resistance to oxidation, carburising, and sulfidation at elevated temperatures. Currently, numerous Ni-Cr alloys with excellent corrosion resistance properties are used in chemical and process industries. Generally, a high concentration of nearly 15% by weight of chromium is required to provide excellent resistance to aqueous corrosion and oxidation. However, the overall concentration can be markedly reduced by producing high chromium or other corrosion resistance layers on the surface for improved corrosion resistance. Thus, one can apply cheaper and stronger substrates with low Cr, Mo, Nb, or Ti content if it can be coated with high Cr, Ti, Mo, etc. base alloy, suitable for providing desirable corrosion resistance. Several surface modification techniques can be used to improve the surface property.

Sputtering techniques [1] such as rf and magnetron have met with increasing acceptance in industrial deposition applications owing to their high deposition rates, minimum substrate heating, and extreme versatility. Since no melting is involved in the sputtering process, it is possible to deposit multi-component films of desired composition by fabricating the sputtering target from alloys of powders. Very few studies have been performed on the corrosion behavior of sputter-deposited coatings. Sputter-deposited amorphous alloys of FeNiCrW [2] and CrB [3] have been reported to exhibit very high resistance to corrosion in NaCl and acid solutions [2]. Recently, Wang and Brimhall [4] have shown that Ni^+ irradiated NiTi amorphous alloys are more resistant to pitting and crevice corrosion than polycrystalline unirradiated alloys, although the improvement was reported to be rather slight because of the excellent behavior exhibited by the ctystalline NiTi alloy.

This paper presents data in support of improved room temperature aqueous corrosion as well as elevated temperature oxidation resistance of several rf-sputtered NiCrTi films.

EXPERIMENTAL

Substrate Ni samples were cut into 1 cm x 1 cm pieces from 0.5 mm rolled Ni sheet stock of 99.997% purity. These samples were polished to mirror finish and cleaned in trichloroethylene, acetone, and deionized water prior to film deposition. rf-sputter deposition was performed by using the deposition system at the University of Illinois, Urbana-Champaign. Both target and subsrate were sputter cleaned for 45 minutes at bias voltages of -1 kV and -120 V, respectively, prior to film deposition using argon discharge. The deposition was made at a base pressure of 5×10^{-7} torr. Homogeneous films of $Ni_{45}Cr_{55}$, NiTiCr, and $Ni_{42}Ti_{30}Cr_{28}$ of thicknesses in the range of 7500°A to 20,500°A were deposited from a composite NiCr and NiTiCr target, respectively, on polished nickel. The overall composition of the film and its homogeneity were determined from Rutherford backscattering measurements. Aqueous corrosion experiments were performed with an EG&G PARC model 332 Soft Corr Corrosion software in conjunction with a PARC model 173 potentiostat/galvonostat, in solutions of 1 N HNO_3 and 0.1 N NaCl solutions at

25°C. The oxidation study was conducted at 750°C in laboratory air. The weight gain of the samples were measured after specified intervals of exposure. After the test, the specimen was sectioned in the middle and the oxide layer was studied under scanning electron microscope (SEM) and x-ray energy spectroscopy (XES).

AQUEOUS CORROSION BEHAVIOR

The anodic polarization curve of sputtered $Ni_{45}Cr_{55}$ in 1 N HNO_3 is shown in Fig. 1. The corrosion

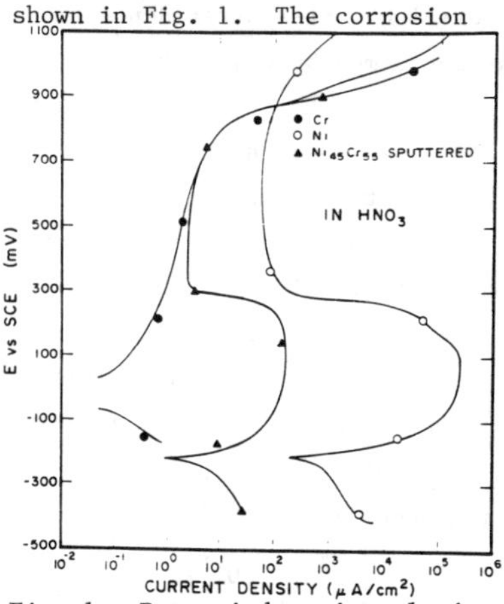

Fig. 1 - Potentiodynamic polarization curves for pure Ni, pure Cr, and sputter deposited $Ni_{45}Cr_{55}$ film on Ni substrate, in .1 N HNO_3.

behavior of Ni and Cr are also shown for comparison. The increased corrosion resistance of $Ni_{45}Cr_{55}$ over pure Ni is very clear. The critical current density is lowered by nearly four orders of magnitude over that of pure Ni. The passivation current is also reduced over an order of magnitude and is similar to that of pure Cr but the overall corrosion behavior still falls short of pure Cr. The addition of Ti to the sputtered layer significantly improves the corrosion behavior in 1N HNO_3 and its performance appears better than even Cr, as evidenced by Fig. 2.

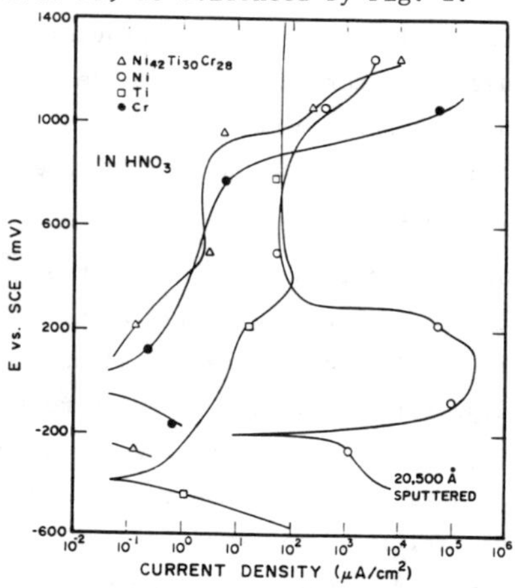

Fig. 2 - Potentiodynamic polarization curves for pure Ni, pure Ti, pure Cr, and sputter deposited $Ni_{42}Ti_{30}Cr_{28}$ film on Ni substrate in .1 N HNO_3.

The performance of $Ni_{42}Ti_{30}Cr_{28}$ in combatting 1 N HNO_3 corrosion is shown to be better than pure Ni, Ti, or Cr. By comparing Fig. 1 and Fig. 2, it can be noticed that the corrosion behavior of $Ni_{42}Ti_{30}Cr_{28}$ sputtered film in 1 N HNO_3 is improved over the binary $Ni_{45}Cr_{55}$ also. Apparently, the addition of Ti helps in reducing the critical current density to a very low value of approximately $3 \mu A/cm^2$ only, and further establishes a passivity near this level of current density.

Aqueous corrosion tests were conducted in .1 N NaCl also to study the improved pitting resistance of the sputtered $Ni_{45}Cr_{55}$ and $Ni_{42}Ti_{30}Cr_{28}$ films. Fig. 3 shows the anodic polarization behavior of $Ni_{45}Cr_{55}$ sputtered film, pure Ni and pure Cr in .1 N NaCl. Obviously,

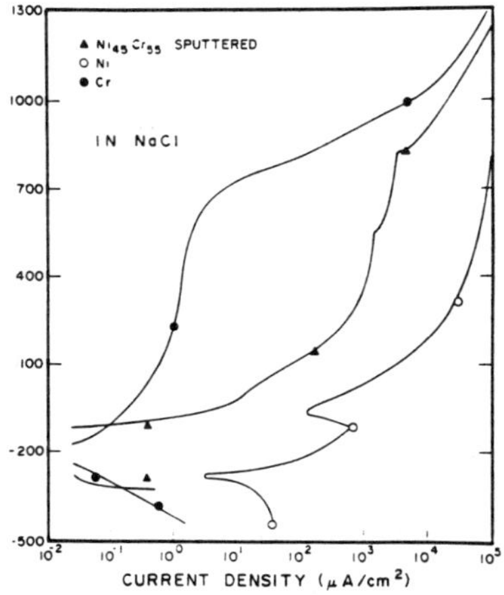

Fig. 3 - Potentiodynamic polarization curves for pure Ni, pure Cr, and sputter deposited $Ni_{45}Cr_{55}$ film on Ti substrate in .1 N NaCl.

pure chromium performs the best; the addition of chromium definitely improves the corrosion resistance of pure Ni in saline water. But it appears that even higher chromium concentration in the sputtered film of NiCr is required to cause passivation.

Fig. 4 shows the effect of the addition of titanium in improving the corrosion resistance of sputtered NiCr film in 1 N NaCl. The influence of titanium is very significant. It is noticed that the corrosion behavior of $Ni_{42}Ti_{30}Cr_{28}$ is better than any of its pure constituents. The corrosion current is tremendously lowered and pitting potential is

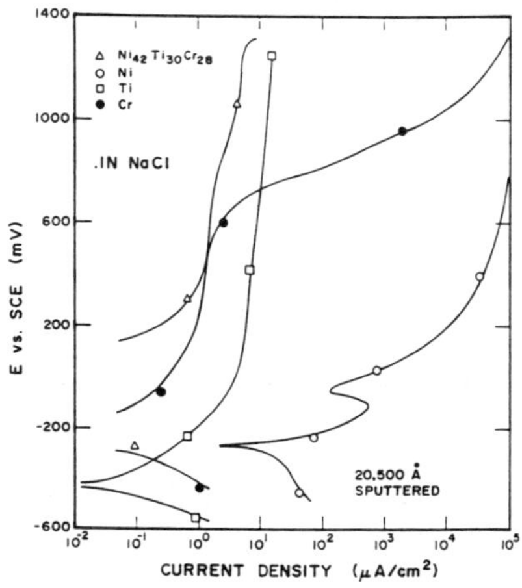

Fig. 4 - Potentiodynamic polarization curves for pure Ni, pure Ti, pure Cr, and sputter deposited $Ni_{42}Ti_{30}Cr_{28}$ film on Ni substrate in .1 N NaCl.

remarkably improved. Overall, the corrosion behavior of sputtered $Ni_{42}Ti_{30}Cr_{28}$ film was found to be excellent, both in .1 N HNO_3 and 1 N NaCl solution. The role of Ti was found to be the most significant in improving the aqueous corrosion resistance of $Ni_{42}Ti_{30}Cr_{28}$ sputtered film. Several other $Ni_xCr_yTi_z$ sputtered films with lower Cr but higher Ni and Ti were also tested for their corrosion behavior, but $Ni_{42}Ti_{30}Cr_{28}$ was found to show the highest corrosion resistance.

OXIDATION BEHAVIOR OF SPUTTERED FILMS

Samples of pure Ni, NiCr, NiTiCr, $Ni_{48}Ti_{34}Cr_{17}$, and $Ni_{42}Ti_{32}Cr_{24}$ were oxidized by exposing them in a box furnace at 750°C. $Ni_{42}Ti_{30}Cr_{28}$ coated specimens were not available for oxidation studies. However, this composition lies within the three different compositions of

$Ni_xTi_yCr_z$ selected for this study. The specimens were periodically taken out to measure the weight gain and observe the oxide build-up. All the samples were exposed for a total period of 125 hours. Fig. 5 shows the oxidation behavior of these materials at 750°C. The oxidation behavior of NiTiCr, $Ni_{48}Ti_{34}Cr_{17}$, and $Ni_{42}Ti_{32}Cr_{26}$, but all of them showed improved oxidation behavior over NiCr sputtered film. A more detailed study with lower concentrations of Ti will enable the determination of the minimum amount of Ti required

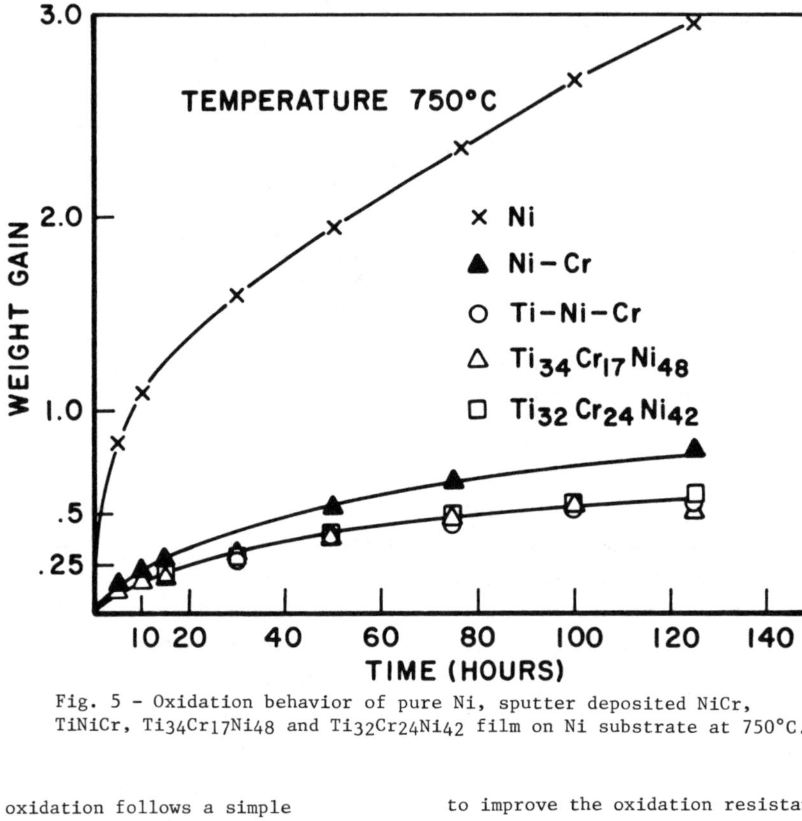

Fig. 5 - Oxidation behavior of pure Ni, sputter deposited NiCr, TiNiCr, $Ti_{34}Cr_{17}Ni_{48}$ and $Ti_{32}Cr_{24}Ni_{42}$ film on Ni substrate at 750°C.

oxidation follows a simple parabolic law. The improvement in the oxidation resistance of NiTiCr over Ni is quite significant. It is surprising to notice that $Ni_xTi_yCr_z$ sputtered film performed better than NiCr, even in its resistance to oxidation. There was hardly any difference noted in the to improve the oxidation resistance of NiCr.

Fig. 6 shows a photomicrograph of a transverse section of an oxidized specimen coated with NiCr. The oxidized section can be divided into three distinct zones. The top layer, nearly 8 µm in thickness above the thin area filled with

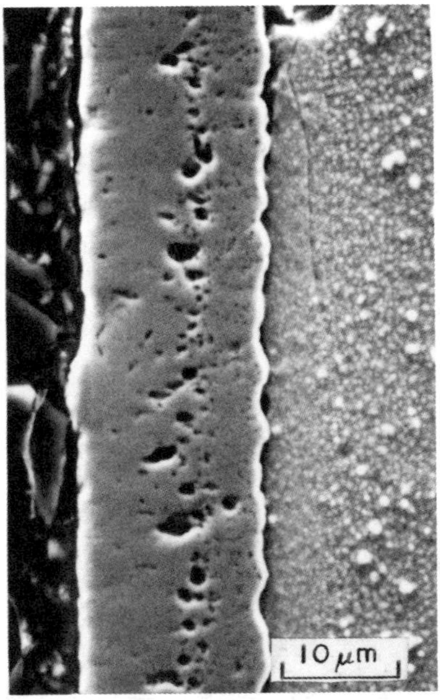

Fig. 6 - Transverse section of the oxidized layer of NiCr sputter deposited film on Ni substrate.

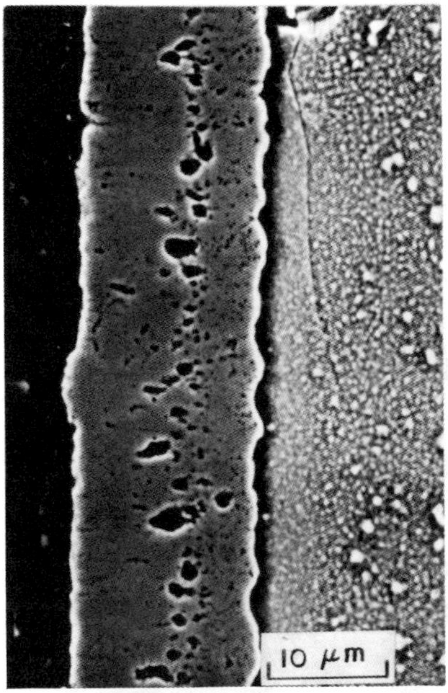

Fig. 7 - Backscattered electron image of the oxidized layer of NiCr sputter deposited film on Ni substrate.

holes, was found to contain mainly nickel oxide with some chromium oxide. The zone just above the substrate contained mainly nickel oxide with traces of chromium. The middle thin layer zone apparently corresponds to the top layer of the coating in the as-received material and appeared pitted. Fig. 7 shows the backscattered image of the same section indicating little difference in the elemental distribution in these zones. However, x-ray spectroscopic analysis showed higher concentration of nickel oxide in the layer next to the substrate with little chromium as compared with the top layer of the oxidized surface.

Fig. 8 is a similar micrograph taken from the oxidized specimen coated with $Ni_{42}Ti_{32}Cr_{24}$. The dark thin band constitutes the middle zone which showed the highest concentration of titania. Nickel and chromium oxides were also present with chromium oxide being the least. The top layer mainly contained nickel oxide followed by a small amount of chromium oxide and traces of titania. The lower zone next to the substrate contained mainly nickel oxide and was depleted of chromium and titanium. Fig. 9 shows the backscattered image of the same section indicating some differences in the element distribution of these elements in these three zones.

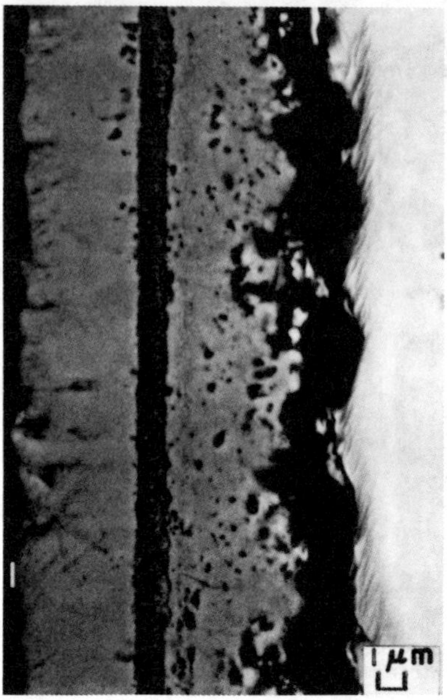

Fig. 8 - Transverse section of the oxidized layer of $Ni_{42}Ti_{32}Cr_{24}$ sputter deposited film on Ni substrate.

It was surprising to find mainly nickel oxide in the top layer with very little titania and chromia. These results indicate that nickel was the easiest to get oxidized. In the case of nickel-chromium coating also, there was hardly any indication of chromium oxidation. The presence of chromium, however, slowed down the oxidation of nickel. The results of Fig. 5 clearly indicate a slower kinetics of oxidation for specimens coated with NiCr as compared with pure Ni. In the case of both NiCr and NiTiCr coated specimens, very small amounts of chromium oxide were found in the oxidized layers.

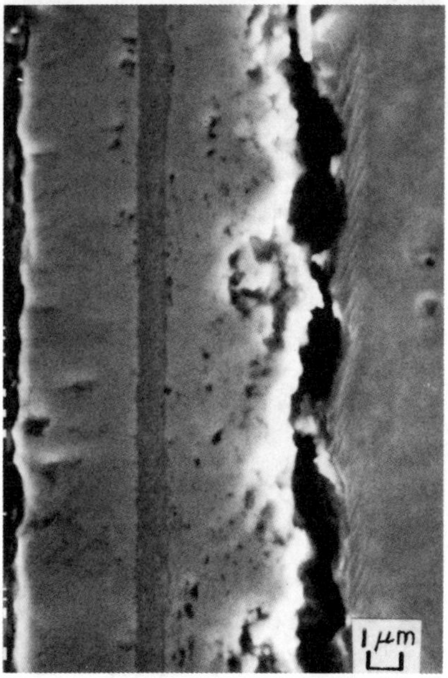

Fig. 9 - Backscattered electron image of the oxidized layer of $Ni_{42}Ti_{32}Cr_{24}$ sputter deposited film on Ni substrate.

In the case of NiTiCr coating, the oxidation was further slowed down, apparently by the presence of Ti. Again, nickel was found to be oxidized first, indicated by the highest concentration of nickel oxide in the top layer. The middle band containing mainly titania acted as a barrier for nickel oxidation. More work is needed to understand the role of titanium in slowing down the oxidation of nickel.

SUMMARY AND CONCLUDING REMARKS

1. NiCr as well as $Ni_xTi_yCr_z$ films provide protection to nickel against aqueous corrosion as well as oxidation.

2. The NiTiCr coating showed excellent performance in resisting aqueous corrosion in both 1 N HNO_3 and .1 N NaCl media.

3. The oxidation resistance of $Ni_xTi_yCr_z$ coated films were found superior to that of NiCr film.

REFERENCES

1. Bunshah, R. F., Ed., Deposition Technologies for Films and Coatings, Noyes Publications, New Jersey, 1982.

2. Wang, R. and Merz, M. D., Corrosion, Vol. 40, No. 6, 1984, p. 272.

3. Ruf, R. R. and Tsuei, C. C., J. Appl. Phys., Vol. 54, 1983, p. 5705.

4. Wang, R. and Brimhall, J. L., Mater. Res. Soc. Symp. Proc., Vol. 27, 1984, p. 729.

M. Khobaib received his Ph.D. in Materials Science from the University of Connecticut. He is currently associated with the University of Dayton Research Institute, and for over ten years is doing research in areas related to the environmental assisted failure of aerospace materials.

Rabi S. Bhattacharya received a M.S. in Solid State Physics from Indian Institute of Technology, Kharagpur and a Ph.D. in Atomic Physics from the University of Calculta in 1975. After spending several years at FOM Institute, Amsterdam, University of Giessen, and Max-Planck Institute for Plasma Physics, Munich as visiting scientist and then at McMaster University, Canada and the University of Florida, Gainesville, as Senior Research Associate, he joined Universal Energy Systems, Inc. (UES) in 1980. He is now a Principal Research Scientist in the Materials Research Division of UES.

Charbel N. Raffoul received his B.S. and M.S. in Mechanical Engineering from the University of Dayton. He joined Universal Energy Systems, Inc. in 1980 where he is currently a Research Scientist working on Materials Research, such as ion implantation and aqueous corrosion.

Amarendra K. Rai received M.S. and Ph.D. in Physics from Banaras Hindu University, Varanasi, India in 1977. He was a research associate at the Department of Materials Engineering, North Carolina State University, Raleigh, North Carolina. He joined Universal Energy Systems, Inc. in 1981. He is currently a Senior Scientist in the Materials Research Division of UES.

2nd International SAMPE Metals Conference
August 2-4, 1988

HIGH TEMPERATURE OXIDATION BEHAVIOR OF Ti_3Al ALLOYS

M. Khobaib[1] and F. W. Vahldiek[2]

[1]University of Dayton Research Institute, Dayton, Ohio 45469

[2]Air Force Materials Laboratory, Wright-Patterson AFB, Ohio 45433

Abstract

High temperature oxidation behavior of two Ti_3Al alloys has been studied at 760 mm Hg oxygen pressure in the temperature range of 700-900°C. Weight changes were obtained from tests conducted at specified time intervals up to 165 hours. Results were compared with data obtained on a commercial high temperature titanium alloy, IMI 834. The weight gain results showed improved oxidation resistance of Ti_3Al alloys as compared with IMI 834. The higher concentration of Nb also had a significant effect in increasing the oxidation resistance. Detailed SEM analysis of the oxidized surfaces was conducted to understand the mechanism of oxidation on these alloys.

1. INTRODUCTION

The titanium aluminide family of alloys has an attractive combination of properties, such as low density coupled with good mechanical properties at high temperatures. Currently, they are being investigated as a candidate material for aircraft structural components. The prime concern is the limited information on the oxidation behavior of these alloys. Resistance of a material to oxidizing environments at high temperature has become a prime design and application consideration. Accordingly, understanding both the fundamental and practical aspects of oxidation of a material is a prerequisite to its recommendation for jet engine structural component application.

Numerous studies [1-8] have been performed on the high temperature oxidation of titanium and titanium alloys in the range of 400-700°C. Depending upon temperature, the oxidation of titanium follows different oxidation laws [9,10]. Oxygen diffusion in titanium is very complex and a titanium surface exposed to air at elevated temperature produces a complex oxide scale whose structure is laminated and composed of several oxides. The oxide compositions range between TiO and rutile TiO_2 represented by the general formula, Ti_nO_{2n-1}, in which $2 < n < 38$ [11-14].

The introduction of alloying elements markedly changes the oxidation kinetics of titanium [13-16]. The particular nature of titanium aluminide base alloys, i.e., their ordered structure and the nature and distribution of other alloying elements, is expected to play a significant role in the kinetics of oxidation.

The present paper mainly deals with the oxidation of Ti_3Al + 5% Nb and Ti_3Al + 11% Nb alloys and a high temperature commercial titanium alloy, IMI 834, at 700-900°C.

EXPERIMENTAL

The titanium aluminides Ti_3Al + 5% Nb and Ti_3Al + 11% Nb were given β-heat treatment (2100°F for one hour in vacuum, followed by forced fan cooling in argon to 1400°F, which was aged for eight hours in vacuum at 1400°F, and finally air cooled to room temperature). IMI 834 was also obtained in heat treated form (solution treated at 1865°F for two hours followed by oil quenching to 1160°F, which was aged for two hours at 1160°F, and finally air cooled to room temperature). Test coupons of 10 mm x 5 mm x 3 mm were polished through 1 - μ diamond and then degreased and cleaned ultrasonically in acetone and methanol to ensure clean starting surfaces.

The oxidation kinetics were determined by exposing the test coupons in laboratory air in a box furnace at 700-900°C. The test coupons were taken out of the furnace after specific intervals of 1, 5, 10, 15, 25, 40, ..., 165 hours and weight gain was recorded within 0.1 mg. Special care was taken to conduct all the tests under similar conditions.

RESULTS AND DICUSSION

The results of oxidation tests conducted in air for Ti_3Al + 5% Nb are shown in Fig. 1. Tests were conducted at 700, 750, 800, 850, and 900°C. Fig. 2 shows similar weight gain vs time plot for Ti_3Al + 11% Nb. The increased oxidation rate for Ti_3Al + 5% Nb at 900°C was found to be very high and scaling was observed. Tests were conducted at 650°C also, where only tarnishing of the sample was noticed. There was no noticeable weight gain over a period of 200 hours. The results of oxidation for both these titanium aluminides are very similar except that Ti_3Al + 11% Nb at 900°C did not show any scaling or increased oxidation. As a matter of fact, the oxidation rates at 900°C were found to be similar to

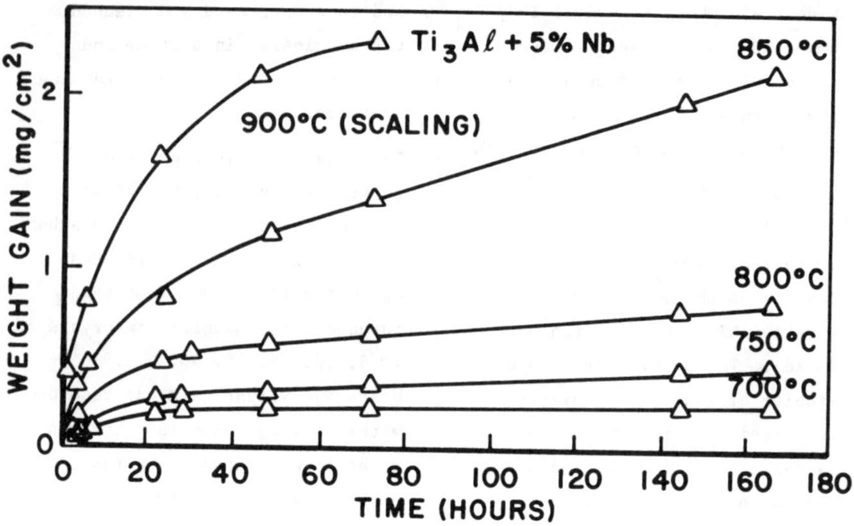

Fig. 1 - Oxidation behavior of Ti_3Al + 5% Nb at 700, 750, 800, and 850°C.

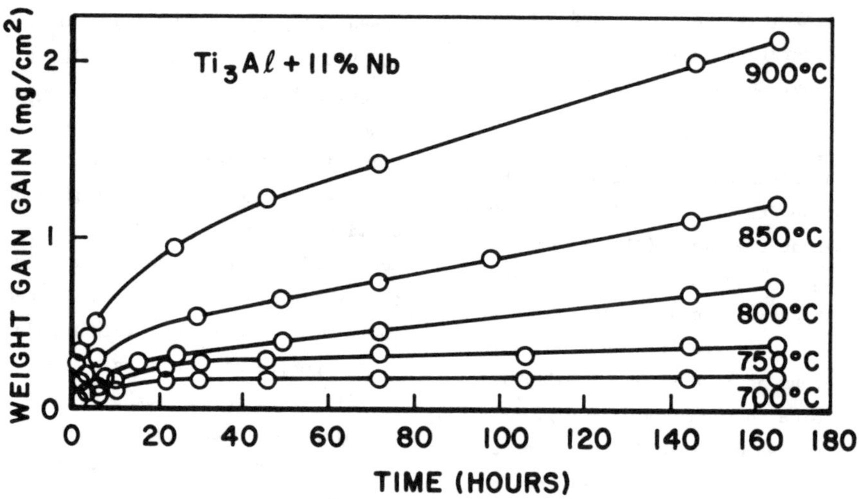

Fig. 2 - Oxidation behavior of Ti_3Al + 11% Nb at 700, 750, 800, 850, and 900°C.

the oxidation rates of Ti_3Al + 5% Nb at 850°C, indicating an increased oxidation resistance capability by nearly 50°C. This increased oxidation resistance could only be the effect of higher Nb concentration. At lower temperatures of 700 and 750°C, the oxidation rates are small and the difference is not that obvious, although very small differences in weight gain over a period of 165 hours were noticed which increased with temperature. The oxidation rate at 800°C for Ti_3Al + 5% Nb in the first 25 hours is noticed to be almost twice as high as that of Ti_3Al + 11% Nb. This difference in rate, however, decreases with exposure. It is quite interesting to notice that the weight gain vs time plot for Ti_3Al + 5% Nb at 850°C and Ti_3Al + 11% Nb at 900°C almost overlap each other, indicating a 50°C improvement in oxidation resistance by 6% addition of Nb. The role of microstructural changes due to higher additions of Nb is discussed later.

Fig. 3 shows the results obtained from a test conducted on commercial high temperature IMI 834 titanium alloy. The oxidation resistance capability is poor with respect to either Ti_3Al + 11% Nb or Ti_3Al + 5% Nb. In other words, the

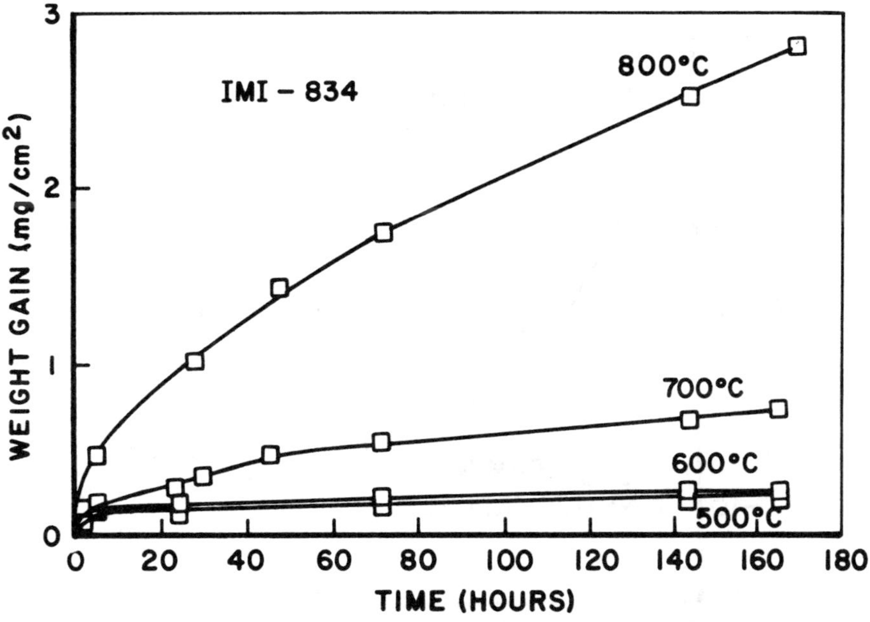

Fig. 3 - Oxidation behavior of IMI 834 at 500, 600, 700, and 800°C.

titanium aluminides show higher oxidation resistance as compared to high temperature titanium alloys. Comparing Fig. 2 with Fig. 3, the oxidation rates of IMI 834 at 700°C is very comparable to the oxidation rates of Ti_3Al + 11% Nb obtained at 800°C. Similar trends are noticed at lower temperatures. This clearly indicates nearly 100°C improvement in oxidation resistance capability of Ti_3Al + 11% Nb over IMI 834.

Fig. 4 shows the parabolic rate constants for oxidation of these three alloys in air as a function of temperature. The K_p values for alumina and titania-forming kinetics compiled from different sources [17-19] are also included in this figure. The K_p values for IMI 834 oxidation indicate predominantly TiO_2-forming kinetics. The K_p values for Ti_3Al + 5% Nb and Ti_3Al + 11% Nb lie almost in the middle. As a matter of fact, K_p values for Ti_3Al + 11% Nb is more closer to alumina-forming kinetics than TiO_2-forming kinetics. In general, it appears that Ti_3Al alloys

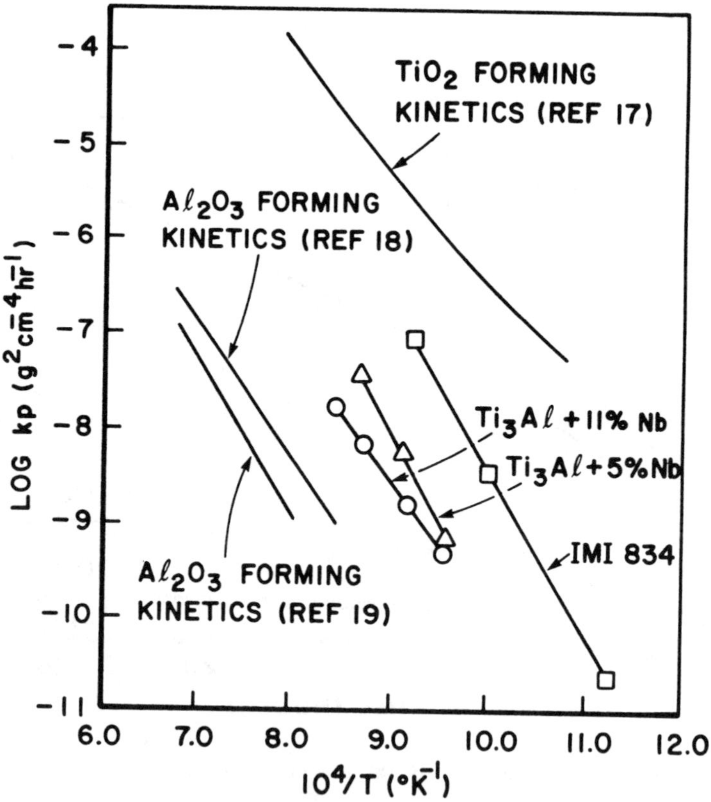

Fig. 4 - Oxidation data plotted as log K_p vs 1/T.

follow a mixed kinetics, resulting from combined oxidation of titanium and aluminum. It appears that higher concentrations of Nb leads to an increase in alumina oxidation resulting in improved resistance to oxidation. A recent study [20] conducted to determine the ratio of the thermodynamic activities of aluminum and titanium in TiAl-base alloys at temperatures of 1400°C and higher have shown an increase in the activity ratio (a_{AL}/a_{Ti}) by addition of Nb. It is difficult to postulate similar effects at 800°C. However, the increased alumina oxidation could occur due to the doping of titania scale by Nb, thereby reducing the concentration of point defects responsible for mass transport through titania. Scanning electron microscope, electron microprobe, and EDAX (energy-dispersive elemental X-ray analysis) studies indicated reduced TiO_2/Al_2O_3 ratio in the oxide scale formed on Ti_3Al + 11% Nb alloy as compared to the scales formed on Ti_3Al + 5% Nb. Fig. 5 is a photomicrograph of a transverse section of the oxidized sample of Ti_3Al + 5% Nb. Fig. 6 shows the back-scattered image of the same area. The oxide layer mainly consists of three zones. The top lighter layer contains mostly rutile (TiO_2), with an indication of a small amount of α-alumina. This zone contains relatively loose oxide and some evidence of cracking was also found. The middle or the

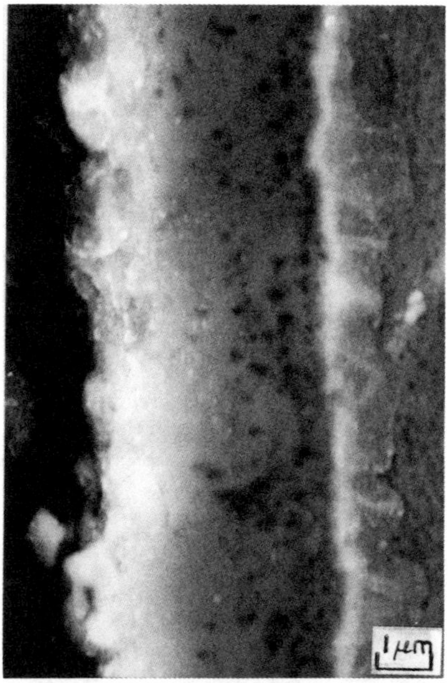

Fig. 5 - Photomicrograph of a transverse section of oxidized Ti3Al+5% Nb at 800°C.

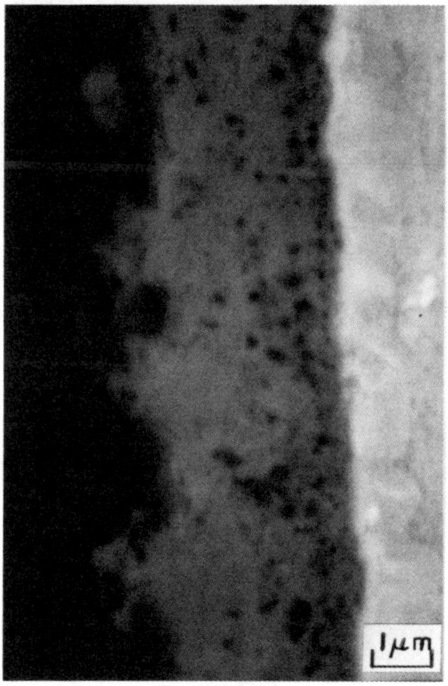

Fig. 6 - Backscattered electron image of oxidized Ti3Al+5% Nb at 800°C.

second layer is again a mixture of titania and alumina, with quite high titania/alumina ratio. The bottom layer is mostly metallic, which is a depleted zone and is most probably not ordered as Ti_3Al. Fig. 7 is the photomicrograph taken

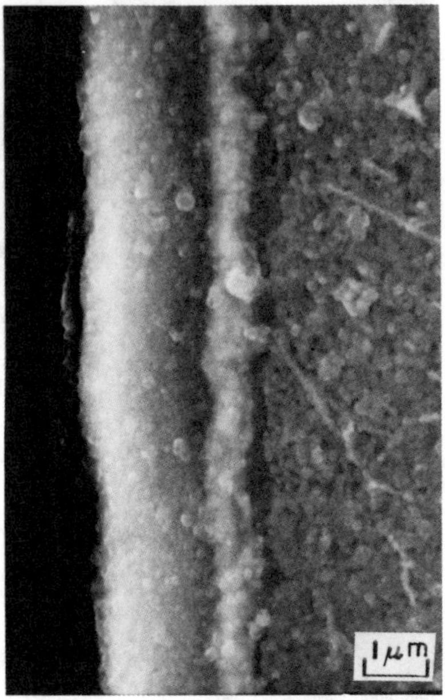

Fig. 7 - Photomicrograph of a transverse section of oxidized Ti3Al+11% Nb at 800°C.

from a transverse section of the oxidized sample of Ti_3Al + 11% Nb. The back-scattered image is shown in Fig. 8 for the sake of comparison. Similarly, three prominent zones are noticed as in the case of Ti_3Al + 5% Nb alloy. The decrease in the thickness of the overall oxidation zone (both alloys were oxidized for 165 hours at 800°C) is quite clear. The top layer was found to be more dense and hardly

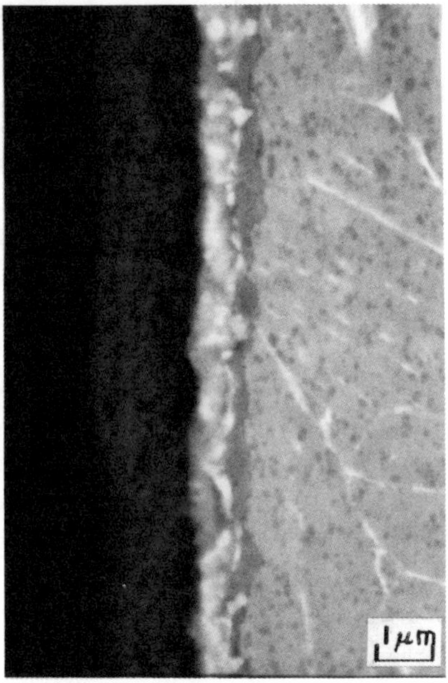

Fig. 8 - Backscattered electron image of oxidized Ti3Al+11% Nb at 800°C.

any evidence of cracking was found in this case. This zone was composed mostly of titania, and although the amount of alumina was small, it was clearly noticed. The second or the middle zone was not as large as in the case of Ti_3Al + 5% Nb and the ratio of titania to alumina was significantly lower, indicating inhibition of titania formation. This is suggested to be the result of higher Nb concentration. The titania could have been doped with Nb. This is supported by the presence of Nb in this zone. The bottom layer, again, is a metallic zone depleted of titanium aluminum, and noibium.

Fig. 9 is the photomicrograph of the oxidized sample of IMI 834. The

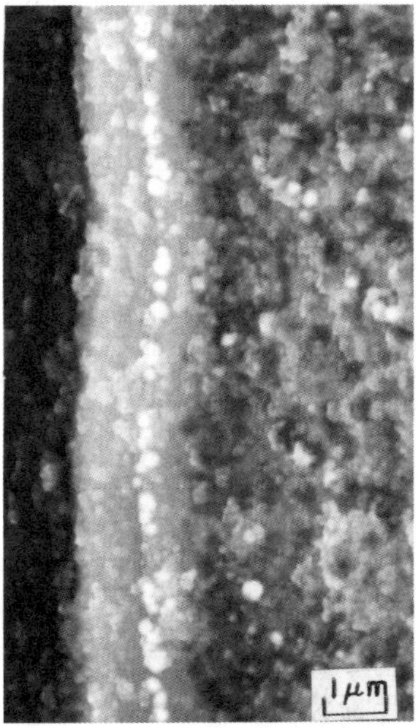

Fig. 9 - Photomicrograph of a Transverse section of oxidized IMI 834 at 800°C.

back-scattered image is shown in Fig. 10. The oxide layer was quite thick as compared to those of Ti_3Al + 5% Nb and Ti_3Al + 11% Nb. Here again, the outermost layer was mainly titania, but was quite loose. The middle zone mainly consisted of titania and alumina. The third zone was small and again, was the depleted zone. In this case, both the top two zones were loosely packed and signs of cracking were found in other sections.

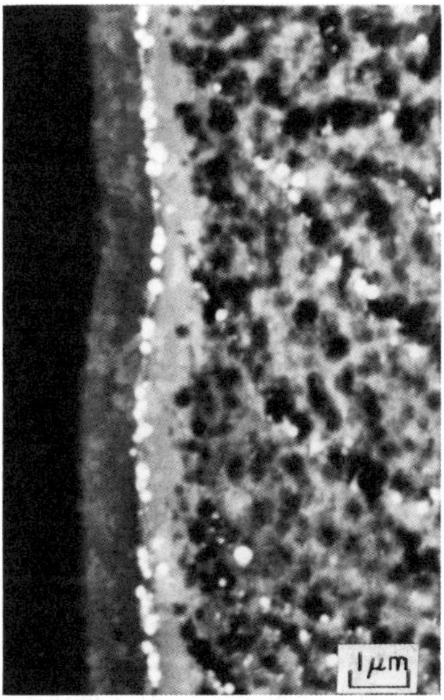

Fig. 10 - Backscattered electron image of oxidized IMI 834 at 800°C.

SUMMARY AND CONCLUDING REMARKS

1. The titanium aluminide-base alloys were found to have better oxidation resistance than the commercially available high temperature titanium IMI 834.

2. The higher niobium concentration improves the oxidation resistance of titanium aluminide. The exact mechanism is not well established. However, the higher niobium concentration was noticed to push the kinetics more towards alumina forming kinetics.

3. Ti_3Al + 11% Nb was found to run out of oxidation resistance at 800°C as compared to 700°C for IMI 834.

REFERENCES

1. Stringer, J., Acta Metall., 8, 1969, p. 758.

2. Hurlen, T., Inst. Metals, 89, 1960, p. 128.

3. Morton, P. H. and Baldwin, W. M., Jr., Trans. ASM, 44, 1952, p. 1004.

4. Jenkins, A. E., J. Inst. Metals, 82, 1953-54, p. 213.

5. Jenkins, A. E., J. Inst. Metals, 84, 1955-56, p. 1.

6. Kofstad, P., J. Less-Common Metals, 12, 1967, p. 449.

7. Sarrazin, P., et al., J. Less-Common Metals, 59, 1979, p. 111.

8. Champin, B., et al., J. Less-Common Metals, 69, 1980, p. 163.

9. Anderson, S., et al., Acta Chem., Scand., 10, 1957, pp. 1641-1657.

10. Bursill, L. A., et al., "On a New Family of Titanium Oxides and the Nature of Slightly-Reduced Rutile," Phil., Mag., Vol. 164, 1969, pp. 347-359.

11. Kofstad, P., "High Temperature Oxidation of Metals," John Wiley & Sons, Eds., 1966.

12. Krol, S. and Gorecki, T., Proceedings of the Fifth International Conference on Titanium, Congress-Center, Munich, FRG, Vol. 4, 10-14 September 1984, p. 2695.

13. Hurlen, T., Journal of Metals, 84, 1960, p. 128.

14. Simnand, M., Spilners, A., and Katz, O., Journal of Metals, 7, 1955, p. 645.

15. Chaze, A. M., Coddet, C., and Beranger, G., J. Less-Common Metals, 83, 1982, p. 49.

16. Chaze, A. M., Beranger, G., and Coddett, C., Proceedings of the Fifth International Conference on Titanium, Congress-Center, Munich, Vol. 4, 10-14 September 1984, p. 2673.

17. Kofstad, P., High Temperature Oxidation of Metals, John Wiley & Sons, Eds., New York, 1966, p. 175.

18. Giggins, C. S. and Pettit, F. S., J. Electrochem. Soc., 118, 1971, p. 1782.

19. Giggins, C. S. and Pettit, F. S., "Oxide Scale Adherence Mechanisms and the Effects of Yttrium, Oxide Particles and Externally Applied Loads on the Oxidation of NiCrAl and CoCrAl Alloys," ARL-TR-75-9234, 1975.

20. Choudhury, N. S., Graham, H. C., and Hinze, J. W., "Oxidation Behavior of Titanium Aluminides," Proceedings of the Symposium on Properties of High Temperature Alloys, Electrochemical Society, Princeton, New Jersey, March 1977, pp. 668-680.

M. Khobaib received his Ph.D. in Materials Science from the University of Connecticut. He is currently associated with the University of Dayton Research Institute and for over ten years is doing research in areas related to the environmental assisted failure of aerospace materials.

Fred Vahldiek is a Materials Engineer at the Air Force Wright Aeronautical Laboratories, Materials Laboratory, Metals Behavior Branch at Wright-Patterson AFB, OH. He has been studying various properties of high temperature materials for about 30 years. Currently, his research is in the areas of oxidation and hot corrosion of aluminides and composites.

2nd International SAMPE Metals Conference
August 2-4, 1988

TITANIUM REQUIREMENTS FOR CURRENT AND FUTURE MILITARY GAS TURBINE ENGINES

E. S. Huron and J. A. Miller
Pratt & Whitney
Operations - Materials Engineering
W. Palm Beach, FL 33410

Abstract

The substitution of titanium for nickel in gas turbine engine components would result in significant weight savings. Currently the use of titanium is limited by the possibility of catastrophic titanium fires and by elevated temperature properties, including creep and damage tolerance. With careful design, the fire problem can be controlled, but mechanical properties must be improved. In this paper, some of the key current and projected requirements for titanium are outlined, and compared with the results of recent alloy developments and future alloy design philosophy.

1. INTRODUCTION

Titanium has played a key role in the performance improvements achieved in gas turbine engines over the past 20 years. Titanium's light weight and high strength and its ability to be used at moderately high temperatures have been key factors in its increased use.

If the useful temperature ranges of titanium could be extended, even greater performance could be achieved. Upgraded derivatives of current engines such as the F100 and J52 with higher thrust-to-weight ratios are in the design stage. These engines will power aircraft such as the F-16 Agile Falcon and A6 derivatives. More advanced engines for aircraft such as the Advanced Tactical Fighter (ATF) and the Light Weight Fighter (LWF) have even higher thrust to weight goals. Titanium alloys in aggressive applications are included in the design of these engines.

Titanium with improved properties could increase the thrust-to- weight ratio by replacing heavier nickel and steel parts with titanium, while providing the same or increased performance. The substitution of titanium is not direct or simple, however. Titanium suffers from limitations in damage tolerance, high temperature creep strength, and fire resistance. Improvements in these properties have resulted from parallel efforts in both alloy and process development. In this paper, some of the use requirements for titanium will be outlined, and the alloy and process development efforts to meet those requirements will be reviewed.

2. BACKGROUND/CURRENT STATUS

2.1 Alloy Development

Titanium was first introduced in gas turbine engines in 1954 in the J57 turbojet in a rotor application. Since that time, significant alloy and process development has occurred.

Subsequent alloy development has resulted in a continuing series of alloys, each with certain advantages and compromises compared to older alloys. Titanium alloys fall into three general classes, categorized by the stable metallurgical phases in the alloy. Alpha alloys consist primarily of hexagonal alpha phase. Beta phase alloys contain alloying additions that stabilize the body centered cubic beta phase. Alpha-beta alloys consist of a mixture of the two phases. Most high temperature titanium alloys have been of the alpha or alpha-beta types. The earliest and most widely used commercial alloy is Ti-6Al-4V. This alpha-beta alloy possessed a good combination of strength, corrosion resistance, and intermediate temperature properties. However, it is limited in temperature to about 400°C (750°F), and to 316°C (600°F) where fatigue loads are high. Ti-6-4 is still used in many disk applications in turbine compressors.

From Ti-6-4, two development paths occurred. To improve creep performance, Ti-6Al-2Sn-4Zr-2Mo and Ti-8Al-1Mo-1V were developed. Ti-8-1-1 is a "super-alpha" alloy with excellent creep strength as high as 538°C (1000°F), but stability limits the useful range of the alloy to 482°C (900°F). Ti-8Al-1Mo-1V has the highest stiffness and lowest density of any commercial alloy, and has good fatigue properties. It is used in fan applications in the forward sections of gas turbines where stiffness and impact resistance are primary considerations. Ti-6Al- 2Sn-4Zr-2Mo possesses higher strength than Ti-8Al-1Mo-1V. Figure 1 compares the creep performance of Ti-6-4, Ti-8-1-1, and Ti-6-2-4-2. This alloy can also be used to 538°C (1000°F), but again, long exposures (greater than 1000 hours) will cause some degradation in properties. A Silicon addition to this alloy resulted in Ti-6-2-4-2 + Si. This alloy is a further improvement over Ti-6-2-4-2, as shown in Figure 2. Maximum use temperature is 566°C (1050°F).

Ti-6Al-2Sn-4Zr-6Mo was developed to improve strength relative to Ti-6-4. This alloy is more beta stabilized than Ti-6-2-4-2 and is hardenable to high strengths in thick sections. Figure 3 shows the relative strengths of Ti-6-4 and Ti-6-2-4-6. Ti-6-2-4-6 has good creep performance up to 427°C (800°F). This alloy is frequently used in disk applications where toughness, damage tolerance and strength are important.

Ti-6-2-4-2 + Si and Ti-6-2-4-6 are used extensively in current gas turbine engines. Newer alloys have been developed, but to date, these alloys tend to be compromises of improvements in some area and reduced properties in other areas. IMI 829 is a recently developed alloy that illustrates this. It has the composition Ti-5.5Al-3.5Sn-3Zr-1Nb-0.3Mo-0.25Si. This alloy was designed to allow improved creep strength without loss of ductility in service (better stability). The Si is added as a solid solution strengthener to improve creep. However, the Si causes silicide precipitation and also promotes short range ordering and Ti3Al formation with loss of ductility (1). The same composition without Si retains ductility after exposure but possesses lower strength.

Ti-1100 is another new alloy which just became available (2). This alloy has the composition Ti-6Al-2.75Sn-4.0Zr-0.4Nb-0.45Si-0.07Ox-0.02Fe. The tight limits on oxygen and iron are specified for optimum creep performance. The alloy is a near-alpha alloy and can be processed in the alpha + beta region or in the beta phase region. The major advantage of this alloy over Ti-6-2-4-2 + Si is improved creep and tensile strength (Fig. 4 and 5), with minimal loss of ductility after creep exposure. The alloy does have lower toughness and higher crack growth rates than Ti-6-2-4-2 + Si (Fig. 6). Whether or not this alloy is an improvement depends on the application. For static parts or rotating parts where creep is life limiting, Ti-1100 may find a use in advanced engines. Where damage tolerance is of prime importance, Ti-1100 may not be preferred.

Figure 1: Relative Creep Performance of Ti–6Al–4V, Ti–8Al–1Mo–1V, and Ti–6Al–2Sn–4Zr–2Mo

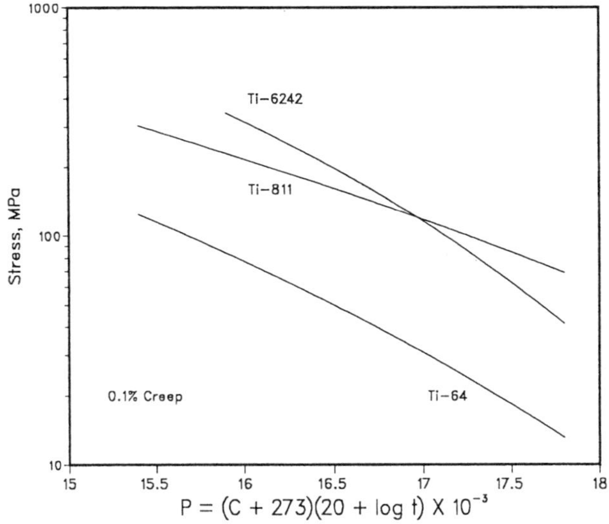

Figure 2: Influence Of Silicon Additions And Processing Routes On Creep Properties of Ti–6Al–2Sn–4Zr–2Mo

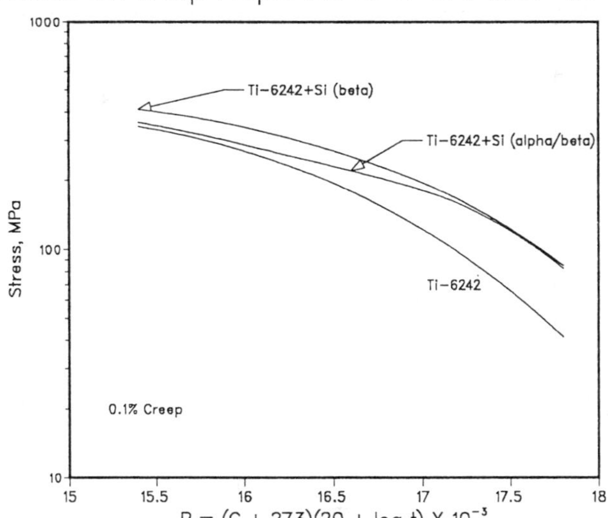

Figure 3: Relative Strength of Ti–6Al–2Sn–4Zr–6Mo And Ti–6Al–4V

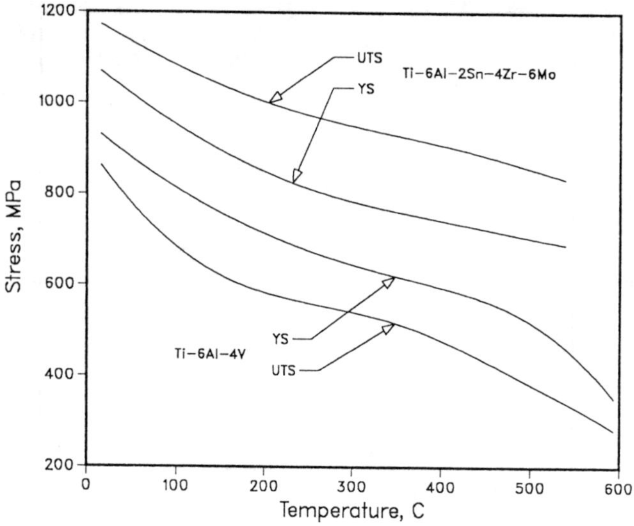

Figure 4: Comparison Of Typical Strength Properties Of Beta–Processed Ti–1100 And Ti–6242–Si

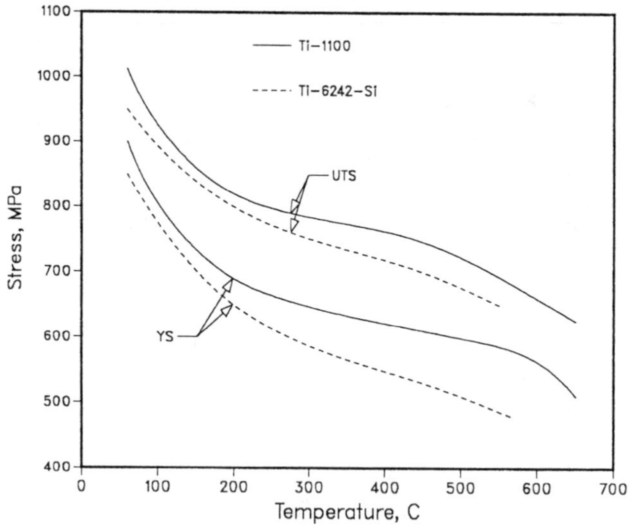

Figure 5: Relative Creep Performance Of Beta–Annealed Ti–6424–Si vs Beta–Processed Ti–1100

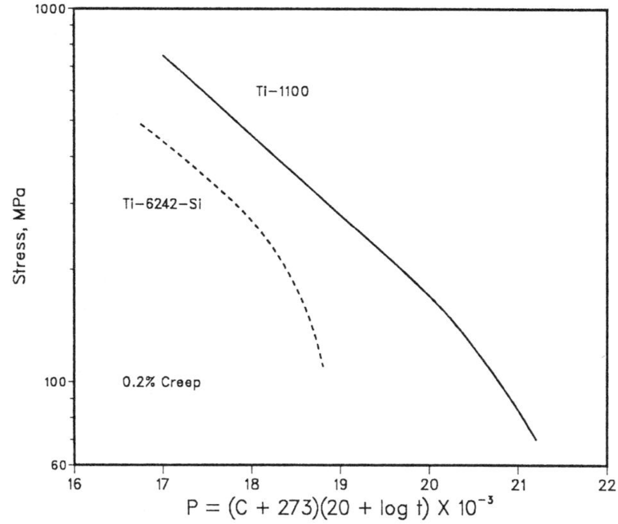

Figure 6: Fatigue Crack Growth Rate Comparison Between Ti–6242–Si And Ti–1100

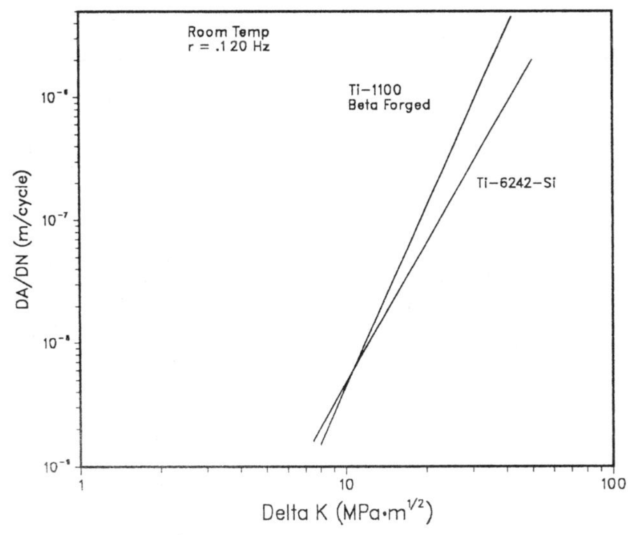

Although the above alloys are currently optimal for gas turbine applications, many other titanium alloys have been developed. The beta alloys, such as Ti-15V-3Al-3Cr-3Sn and Ti-13V-11Cr-3Al, are easily formed into sheet, have good room and elevated temperature strength, and have been used in structural airframe applications. However beta alloys have been limited in high temperature creep performance and stability and to date have not been widely used as engine components.

2.1 Process Development

Improved titanium ingot metallurgy methods have resulted in continually increasing quality. Titanium is particularly sensitive to internal defects. Titanium defects fall into three categories - Type I defects, oxygen or nitrogen stabilized alpha-phase regions, Type II defects, from chemical segregation of alloying elements, and High Density Inclusions (HDI's), which are typically unmelted refractory metals or their oxides or carbides. Figure 7 shows an example of a Type I defect. These defects have caused severe component failures, and continual efforts have been made to reduce their occurrence. In the very early years titanium was produced using single melt, non-consumable tungsten electrode practice. Severe problems with oxygen-stabilized (Type I) defects and tungsten inclusions (HDI's) were encountered. The next step was double consumable melting under vacuum, termed vacuum are remelting (VAR). This was later supplemented with triple melting for some applications. Failures still occurred occasionally, however. Improvements were made in scrap control and sample inspection through sonic methods and macroetching. Process control at the vendors was improved to the point that failures from Type I defects are now very rare. In addition, cryogenic spin testing was developed as a proof test. In this test, the disk is rotated at high speeds, to high loads, at cryogenic temperatures. The strength is increased, but toughness is decreased at the low temperatures, causing flaws that may otherwise not be detected to show up dramatically (Fig. 8).

Improved control of forging, particularly beta forging, and heat treatment has resulted in some benefits. Titanium forging may be performed above the beta transus (beta forging) or below the beta transus (alpha-beta forging). In general, alpha-beta forging leads to better fatigue properties while beta forging leads to better creep properties and damage tolerance (3). Both forging techniques are difficult due to a narrow temperature window: beta forging is limited by the beta transus, and alpha-beta forging is limited on the low side by tooling forces and material flow stress and on the high side by the beta transus. In traditional cold-die forging, this led to slow forging rates, frequent reheating of parts, and inconsistent temperature control. The development of isothermal forging using heated dies and controlled atmospheres was a significant improvement (4). Isothermal forging in the beta range coupled with close control of the cooling rate allows achievement of optimum properties. Beta forging without close control results in a larger beta grain size and a coarse, aligned Widmanstatten alpha structure. This results in poor toughness. With increased work in the beta range (made easier by isothermal forging) and control of cooling rate, the grain size can be refined if desired and a basket-weave transformed beta structure can be achieved. The choice of forging route influences which properties are optimized.

Figure 7: Type I Defect In Ti-6-2-4-6 Disk

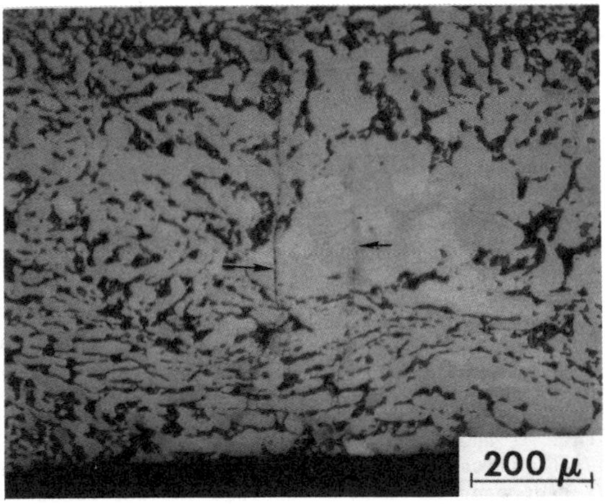

Figure 8: F100 Ti-6-2-4-6 Disk Which Failed During Cryospin Proof Test

Figure 9 illustrates the microstructural effects on low cycle fatigue initiation and propagation. While it is generally found that alpha-beta microstructures provide superior life to crack initiation, beta microstructures result in superior propagation characteristics (slower fatigue crack growth rate). In today's design environment which stresses structural integrity and a "retirement-for-cause" philosophy, crack propagation is becoming a dominant selection criteria. Hence in several applications beta structures are preferred.

2.3 The Current State of The Art - the 538°C (1000°F) Barrier

At the present state of the art, titanium usage seems limited by a temperature barrier of approximately 538°C and compromises in alloy capabilities. The designer has the choice of high strength alloys such as Ti-6-2-4-6 or higher creep performance of Ti-6-2-4-2. Processing can be used to some extent to tailor properties to each application, with recent emphasis being placed on fatigue crack propagation resistance. Also, beyond this temperature, environment factors (oxidation and the pyrophoric nature of currently available alloys) become a major concern.

3. FUTURE PROPERTY REQUIREMENTS

In general, the overall goal of advanced engines is to reduce weight and increase performance. An "easy" way to do this is to replace nickel components with titanium. This requires that the titanium alloys used have the capability for service at higher temperatures, because in general the limit of titanium application has been set by the operating temperature of the particular component. The payoff for such a substitution is huge. For example a rough calculation based on a J52 engine (a 12,000 lb.-class thrust engine used in the A-6) showed that a direct substitution of Ti for IN-718 and A-286 parts of the same size would save over 60 lb. of engine weight. This does not include further reductions due to decreases in part sizes made possible by lower rotating weight. The present state of the art has been discussed, and alloy capability might appear to be limiting factor. However, development continues towards aggressive property goals.

Figure 9: Titanium Microstructural Effect On Fatigue Crack Initiation And Propagation

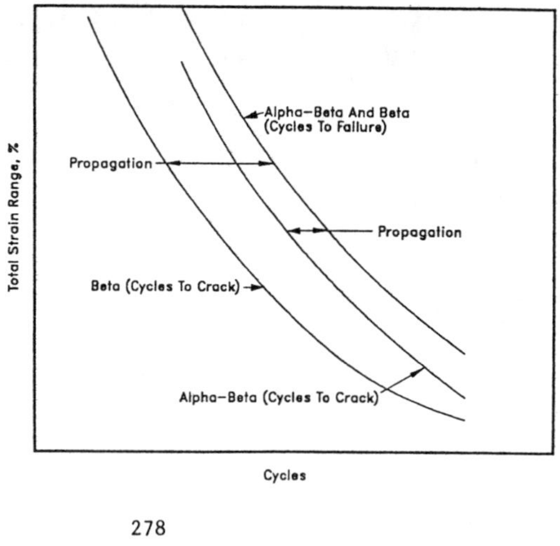

In order to substitute for nickel, new titanium alloys must meet four general requirements. First, they must have higher temperature capability in strength and creep. As an example, Figure 10 compares creep properties of Ti-6-2-4-2 + Si and the aluminides Ti_3Al and TiAl with goal properties. The goal properties represent a 110°C (200°F) improvement at a stress of 138 MPa (20 ksi). Another requirement is improved damage tolerance (lower crack growth rates), at higher temperatures. Increasing temperature causes decreased crack propagation resistance because of more environmental damage. New titanium alloys should possess the same (or better) crack propagation resistance as current alloys, but at higher temperatures. Closely related to damage tolerance at higher temperatures is oxidation resistance. Oxidation resistance has not been a severe problem to date because titanium has been limited by mechanical strength limitations to temperature ranges below where oxidation becomes critical. These considerations taken together indicate that a 110°C (200°F) improvement in alloy capability is necessary for titanium to replace nickel.

The fourth requirement is that titanium ignition and burn resistance must be understood and improved. When exposed to a rapid heat input, conventional alpha and alpha-beta alloys are subject to ignition and self-sustained combustion, with damage ranging from slight erosion of part surfaces to catastrophic loss of the engine and airframe. Most titanium ignition cases have been caused by one of four major areas (5) - blade or vane tip rubs, root rubs, impact from foreign or internal damage debris (FOD/IOD), and aerodynamic stall. Rub events account for most of the titanium fires experienced to date. Rubs occur because of component breakage, excessive rotor movement from high G-maneuvers, or bearing failures. The net effect is that above a certain temperature and pressure combination, titanium alloys can't be used because should a rub occur, a fire would likely start and be self-sustaining, because of the high reactivity of titanium. Burn resistance is closely related to oxidation.

Figure 10: Creep Goal Properties Compared To Conventional Ti–6Al–2Sn–4Zr–2Mo And The Aluminides

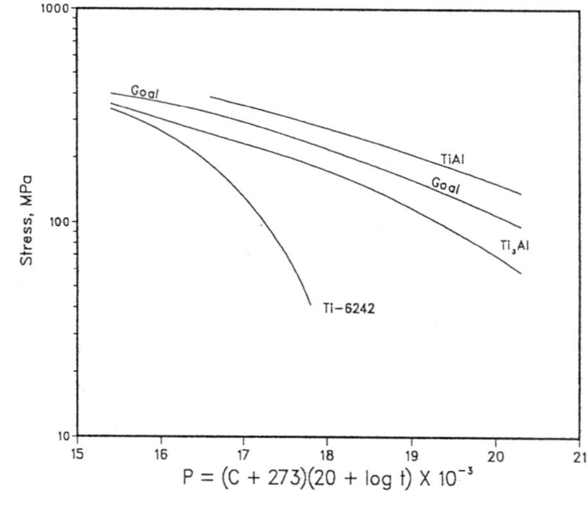

The fire problem restricts the use of titanium in some cases even when the mechanical properties of the titanium would be adequate at that temperature. Standards for fire resistance and its effect on alloy usage and engine design are proprietary, and this subject has not been widely discussed in the open literature. At the present time the "requirements" and applications for titanium relative to burn resistance are determined primarily through service experience, and quantitative "goals" are not available. Standardized testing of titanium combustion for gas turbine applications does not exist. Standardized testing of other metals for gaseous oxygen applications is being developed (6). Since the fire problem is a major barrier to increased usage, further development and the establishment of test procedures will surely occur.

4. DEVELOPMENT EFFORTS TO MEET GOALS

4.1 Alloy Development

Ti-8-1-1, Ti-6-2-4-2, and Ti-6-2-4-6 have met most of the requirements for engine components for many years. Alloy development of conventional alpha and alpha + beta alloys is limited by the metallurgy of titanium. Excessive alpha stabilizers, such as aluminum and tin, lead to embrittlement by Ti_3Al formation. Increased beta stabilizer additions, such as Mo, Zr, or Cr, lead to poor toughness and may lead to loss of ductility after exposure. Additions of precipitate formers such as Si are limited by loss of ductility due to excessive amounts of the precipitate and a tendency for preferential precipitation at grain boundaries. As previously discussed, recently developed alloys reflect these limits in that they usually involve a compromise between an improvement in one area with a debit in another. The designer will have to choose alloys based on the particular limiting property of each application.

Alloy development based on conventional alloys will continue. In addition, there are alternative systems that show promise. The titanium aluminides, TiAl ("gamma") and Ti_3Al ("alpha-two") have been the subject of many investigations (7,8). The aluminides in general offer improved creep performance and high strength on a density-corrected basis, along with good oxidation characteristics, but suffer from poor toughness, impact resistance, and ductility. TiAl-type alloys in particular offer excellent high temperature capabilities with densities around 3.88 g/cm3 (0.14 lb/in3). Figures 11 and 12 compare the specific tensile and creep strength of TiAl relative to IN-718. TiAl obviously is very attractive, but does not show consistent ductility. Programs to improve the toughness and ductility of TiAl are underway, but it may still require a change in design philosophy to allow use of this material in rotating hardware.

Alloy development specifically for burn resistance has not been widely attempted or discussed. Most of the available information comes from programs conducted by Pratt & Whitney with the U.S. Navy and U.S. Air Force in the 70's and early 80's (9-11). P&W has developed an in-house burn testing capability using a small window and a laser heat source to simulate ignition and combustion in small titanium samples (Figure 13). This device was used primarily for characterizing combustion limits of conventional alloys. Figure 14 shows samples which burned completely and samples which only burned to the limit of the igniter, but did not sustain combustion. The ease of ignition and degree of combustion are different in the various titanium alloy classes. For example, the aluminides Ti_3Al and TiAl were found to be more burn resistant than Ti-8-1-1 and Ti-6-4. Binary alloys such as Ti-13Cu and Ti-20Ni were found to be non-burning in the P&W rig. The extent of alloying was related to burn severity.
The commercial alloy Ti-13V-11Cr-3Al was also found to be more

Figure 11: Specific Tensile Strength Of TiAl vs IN–718

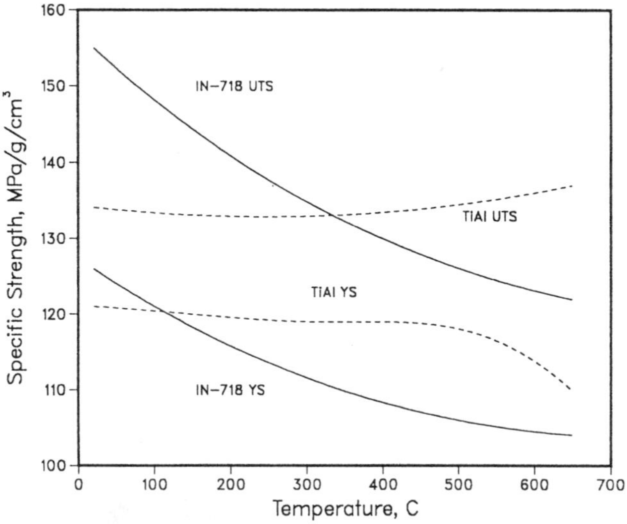

Figure 12: Specific Creep Strength Of TiAl vs IN–718

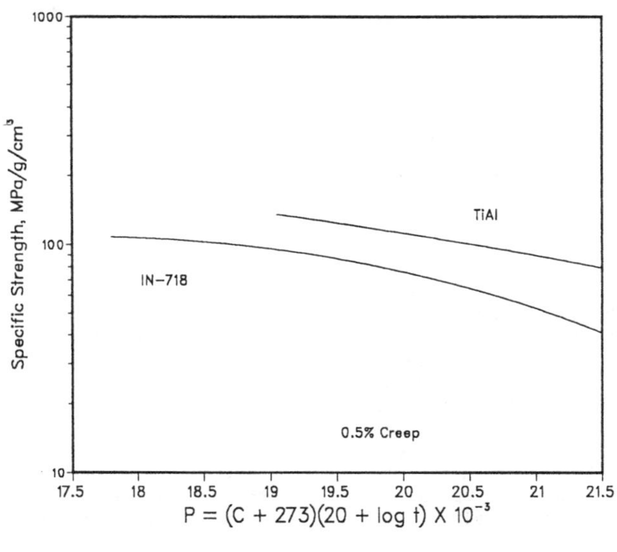

Figure 13: Schematic Of Burn Test Rig

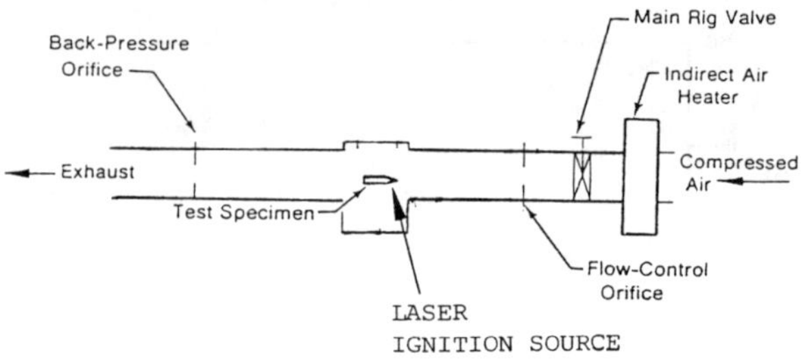

Figure 14: Comparison Of Burn Test Samples

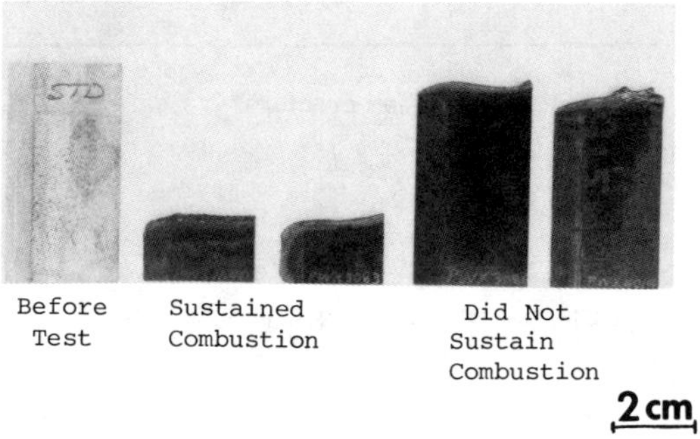

burn resistant than the alpha and alpha-beta alloys. The general trend seen was that heavily alloyed compositions such as the aluminides or the beta alloys were more burn resistant. The burn resistance of the aluminides is another factor in addition to their creep capability supporting programs to improve their poor toughness. The beta alloys in general are limited to low temperature usage where fire considerations are not important anyway. Still, some recent alloy development in beta systems is being done (12). These studies are attempting to add strength by dispersion strengthening and rapid solidification processing.

4.2 Process Development

Titanium is very sensitive to failure at internal defects, a fact that has been long recognized and has led to the improvements in ingot metallurgy described earlier. The need for further improvement remains. Type I defects are particularly troublesome because they are difficult to find using standard ultrasonic methods.

Two recent melting innovations offer hope of reducing the occurrence of Type I, Type II, and HDI defects. Both are a variation of Cold Hearth Melting (CHM). In one method an electron beam (EB) provides the melting power, under a vacuum atmosphere. In the other, a plasma arc (PA) provides the melting power, under a vacuum or inert gas atmosphere. These processes allow for improved dwell time in the liquid state, longer solution periods, and more complete mixing. Any unmelted refractory metals or oxides tend to sink to the bottom of the hearth instead of going into the ingot. Programs to make production quantities of material by both methods and evaluate properties are underway. Included in these programs are development efforts for improved sonic detection methods. Although these programs may not provide a direct increase on mechanical strength, they will ensure a greater degree of safety and reduce the chance of catastrophic, premature failure as titanium is used in components at increased loads and temperatures.

Rapid solification processing (RSP) of titanium alloys is an area that shows promise in improving properties. For titanium this can provide several benefits. Titanium is a universal solvent when molten, but on cooling, many alloying additions are prone to segregation. For this reason, titanium is usually produced as very large ingots, then hot worked and heat treated extensively to break up and homogenize the segregated areas. RSP allows addition of elements in higher amounts while still achieving a homogenous end product. Dispersion-forming elements can be added to form distributions of strengthening particles not possible through conventional ingot metallurgy. Further, the powder particles can be screened to fine powder sizes, effectively placing an upper limit on the maximum defect in the final component.

There are several methods of RSP. Examples include melt spinning, crucible melting plus rotary atomization, and consumable/non-consumable skull melting plus rotary or gas atomization. Pratt & Whitney has made RSP titanium using all three methods. The crucible melt plus rotary atomization route has been used most extensively, and has been used to make over 1800 kg (4000 lb.) of titanium powder. This process may involve an interaction between the crucible material and the melt, which may be a problem depending on the alloy. To eliminate the interaction a system using skull melting has been brought on line. Initial production using a $3 million facility began in late 1987, and production is nearing the 450 kg (1000 lb.) mark.

Alloying additions possible using RSP include rare earths for oxide dispersion strengthening (12-14), dispersion formers such as Si and B, and high amounts of segregation-prone beta stabilizing elements. Most investigations have concentrated on dispersion strengthening through rare earth additions. Erbium has been reported to provide optimum dispersions and improved creep strength in binary alloys (13-15). However, dispersion coarsening occurs above the beta transus (16), which restricts heat treatments in alpha alloys and will lead to unavoidable coarsening in alpha-beta and beta alloys. Grain boundary coarsening of dispersoid in RSP-processed Ti-6-2-4-2, Ti-6Al-3Sn-3Zr, and Ti-6Al-6Zr has been observed (17). To date, no conclusive evidence of benefits in creep properties from dispersion strengthening has been shown, but development is continuing. Studies to identify more stable dispersions such as rare earth-oxygen-sulfur compounds are underway (18). Effective use of dispersions in titanium alloys will involve the combination of RSP processing with low temperature consolidation and processing.

As mentioned before, burn resistance is an additional and important criteria for more aggressive application of titanium. However, reports on alloy development in this area are limited to date. A general trend observed is that heavily alloyed compositions have increased burn resistance, such as the aluminides and the beta alloys. Should further development and testing for burn resistance identify these systems as the best candidates, RSP processing may be necessary because of difficulty in achiving homogenous ingots with extensive alloying additions.

5.0 SUMMARY

The payoff from the use of titanium in gas turbine engines in terms of thrust/weight ratio improvements has been large. Alloy development and processing innovations over the years have provided a stable of alloys which are used with confidence and a high level of reliability. Current state-of-the-art, however, limits application of this technology to a temperature limit of approximately 538°C (1000°F). To move beyond this limit, general requirements with respect to strength, damage tolerance and environmental resistance have been defined. Notable progress has been made in understanding this temperature limit and in defining alloys and innovative processes which can surpass the barrier. Much remains to be done.

6.0 REFERENCES

1. Woodfield, A.P., et al., Acta Met 36, 507-515 (1988).

2. Bania, P.J., J. Metals 40, 20-22 (1988).

3. Dadras, P. and Thomas, J.F., Jr., Met. Trans. 12A, 1867-1876 (1981).

4. Chen, C.C. and Coyne, J.E., Met. Trans. 7A, 1931-1941 (1976).

5. Elrod, C.W., AFWAL TR-81-4019, 2, 55-92 (1981).

6. Benning, M.A., ed. Flammability and Sensitivity of Materials in Oxygen-Enriched Atmospheres: Second Volume, ASTM STP 910, American Society for Testing and Materials, Philadelphia (1986).

7. Lipsett, H.A., Schechtman, D., and Schafrik, R. E., Met. Trans. 11A, 1369-1375 (1980).

8. Graves, J.A., et al, Scripta Met. 21, 567-572 (1987).

9. Anderson, V.G., and Manty, B.A., NADC-77153-30, (1978).

10. Glickstein, M.R., AFML-TR-79-4001 (1978).

11. Glickstein, M.R., et al, AFWAL-TR-81-2092 (1981).

12. Rowe, R.G., et al, in Rapidly Solidified Materials: Properties and Processing, edited by P.W. Lee and J.H. Moll, ASM International, Metals Park, Ohio (1988), pp. 99-106.

13. Sastry, S.M.L., Meschter, P.J., and O'Neal, J.E. Met. Trans. *15A*, 1451-1463 (1984).

14. Sastry, S.M.L., Peng, T.C., and Beckerman, L.P., Met. Trans. *15A*, 1465-1474 (1984).

15. Konitzer, D.G., et al, Acta Met. *34*, 1269-1277 (1986).

16. Lofvander, J.P.A., et al, Scripta Met. *21*, 859-861 (1987).

17. Rowe, R.G., et al, in *Rapidly Solidified Materials*, edited by P.W. Lee and R.S. Carbonara, ASM, Metals Park, Ohio (1985), pps. 107-113.

18. Rowe, R. G., and Koch, E.F. in *Rapidly Solidified Materials*, ed. by P.W. Lee and R.S. Carbonara, ASM, Metals Park, Ohio (1985), pp. 115-120.

7. BIOGRAPHY

Eric S. Huron received a Bachelor's Degree in Metallurgical Engineering from the University of Cincinnati and a Master's Degree in Metallurgy from Georgia Institute of Technology. As an undergraduate student he participated in a co-operative education program and was employed by the Timken Company. Mr. Huron is now a Senior Materials Engineer with Pratt & Whitney, Operations, and is responsible for titanium alloy development and rapid solidification processing.

John A. Miller received a Bachelor's Degree in Mechanical Engineering (Materials Option) from Marquette University. Through his career Mr. Miller has worked in the areas of titanium alloy and process development, powder metallurgy processing (nickel and titanium) and turbine airfoil alloy and process development. He is currently Supervisor, Advanced Materials Processing with Pratt & Whitney, Operations.

2nd International SAMPE Metals Conference
August 2-4, 1988

Ti-1100: A NEW ELEVATED TEMPERATURE TITANIUM ALLOY

Paul J. Bania
TIMET Henderson Technical Lab
P.O. Box 2128
Henderson, NV 89015

Abstract

A new elevated temperature titanium alloy has been developed which offers roughly a 55°C creep and stress rupture advantage over beta processed Ti-6242-Si. This alloy, designated Ti-1100, has a nominal aim composition of Ti-6Al-2.8Sn-4Zr-.4Mo-.45Si-.07O_2-<.03Fe. Preliminary studies reported in this paper show that this creep advantage over Ti-6242-Si is achieved without a significant trade-off of strength, toughness, crack growth resistance or fatigue properties. Although the optimum processing sequence appears to be beta forge plus direct stabilization age, the alloy has not demonstrated a strong creep or tensile sensitivity to processing parameters such as forge temperature, cooling rate or final age temperature. This could provide for a relatively wide processing window and greater uniformity of properties compared to other process sensitive alloys. The Ti-1100 alloy has also been shown to exhibit excellent bulk stability after long exposures, although a coating may be required for long term exposures over roughly 540°C.

1. INTRODUCTION

The historical development of high temperature titanium alloys has been reviewed[1,2] and shown to be a progressive evolution. In the mid-1950's when the Ti-6Al-4V alloy was developed, it had a maximum use temperature of approximately 350°C (660°F). Towards the late 1950's and early 1960's, two additional alloys emerged with a maximum temperature capability of roughly 400°C (750°F). These were the Ti-7Al-4Mo and Ti-8Al-1Mo-1V alloys. Although these alloys offered increased temperature capabilities, they were somewhat limited because of their stability - ie, they became embrittled after long exposures at elevated temperature. In the mid-1960's, the alloys Ti-679 (Ti-11Sn-5Zr-1Mo-2.5Al-.2Si) and Ti-6Al-2Sn-4Zr-2Mo emerged, pushing titanium's temperature limit even further The Ti-679 alloy had a useful limit up to roughly 450°C (840°F) while the Ti-6242 alloy was useful up to 480°C (900°F). In the late 1960's, Ti-685 (Ti-6Al-5Zr-.5Mo-.25Si) pushed the limit to around 510°C (950°F) and in the early 1970's, a .1% silicon addition to Ti-6242, now called Ti-6242-Si, extended that alloy to the 510°C range and even as high as 540°C (1000°F) in some cases. Although other alloys have been introduced with varying degrees of success, none have increased the temperature capability of conventional titanium alloys beyond that 540°C (1000°F) barrier. It should be noted, however, that other non-conventional alloys such as alpha-two and gamma titanium aluminide compounds do offer significant promise in this regard. However, the limited ductility and toughness of those alloys will probably limit their use.

2. ALLOY REQUIREMENTS

As is often the case in alloy development programs, a new design (in this case a gas turbine) in the early 1980's identified the need for yet another temperature improvement. A titanium alloy was required which provided a creep resistance improvement of up to 40°C (75°F) over the best alloy available at the time - beta processed Ti-6242-Si. An alloy development program was initiated in order to satisfy this requirement. Although creep resistance was the primary goal, the improvement was to be achieved without a significant sacrifice in strength or other design properties such as stability, toughness, fatigue and fatigue crack growth.

3. ALLOY DEVELOPMENT

The alloy development program began with the screening of over 250 compositions of button (250-gm) heats. These studies identified compositions which were scaled to 45 kg (100-lb) heats in order to provide forged product for evaluation. The most promising of these alloys was then scaled to several 816 kg (1800-lb) heats for melting and conversion studies as well as thermomechanical processing (TMP) studies. This successful progression culminated with the production and evaluation of a production-sized 3,630 kg (8,000-lb) ingot.

The outcome of this alloy development study was a composition consisting of Ti-6Al-2.75Sn-4Zr-.4Mo-.45Si-.07O_2-.02Fe. This alloy is clearly a modification of the Ti-6242-Si alloy which is so widely used today. Although the chemistry differences would appear to be subtle, they are quite dramatic in their effect on creep response. Examples are shown in Figure 1. Figure 1(a) shows the effect of silicon in the new alloy. This plot shows that creep resistance is significantly enhanced up to ~.5% silicon but beyond that point post creep ductility (stability) is compromised with no further creep enhancement. A similar plot could be shown for tin, with stability sacrificed above the 3% level. A plot of the iron effect in this base chemistry is shown in Figure 1(b). This plot demonstrates a strong iron effect on time to .2% creep strain at the 510°C test condition cited, necessitating iron levels well below those typically encountered in the Ti-6242-Si alloy. The iron effect in this base is somewhat surprising in light of the Seagle and Ankem(3) results which showed little or no effect of iron at these levels on creep properties in beta processed Ti-6242-Si at a similar test temperature. This is clearly a topic which requires further exploration. One final comment on the iron effect should be made. If creep tests are run at higher temperatures, eg at 565°C (1050°F) or above, the iron effect disappears. This is likely due to a change in creep mechanism at the higher temperatures as demonstrated by Bania and Hall(4) in earlier work on Ti-6242-Si. It is not clear at this time why this modified alloy is so much more sensitive to iron content than its' predecessor.

The aluminum level in the new alloy was kept at 6% due to stability problems at higher levels and strength problems at lower levels. Zirconium was kept high in order to promote a uniform distribution of silicides in light of the high silicon level. Thus, the chemistry of this alloy was optimized not only for creep strength, but also for stability, strength and uniformity.

As will be shown in a subsequent section, this critically controlled chemistry will exhibit roughly a 55°C (100°F) advantage over the beta processed Ti-6242-Si alloy, thus extending the useful range to roughly 593°C (1100°F). This 1100°F alloy has been named Ti-1100.

4. PROCESSING

4.1 Melting and Conversion

The Ti-1100 alloy, due to its extremely low iron limit, requires a select grade of titanium sponge. However, sponge containing roughly

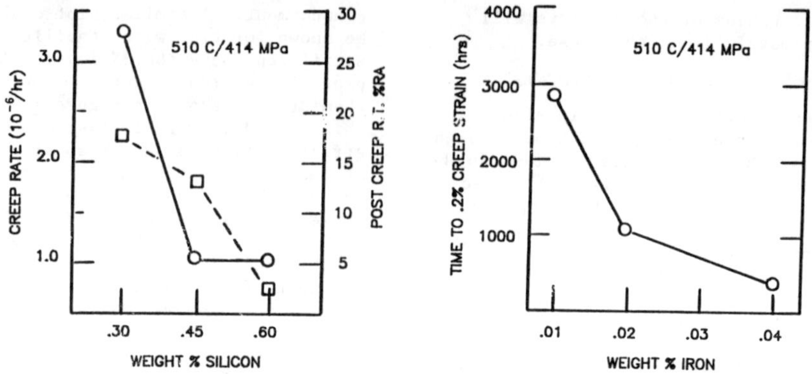

(a) (b)

Figure 1: Chemistry Effects in Ti-6Al-2Sn-4Zr-.4Mo-xSi-.XFe base.

100 ppm Fe (.01%) has been produced on a commercial basis and hence no problems exist concerning raw material supply.

In terms of melting, the high Si content of this alloy calls for special controls during VAR melting, especially on the third and final melt. However, Ti-550 (Ti-4Al-2Sn-4Mo-.5Si) has a comparable silicon content and this alloy has been successfully melted for several years. Early indications suggest that Ti-1100 will behave similar to Ti-550.

The Ti-1100 alloy has been successfully processed to billet, bar, sheet and weld wire. Other product forms such as plate, castings and foil are currently being evaluated.

4.2 Forging

The Ti-1100 alloy was developed primarily to optimize creep resistance in forgings such as compressor discs and impellers. It is well established[5-7] that for alpha-beta or near-alpha alloys, the beta processed microstructure is the best in terms of creep resistance. Since this alloy was optimized for creep resistance, the beta processed microstructure was chosen.

Once beta processing is decided upon, there are still a number of questions to be answered. For example, will the forging be beta forged (ie, forged above the alloy's beta transus) or alpha-beta forged followed by a beta anneal? How far above the transus should one forge/anneal? What will the cooling rate be? What about final age temperature? Obviously, the answers to these questions should be arrived at based on the impact of these parameters on resultant properties.

Since it has been shown in a prior study[8] that beta forging plus direct aging resulted in improved post-creep ductilities (ie, stability) than beta annealed material, the beta forge + direct age route was selected.

The next step was aimed at evaluating forging temperature. While most forgers would relate forging temperature of titanium alloys to the beta transus, there is another consideration for this high silicon alloy - the silicide solvus. Since it was assumed that processing below the silicide solvus would harm creep resistance by coarsening silicides, it was decided to process above the silicide solvus. This was determined metallographically as approximately 1050°C (1925°F). This is 35°C (65°F) above the alloy's average beta transus of 1015°C (1860°F). In order to determine if forging temperature over the silicide solvus was

Table I. Cooling Rate Study (1)

Cooling Rate	Location(2)	YS (MPa)	UTS (MPa)	% El	% RA
		R.T. Tensile Results			
Oil Quench	A	901	996	10.0	18.3
	B	874	958	10.9	19.7
Air Cool	A	862	949	11.5	19.3
	B	860	960	11.0	17.9
Slow Cool	A	831	921	11.3	21.6
	B	824	927	12.3	18.1

Cooling Rate	Location	Creep Exposure °C/MPa	$\dot{\varepsilon}$	t(.2)	RA'
			Creep Results (3)		
Oil Quench	A	510/414	1.6	660	15.9
"	B	"	1.2	1300	12.0
Air Cool	A	"	0.9	980	18.2
"	B	"	1.5	820	17.6
Slow Cool	A	"	1.9	690	13.5
"	B	"	1.6	850	15.6
Oil Quench	A	538/276	2.9	380	5.6
"	B	"	2.5	590	9.1
Air Cool	A	"	1.8	800	17.4
"	B	"	2.2	350	11.6
Slow Cool	A	"	2.4	400	10.5
"	B	"	3.3	300	10.5
Oil Quench	A	593/165	2.4	550	6.6
"	B	"	1.6	1140	5.8
Air Cool	A	"	1.6	980	9.2
"	B	"	1.6	750	12.1
Slow Cool	A	"	1.5	880	----
"	B	"	1.5	1100	11.5

(1) All samples forged 15.2 cm → 7.6 cm from 1093°C.

(2) A = 1.2 cm from forging surface, B = from mid section.

(3) $\dot{\varepsilon}$ = steady state creep rate (%/hr x 10^{-4}); t (.2) = hrs to .2% creep strain, (extrapolated if over 500 hrs); RA' = room temp redn in area after ≥500 hrs creep exposure (surface removed). Single test results.

---

significant, a brief study was conducted comparing forging at 1093°C (2000°F) vs 1149°C (2100°F). Warm (~300°C) dies were used. The results showed no significant differences in creep or tensile properties.

Next was a study to evaluate cooling rate effects after the beta forge operation. Table I contains the results of this study. In this case, all material was beta forged from 1093°C (2000°F), again using warm dies. After beta forging, the 7.6 cm (3-in) thick x 15.2 cm (6-in) diameter pancakes were either oil quenched, air cooled or slow cooled off the press. Slow cooling involved a wrap in an insulating material. All material was given a post-forge stabilization of

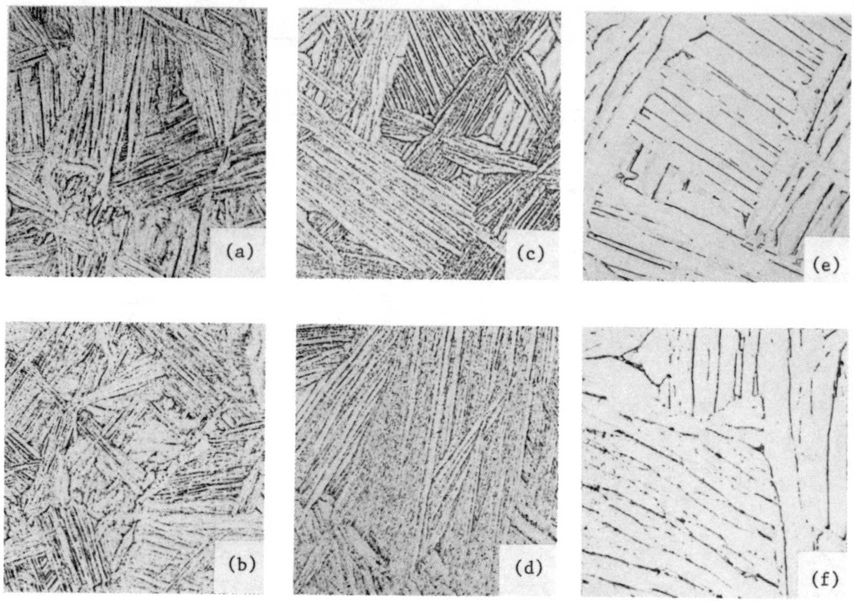

Figure 2: Microstructures of 7.6 cm thick pancakes (see Table II). Top row represents near surface, bottom row mid-section. (a) and (b) = oil quench; (c) and (d) = air cool; (e) and (f) = slow cool.

600°C (1100°F)/8 hr. The resultant microstructures, taken from approximately 1.5 cm below the surface and at mid-thickness are shown in Figure 2. These microstructures show that as cooling rate decreases, alpha-platelet thickness increases. Also, the faster cooling rate tends to produce a more basketweave type of structure as opposed to the larger colony structure associated with the slower cooling rates. The differences between the middle of the forging and the surface are minor. These microstructural trends are consistent with those observed in similar near-alpha alloys.

Comparing the properties noted in Table I, it is apparent that the oil quench resulted in about a 50 MPa increase in near surface strength over the air cool and a 70 MPa increase compared to the slow cool. However, the differences diminish to 14 MPa and 50 MPa respectively at the center. These are rather small effects suggesting that the alloy has very limited hardenability. The creep values also show quite minor effects, considering the large scatter inherent in such tests. It is thus concluded that cooling rate off the press is not a critical factor in determining strength and creep properties of this alloy. However, other factors such as fatigue, crack growth and toughness may be more affected by these microstructural changes.

4.3 Heat Treatment

Another important processing consideration is post-forging heat treatment. Table II summarizes the results of a variety of post forging treatments, ranging from a beta-anneal (Cond B) to a high alpha-beta anneal (C and D) to mill anneal type heat treatments (Conds E and F). All treatments were followed by a 600°C/8 hr stabilization. Since all were air cooled off the press, most yield strengths were within a 40 MPa range with only the long term high

Table II. Post Forge Heat Treat Study(1)

Post Forge Heat Treat(2)		R.T. Tensile Results YS (MPa)	UTS (MPa)	% El	% RA
A]	None	900	997	11.1	20.9
B]	1106°C/.5 hr/AC	882	977	7.5	15.9
C]	995°C/1 hr/AC	850	956	10.7	24.5
D]	870°C/24 hr/AC	870	958	9.8	15.9
E]	700°C/4 hr/AC	917	1010	10.1	13.6
F]	790°C/4 hr/AC	887	961	7.2	14.1

Post Forge Heat Treat	Creep Exposure °C/MPa	Creep Results(3) $\dot{\varepsilon}$	t(.2)	RA'
A]	510/414	1.0	1825	20.8
B]	"	0.6	2615	14.0
C]	"	0.9	1600	22.7
D]	"	7.9	175	14.2
E]	"	2.2	515	13.9
F]	"	3.1	450	13.1
A]	538/276	3.1	350	14.8
B]	"	2.0	735	18.4
C]	"	4.8	265	15.2
D]	"	21.5	70	13.2
E]	"	3.2	495	15.5
F]	"	8.0	175	11.1
A]	593/165	1.7	1080	8.2
B]	"	0.9	2270	4.5
C]	"	2.4	550	14.7
D]	"	2.9	425	11.5
E]	"	1.6	1125	11.0
F]	"	0.9	1970	9.5

(1) All material forged 7.6 cm → 1.9 cm from 1150°C/air cool (AC) off press.

(2) All given 593°C/8 hr after indicated heat treatment.

(3) $\dot{\varepsilon}$ = steady state creep rate (%/hr x 10^{-4}); t (.2) = hrs to .2% creep strain, (extrapolated if over 500 hrs); RA' = redn in area after ≥500 hrs creep exposure (surface removed). Average of 2 tests.

---

alpha-beta anneal (Cond D) falling slightly below this range.

Examining the creep results in Table II, several observations can be made:

a) The beta anneal treatment (Cond B) resulted in the best overall creep performance. However, it also resulted in the lowest post-creep ductility. This is consistent with the earlier experiment[8] which showed beta forge to result in better post-creep ductility than beta anneal.

b) While the high alpha-beta anneal treatment (Cond C) reduced creep resistance slightly, the long term anneal dramatically reduced creep resistance. This is likely due to silicide coarsening.

c) The mill anneal cycles (Cond E and F) in general also lowered creep resistance except for the 593°C creep tests.

The information contained in Table II suggests that, from a tensile and creep standpoint, a direct low temperature age after forging is the optimum practice.

Table III. Stabilization Age Study(1)

8 Hr Age Cycle	YS (MPa)	UTS (MPa)	% El	% RA
		Tensile Results		
538°C	896	986	12.5	19.6
593°C	930	1006	10.0	21.7
648°C	889	986	10.0	19.9

8 Hr Age Cycle	Creep Exposure °C/MPa	$\dot{\varepsilon}$	t(.2)	RA'
		Creep Results (2)		
538°C	510/414	0.8	1900	23.0
593°C	"	0.9	1520	22.3
648°C	"	1.7	530	21.1
538°C	538/276	3.6	420	17.1
593°C	"	2.4	590	15.4
648°C	"	4.2	250	16.7
538°C	593/165	1.3	1020	13.7
593°C	"	1.8	780	16.3
648°C	"	0.9	1560	15.8

(1) All material forged 6.4 cm → 1.9 cm from 1150°C/air cool off press.
(2) $\dot{\varepsilon}$ = steady state creep rate (%/hr x 10^{-4}); t (.2) = hrs to .2% creep strain, (extrapolated if over 500 hrs); RA' = redn in area after ≥ 500 hrs creep exposure (surface removed). Single test results.

A final study was conducted to evaluate final stabilization temperature with the results shown in Table III. These results show only very minor effects associated with the various stabilization age cycles.

Considering all of the data in Tables I, II and III, it is clear that a beta forge plus direct age at 600°C/8 hr is an optimum cycle for Ti-1100 from a strength and creep standpoint. However, since the properties are not highly sensitive to forging temperature, cooling rate off the press or final heat treatment, there should be considerable latitude in processing parameters - a distinct advantage when considering reproducibility.

5. ALLOY PROPERTIES

In cases where a comparison is made in this section between Ti-1100 and Ti-6242-Si, each alloy will be referenced to the beta processed microstructure, similar to that shown in Figure 2(c). All Ti-1100 properties are for material beta forged to 1.9 cm thickness at 1093°C, air cooled off the press, followed by a 600°C/8 hr stabilization.

5.1 Physical Properties

The physical properties of Ti-1100 from room temperature up to 760°C are summarized in Table IV. These properties are comparable to similar alloys such as Ti-6242-Si. The density of Ti-1100 is 4.51g/cc (.163 lb/cu-in).

5.2 Strength

Figure 3 compares the strength properties of Ti-1100 vs. beta annealed Ti-6242-Si. These results show approximate parity with beta annealed Ti-6242-Si, with an advantage in yield strength at the higher temperatures.

Table IV: Ti-1100 Physical Properties

Property	Equation (T = °C)
Specific Heat C(cal/gm°C)	$C = .117 + 6.7 \times 10^{-4} T$
Conductivity (w/m°k)	$Q = 6.62 + 1.27 \times 10^{-2} T$
Resistivity (10^{-8} ohm-m)	$R = 178 + 0.57T - 5 \times 10^{-5} T^2$
Thermal Expansion (10^{-6}/°C)	$\alpha = 8.12 + 8.17 \times 10^{-3} T - 1.37 \times 10^{-4} T^2 + 10^{-8} T^3$

A more important comparison might be of Ti-1100 to a wrought nickel-based alloy such as IN718. The following is a strength comparison at 593°C:

Alloy	UTS(MPa)/ g/cc		Specific Strength
Ti-1100:	674/4.51	=	149
IN718[9]	1251/8.19	=	152

At the maximum anticipated use temperature of Ti-1100, it is clearly competitive with IN718 on a specific strength basis.

5.3 Creep Resistance

Perhaps the most outstanding advantage of Ti-1100 over beta processed Ti-6242-Si is its' creep resistance. Table V provides a comparison of creep properties for Ti-1100 and Ti-6242-Si under three creep conditions. At the lower test temperature, the Ti-1100 alloy shows a reduction in creep rate by a factor of 7 whereas at the higher temperature a reduction factor of 25 is demonstrated. It should be emphasized that the steady state creep rates cited in Table V were obtained over a period of typically

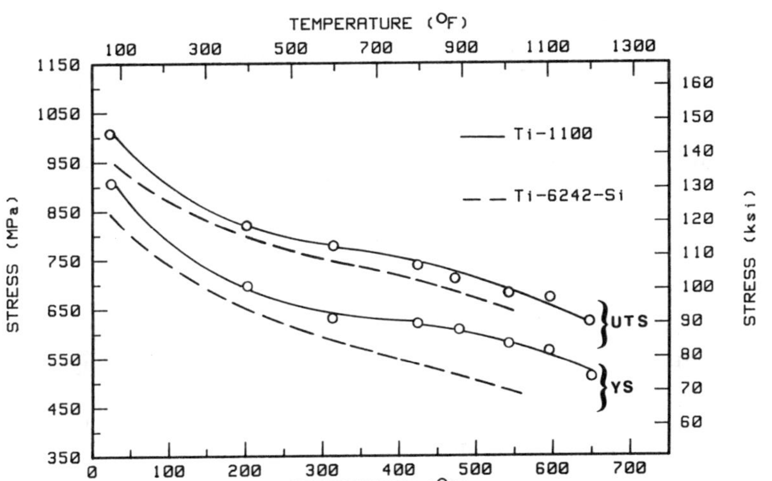

Figure 3: Strength Comparison of beta forged + aged Ti-1100 to beta annealed Ti-6242-Si.

Table V: Creep Comparison

Alloy	Creep Exposure 510°C/414 MPa $\dot{\varepsilon}$	510°C/414 MPa t(.2%)	565°C/275 MPa $\dot{\varepsilon}$	565°C/275 MPa t(.2%)	600°C/165 MPa $\dot{\varepsilon}$	600°C/165 MPa t(.2%)
Ti-1100	0.8	2200	3.0	430	1.6	1025
Ti-6242	5.5	150	65	40	41.0	40

$\dot{\varepsilon}$ = steady state creep rate (%/hr x 10^{-4}) t(.2%) = hrs to .2% creep.

350-400 hrs, with the remaining 100 to 150 hrs. of the 500 hr. creep test being consumed by Stage I or primary creep.

Another way of looking at creep, one more commonly used in design, is a Larson-Miller stress rupture plot. Such a plot is shown in Figure 4. This plot shows the 55°C advantage over Ti-6242-Si. It also shows that even compared to the Ti-14Al-21Nb titanium aluminide, the Ti-1100 is attractive up to ~620°C (1150°F) or slightly beyond.

5.4 Stability

One of the most critical properties of an elevated temperature titanium alloy is its' ability to retain acceptable ductility (at room temperature) after an extended creep exposure. In terms of bulk stability - ie, room temperature tensile ductility measured after creep exposure and removal of the surface contaminated layer, Ti-1100 meets the standard goal of greater than 10% reduction in area after 500-hr exposure up to as high

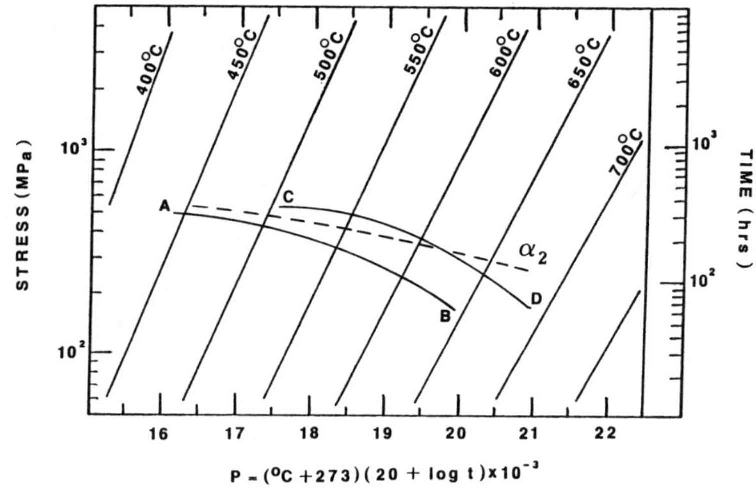

Figure 4: Larson-Miller Stress Rupture Time to Failure Comparison. A-B = Ti-6242-Si; C-D = Ti-1100.

as 650°C. However, if the surface layer is left intact, the %RA values are roughly halved. This suggests that coatings will likely be required for long term exposures above ~540°C (1000°F). It has been shown[8] that a thin sputtered Pt coating can be effective in reducing the loss associated with air contamination.

5.5 Fracture Toughness

The room temperature fracture toughness of Ti-1100 has been measured as 67 MPa-m½ while at 480°C the toughness is 59 MPa-m½. Comparable values for beta annealed Ti-6242-Si would be 85 and 75 MPa-m½ respectively. Although the values for Ti-1100 are lower, they are nonetheless quite sufficient for most anticipated uses. These beta processed Ti-1100 toughness values are also well above those reported for similar alloys in an alpha-beta processed condition.

5.6 Crack Growth Resistance

Figure 5(a) shows the crack growth behavior of Ti-1100 at room temperature for two r-ratios (K min/K max = r). Also shown for comparison purposes is a typical curve for beta annealed Ti-6242-Si. This plot shows that the Ti-1100 alloy is roughly equivalent to the Ti-6242-Si alloy, particularly in the more important low delta-K regime. The separation at the higher delta-K region is a reflection of the higher toughness of the Ti-6242-Si alloy. Nonetheless, these rates are exceptionally good from a damage tolerance standpoint, and highlight the advantage of the beta processed microstructure. Crack growth rates for alpha-beta processed micro-structures would be at least 2-3 times faster.

Figure 5(b) shows the crack growth behavior at 400°C and two r-ratios for a much slower frequency than the previous case (10 cpm vs 20 Hz). While direct comparison data is not available for Ti-6242-Si, it is clear that the temperature and frequency effects are not very prominent.

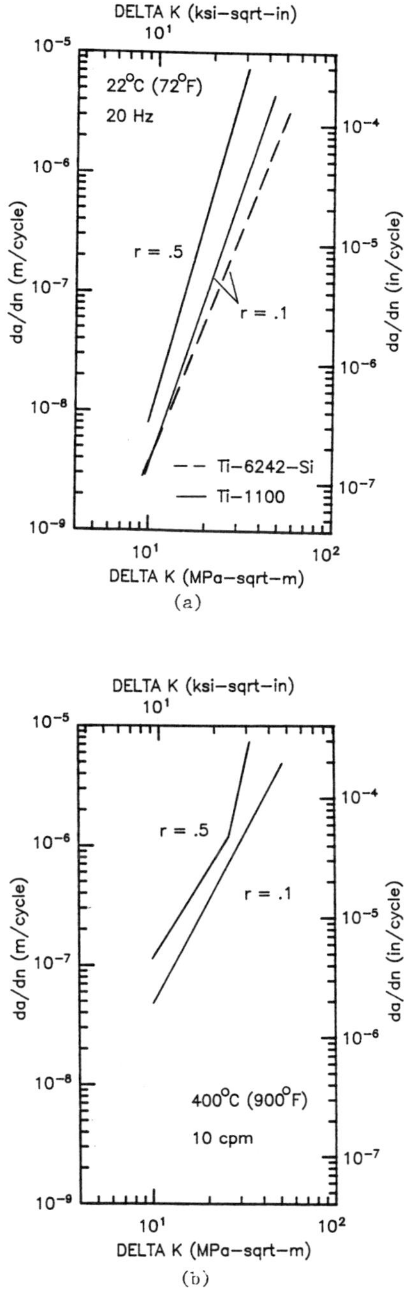

Figure 5: Fatigue Crack Growth Properties of Ti-1100.

Table VI: Ti-1100 High Cycle Fatigue

Temperature	K_t	10^7 Cycle Fatigue Strength
22°C	1.0	655 MPa
22°C	3.0	250 MPa
480°C	1.0	500 MPa
480°C	3.0	235 MPa

NOTE: f = 30 Hz, r = .1 for all tests

5.7 Fatigue Resistance

The high cycle fatigue strength of the Ti-1100 alloy is summarized in Table VI. This data demonstrates that, as expected, Ti-1100 shows no notch sensitivity in fatigue. These values for endurance strength are roughly comparable to Ti-6242-Si. This is supported by the work of Buttrill et al [10] wherein the high cycle fatigue strength of Ti-1100 was shown to be comparable not only to alpha-beta processed Ti-6242-Si, but also to alpha-beta processed IMI834, another near alpha high temperature titanium alloy.

Another interesting feature of Ti-1100 fatigue resistance was reported by Buttrill et al. One of the only drawbacks assumed to be associated with the beta processed microstructure was a reduced low cycle fatigue strength. However, the Buttrill work showed that beta processed Ti-1100 is comparable to alpha-beta processed Ti-6242-Si and IMI834 when tested at 540°C.

6. SUMMARY

1) A new alloy, designated Ti-1100, has been developed which offers approximately a 55°C creep advantage over beta processed Ti-6242-Si at roughly equivalent strength.

2) The Ti-1100 aim chemistry is: Ti-6Al-2.8Sn-4Zr-.4Mo-.45Si-.07O_2-<.03Fe.

3) The creep behavior of Ti-1100 is strongly affected by iron content when testing below roughly 565°C, but not above this temperature. The mechanism producing the iron sensitivity is not understood at this time.

4) Ti-1100 will be melted and converted by conventional practices.

5) Although beta-forge plus direct stabilization age was shown to be an optimum forging sequence for Ti-1100, the tensile and creep behavior of the alloy is not highly sensitive to processing parameters. However, other properties such as toughness, crack growth and fatigue may be more sensitive.

6) Beta forge plus direct age enhances post-creep ductility compared to beta anneal plus age.

7) When comparing processing effects, one should be careful to evaluate a range of creep test temperatures. Beneficial effects under low test temperatures could show detrimental effects at high temperatures or vice versa.

8) Ti-1100 possesses excellent bulk stability up to 600°C. However, due to surface interaction, a coating may be required for long term exposures in this temperature regime.

9) The damage tolerance properties of beta processed Ti-1100 are roughly comparable to beta processed Ti-6242-Si and in most cases superior to similar alpha-beta processed alloys.

7. REFERENCES

1. Rosenberg, H. W., "Critical Review: High Temperature Alloys", Titanium Science and Technology, Vol 4, Ed by Jaffee and Bunte, Plenum Press, 1973.

2. Blankinsop, P.A., "Critical Review: Developments in High Temperature Alloys", Titanium Science and Technology, Vol 4, Ed by Litjering et al, DGM, 1984. p 2323.

3. Ankem, S. and Seagle, S. R., "The Detrimental Effects Of Iron On Creep Of Ti-6242S Alloys", Titanium Science and Technology, Vol 4, Ed by Lutjering et al, DGM, 1984. p. 2411.

4. Bania, P. J. and Hall, J. A., "Creep Studies of Ti-6242-Si Alloy", Titanium Science and Technology, Vol 4, Ed by Bunk et al, DGM, 1984. p 1271.

5. Chen, C. C. and Coyne, J. E., "Relationships Between Microstructure and Properties of Ti-6242-Si Alloy Forgings", Titanium '80 Science and Technology, Vol 2, Ed by Kimura and Izuni, TMS-AIME Publications, 1988. p 1197.

6. Eylon, D., Hall, J. A., Pierce, C.M., and Ruckle, D. C., "Microstructure and Mechanical Properties Relationships in Ti-11 at Room and Elevated Temperatures", Met Trans. A, 7A, 1976. p 1817.

7. Vassel, A. et al, "Influence of Processing and Heat Treatment Variables on the Mechanical Properties of Two Advanced High Temperature Titanium Alloys", Titanium Science and Technology, Vol 4, Ed by Lutjering et al, DGM, 1984. p 515.

8. Bania, P. J., "Ti-1100: A New High Temperature Titanium Alloy", presented at Sixth International Titanium Conference, Cannes, France, June, 1988.

9. Aerospace Structural Metals Handbook, MCIC, Battelle Columbus, 1988 edition, Vol 4, Code 4103.

10. Buttrill, W., Hall, J. A., and Park, J., "Microstructure/Property Relationships in New Near-Alpha Titanium Alloys", presented at Westec '88 Conference, Los Angeles, CA, March, 1988.

8. BIOGRAPHY

Dr. Bania has been working in the titanium field for 17 years, commencing with his M.S. and PhD work at the University of Cincinnati.

In 1974, he joined the Air Force Materials Laboratory at Wright Patterson Air Force Base as a Project Engineer in the Titanium Group. In four years at AFML, he directed in-house and contractural programs dealing with microstructure/property relationships. In 1977, Dr. Bania joined Avco Lycoming where he later became Chief of the Materials Development Department. In this capacity, he directed alloy and process development programs for iron, nickel and titanium-based alloys applied to gas turbine engines. In 1981, Dr. Bania joined TIMET where he currently serves as Manager of Product Development. In this capacity, he has responsibility for research and development programs in the areas of melting, alloy development and process development. Dr. Bania is a member of TMS-AIME and ASM.

2nd International SAMPE Metals Conference
August 2-4, 1988

FAME - An Expert System For Failure Analysis Of MEtals

R. K. Elsley and W. A. Guthmiller
Rockwell International Science Center
P.O. Box 1085
Thousand Oaks, CA 91360
(805) 373-4155

S. M. El-Soudani and J. D. Young
Rockwell International North American Aircraft Operations
201 North Douglas Street
Los Angeles, CA 90245
(213) 647-6411, -6945

Russell Henderson
Air Force Systems Command
WPAFB, OH 45433-6533

ABSTRACT

The development of a diagnostic expert system for failure analysis of metal components is described. Through a consultation with the user, the expert system determines the failure mechanism, suggests causes of failure, and makes recommendations for corrective actions.

The expert system contains a knowledge base of rules relating failure descriptors to failure mechanisms, causes and corrective actions. It also contains a database of material properties and software for performing fracture mechanics and other calculations.

The user provides identifying information and the results of measurements via fill-in-the-blank forms which appear on the screen of the computer. As the expert system reasons, it asks for additional information (data and measurements) based on the failure hypotheses it is considering. During the course of the reasoning, it gradually increases its confidence in its conclusions. After the consultation, it prints a summary report. The user can also inquire as to the conclusions reached and the data used to reach them.

This system, when completed, will allow senior failure analysts to more quickly analyze and document failures and will allow junior analysts to perform more reliable analyses with confidence that relevant data and failure knowledge have not been overlooked.

1.0 INTRODUCTION

PERFORMING A FAILURE ANALYSIS involves a complex reasoning process which requires bringing to bear years of experience in a variety of disciplines. It is also a time consuming process which requires considering a large amount of detail information in order to avoid reaching erroneous conclusions.

Expert systems are computer programs which carry out reasoning processes based on large numbers of rules which have been derived from the experience of human experts. The user of the expert system participates in a dialog with the computer in order to enter the data which the expert system needs to carry out its reasoning. The expert system draws as many conclusions as it can from the data and reports them to the user.

The use of an expert system as an aid to a failure analyst offers several advantages. For the senior failure analyst, it speeds the process and frees his time for working on the most difficult aspects of the analysis. It also provides automatic report generation at the end of the analysis. For junior failure analysts, the expert system provides a tool which will methodically check a wide variety of possibilities and ensure more reliable analyses.

The kinds of information provided by a failure analysis are summarized schematically in Figure 1. First, the failure mechanism(s) are determined. For example, the failure mechanism may be high cycle fatigue. Next the cause(s) of the failure are determined. For example, a design exceedance may have occurred during service. Then corrective actions are proposed, such as derating the component or redesign. Eventually, policy decisions may be made, such as grounding a fleet.

Human failure analysts perform a portion of each of these tasks, with the remainder being performed by others who have the required expertise. In the first area (failure mechanism), the failure analyst performs the entire task. In the second area (cause of failure), other experts are brought in to provide additional input. In the third area (corrective actions), the failure analyst makes some recommendations, but much of the work (such as redesign) is performed by experts in other areas. Finally, policy decisions are usually made not by failure analysts, but by those with policy responsibility.

This same pattern holds true for the failure analysis expert system. It will comprehensively cover failure mechanisms. It will do what is within its scope to determine causes and it will make what recommendations it can for corrective actions.

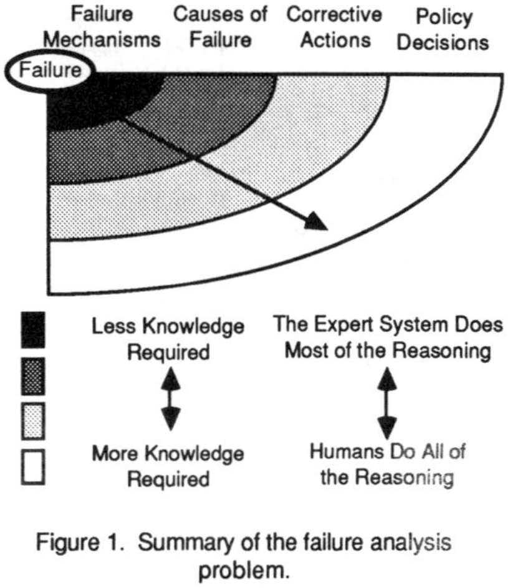

Figure 1. Summary of the failure analysis problem.

2.0 FAILURE ANALYSIS METHODOLOGY

In order to develop an expert system which can perform a failure investigation the way a senior failure analyst would, it is first necessary to codify the failure analysis procedures, logic and experience of the human expert. Figure 2 is a schematic representation of the analysis sequence upon which the expert system is based.

The organization of a failure investigation for an expert system is shown in Fig. 3 in which a conventional "Failure Analysis Methodology"[1-7] (shown in Table A) is divided into two tasks;

1) an examination of those variables which led up to the failure, designated "Fracture Contributors/Drivers," and

2) the post mortem investigation and testing of the failed component (evidence) to determine the "Fracture Topography Descriptors."

TABLE A
FAILURE ANALYSIS METHODOLOGY

- Aircraft/Failed Component Data
- Drawing Requirements
  - Part
  - Material
    - Material Description
    - Chemistry (bulk)
    - Microstructure
    - Properties
    - Susceptibility
  - Geometry (A_F, K_T, K_S), Fracture
  - Processing/Heat Treat
  - Processing/Surface Maching
  - Inst.I/Ass'y.
  - NDI
  - Drawing Error
- Service Environment
  - Loads
  - Thermal
  - Chemical
  - Maintenance
  - Exceedence
- Failure Description (Macroscopic)
  - Fracture
  - Surface
  - Part
- Microstructural Details (Microscopic)
  - Fracture
  - Surface
  - Base Metal
- Chemical Details (Microscopic)
  - Fracture
  - Surface
  - Microstructure

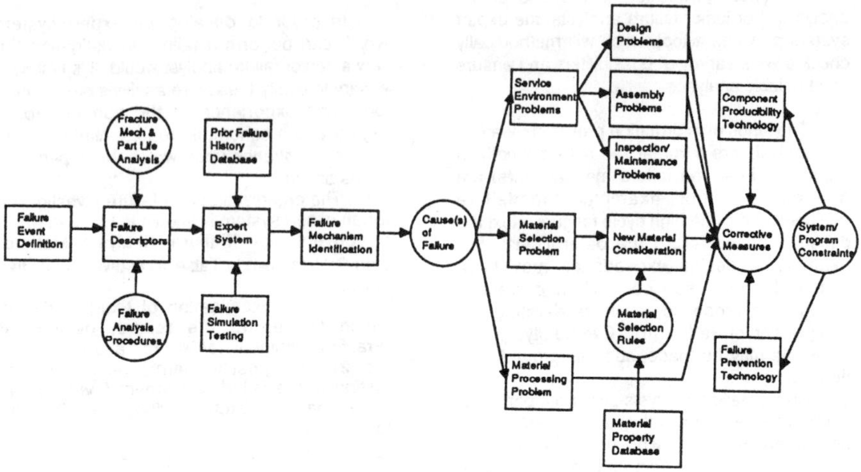

Figure 2. Flow Diagram for Failure Analysis

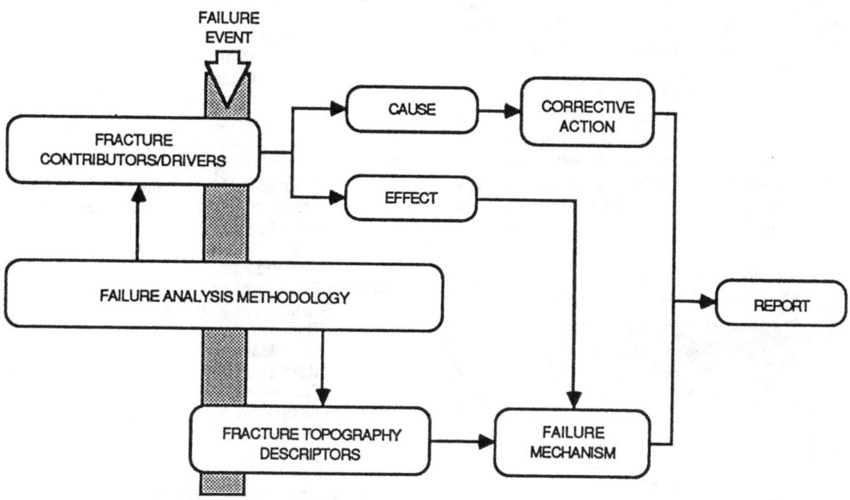

Figure 3. Failure Analysis Methodology Organization For an Expert System

Fracture Topography (Microscopic)
- Origin
- Zone II
- Zone III
- Zone IV

Laboratory Failure Simulation
- Test
- Results

Fracture Mechanics/Stress Analysis
- Stress Analysis
- Fracture MechanicsFailure Mechanism

The "Fracture Contributors/Drivers," shown in Table B,evaluate the "Cause of Failure," as shown in Table E, while the"Fracture Topography Descriptors" shown in Table D determine the "Failure Mechanism" listed in Table C.

TABLE B
FRACTURE CONTRIBUTORS/ DRIVERS

Design/Stress
- Stress Analysis
- Fracture Mechanics
- Service Environment Assumptions
- Material/Processing Selection
- Configuration

Geometry
- A_F(Area of fracture)
- K_T /K_S (Stress conc. factor)

Material
- Properties
- Susceptibility
- Chemistry
- Microstructure
- Product Form

Processing
- Heat Treat
- Surface/Machining
- Forming
- Welding
- Mill

NDI
- Defect (type/size/distribution)
- Geometry

Installation/Assembly
- Preload/Overtorque
- Interference/Residual Stress
- Galvanic Couple
- Joining

Service Environment
- Loads
- Thermal
- Chemical
- Exceedence
- Maintenance/Inspection

Drawing Error(s)

Sqawk(s), NDR(s), EO(s)

TABLE C
FAILURE MECHANISMS:

Overload - Tension (Low-Int. de/dt)
- HI K_{1C} material
- LO K1C material

Overload - Tension (HI de/dt,Low T)
- HI K_{1C}material
- LO K_{1C}material

Overload - Tension (elevated T)

Overload - Bending (LO-INT. de/dt)
- HI K_{1C}material
- LO K_{1C}material

Overload - Bending (HI de/dt)
- HI K_{1C}material
- LO K_{1C}material

Overload - Shear

Fatigue - Stage I

Fatigue - Stage II

Fatigue - Stage III

Processing - Cold/Warm Forming Cracks

Processing - Hot Forming Cracks

Processing - Extrusion

Processing - Machining

Creep/Stress Rupture

Low Temp. - Ductile/Brittle Transition

Processing - Welding Hot Tear

Processing - Grinding Cracks

Heat Treat - Quench Cracks

Creep Fatigue

Thermal Fatigue

T/M Processing - Overheat

Thermal Cracking

Stress Corrosion Cracking

Corrosion Fatigue

Hydrogen Embrittlement

Heat Treat/Service - Temper Embrittlement

Solid/Liquid Metal Embrittlement

Heat Treat - Underaged/Undertempered

Processing - Coating/Surface Hardening

Each "Failure Mechanism," while being defined by a set of"Fracture Topography Descriptors" is driven by a specific combination of "Fracture Contributors/Drivers" which may take the form of

1) 'Told' load, thermal, &/or chemical service environment 'Fracture Drivers,' which may, by themselves, drive a 'Failure Mechanism' &/or

2) a discrepancy/deficiency ("Fracture Contributor") which can act by itself, or in concert with a service environment condition to drive a failure mechanism.

Each "Fracture Contributor/Driver" results in a cause/effect relationship for a given "Failure Mechanism," whether 'Told' *a priori* from the service environment, or determined *in situ* from the failure investigation. Any discrepancy in

1) design/stress analysis
2) geometry
3) material
4) processing
5) NDI

6) installation/assembly &/or
7) service environment

can be a "cause" of failure. For example, "material microstructure" [too coarse grain size] may be a "cause of failure," which specifies a "corrective action" shown in Table F.

TABLE D
FRACTURE TOPOGRAPHY - DESCRIPTORS:

Surface Origin
Subsurface (s/s) origin
Multiple Origins or Ratchet Marks
Shear Lip/Slant Fracture
Flat Fracture
Mixed Mode
Intergranular (I/G)
Transgranular (T/G)
Microvoids
Cleavage
Striations
Tearing /Tear Ridges
Oxidation
Corrosion Products
Pitting
Thumbnail
Radial Growth
Featureless/Slip Plane Fracture
Featureless/Intergranular
Crack Branching /Secondary Cracks
Fan Pattern
River Marking
Elastic
Elastic/Plastic
Plasticity/Plastic Def.
Necking
Brittle/Ductile
Grain Facets
Ligament
Compression Hinge
Chevron Pattern
Growth Rings/Conchoidal Markings
Zone
Path

The "Effects" resulting from these "causes" (shown in Table G) can be either, 1) degraded part/material or, 2) increased severity of service environment, with specific properties/conditions involved. Each "Failure Mechanism" can be driven only by specific 'Effects.'

The specific relationship between "Fracture Topography Descriptors," "Fracture Contributor/Driver Cause/Effects," and"Failure Mechanisms" reside in the expert system as fault tree matrices defined by "If/Then" rules.

3.0 ARCHITECTURE

The architecture of the expert system is shown in Figure 4. All of the system resides in a PC/AT class computer. The MS Windows[8] environment is used in order to allow the expert system software to interact with the other software components. The expert system itself is built using the Nexpert/Object[9] expert system building shell. It interacts with the user through a keyboard- and mouse-based interface, and with a materials property database and with external programs for performing numerical calculations.

TABLE G
EFFECTS

DEGRADED PART/MATERIAL
- Embrittled/Increased Susceptability.
  - K_{1SLC}
  - Susceptability
  - S/t
  - da/dt
- Lowered Mechanical Properties
  - F_{tu}
  - F_{tys}
  - el
  - R.A.
  - H_R
- Lowered Toughness
  - CVN
  - K_{1c}
  - K_{init} @ a_i
  - $K_{crit}max$ @ a_c
- Lowered Fatigue Properties
  - S_E, n_F
  - 'R' Factor
  - da/dn
  - K_{TH}
- Introduced Initial Flaw
  - a_i

INCREASED SEVERITY OF SERVICE ENVIRONMENT
- Increased Operating Stress
  - A_F
  - K_T
  - K_S
  - S^S_{nom}
  - S^D_{max}
  - dS/dx
  - S_{RES} (+)
  - de/dt
- Increased Thermal Exposure
  - dT/dt
  - T_{opr}
  - dT/dx
- Increased Corrosion Exposure
  - DELTA EMF
  - pH
  - E_{corr}

TABLE E.
CAUSE(S) OF FAILURE:

Design

Fracture mechanics or stress analysis too conservative.
Service environment assumptions were wrong.
Material and/or processing selection not optimized for service enviornment.

Geometry

Area of fracture (A_F) inadequate for loads.
Stress concentration factors (K_T,K_S) (SPECIFY) too severe.

Material

Material properties not optimized for service environment/design.
Material selected susceptible to (SPECIFY).
Chemical composition did not meet specification, impacting properties (SPECIFY).
Microstructure is discrepant, indicating incorrect processing (SPECIFY) and impacting properties (SPECIFY).
Wrong product form used, was (SPECIFY), should be (SPECIFY).

Processing

Incorrect heat treat used, resulting in discrepant properties (SPECIFY).
Discrepant surface treatment/machining (SPECIFY DEFECT/ANOMALY).
Weld not performed in accordance with specification, (SPECIFY DISCREPANCY).
Part exhibits positive residual stress profile at fracture plane.

TABLE F.
CORRECTIVE ACTION

Design

If fracture mechanics or stress analysis were too conservative,
Then redesign the part with revised loads, or
reduce the severity of service environment by part isolation or flight restriction.

If the design service environment assumptions were wrong,
Then input the correct service environment conditions and redesign the part.

If material and/or processing selection was not optimized for actual service environment,
Then re-evaluate material and processing selection based on actual/correct service environment, recommend new material and processing.

Geometry

If area of fracture (A_F) is inadequate because of machining error,
Then inspect fleet. limit part life, and replace with part machined to drawing.

If (A_F) is inadequate because of design error,
Then limit existing part life, insepct fleet for cracked parts and retrofit fleet with redesigned parts.

If K_T and K_S are too severe because of machining errors,
Then rework existing parts to meet drawing or removed and replace existing parts with correctly machined parts.

3.1 USER INTERFACE - The expert system interacts with the user in two ways. Initially, the user provides preliminary information by means of a fill-in-the-blank form displayed on the screen. (See "Forms-based Input" below.) During the reasoning process, the expert system asks the user for further information that it requires. This may include the results of tests on the specimen, service environment data, etc. (See "Reasoning-based Input" below.) Finally, the expert system produces a report summarizing the analysis. (See "REPORTING" below.)

3.1.1 Forms-based Input - Initial data is acquired from the user by means of a fill-in-the-blank form displayed on the screen. Figure 5 shows a portion of the form as it appears to the user. This data includes part identification information, design information, and preliminary failure data. Any information which is unknown may be left blank and the expert system will either ask for it later or do without it.

3.1.2 Reasoning-based Input - During the course of the reasoning process, the expert system will require additional information. This information may include visual observations by the operator, the results of tests on the specimen, service environment data, etc. For example, in order to determine what type of fatigue has occurred, it may ask the user whether striations are fine or coarse.

Figures 6-8 are examples of three types of questions which the expert system may ask during the reasoning process. In Figure 6, the question is one which should be answered with either a yes (true) or a no (false). The expert system offers these two options and the operator need only click the mouse in the appropriate box. In addition, two other boxes are present. "Why?" provides information on why this question is being asked and "Apropos" provides additional information to explain the question. The user can also specify "NOTKNOWN" and let the expert system continue its reasoning without this piece of information.

Figure 7 asks for the value of a quantity for which there are a set of possible answers. In this case, there are two geometries about which the system knows how to reason. The possible answers are offered to the user in a list and he selects the appropriate one.

Figure 8 asks for the value of a numeric quantity. The user types in the appropriate value.

3.2 DATA - The expert system uses a number of types of data to carry out the reasoning process. These include:

Identifying information about the part being analyzed.

Design and service environment data.

Observations of the part by the user.

The results of tests performed on the part.

Hypotheses which the system is considering or has confirmed.

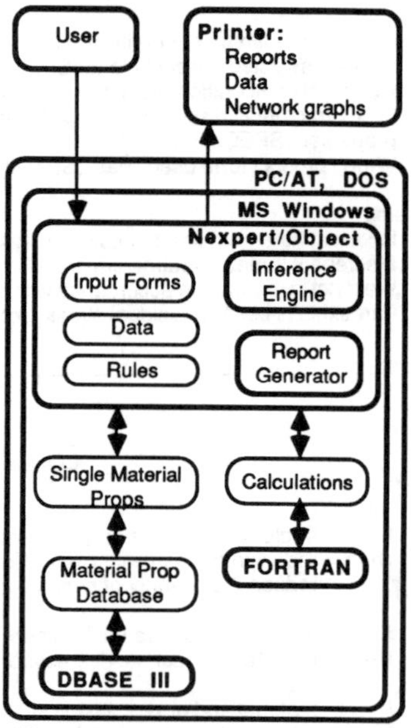

Figure 4. Architecture of the expert system.

3.3 RULES

The expert system for failure analysis of metals uses three classes of "If/Then" rules:

1) "Fracture Contributor/Driver" cause and effect rules,
2) "Fracture Topography Descriptor" rules and,
3) Operator/Tell User rules

Examples of each of these rules follow.

- Fracture Contributor/Driver:
  - If - the Aircraft is based in a "Hostile" environment,
  - Then - "Service Environment
    - Chemical"
    - cause (basing, hostile) can result in
    - effect [Incr. Corr. Exp. - HI Ecorr, &/or HI WEMF &/or Acidic/basic pH] and can be a Fracture Contributor/Driver.

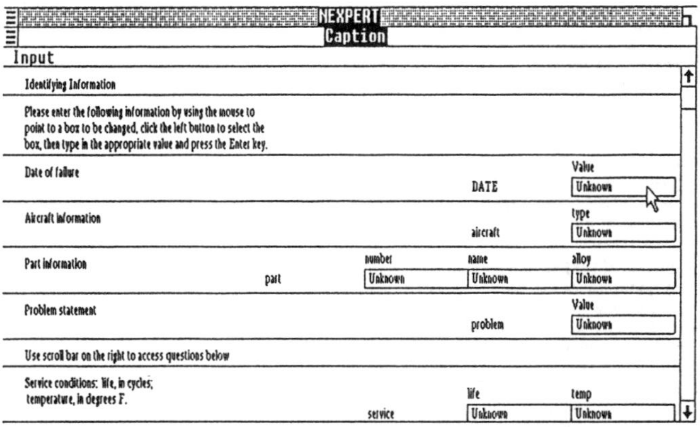

Figure 5. Fill-in-the-blank forms-based input which is presented to the user at the beginning of the consultation.

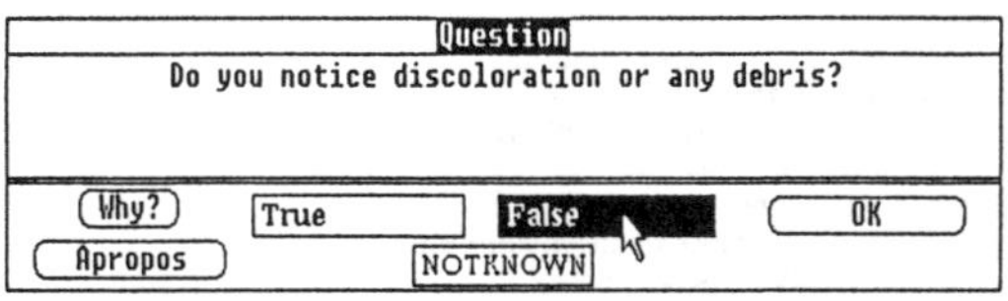

Figure 6. The expert system asks the user a yes-no (true-false) question.

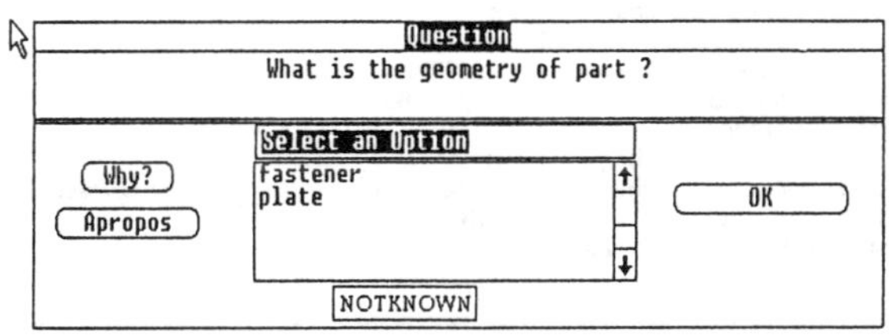

Figure 7. The expert system asks the user which of a set of possible answers to a question is the right one.

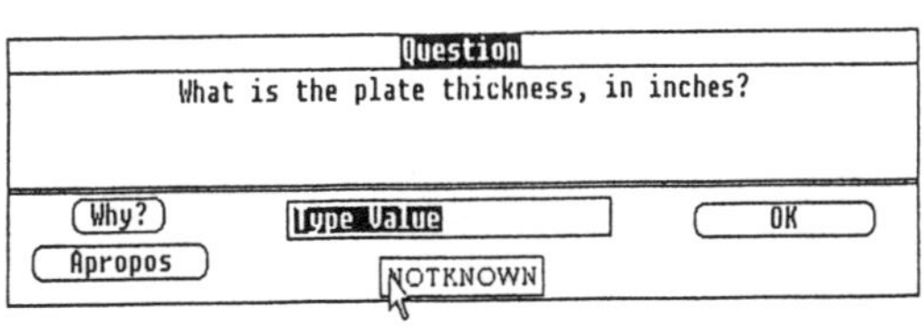

Figure 8. The expert system asks the user for the value of a numeric variable.

Table H - Summary of Consultation Between User and Expert System

Forms Input values entered for demonstration:

DATE = **3/30/87**
Aircraft type = **F99**
Part information:
part number = **6642**
part name = **stripcrak**
alloy = **7475**
Problem statement = **strip failed after 10^6 cycles but designed for infinite life.**
Service conditions:
life, in cycles = **1.1e6**
temperature, in degrees F = **75**
Design specifications:
max. load, in 1000 lbs. = **18**
temperature, in degrees F = **75**

Questions asked and text displayed during demonstration session:

What is the geometry of the part? (fastener, plate) **plate**
Is the crack plane normal to the uniaxial tensile direction? **yes**
What is the plate thickness, in inches? **.1**
What is the plate width, in inches? **10**
Please perform a visual examination of the failed part.
Do you see any thumbnail, river markings, or beach markings? **no**
Do you notice discoloration or any debris? **no**
What kind of granular cracking is there? (both_inter_&_trans, intergranular.cracking, transgranular.cracking) **NOTKNOWN**
Please perform a SEM inspection of the failed part.
What kind of cracking is there at the origin? (intergranular, transgranular) **transgranular**
How are the transgranular facets oriented? (inclined, parallel_to_normal_surface) **parallel_to_normal_surface**
Is there micro structure on the fracture surface? **no**
Is secondary cracking noticeable? **no**
Is there any evidence of striations? **yes**
How would you classify the striations? (coarse, fine) **coarse**
Failure mechanism determination phase is complete.
Determining cause of failure and suggesting possible corrective actions is next.
Is the microstructure normal? **yes**
Is chemical analysis within specifications? **yes**
Is the part hardness within specifications? **yes**
What is the part's Yield strength, in ksi? **56**
What is the part's Ultimate Tensile Strength, in ksi? **60**
What is the part's fracture toughness, in ksi*sqrt(in)? **65**
Are there any defects at the fatigue crack origin? **no**
A fracture mechanics analysis will now be performed.
From the SEM analysis, select an intermediate crack radius (a) and determine the corresponding da/dN.
What intermediate crack radius did you select? **.5**
What is the value of da/dN? **.0000125**
End of Session.

• Fracture Topography/Failure Mechanism - Domain:

If - the origin/Zone I Fracture Topography is [Transgranular Microvoid Coalescence w/open voids facing the same direction]
Then - the Failure Mechanism is [overload, bending].

• Operator:

If a stress analysis was not conducted,
Then conduct a stress analysis and Determine [σ^S_{nom} and σ^D_{max}].

The expert system for failure analysis of metals uses three "Master" rules which establish the criteria for the "Fracture Contributor/Driver" vs "Failure Mechanism" rules, shown below:

A.
If the part/material meets the drawing, and
If the drawing contains no errors, and
If the Fracture Topography and the Fracture Drivers are satisfied either directly from the (Told) "Service Environment" or by combined effect of (Told) "Service Environment" and (observed) Fracture Contributors,
Then a Failure Mechanism and Cause of Failure are established.

B.
If Failure is caused by Fracture Drivers (only) and
If the source of those Fracture Drivers is (Told) "Service Environment" (only),
Then "Design/Stress" are the Cause of Failure and part Redesign is required.

C.
If Failure is caused by Fracture Drivers and Fracture Contributors,
Then The Fracture Contributors are the Cause of Failure and dictate the Corrective Action(s).

3.4 MATERIAL PROPERTY DATABASE - The expert system must have access to data on the properties of the materials about which it is reasoning. Figure 9 shows the structure of the materials property database which is being developed. The master database contains a table of a variety of properties for each of the materials which the expert system will encounter. This database will be maintained using the dBase III database program, which is also resident in the computer. The expert system extracts a single record or line from the database file for each material it is considering and loads it into its knowledge base. Each record contains the properties of one material and may include the items listed in Table E. Not all items need to be stored for every material,only those which are meaningful for that material.

3.5 CALCULATIONS - At a number of points in the reasoning process, it is necessary to perform calculations. These include very simple ones such as calculating the area of a plate by multiplying its width by its thickness, and more complex ones, such as calculating applied stresses and expected lifetimes. The very simple calculations are performed directly within rules in the expert system. More complex calculations can be performed externally to the expert system (but still within the same computer) using Fortran, a language much better suited to such calculations.

Very complex calculations, such as finite element calculations required for redesign, are outside the scope of this expert system and it will only suggest to the user that they be performed externally where appropriate.

3.6 REPORTING - When the analysis is complete, the expert system produces a report summarizing its results. An example of this report is shown in Figure 10. Much of the identifying information provided during the forms-based-input at the beginning of the analysis (Figure 5) has been reproduced. The conclusions reached are also listed. The format of this report is easily changed and multiple reports with different formats can be prepared as deemed necessary.

3.7 REASONING PROCESS - The overall reasoning process carried out by the expert system is as follows:

3.7.1 Forms-based Input of Data - Initial data is acquired from the user by means of fill-in-the-blanks forms displayed on the screen. As described above (see Figure 5), this data includes part identification information, design information, and preliminary failure data. Any information which is unknown may be left blank and the expert system will either ask for it later or do without it.

3.7.2 Rule Evaluation - Once the user indicates that he is finished inputting data on the form, the expert system suggests the hypothesis that a failure analysis result should be determined and then invokes a backward chaining analysis of the knowledge base to try to establish this hypothesis. This causes the expert system to inspect all rules which have as their conclusion that an analysis has been completed. It attempts to determine if the conditions which must be true for these rules to be true can be verified. Since these conditions contain intermediate conclusions which may be the result of firing other rules, a search through many levels of rules in the knowledge base is thereby initiated.

During the course of this search, the expert system will need to determine the values of numerous pieces of data. If these data have not already been input, the expert system tries one of several other ways of determining the values. For some data, it asks the user by means of one of the dialog boxes shown in Figures 6-8. For others, it may perform calculations or look in the material property database. In any of these cases, if it cannot find a value, it simply skips whatever reasoning it would have been able to do if it had found a value.

3.7.3 Reporting - When the analysis has been completed, a report (Figure 10) is automatically generated.

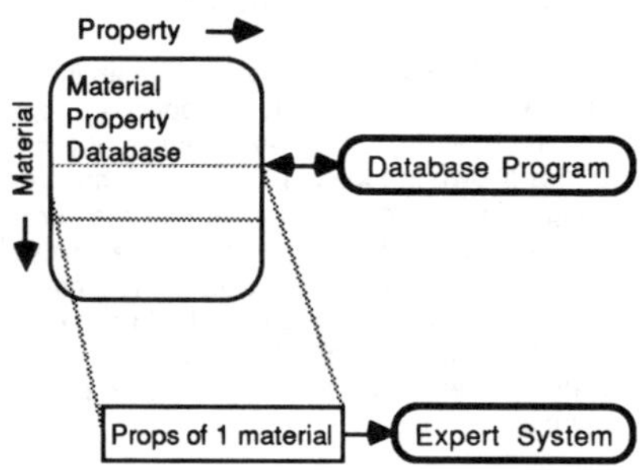

Figure 9. A single line of the material database is used by the expert system when analyzing a particular material.

4.0 A FAILURE ANALYSIS CASE STUDY

The preliminary version of the expert system has been used to perform a failure analysis on a specimen which suffered a premature fatigue failure. A load carrying member in the form of a 10 inch wide X 0.1 inch thick strip failed prematurely after experiencing 1.1 x 10^6 fatigue cycles. The part was designed for a maximum service stress range of 18 ksi (zero to max stress) and had an endurance limit of 24 ksi (at R-ratio=0), thereby providing for anticipated infinite design life (10^7 cycles). The part had no pre-existing defects, and its microstructural and chemical analyses, hardness, tensile and fracture toughness test results all met the specification requirements.

Table H gives a compact summary of the consultation. The first part lists the information which is asked for on the input form (Figure 4) initially presented to the user. **Boldface** entries are the user's replies. The second part contains the questions asked, **answers given** and informational comments made by the expert system during the course of the reasoning process. (The questions were asked in the form of one of Figures 5-7, and are shown in single line form in Table H only to provide a concise display.)

FAME first determines the failure mechanism by asking for the necessary data and evaluating possible mechanisms. It then performs fracture mechanics and fatigue life analyses to establish the failure cause.

The expert system first verifies that the failure should not have occurred, since the design stress level was below the endurance limit. It then examines the service environment. It uses a suitable crack model (in this case center-cracked sheet subject to remote tension), and given fatigue crack growth rates (da/dN) from striation spacings, and from the material data base (da/dN curves) to calculate the applied stress and verify that this stress estimate is consistent with the observed critical crack length .

The cause is determined to be fatigue overload at an applied stress range (zero to σ_{max}) of 27 ksi, which accounted fully for the observed reduction in fatigue life from 10^7 cycles to 1.1 X 10^6 cycles.

Finally, FAME uses its corrective action rules to prescribe corrective measures and includes these in the report.

5.0 CONCLUSIONS

This paper presents a progress report on the development of FAME, a diagnostic expert system for use by failure analysts in diagnosing failures of metallic components. Based on a dialog with the user, FAME determines failure mechanisms, causes of failure and recommends corrective actions. It is built from commercially available software tools and has an architecture which can be extended to other types of failure analysis. For the senior failure analyst, FAME will speed the analysis process and free his time for working on the most difficult aspects of an analysis. It also provides automatic report generation at the end of the analysis. For junior failure analysts, FAME will provide a tool which will methodically check a wide variety of possibilities and ensure more reliable analyses.

6.0 ACKNOWLEDGEMENTS

This work has been supported by the Air Force Systems Command under Contract No. F33615-86-C-5078, and by Rockwell International IR&D funds.

7.0 REFERENCES

1. Metals Handbook, Vol. 12, 9th Edition, "Fractography", ASM International, 1987.

2. Metals Handbook, Vol. 11, 9th Edition, "Failure Analysis and Prevention", ASM International, 1986.

3. Damage Tolerant Design Handbook, MCIC-HB-01, Battelle Columbus Laboratories, Dec 1972.

4. First Supplement to the Damage Tolerant Design Handbook, MCIC-HB-01S1, Battelle Columbus Laboratories, Dec. 1973.

5. Second Supplement to the Damage Tolerant Design Handbook, MCIC-HB-01S2, Battelle Columbus Laboratories, Jan. 1975.

6. SEM/TEM Fractography Handbook, MCIC-HB-06, Dec. 1975, Battelle Columbus Laboratories, Dec. 1975.

7. Electron Fractography Handbook, MCIC-HB-08, June 1976, Battelle Columbus Laboratories, June 1976.

8. MS windows, Microsoft Corporation, 16011 NE 36th Way, Box 97017, Redmond, WA 98073-9717.

9. Nexpert/Object, Neuron Data, Inc., 444 High St., Palo Alto, CA 94301.

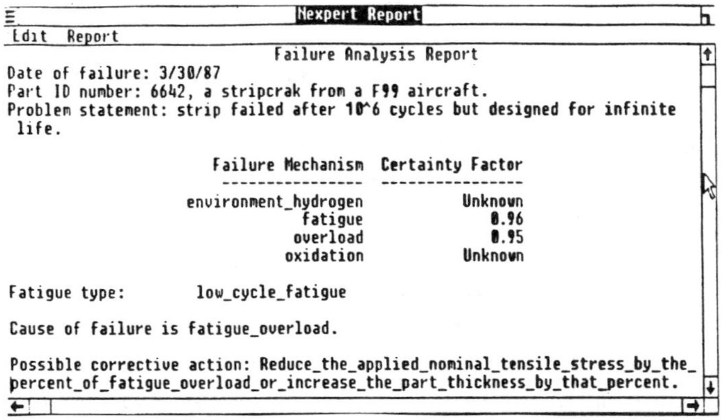

Figure 10. The report generated at the end of the consultation includes identifying information supplied by the operator and the conclusions of the reasoning process.

2nd International SAMPE Metals Conference
August 2-4, 1988

HYDROGEN AND ADVANCED AEROSPACE MATERIALS

Howard G. Nelson
NASA-Ames Research Center
Moffett Field, CA 94035

Abstract

The hydrogen embrittlement is briefly reviewed and discussed in terms of specific structural materials considered for use on a generic, hydrogen-fueled, hypersonic aerospace vehicle. A few unusual hydrogen-material incompatibility concerns are identified and some solution methodologies are discussed that could potentially lessen these concerns.

1. INTRODUCTION

Recent advances in aerospace technology have brought materials to the forefront. More frequently, we no longer have the luxury of selecting one of a number of materials to perform a specific and often critical function. Instead, materials technology is beginning to pace the advancement in aerospace technology --a specific material with a specific set of properties is often required to meet the mission objectives.

Demands on materials take many forms; lighter weight, higher strength, higher stiffness, reasonable low temperature behavior, good high temperature properties, etc. All of these requirements are combined with the need for acceptable long-term durability. Durability often translates into chemical compatibility with the sometimes unique operating environments of the advanced vehicles. One such unique operating environment is that associated with the advanced, hydrogen-fueled, hypersonic, aerospace vehicles such as the National Aero-Space Plane. In this application, materials are required to perform over the long term in hydrogen environments never before considered --hydrogen pressures up to about 2000 psi and hydrogen temperatures up to about 1800 F and even above. Add to this the fact that for many applications both pressure and temperature may be frequently cycled and the chemical environment can become extremely severe.

The purpose of this paper is to briefly look at some of the potential problems associated with the use of advanced, structural materials in the high pressure, high temperature hydrogen environment of a generic, hydrogen-fueled, hypersonic, aerospace vehicle. We will review our understanding of the hydrogen embrittlement phenomena and then, apply this relatively limited understanding in an attempt to identify potential areas of hydrogen-material incompatibility. We will conclude by identifying a few approaches that could be taken to lessen the susceptibility of advanced structural materials to hydrogen degradation.

2. HYDROGEN EMBRITTLEMENT PHENOMENA

Virtually every material can be influenced by the presence of hydrogen. Under the right conditions, the mechanical and/or fracture behaviors of most structural materials can be significantly degraded. Severity of degradation will depend on (1) the magnitude and extent of the hydrogen interaction with the materials (forms of degradation available in the material) and (2) the ease of hydrogen transport from the environment to the point in the material where it can interact. Because most materials will interact in some way with hydrogen, the degree to which a material is embrittled is often determined solely by the hydrogen transport processes--whether hydrogen can become available to interact.

Hydrogen in the liquid state is generally benign to most materials since insufficient thermal energy is available to dissociate the hydrogen molecule. Depending on the catalytic properties of the material, hydrogen-material incompatibility will begin to develop as the temperature is raised above about -150 F. At these temperatures, hydrogen dissociation and the various other hydrogen transport processes begin to occur at significant rates. Because many of these transport processes are thermally activated, the higher the temperature, the greater the rate of hydrogen transport and the greater the magnitude of degradation.

The process of hydrogen degradation of a material is not a single phenomenon but an assembly of a number of phenomenon. Over the years it has become accepted that most of the potential embrittlement interactions identified to date can be fit into the following categories:

2.1 <u>Hydrogen Attack</u>

Hydrogen attack is a high temperature phenomenon associated with the formation of a volatile, immobile reaction product between hydrogen and some lesser alloying element in the host material; an example of which is the formation of methane from hydrogen and carbon in steel. High temperature is required since both reactants (including carbon) must diffuse to an internal interface where they can combine and nucleate a cavity. The continued accumulation of reaction product can result in high pressure gas pockets within the material.

2.2 Internal Hydrogen Pressure Formation

This phenomenon is similar to hydrogen attack in that it involves the formation of a reaction product, but here the reaction product is molecular hydrogen. Molecular hydrogen is immobile in most materials and will precipitate at internal interfaces develop high gas pressures, depending upon the degree of protonic hydrogen supersaturation in the material lattice. Because both reactants (protonic hydrogen) diffuse readily in most materials, this form of degradation can occur at room temperature and below.

2.3 Dislocation Interaction

Depending upon the specific materials, the presence of hydrogen can either enhance plastic deformation or make it more difficult, and can strongly influence those properties dependent upon plastic behavior, such as many forms of crack initiation and crack growth. This phenonmenon can occur over a broad range of temperatures and does not require protonic hydrogen supersaturation in the material.

2.4 Hydride Formation

Protonic hydrogen at concentrations in excess of saturation can interact with some materials to form a hydrogen rich second phase known as a metal hydride. Hydrides are less dense than the host material and will preferentially precipitate along selected planes and at local regions of high strain. In addition, hydrides are brittle and can be easily fractured.

2.5 Lattice Bond Interactions

Hydrogen on or within a material can interact with some materials to reduce its bond strength. A reduced bond strength will facilitate separation of the host material along crystallographic planes and at grain boundaries.

In summary, it is important to realize that any of the above hydrogen interaction phenomena, can occur in most materials; the exact interaction mechanism that will be the primary cause of embrittlement will depend both on the specific material and on the amount and location of available hydrogen in the material.

3. HYDROGEN-MATERIAL INTERACTIONS

The vast majority of the hydrogen-material interaction studies performed to date have involved the more common, moderate temperature, structural materials. Unfortunately, this limited understanding has only narrow applicability to advanced, hydrogen-fueled, hypersonic vehicles. Material requirements for these vehicles will range from the commonly used, commercially available metals alloys, such as titanium-, aluminum-, nickel-base alloys and autenitic stainless steels, to the more exotic, advanced structural materials, such as titanium- and nickel aluminides and lithium containing aluminum alloys in monolithic form and in silicon-carbide and/or carbon fiber reinforced composite forms. This section will briefly review our understanding (or lack of understanding) of the hydrogen interaction and the resulting degradation associated with these materials.

3.1 Titanium-Base Alloys

Most of the commonly used, commercial titanium alloys are highly susceptible to hydrogen degradation. These alloys consist of combinations of the alpha-phase (stable at room temperature) and the beta-phase (normally stable at elevated temperatures). The extent, location, and configuration of each phase is controlled by the combination of alloy additions and heat treatment.

Alpha titanium has a low hydrogen solubility at room temperature and readily forms titanium hydride. Hence, it is susceptible to hydrogen degradation by Hydride Formation. When a hydride forms, there is about an 18% volume increase. This volume increase must be accommodated by the host lattice and can result in sizable lattice strains. Likewise, localized strains in the material will enhance the nucleation and growth of the hydride. Finally, hydrogen transport by lattice dif-fusion is slow in alpha-titanium.

In contrast to the alpha-phase, beta-phase of titanium can hold large quantities of hydrogen in solid solution (well over 30 at.% at room temperature) without the formation of a metal hydride phase. Additionally, hydrogen transport by lattice diffusion in beta-titanium is much more rapid than in alpha-titanium.

Hydrogen embrittlement of the commercial titanium alloys, although most often associated with a single form of degradation (Hydride Formation), can be very complex. These complications arise from the large variations in hydrogen solubility and transport behavior that can develop in an alloy, depending on the relative amount and distribution of the alpha and beta phases present. When embrittlement occurs as the result of an external hydrogen environment, alloys containing mostly alpha-phase will degrade by the repeated formation and fracture of the brittle hydride phase formed at or very near the gas-metal interface [1]. When a continuous beta-phase is present throughout the alpha + beta microstructure, hydrogen can be transported well within the alloy where it can interact with the neighboring alpha-phase at the alpha/beta boundaries to form a brittle hydride film [2].

The advanced titanium alloys considered for elevated temperature applications on the generic, hypersonic vehicles differ from those discussed above in that they are comprised primarily of intermetallic compounds--either Ti_3Al (alpha-2) or TiAl (gamma)--often together with significant amounts of the alpha and/or beta phases as well. Although very little understanding presently exists of the hydrogen interaction with the titanium-aluminide phases, significant degradation can be anticipated from the rapid transport of hydrogen in the beta-phase constituent and the formation of a hydride in the alpha-phase, if not in the alpha-2 and gamma phases themselves.

3.2 Aluminum-Base Alloys

Hydrogen appears to degrade aluminum alloys only under unique or extreme conditions. Gaseous hydrogen at temperatures up to at least room temperature does not appear to effect most aluminum alloys even at very high pressures. However, hydrogen-induced surface cracking has been observed in a 7075 alloy exposed to an ionized hydrogen environment for four days under an accelerating voltage of 1.5 kV while being strained to 75 percent of its failure load [3]. Although these conditions are extremely severe, it does offer unequivocal proof of the potentially damaging effect of hydrogen in aluminum. Degradation has also been observed in aluminum alloys when cathodically charged with hydrogen. Degradation under these conditions is most severe at around -100F and is essentially non-existent at room temperature [4].

The exact forms of hydrogen degradation operative in aluminum alloys is still open to question. Lattice Bond Interaction has been suggested as has Hydride Formation. The latter appears more associated with specific alloying elements rather than aluminum itself; a severely charged magnesium containing aluminum alloy was found to exhibit a fragmented layer (most probably magnesium hydride) on its intergranular fracture facets [5].

The behavior of the advanced Al-Li alloys in hydrogen is yet to be completely defined. Because of the known reactivity of lithium with hydrogen--readily forming lithium hydride--a concern exists for the possible degradation of Al-Li alloys by hydride formation in much the same manner as indicated above for the magnesium containing aluminum alloy.

3.3 Nickel-Base Alloys

Hydrogen interaction with nickel alloys can be extensive. Hydrogen can associate itself with dislocations and influence deformation behavior. Hydrogen can also reduce the ductility of nickel alloys by decreasing its flow stress and enhancing planar slip. Severe degradation of hydrogen charged nickel suggests Lattice Bond Interactions may be operative, either along grain boundaries or through grains. Finally, nickel alloys can be degraded by Hydride Formation. Like the titanium hydride, nickel hydride requires a lattice expansion, is brittle, and is generally stable about room temperature and below. In fact, some investigators believe observations attributed to Lattice Bond Interactions in nickel are actually the result of localized, strain-induced, hydride formation, which decomposes after fracture. Embrittlement of nickel alloys in an external hydrogen environment has been observed to extend to temperatures as high as 1100 F [6].

3.4 Austenitic Stainless Steels

Stable austenitic stainless steels such as 316 or 321 are far less susceptible to hydrogen degradation than most other structural materials. The austenitic phase of iron shows no evidence of degradation by Lattice Bond Interaction or by Hydride Formation. Thus, degradation will occur only through Dislocation Interaction or by Molecular Hydrogen/Internal Pressure Formation.

Although austenite is less prone to hydrogen degradation, many so called austenitic stainless steels, such as 304, are relatively unstable and can be partially transformed to martensitic-type phases by plastic deformation. These martensitic phases are highly susceptible to hydrogen degradation and thus, significant embrittlement of these unstable austenitic stainless steels by mechanisms other than Internal Pressure Formation and Dislocation Interaction is possible.

3.5 Silicon Carbide

Although data in the literature is limited, significant degradation of sintered alpha-silicon carbide has been observed at hydrogen pressures of about one atmosphere and at temperatures as low as 1800F [7]. The hydrogen interaction produces significant grain boundary attack with an associated reduction in fracture strength. In the same study, thermodynamic calculations were performed on the hydrogenation of SiC over the range of temperature from 700C to 1700C. The predominant carbon-based reaction product was methane, with its equilibrium concentration remaining essentially constant over the entire temperature range investigated. These calculations suggest that degradation is controlled by transport kinetics and may be significantly more severe at higher hydrogen pressures.

3.6 Carbon and Graphite

Molecular hydrogen at temperatures as low as about 600F can interact relatively rapidly with carbidic carbon to produce methane gas and other hydrocarbons [8]. This reaction appears to be kinetically controlled and would be expected to increase rapidly as temperature is increased.

The hydrogenation of graphitic carbon in a molecular hydrogen environment occurs at a much slower rate than with carbidic carbon [8], and is essentially unmeasurable at low hydrogen pressures even at 1800F [9]. This is not the case, however, when graphite is exposed to atomic hydrogen. Atomic hydrogen will react rapidly with graphite even at room temperature to form methane gas [10]. Assuming the ideal equilibrium between atomic hydrogen and molecular hydrogen in the gas phase, sufficient atomic hydrogen may be present at a hydrogen pressure of 2000 psi and a temperature of 1800 F to form significant amounts of methane from graphitic carbon.

3.7 Metal Matrix Composites

Little information is presently available relative to the influence of hydrogen on fiber reinforced, metal matrix composites. Areas of concern are the potential for accelerated hydrogen transport along the fiber/matrix interface, the effect of residual stresses (inherently present in most composites) in concentrating the hydrogen, and the increased ease of molecular hydrogen precipitation at the weaker and readily available fiber/matrix interface. Additionally, graphite fibers and many silicon carbide fibers, which contain significant surface carbon and carbon filled defects, may readily react with hydrogen to form cavities of high pressure methane gas in much the same manner as Hydrogen Attack observed in carbon steels. Finally, both graphite and SiC fibers (themselves) may be significantly degraded by the presence of hydrogen at high concentrations in the host metal matrix.

4. UNUSUAL INCOMPATIBILITY CONCERNS

Materials to be used on a generic, hydrogen-fueled, hypersonic vehicle will see a variety of hydrogen environments, ranging from liquid hydrogen at cryogenic temperatures and low pressures to very high hydrogen temperatures and pressures. Without question, the greatest overall concern stems from our general lack of understanding and the lack of demonstrated compatibility between those advanced, elevated temperature materials deemed critical to hypersonic flight and the extremely varied and severe hydrogen operating environments. Certainly, it is reasonable to expect that many of the hydrogen degradation mechanisms previously identified to be active in the more common materials about room temperature could be active in these advanced materials at elevated temperatures as well. In addition, the unusual mission requirements associated with a hypersonic vehicle could develop degrading interactions never before considered.

Most structural materials will rapidly become staturated with hydrogen when exposed to high hydrogen pressures at temperatures above about 1200 F. When this occurs, the character of the material is changed and the material must be considered "alloyed" with hydrogen. Its mechanical and fracture behaviors may be markedly different. Under these conditions, the hydrogen transport processes discussed previously will have little or no influence on the severity of degradation. Sufficient hydrogen will most probably be available and transport will be sufficiently rapid that, if degradation does occur, it will be determined solely by the hydrogen-material interaction processes. Such properties as creep, ductility, and sustained load cracking of the hydrogen-alloyed material must be defined. Additionally, hydrogen has been considered for active-cooling of hot structures. Under such conditions, the material's surface in contact with the hydrogen environment may develop large residual tensile stresses and strain which could dramatically change both the solubility of hydrogen in the material and the severity and extent of the hydrogen-material interaction.

Thermal cycling can have a strong influence on the degree of hydrogen-degradation. If the temperature of a hydrogen-alloyed material is lowered at a sufficient rate, hydrogen will be retained in the material. The hydrogen host-phase may become supersaturated with hydrogen and various hydrogen-rich phases such as hydrides or volitile reaction products such as methane, water, or molecular hydrogen may be created, each having a potentially degrading effect on the material.

Another aspect of thermal cycling is the potential degradation that could occur when a material containing hydrides is heated. If the temperature rise is sufficiently rapid or if the diffusion of hydrogen away from the hydride is sufficiently slow, a high localized hydrogen supersaturation can develop as the hydride goes back into solution [11].

Finally, for some materials, a rapid change in temperature may not be required to form the potentially degrading hydrogen-rich phases such as a hydride or methane. Such hydrogen reactions do not require a high hydrogen supersaturation and a simple temperature gradient along a structure may be sufficient. In the case of a hydride former, if the low temperature end of the gradient is below the hydride eutectic, hydride precipitation will occur. The precipitation of a hydride will in turn reduce the hydrogen solid solution concentration in the host-phase with a resulting flow of hydrogen down the temperature gradient. This in turn will result in the further precipitation of the brittle metal hydride.

5. POSSIBLE SOLUTION METHODOLOGIES

Based on our present knowledge of the hydrogen degradation phenomena, possible solution methodologies to a hydrogen-materials incompatibility problem must address either the hydrogen-material interaction or the hydrogen transport processes. Obviously, if either of these can be reduced or eliminated, the incompatibility problem will be reduced or eliminated as well.

Reducing the hydrogen-material interaction would require a change in material. For most critical applications, this may be impractical, if not impossible. Consequently, solution methodologies must address the hydrogen transport processes. One approach is to develop a material system in which the hydrogen transport processes are slow or nonexistent or are extremely rapid. The latter would allow hydrogen to rapidly move in and out of the material, lessening the chance of degradation as the result of a large hydrogen supersaturation. The most complete approach would be the use of a hydrogen barrier system that would lessen or eliminate hydrogen entrance into the material.

Considerable experience has been gained in the use of barrier coatings as a secondary protection system in specific applications. Unfortunately, however, they have seldom been used as the primary method of protection. Surface barriers are susceptible to damage during assembly, operation, and testing, and their service life can be affected by thermal cycling, temperature gradients, and time dependent microstructural changes. Additionally, it is often difficult to inspect a barrier coating to insure even its initial integrity.

The two primary types of barrier coatings are diffusion barrier coatings and catalytic or non-catalytic coatings. The former slows hydrogen entrance by slowing diffusion and the latter by influencing the rate of hydrogen dissociation and/or recombination.

The use of diffusion barrier coatings to protect against structural degradation appears reasonable only at lower temperatures. At elevated temperatures, hydrogen transport is so rapid in most materials that coating thicknesses become significant and minor defects in the coating permit rapid, local hydrogen entrance. Additionally, a diffusion barrier makes it difficult for hydrogen to exit a material once it enters.

The catalytic-type barrier coatings appear to offer the greatest potential. The significant advantage of such coatings is that theireffectiveness is generally independent of coating thickness. An optimum catalytic-type barrier is one that hinders molecular hydrogen dissociation, accelerates atomic hydrogen recombination, permits hydrogen diffusion, is easily applied to all structural configurations (a gas-phase application), adds little or no weight, and guarantees complete coverage (a surface-conversion coating).

6. SUMMARY

Hydrogen degradation of the mechanical and fracture behavior of a material can be a complex process, influenced both by hydrogen transport and the form and extent of the hydrogen-material interactions. Without question, our greatest concern with advanced, high temperature, structural materials lies with the general lack of understanding and the lack of demonstrated compatibility between those materials and the extremely varied and severe hydrogen operating environments anticipated to be encountered in advanced vehicles. Although hydrogen-materials incompatibility problems will most probably exist, possible solution methodologies also exist which, if developed, could significantly lessen or eliminate these problems. It is believed, however, that unless extreme care is taken to characterize and understand the potential hydrogen interactions, to identify the potential hydrogen-materials incompatibility issues, and to develop and evaluate the required solution methodologies, severe limitation on the life and operating capability of an advanced, hydrogen-fueld, hypersonic, aerospace vehicle could result.

7. REFERENCES

(1) Williams, D. P. and Nelson, H. G. Met. Trans. 3, 2107 (1972).

(2) Nelson, H. G., et al. Met. Trans. 3, 469 (1972).

(3) Koch, G. H. Corrosion 35, 73 (1979).

(4) Gest, R. J., and Troiano, A. R. Corrosion 30, 275 (1974).

(5) Ciaraldi, S. W. et al. Hydrogen in Metals AIME, Warrendale, PA, 1981.

(6) Frick, V. et al. Space Shuttle Materials 3, 597 (1971).

(7) Hallum, G. W., and Herbell, T. P. Sintered Silicon Carbide, NASA TM 88819, 1986.

(8) Breisacher, P., and Marx, P. C. J. Amer. Chem. Soc. 85, 3518 (1963).

(9) Bonzel, H. P., and Krebs, H. J. Solid Surfaces and Surface Science 411 (1980).

(10) Wood. B. J., and Wise, H. Gaseous Hydrogen with Graphite, TR SRI-26-PU, Purdue, 1968.

(11) Pesch, W., et al. Met. Trans. 11A, 1821 (1980).

2nd International SAMPE Metals Conference
August 2-4, 1988

FATIGUE BEHAVIOR OF POWDER METALLURGY Ti-6Al-4V

J.W. Bohlen, J.A. Walsh, A.A. Sheinker, and G.R. Chanani
Northrop Corporation, Aircraft Division
Hawthorne, CA 90250-3277

1. ABSTRACT

The fatigue properties of standard grade Ti-6Al-4V powder metallurgy (PM) products were evaluated. Prealloyed Ti-6Al-4V powder was made by (a) the rotating electrode process (REP) and (b) the plasma rotating electrode process (PREP). The powder produced from the REP was compacted into various shapes by hot isostatic pressing (HIP) powder in ceramic molds. The powder produced from the PREP was consolidated into round bars in steel cans by HIP and rapid omnidirectional compaction (ROC). A hot rolled plate of standard grade Ti-6Al-4V annealled to the same strength level as the PREP PM bars was evaluated as an ingot metallurgy (IM) baseline for comparison. Both strain-controlled and load-controlled fatigue tests were run. In general, fatigue lives for the REP powder products were lower than those for IM products and typical standard grade Ti-6Al-4V forgings. Strain-fatigue lives of the PREP powder products were equivalent to those of the IM plate because there was less contamination in the PREP products than in the REP products.

The primary fatigue crack initiation sites in most of the REP specimens were both internal and at the surface while most of the PREP specimens were located at the surface. However, several of the HIP PREP specimens had the primary initiation sites located internally. Several of the IM plate specimens also had internal primary initiation sites associated with favorably oriented grains. Therefore, internal initiation in Ti-6Al-4V PM products does not necessarily indicate inferior fatigue resistance compared to Ti-6Al-4V IM products.

2. INTRODUCTION

Powder metallurgy (PM) Ti-6Al-4V aircraft components are being considered as replacements for wrought, ingot metallurgy (IM) Ti-6Al-4V parts to reduce manufacturing costs[1-12]. The Ti-6Al-4V alloy is used in critical load-bearing applications on multirole tactical aircraft because of its high strength-to-weight ratio, high modulus of elasticity and elevated temperature strength. To qualify for this usage, the mechanical properties of PM parts must be at least equivalent to those of the IM parts that they would replace. These critical applications require high damage tolerance, which includes good strain-life fatigue resistance, a key design parameter.

Earlier Ti-6Al-4V powder was made by the rotating electrode process (REP) which employed a tungsten electrode. Work on REP powder products was conducted on test blocks consolidated by hot isostatic pressing (HIP) from prealloyed powders in ceramic molds. Data showed that the static mechanical properties, as well as notched fatigue and fatigue crack growth properties, were comparable to standard IM Ti-6Al-4V alloys. Therefore, the PM alloy could potentially replace the IM alloy. However, the smooth fatigue properties were inferior to the fatigue properties of standard IM Ti-6Al-4V stock;[13] and, therefore, rendered the REP PM inadequate as a potential replacement product. The results of the REP powder did provide a good understanding of the effects of contaminants on the fatigue behavior of PM products.

The REP process has now been superseded by the plasma rotating electrode process (PREP), which employs a plasma arc instead of a tungsten electrode, thereby eliminating the source of tungsten contamination.[14] Further improvements also have been made in the powder handling procedures to reduce contamination. Extensive fatigue testing and fractography were performed on both the REP and PREP powder products. The technical approach involved predominantly (1) strain-controlled, and some stress controlled, fatigue testing of smooth and hourglass specimens from consolidated Ti-6Al-4V PM products and (2) fractographic analysis of failed specimens to determine the cause of fatigue failure initiation.

3. EXPERIMENTAL

The REP and PREP Ti-6Al-4V powders were made from standard grade bar stock and were consolidated as indicated in Table 1. The cylinders (1) were as-HIP, and the witness blocks (2), which were

TABLE 1. MATERIAL PROCESS AND FORM

POWDER	FORM
REP	1. Three cylinders (6.4 cm DIA x 26.7 LONG)
	2. Two witness blocks (7.6 x 12.7 x 12.7 cm)
	3. Near net shape in the form of an arrestor hook support fitting: AHSF I-1 (140 cm LONG x 12.4 cm WIDTH)
	4. Three additional AHSF shapes, designated AHSF I-2, AHSF II-1, and AHSF II-2 produced from single lot of clean-room quality powder (140 cm LONG x 12.4 cm WIDTH)
PREP	5. HIP in steel cans, (3.3 cm DIA x 50.8 cm LONG)
	6. Rapid omnidimensional compaction (ROC), (1.9 cm DIA x 13.2 cm LONG)

part of a heat treatment study,[15] included as-HIP, beta anneal, recrystallize anneal, and mill anneal conditions. Near net shapes and bars, (3), (4), (5), and (6), were vacuum annealed at 700 C for 4 hours prior to mechanical testing. All products were inspected radiographically to verify the quality prior to the fabrication of mechanical test specimens. The REP PM forms were documented metallographically to show that most had fully compacted sections. One specimen which was not fully compactd was tested to determine the effect of poor compaction.

A large and a small fatigue specimen with 12.9 mm DIA x 30.7 mm gage, and 6.35 mm DIA x 15 mm gage, respectively were used. The large specimens had eight times the volume of the small specimens and one-half the surface-to-volume ratio in the gage section. In some of the strain-fatigue life tests utilizing complete reversals (R=-1), an hour-glass specimen (6.35 mm minimum DIA, 51 mm radius of curvature) was used to prevent buckling.

The fatigue tests were conducted in strain control, at a strain ratio of -1, and K_t = 1.0, in 50% RH laboratory air, at room temperature. Fractographic analyses were conducted on many specimens to determine the cause of crack initiation using a stereo scanning electron microscope (SEM) and the associated Energy Dispersive X-Ray Analyzer (EDXA) for chemical analysis.

4. RESULTS

4.1 REP Powder Products

The fatigue properties of the PM materials were compared to the properties of wrought IM products (plates and forgings). Strain life fatigue data was available for the Ti-6Al-4V ELI (extra low interstitial) plate, evaluated for use in aircraft design. Figure 1 shows the strain-life (R=-1.0) performance of specimens from AHSF I-2 and AHSF II-2 relative to the ELI plate.[16] The strain life performance of the PM material was similar to that indicated by the curve for ELI plate, but tended to be inferior to the ELI plate at high strain amplitudes. At R=0 the strain-life performance (not shown) of a variety of PM materials was also tested at moderate strain amplitudes of 0.005 and 0.00225. For these strain amplitudes, the strain-life performance was comparable to that for ELI plate.

The data from a series of load controlled tests using the small smooth specimens are shown in Figure 2. The smooth fatigue behavior of the PM specimens from AHSF I-2 and AHSF II-1 tested at R=0.1 were averaged to determine the curve shown in Figure 2. The failure lives for the PM ELI specimens were shorter than is typical for specimens from forgings of Ti-6Al-4V, standard grade[17] and were significantly shorter at the higher maximum stresses.

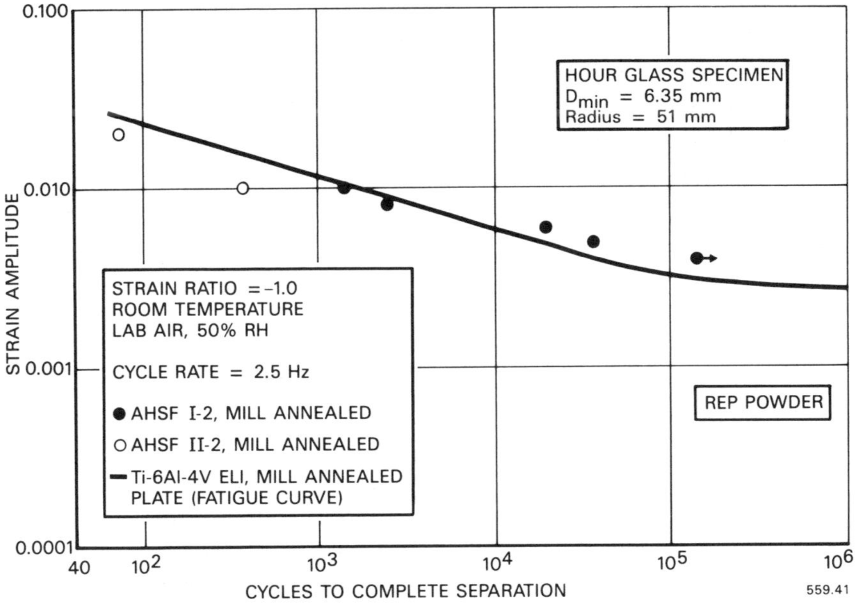

FIGURE 1. Strain-Life Fatigue for AHSFs vs Ti-6Al-4V ELI Plate at R = −1

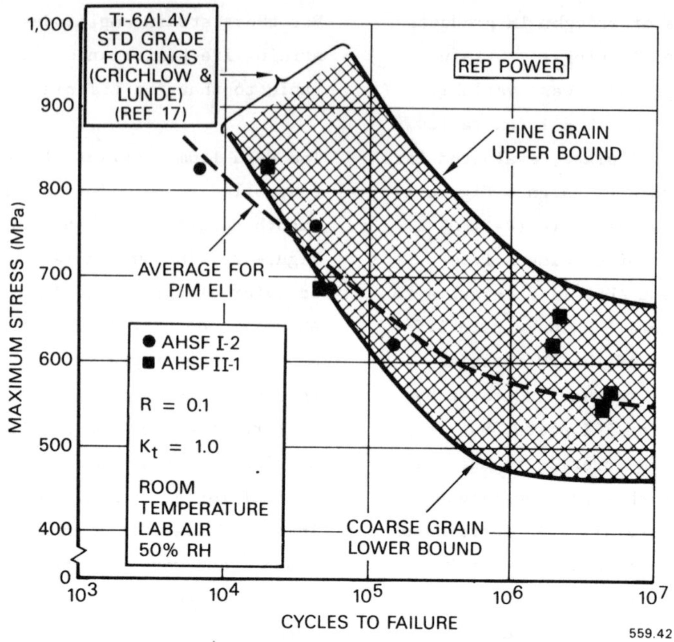

FIGURE 2. Fatigue Behavior of Smooth Specimens From AHSF Material (Mill Annealed)

Fractography was performed on 37 strain controlled and 9 load-controlled fatigue specimens. Inclusions from the AHSF generally contained silicon either singly or in combination with sulfur, calcium, chlorine, magnesium and/or yttrium. These inclusions were often associated with failure initiation. Figure 3 shows a typical combination of macro fractography and high magnification fractography showing the primary and secondary initiation sites caused by contamination. Of the fractures examined, 12 initiated from subsurface inclusions, 10 from near surface inclusions or pores and 23 initiated at the specimen surface, including three which initiated at inclusions which intersected the surface. Therefore, 25 out of 46 smooth fatigue specimens had defect initiated failures. Failures initiated at inclusions in 19 cases, at pores in 5 cases, and a lack of compaction in one case. Three other coupons had subsurface or obscured initiation sites which were not attributed to a specific defect.

Comparisons were made between performance of individual fatigue specimens and reference curves to establish the consequences of each of the different locations for

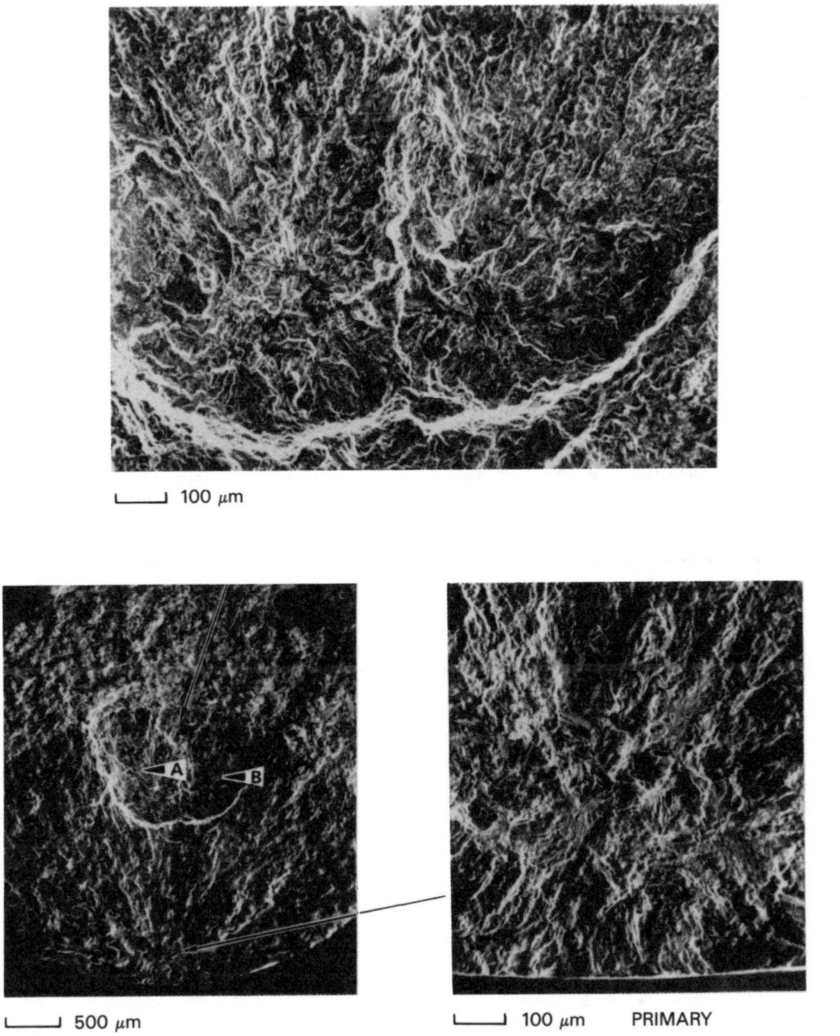

FIGURE 3. SEM Fractographs Showing the Primary Sub-Surface Initiation in a Specimen From AHSF-I-1. Two Secondary Initiation Sites, "A" and "B" Which Are in a Different Plane, are Also Present on the Fracture Surface.

crack initiation relative to the specimen surface. The performance relative to the PM average for both S-N and ε-N tests of smooth specimens is shown in Figure 4 versus the initiation location. A similar curve, not shown, was calculated as percentages of the typical ELI curve. In three out of four cases where initiations were associated with edge inclusions, lives were below the typical ELI or PM average. In one case, a large tungsten inclusion at the surface reduced the expected fatigue life by one-half. Previous work showed that the tungsten electrode provided a source of contamination.[18] Over 60% of the failures initiating at subsurface inclusions reduced the fatigue life below typical ELI (not shown) or average PM (Figure 4). Therefore, incomplete compaction and inclusions can be concluded to be somewhat detrimental to fatigue life. The fatigue results of material from industrially produced powder versus clean-room produced powder showed no significant differences. Thus, contaminants had to be removed through improved processing; and therefore, PREP was established.

4.2 PREP Powder Products

Prior to fatigue testing, the oxygen and hydrogen contents were

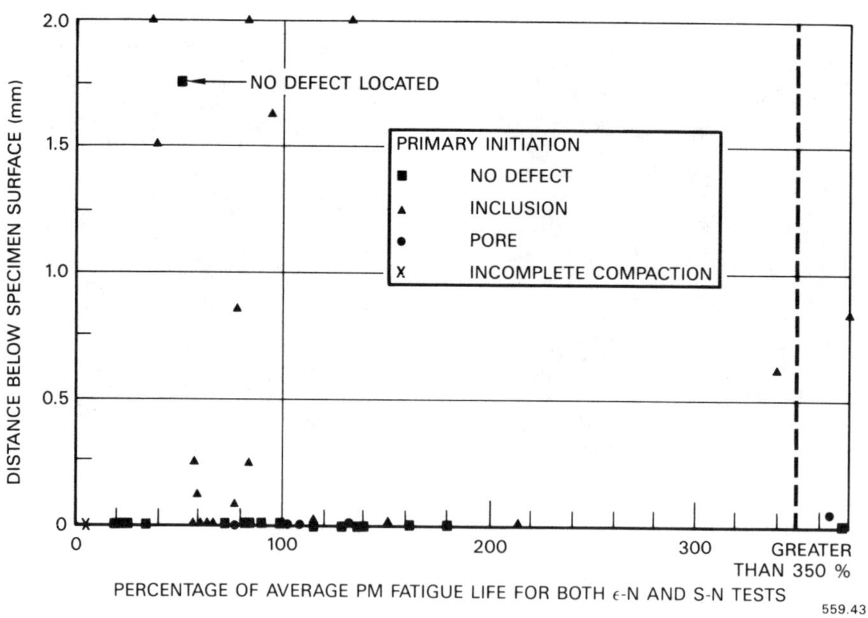

FIGURE 4. Effects of Inclusion or Pore Location on Smooth Specimen Fatigue Life Relative to Average PM Fatigue Life

determined. The oxygen content of the rapid omnidimensional compaction (ROC) bars was average; and the content of the HIP bars was above average, but still below the maximum limit for the standard grade. The oxygen contents were essentially unchanged by the vacuum mill anneal. The hydrogen content of the ROC bars was average; and the content of the HIP bars was slightly above average, but well below this maximum limit. The vacuum mill anneal was effective in removing most of the hydrogen. The mechanical properties: TYS, UTS, %e, and %RA, of the vacuum-mill annealed ROC and HIP bars were equivalent to or better than the mill annealed IM plate.

The strain-life fatigue results for the ROC and HIP PM bars and the IM plate are shown in Figure 5. Both small and large specimens were tested from the HIP PM and IM products while only the small specimens were tested from the ROC PM bar. The strain-life data for the small and large specimens of the IM plates were about the same. The fatigue lives of the small specimens from both the HIP bars and the ROC bars were generally longer than those of the IM plate. No significant difference was evident in this plot between the HIP and the ROC methods. The fatigue lives of the large specimens from the HIP PM bars were equal to those of the IM plate. The fatigue lives of the large HIP PM specimens were all shorter than those of the small PM specimens (both HIP and ROC), except for one specimen tested at the lowest strain amplitude, 0.004, where they were nearly equal, suggesting a size effect in the HIP PM materials.

To verify this size effect, pairs of one large and one small specimen were excised from adjacent areas in each of five of the HIP PM bars, and both specimens in each pair were tested at the same strain amplitude. For two specimen pairs tested at a strain amplitude of 0.010 and two specimen pairs tested at a strain amplitude of 0.013, the fatigue lives of the large specimens were shorter than those of the small specimens. The fifth specimen pair was tested at a strain amplitude of 0.007, and the fatigue lives of the large and small specimens in this pair were about equal. Thus, there was a size effect for the strain-life fatigue properties of the HIP PM bars. This effect was probably due to the higher probability of crack initiation from the presence of defects or contaminants in the large specimens as compared to the small specimens.

A summary of the fractography results is shown in Table 2. No

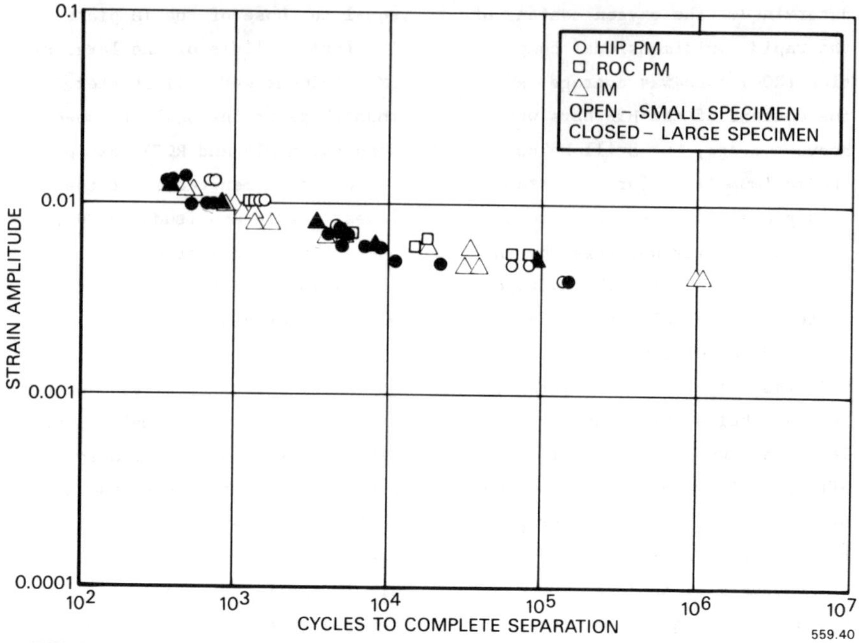

FIGURE 5. Strain-Life Fatigue Results for Specimens From PREP HIP and ROC PM Ti-6Al-4V Bars and IM Ti-6Al-4V Plate

TABLE 2. FRACTOGRAPHY RESULTS OF PREP POWDER AND IM SPECIMENS

Material	Specimen Size	Location: Primary Initiation Site	No. Inspected
IM	S+L	50% Surface (scratches)	1L, 2S
		50% Internal (grain orientation): S-N*	1L, 2S
HIP PM	S	87.5% Surface (scratches)**	7
		12.5% Internal (voids)	1
ROC PM	S	100% Surface (misc)	10
HIP PM	L	33% Surface (scratches)**	3
		67% Internal (33% voids, 34% misc)	6

*All else were ε-N.
**In addition to primary sites, there were multiple internal secondary sites.

contaminants were detected at any of the initiation sites in the large and small fatigue specimens from the IM plate or the small specimens from the HIP and ROC PM bars. The large specimens from the HIP PM bar, however, showed primary fatigue crack initiation associated with contaminants; contaminants on the surface of internal voids twice included sulfur-calcium, and once magnesium-potassium. Secondary crack initiation sites associated with voids containing silicon-phosphorphous-sulfur were also found.

The cause of the voids and contaminants in these specimens was tentatively attributed by the powder producer to the decomposition of an organic compound originating from a tube used to fill the steel HIP cans. To eliminate this source of contamination, the producer replaced the tube with a stainless steel tube. Nevertheless, the defects associated with these contaminants were not severe enough to degrade the strain-life fatigue properties of the HIP PM bars below that of the IM wrought plate, but were significant enough to be manifested as a specimen size effect.

5. CONCLUSIONS

The following conclusions were drawn from an investigation of the cylindrical and hourglass gage-section fatigue properties of standard grade Ti-6Al-4V powder products produced by the rotating electrode process (REP) and more state-of-the-art plasma rotating electrode process (PREP):

1. The strain-life fatigue properties of standard grade Ti-6Al-4V PM products made from REP powder were inferior to IM products and those made from PREP powder were equivalent to the IM products.

2. Differences in fatigue lives between REP PM and IM products and small and large PREP PM products disappeared at low strain amplitudes. This result indicates less influence of stress/strain concentrators, such as inclusions, at low applied strains.

3. Small (11 $\mu m \leq$ width < 80 μm,) Si-rich and Fe-rich, inclusions were found to initiate subsurface fatigue cracks in the earlier, more-contaminated REP products.

4. The subsurface fatigue cracks, which initiated at the small inclusions present in PM Ti-6Al-4V ELI from REP powder, tended to decrease smooth fatigue life, with the greatest reduction of life occurring in the low-cycle range (high stress or strain condition). Elimination of these particles improved low-cycle fatigue life as shown with the PREP PM products.

5. The primary fatigue crack initiation sites in PREP PM products

specimens were usually located at the surface indicating a sound metallurgical structure. Those sites located internally were attributed to contaminants originating from the powder handling equipment. The source of these contaminants was eliminated by the powder producer.

6. Fatigue crack initiation can occur internally in Ti-6Al-4V IM products such as plate. Therefore, internal fatigue crack initiation does not necessarily indicate inferior fatigue resistance in Ti-6Al-4V PM products.

6. ACKNOWLEDGEMENTS

The authors thank E.J. Kosinski of Nuclear Metals Inc. for supplying the HIP bars and C.A. Kelto of Kelsey-Hayes for supplying the ROC REP PM cylindrical bars. The authors acknowledge valuable discussions with J.S. Bruce of the Naval Air Systems Command, L.D. Parsons of the Air Force Wright Aeronautical Laboratories/Materials Laboratory, and J.H. Moll, V.C. Petersen of the Crucible Research Center, and E.J. Kosinski. The authors thank Northrop's testing and microstructural labs for their invaluable services. This work was performed under Northrop Independent Research and Development funding.

7. REFERENCES

1. J.N. Fleck and L.P. Clark, "Advancements in Titanium Powder Processing," SAMPE Quarterly, Vol. 7, October 1976, pp. 10-20.

2. F.H. Froes, D. Eylon, G.E. Eichelman, and H.M. Burte, "Developments in Titanium Powder Metallurgy," Journal of Metals, Vol. 32, No. 2, February 1980, pp. 47-54.

3. C.A. Kelto, B.A. Kosmal, D. Eylon, and F.H. Froes, "Titanium Powder Metallurgy - A Perspective," Powder Metallurgy of Titanium Alloys, F.H. Froes and J.E. Smugeresky, eds., The Metallurgical Society of AIME, 1980, pp. 1-19.

4. V.C. Petersen, V.K. Chandhok, and C.A Kelto, "Hot Isostatic Pressing of Large Titanium Shapes," Powder Metallurgy of Titanium Alloys, F.H. Froes and J.E. Smugeresky, eds., The Metallurgical Society of AIME, 1980, pp. 243-254.

5. R.H. Witt and W.T. Highberger, "Hot Isostatic Pressing of Near-Net Titanium Structural Parts," Powder Metallurgy of Titanium Alloys, F.H. Froes and J.E. Smugeresky, eds., The Metallurgical Society of AIME, 1980, pp. 255-265.

6. G.R. Chanani, W.T. Highberger, C.A. Kelto, and V.C. Petersen, "Application of Titanium Powder Metallurgy for Manufacture of a Large and Complex Naval Aircraft Component," Powder Metallurgy of Titanium Alloys, F.H. Froes and J.E. Smugeresky, eds., The Metallurgical Society of AIME, 1980, pp. 279-290.

7. R.H. Witt and J. S. Bruce, "Progress on Hot Isostatic Pressing of Titanium," Materials and Processes - Containing Innovations, SAMPE, Vol. 28, 1983, pp. 422-435.

8. J.S. Bruce, A.A. Sheinker, J.W. Bohlen, and G.R. Chanani, "Evaluation of Ti PM Technology for Naval Aircraft Components," Material and Processes - Containing Innovations, SAMPE, Vol. 28, 1983, pp. 1454-1463.

9. D. Eylon, F.H. Froes, and L.D. Parsons, "Titanium PM Components for Advanced Aerospace Applications," Metal Powder Report, Vol. 38, No. 10, October 1983, pp. 567-571.

10. F.H. Froes and J.R. Pickens, "Powder Metallurgy of Light Metal Alloys for Demanding Applications," Journal of Metals, Vol. 36, No. 1, January 1984, pp. 14-28.

11. F.H. Froes and D. Eylon, "Titanium Powder Metallurgy - A Review," Titanium Net Shape Technologies, F.H. Froes and D. Eylon, eds., The Metallurgical Society of AIME, 1984, pp. 1-20.

12. R.H. Witt and I.G. Weaver, "Titanium PM Components for Airframes," Titanium Net Shape Technologies, F.H. Froes and D. Eylon, eds., The Metallurgical Society of AIME, 1984, pp. 29-38.

13. G.V. Scarich, G.R. Chanani, V.C. Petersen and D.M. Weaver, "Relationship Between Mechanical Properties, Contaminants, and Fracture Topography of HIP Ti-6Al-4V Powder Metallurgy Products," Powder Metallurgy of Titanium Alloys, F.H. Froes and J.E. Smugersky, eds, The Metallurgy Society of AIME, 1980, pp. 103-114.

14. E.J. Kosinski, "The Mechanical Properties of Titanium PM Parts Produced from Superclean Powders," Progress in Powder Metallurgy 1982, J.G. Bewley and S.W. McGee, eds., MPIF/-APMI, Princeton, NJ, 1982, pp. 491-502.

15. G.R. Chanani, W.T. Highberger, C. Kelto, and V.C. Peterson, "Application of Titanium Powder Metallurgy of Manufac-

ture of Large and Complex Naval Aircraft Component" presented at the 109th Annual AIME Meeting, Las Vegas, February 1980.

16. P.G. Porter, "Basic Fatigue Data Test Program F-18A," NOR 79-317, January 1982, p. 38.

17. W.J. Crichlow and T. Lunde, "High Cycle Fatigue Properties of Titanium in Aircraft Applications," Titanium Science and Technology, R.I. Jaffee and H.M. Burte, eds., Plenum Press, New York, 1973, p. 1257.

18. P.R. Roberts and P. Lowenstein, "Titanium Alloy Powder Made by the Rotating Electrode Process," Powder Metallurgy of Titanium Alloys, F.H. Froes and J.E. Smugeresky, eds, The Metallurgical Society of AIME, 1980, pp. 21-35.

8. BIOGRAPHIES

James W. Bohlen is a Senior Technical Specialist in the Materials and Process Research and Development Organization of the Northrop Corporation Aircraft Division. Mr. Bohlen has an MS in Materials Science from Cornell University and a BS in Metallurgical Engineering from the University of Illinois. Mr. Bohlen has been involved in PM Ti-6Al-4V research at Northrop since 1980.

Janet A. Walsh is an Engineer in the Materials and Processes Research and Development Organization of the Northrop Corporation Aircraft Division. Miss Walsh has an MS in Materials Science and a BS in Mechanical Engineering, both from the University of Virgina.

A. A. Scheinker was an Engineering Specialist in Materials and Process Research and Technology at Northrop, where he conducted research in titanium powder metallurgy and fatigue and fracture behavior of airframe structural metals. He received his PhD degree in Metallurgy and Materials Science from Lehigh University in 1972. Dr. Sheinker joined Northrop in 1982 and previously conducted research in fatigue, fracture mechanics, stress corrosion cracking, hydrogen embrittlement, and powder metallurgy. He is currently with Rockwell

Govind R. Chanani is the Manager of the Metallic Materials and Processes Research and Development organization of the Northrop Corporation, Aircraft Division where he directs a broad range of research and development activities dealing with airframe structural metals. Dr. Chanani has a PhD and MS from the University of California, Berkeley in

Metallurgical Engineering, and a BS from Banaras University, India also in Metallurgical Engineering. Dr. Chanani has more than 10 years experience investigating PM Ti-6Al-4V alloys.

2nd International SAMPE Metals Conference
August 2-4, 1988

FATIGUE CRACK GROWTH BEHAVIOR OF A SINGLE CRYSTAL ALLOY AS OBSERVED THROUGH AN IN SITU FATIGUE LOADING STAGE

Jack Telesman and Peter Kantzos*
National Aeronautics and Space Administration
Lewis Research Center
Cleveland, Ohio 44135

Abstract

An in situ fatigue loading stage inside a scanning electron microscope (SEM) was used to determine the fatigue crack growth behavior of a PWA 1480 single-crystal nickel-based superalloy. The loading stage permits real-time viewing of the fatigue damage processes at high magnifications.

The PWA 1480 single-crystal, single-edge notch specimens were tested with the load axis parallel to the (100) orientation. Two distinct fatigue failure mechanisms were identified. The crack growth rate differed substantially when the failure occurred on a single slip system in comparison to multi-slip system failure. Two processes by which crack branching is produced were identified and are discussed. Also discussed are the observed crack closure mechanisms.

1. INTRODUCTION

Single-crystal nickel-based superalloys are being utilized in advanced turbine engine applications and are also candidate materials for the use in space shuttle main engines. Due to the current emphasis on the damage tolerant design for turbine engine components, the understanding of fatigue crack growth (FCG) behavior of these alloys has become increasingly more important.

Single-crystal alloys have also been used for fundamental studies of the fatigue damage mechanisms. These alloys are ideal candidates for such studies since they eliminate number of problems encountered with polycrystalline alloys. In polycrystalline alloys, the fatigue crack front encompasses many grains of various unknown orientations, with each grain individually contributing to the overall FCG behavior. The measured fatigue crack growth rate is an average of these contributions and is also effected by the interaction of the crack front with a myriad of grain boundaries. These problems are not encountered when the fatigue behavior of the single-crystal alloys are studied.

In order to obtain a better understanding of the fatigue damage mechanisms, a fatigue loading stage inside a scanning electron microscope (SEM) was developed at NASA Lewis. The small loading stage consists of a specially designed closed loop, servohydraulic load frame mounted inside a SEM. This unique device permits real-time

*NASA Resident Research Associate.

viewing of the fatigue damage processes at high magnifications. An attached video tape recorder and monitor allow repeated test viewing for a more detailed evaluation of the pertinent fatigue mechanisms. The loading stage is detailed elsewhere,(1) however a summary of the operational characteristics is shown in Table 1.

The fatigue loading stage was used to determine the fatigue crack growth damage mechanisms of PWA 1480, a single-crystal nickel-based superalloy.

2. EXPERIMENTAL PROCEDURE

Cast billets of the PWA 1480 single-crystal alloy were obtained in a near (100) orientation. The nominal composition of the alloy is shown in Table 2. Single-edge notched (SEN) specimens of the geometry shown in Fig. 1 were machined using electrical discharge machining. The faces of the SEN specimens were nearly parallel to the (001) orientation. Prior to testing, the specimen faces were polished through a 1-μm diamond paste. Two of the four specimens were then etched to reveal the precipitate structure. Testing was performed in the loading stage. Three of the four specimens were tested at a load ratio R (minimum load/maximum load) of 0.1 with the maximum load being 3110 N (700 lb). The other specimen was tested at R = 0.5 with the maximum load being 2220 N (500 lb). The testing was conducted at room temperature at a frequency of 10 Hz. The vacuum inside the SEM ranged from 1x10-7 to 5x10-7 torr. Tests were stopped periodically to obtain high magnification micrographs of the crack front. Crack growth measurements were obtained for the R = 0.5 specimen and one of the R = 0.1 specimens, while the other two specimens were used to make general observations of the fatigue processes. In all cases the crack length was taken to be the projection of the crack on to a plane perpendicular to the loading axis. To ease the determination of the crack tip location, the micrographs were obtained typically at the maximum load. The crack tip was taken to be the furthest point where displacement between the two surfaces was observed. Some post failure fractography was also performed.

3. RESULTS AND DISCUSSION

3.1 Crack Propagation Mechanisms

The surface observations revealed two distinctive fatigue failure mechanisms. When viewed at low magnifications, the crack growth was either in the plane perpendicular to the loading direction ("straight" crack shown in Fig. 2(a)), or at an inclined angle to the loading axis (Fig. 2(b)). Interestingly, both modes of crack growth were sometimes encountered in the same specimen. For instance, on one side of specimen SC6 tested at R = 0.5, the crack grew perpendicular to the loading direction, while on the other side an inclined crack of approximately 60° to the loading axis was present.

Higher magnification views of the "straight" crack reveal that in reality the crack for the most part propagated on two competing, distinct slip systems. This process has also been observed by Neumann(2) in a copper single crystal. The process consists of alternating shear decohesion at the crack tip on two intersecting (111) type planes creating an "unzipping" effect.(3) This failure mechanism results in activation of large number of slip planes at the crack tip as shown in Fig. 3. As a result, a large deformation zone was formed in the crack wake.

The process through which the fatigue crack growth on an inclined plane occurs is somewhat different. In this case only one (111) type slip system was activated and the crack tended to follow it as shown in Figs. 4(a) and 4(b). Very few slip planes were activated ahead of the crack tip, thus the damage zone was very small. The actual crack growth mechanism appears to be a combination of slip damage on the active

(111) plane due to the resolved shear stresses, combined with the tensile stresses normal to this plane which separate the surfaces on each side of the slipped and damaged material. When this type of mechanism is operative, it becomes difficult to locate precisely the location of the crack tip due to the indistinguishability between the areas where only slip damage has occurred versus areas where the separation of the surfaces has taken place (Figs. 4(a) and 4(b)). At times, the crack-advancement mechanism also involved linking of cracks on parallel slip bands, as shown in Fig. 5, and also documented by Wu and Hoeppner.[4]

The surface observations revealed that the failure mechanisms were mostly confined to the damage on the (111) type slip planes in agreement with findings of others[5,6] for nickel-based single crystals tested at room temperature. However, a post failure analysis revealed the presence of another failure mechanism. A fracture surface of one of the failed specimens was etched and analyzed under the SEM. The evaluation revealed the presence of cuboidal facets of the strengthening precipitates on the fracture surface which is indicative of (100) type failure (Fig. 6(a)). These regions of probable (100) failure were present only in the mid-thickness of the specimen tested under relatively low stress intensities. As the stress intensity increased, the failure mechanism changed in the mid-thickness to the (111) type as indicated by the presence of the (111) facets (Fig. 6(b)).

3.2 Effect of Failure Mode on FCG Rate

The crack length versus cycles plots of specimens tested at a R ratio of 0.1 and 0.5 are shown respectively in Figs. 7(a) and 7(b). Specimen SC2, tested at R = 0.1, exhibited macroscopic crack growth perpendicular to the loading axis on both sides of the specimen. The FCG behavior on each side of this specimen was similar. Specimen SC6 tested at R = 0.5 exhibited macroscopically "straight" crack growth on one side and an inclined crack growth on the other side. At first, crack growth was only observed on the side exhibiting inclined growth (Fig. 7(b), Side A). However, once the "straight" crack appeared on the other surface, it quickly began growing more rapidly than its counterpart on Side A (Fig. 7(b), Side B). FCG rates were calculated on each side of this specimen as a function of the crack length and are shown in Fig. 8. The FCG rates on the side which exhibited "straight" crack growth and deformation on the multislip systems, are faster than the crack growth rates observed on the other side where the crack growth was at an inclined angle and the damaging slip was mostly limited to one slip system. These results imply that damage mechanisms can have a significant effect on FCG behavior. However, since only surface observations and measurements were performed, it is not clear how the subsurface interaction between the two active failure mechanisms influences the observed surface FCG behavior. More work is required to fully comprehend the role of these failure mechanisms on FCG behavior. Ideally, a direct comparison of behavior of a specimen exhibiting entirely inclined crack growth with one exhibiting only "straight" crack growth is desired.

3.3 Mechanisms of Crack Branching and Crack Closure

The observed mechanisms of crack branching can be divided into two categories. One category consisted of a secondary crack developing at or slightly behind the crack tip as shown in Fig. 9. These cracks were usually formed on the slip bands. The usual sequence of events consisted of slowing down of the crack growth rate of the main crack, followed by the initiation of another crack slightly behind the crack tip on the other competing slip system. The new crack then started growing rapidly, leaving in its wake the original crack tip which became in effect the branched crack segment.

The other type of crack branching mechanism encountered consisted of a crack nucleating ahead of the main crack, followed by the coalescence of the two cracks slightly behind the original crack tip. This sequence of events is illustrated in Fig. 10. As shown, crack nucleation was preceded by formation of a slip band (Fig. 10(a) and 10(b)), followed by an initiation of another crack ahead of the crack tip on a parallel slip band (Fig. 10(c)), and finally the coalescence of the three cracks (Fig. 10(d)). Post failure evaluations of some of the events described above, revealed that the surface observations can at times be misleading. What may appear on the surface to be the initiation of an entirely new crack, may in fact be the surfacing of a branched, subsurfaced, region of the main crack.

Three different mechanisms of crack closure were observed to occur in this nickel-based superalloy single crystal. Due to the high degree of tortuosity of the crack path, interlocking of asperities occurred in the crack wake as demonstrated in Fig. 11(a) (roughness induced closure). Another method by which crack closure was found to occur was through the lodging of debris, formed during the fatigue processes, between the two cracked surfaces (Fig. 11(b)). Crack closure was also found to depend upon the mode of loading in the crack segments near the crack tip. This was probably the most common and important mechanism of crack closure. Segments of the crack which experienced mostly Mode I loading (tensile type), exhibited the largest crack tip opening displacements (CTOD) and tended to remain open throughout most of the loading cycle. Other segments, subjected to mixed-mode loading (combination of tensile and shear), exhibited much smaller CTOD and were closed for considerably longer portions of the loading cycle. This behavior is clearly shown in Figs. 11(c) and 11(d). The larger magnitude of crack closure of the inclined crack subjected to mixed mode loading, may have contributed to its lower observed FCG rates since the crack tip was closed for longer portions of the loading cycle.

4. SUMMARY

The in situ fatigue loading stage inside a SEM has proven to be a powerful tool in identifying the fatigue failure mechanisms. Using this tool, relationships between the measured FCG rates and the individual fatigue failure mechanisms were established for the PWA 1480 single-crystal nickel-based superalloy. The crack growth rate was substantially lower when the failure occurred on a single (111) slip system in comparison to the multislip failure. The details of two different processes resulting in formation of crack branching were observed. Three different crack closure mechanisms were identified.

The in situ loading stage is capable of providing qualitative information such as the mode of failure, as well as quantitative information such as CTOD and the magnitude of crack closure load. Such information is needed to improve the fundamental understanding of the fatigue damage processes as well as life prediction methodology of engineering structures.

5. REFERENCES

1. J. Telesman, P. Kantzos, and D.M. Fisher, "Fatigue Loading Stage Combined with Scanning Electron Microscope," NASA TM-100822, 1988.

2. P. Neumann, "New Experiments Concerning the Slip Processes at Propagating Fatigue Cracks - I," Acta Met., Vol. 22, No. 9, Sept. 1974, pp. 1155-1178.

3. C.Y. Yang, and H.W. Liu, "The Application of the Unzipping Model of Fatigue Crack Growth to Micro-Cracks," Scr. Met., Vol. 14, No. 7, July 1980, pp. 785-790.

4. D.C. Wu, and D.W. Hoeppner, "Observations and Characterization Considerations of Fatigue Crack Growth in a Single Crystal Nickel-Base Superalloy," Scr. Met., Vol. 19, No. 4, Apr. 1985, pp. 493-498.

5. J.S. Crompton, and J.W. Martin, "Crack Growth in a Single Crystal Superalloy at Elevated Temperatures," Metall. Trans. A., Vol. 15, No. 9, Sept. 1984, pp. 1711-1719.

6. C. Howland, and C.W. Brown, "The Effect of Orientation on Fatigue Crack Growth in a Nickel-Based Single-Crystal Superalloy," Report No. PNR-90207, Rolls Royce Ltd., Derby, England, 1984.

TABLE I. - SEM LOADING STAGE CHARACTERISTICS

Maximum cyclic load, N (lb)	4450 (1000)
Maximum cyclic frequency, Hz	30
Maximum temperature (designed), °C (°F)	1100 (2000)
Working distance (room temp. stage), mm (in.)	25 (1)
Working distance (high temp. stage), mm (in.)	55 (2.2)
Useful maximum magnification (room temp. stage)	10 000X
Useful maximum magnification (high temp. stage)	3000X

TABLE 2. - PWA 1480

NOMINAL COMPOSITION

Element	Weight percent
Co	5
Cr	10
W	4
Ta	12
Al	5
Ti	1.5
Ni	62

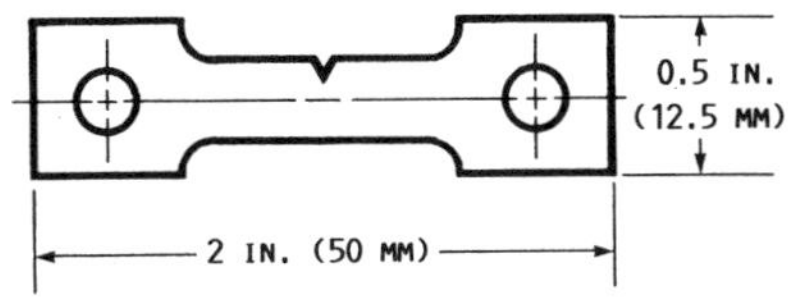

FIGURE 1. - SINGLE-EDGE NOTCH SPECIMEN.

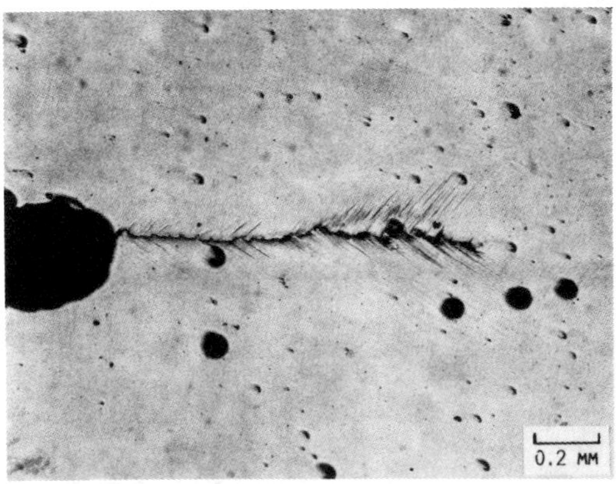

(a) NORMAL TO THE LOADING DIRECTION.

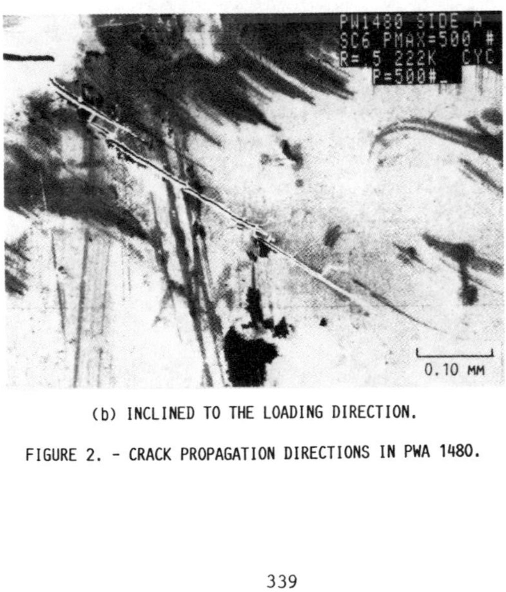

(b) INCLINED TO THE LOADING DIRECTION.

FIGURE 2. - CRACK PROPAGATION DIRECTIONS IN PWA 1480.

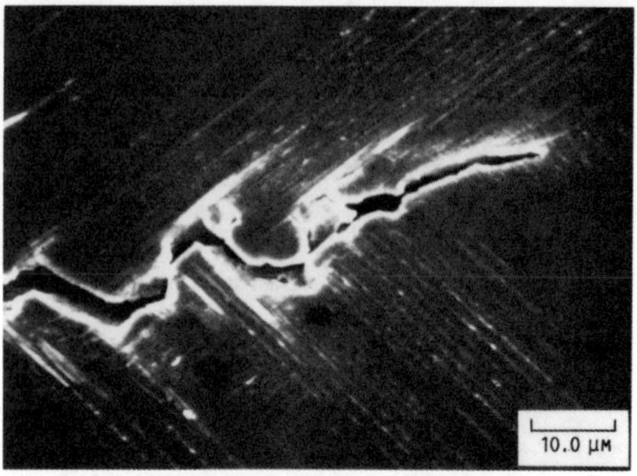

FIGURE 3. - HIGH MAGNIFICATION OF THE MACROSCOPICALLY STRAIGHT CRACK. TWO SLIP SYSTEMS AND A LARGE NUMBER OF SLIP PLANES WERE ACTIVATED.

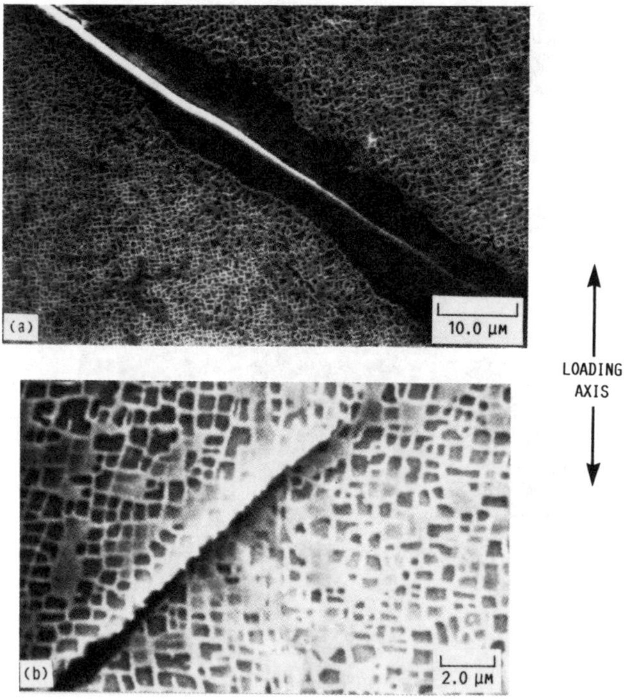

FIGURE 4. - GROWTH OF AN INCLINED CRACK, PARALLEL TO ONE SLIP SYSTEM. VERY FEW SLIP PLANES WERE ACTIVATED AND SLIP WAS RESTRICTED PRIMARILY TO ONE (111) SLIP SYSTEM.

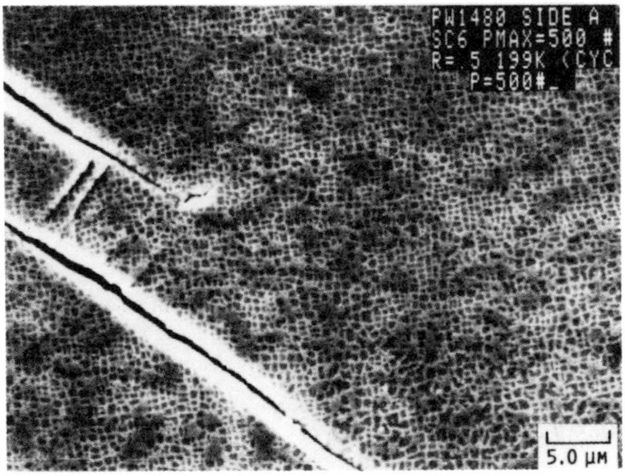

FIGURE 5. - LINKING OF TWO SEGMENTS OF AN INCLINED CRACK.

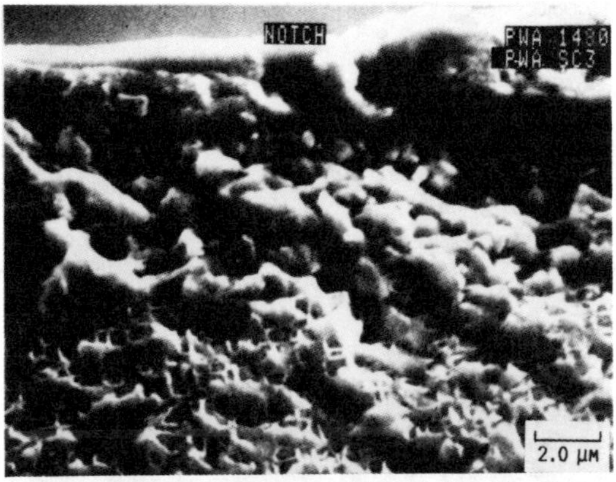

(a) LOW ΔK REVEALED AREAS OF (100) FAILURE.

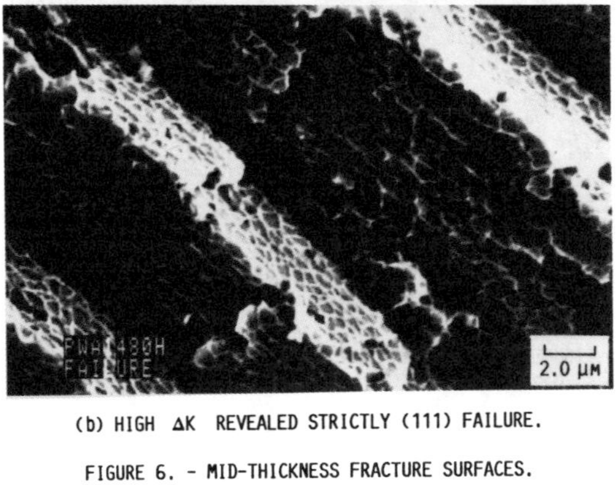

(b) HIGH ΔK REVEALED STRICTLY (111) FAILURE.

FIGURE 6. - MID-THICKNESS FRACTURE SURFACES.

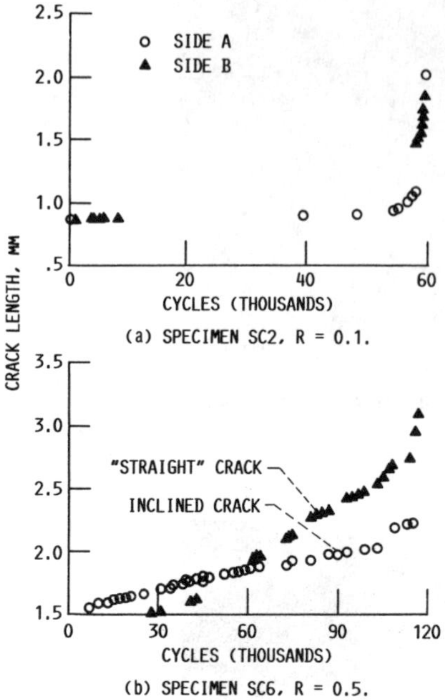

FIGURE 7. - CRACK LENGTH VERSUS CYCLES FOR TWO PWA 1480 SPECIMENS.

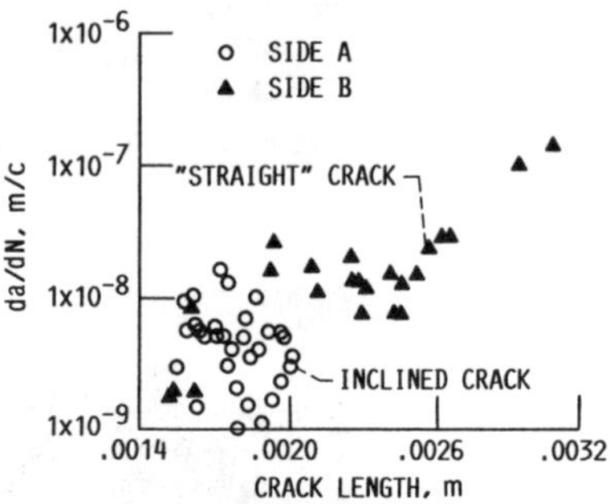

FIGURE 8. - CRACK GROWTH RATE VERSUS CRACK LENGTH FOR SPECIMEN SC6 WHICH EXHIBITED TWO TYPES OF CRACKING PROCSSES.

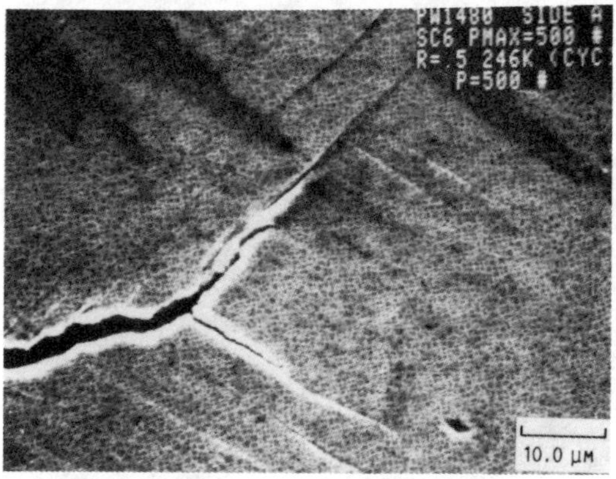

FIGURE 9. - CRACK BRANCHING NEAR THE CRACK TIP (MAXIMUM LOAD).

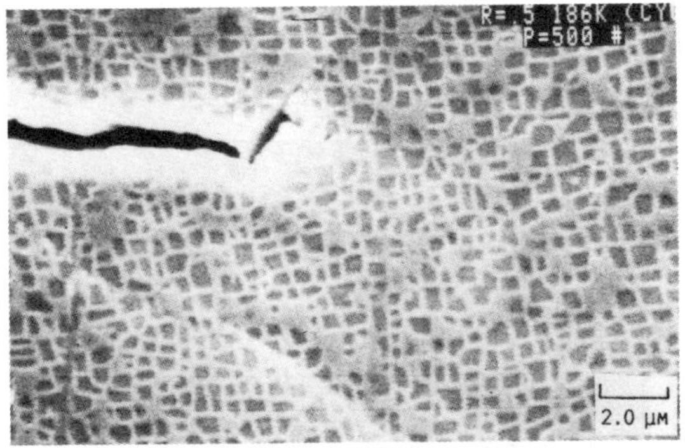

(a) 186 000 CYCLES.

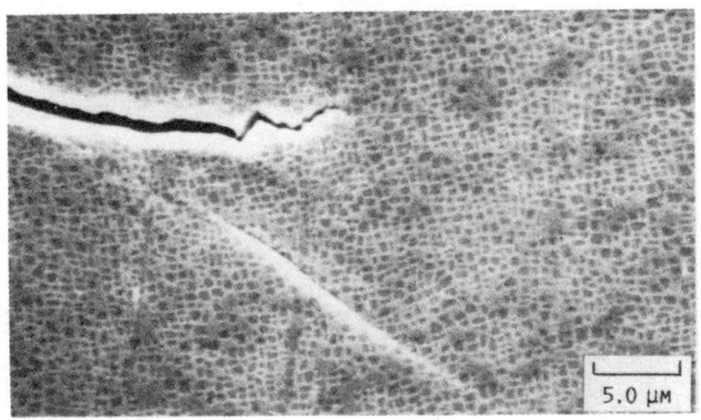

(b) 188 000 CYCLES.

LOADING
AXIS

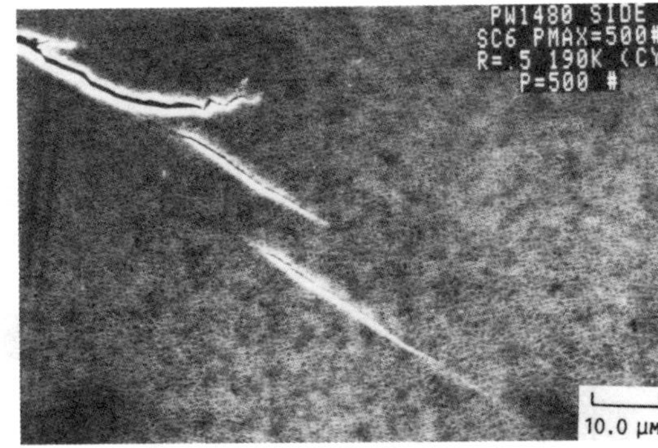

(c) 190 000 CYCLES.

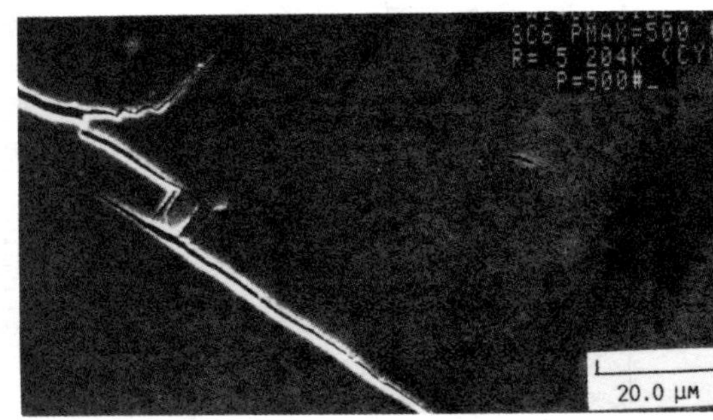

(d) 204 000 CYCLES.

FIGURE 10. - THE SEQUENCE OF CRACK INITIATION AND LINKUP OF SLIP BAND CRACKS WITH THE MAIN CRACK (MICROGRAPHS TAKEN AT MAXIMUM LOAD).

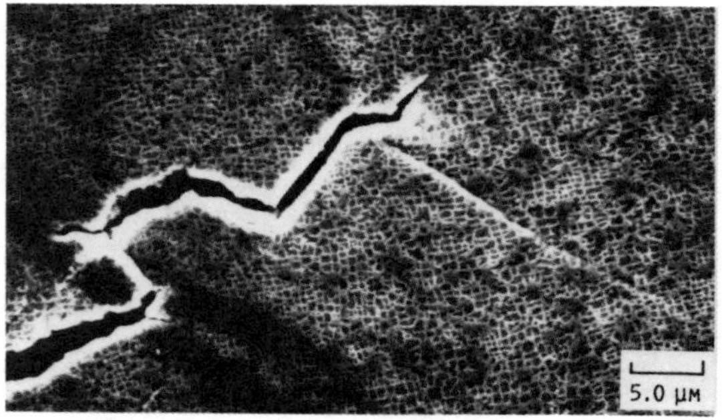

(a) 2220 N (500 LB).

(b) 2220 N (500 LB).

LOADING AXIS

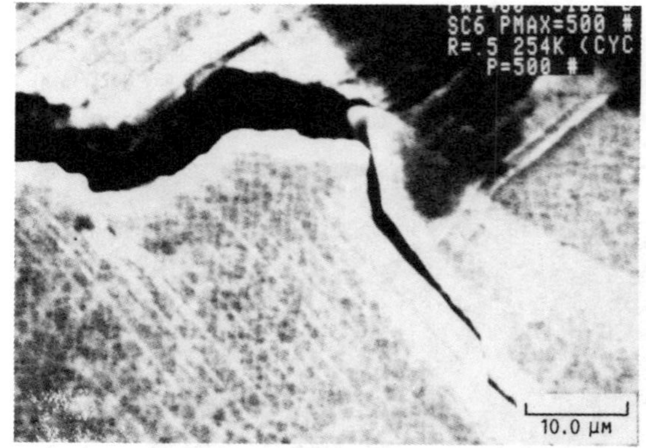

(c) 2220 N (500 LB).

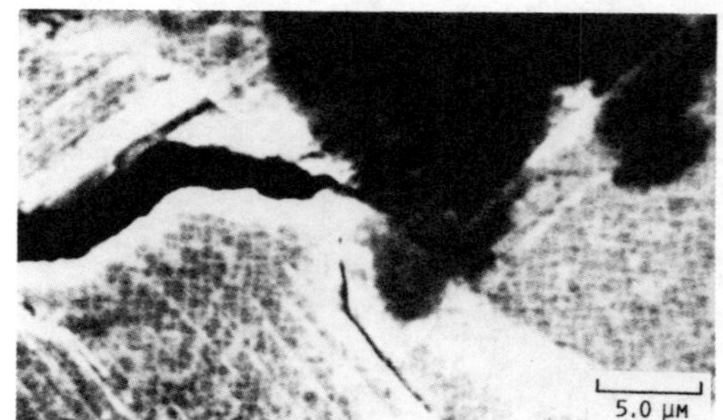

(d) 1335 N (300 LB).

FIGURE 11. - THREE MECHANISMS OF CRACK CLOSURE ARE: ROUGHNESS INDUCED (a), DEBRIS FORMATION (b), AND THE EFFECT OF MIXED MODE LOADING (c) AND (d).

2nd International SAMPE Metals Conference
August 2-4, 1988

ADVANCED MAGNESIUM ALLOYS FOR AEROSPACE APPLICATION

C. F. Chang, S. K. Das, D. Raybould, R. L. Bye, N. J. Kim
Metals and Ceramics Laboratory
Allied-Signal, Inc.
Morristown, NJ 07960

Abstract

Magnesium alloys are attractive for high performance aerospace structural applications because of their light weight, high specific strength, and good machinability. Improvements in corrosion resistance and mechanical properties of magnesium alloys by the alloy and process development can further improve performance and expand the usage of magnesium alloys, allowing significant weight savings. Mg-Al-Zn-Rare Earth (RE) alloys made by rapid solidification processing (RSP) possess an excellent combination of corrosion resistance and room temperature mechanical properties in both extrusions and forgings. These alloys have excellent formability at elevated temperatures which are critical for near-net shape forging. Conventionally cast Mg-Al-Zn-RE alloys have good mechanical properties and improved corrosion resistance compared to other cast magnesium alloys.

1. INTRODUCTION

One limitation of aircraft performance is the weight of the aircraft structure. Improvements in performance can be achieved by the use of magnesium alloys due to their low density. However, their lower mechanical strength and poor corrosion resistance compared to aluminum alloys have limited their use. Improvements in strength and/or corrosion resistance would probably greatly increase magnesium alloy usage. The application of RSP in metallic systems results in the refinement of grain and intermetallic particle size, extended solid solubility, and improved chemical homogeneity. A significant improvement in the mechanical properties [0.2% yield strength (Y. S.) up to 435 MPa, ultimate tensile strength (U. T. S. up to 476 MPa, elongation (El.) up to 14 %] has been achieved in rapidly solidified Mg-Al-Zn-Si-Mn and rare earth containing Mg-Al-Zn alloys(1-4). The rapidly solidified Mg-Al-Zn-RE alloys have also been found to have outstanding corrosion resistance, better than that of any commercially available magnesium alloy and equivalent to that of some aluminum alloys(2,3). The alloys show a great potential application in automotive and aerospace industries. However, to ensure the successful application of RSP Mg-Al-Zn-RE alloys in the aerospace industry, it is necessary to characterize the alloys in detail. In addition, practical aspects of magnesium alloy: such as the ability to produce RSP alloys in large quantities, and the formability of RSP alloys need to be addressed. There have been indications that the rare earth containing alloys need not be rapidly solidified in order to achieve superior corrosion resistance and that conventional castings made from these alloys may exhibit enhanced corrosion behavior. These castings may be used for applications where strength is not a critical factor and forgings or extrusions are not practical. In the present paper we will examine the developments in magnesium alloy rapid solidification processing, fabrication techniques, and the engineering properties achieved in Mg-Al-Zn-RE alloys. The properties of chill cast alloys will also be compared with those obtained for commercial magnesium alloys.

2. RAPID SOLIDIFICATION PROCESSING

Rapid solidification processes that show evidence of being commercially viable can be broadly classified into atomization and melt spinning methods. The reactivity of molten magnesium metal increases rapidly with temperature, such that above 850°C a freshly exposed surface spontaneously bursts into flame. Therefore, air atomization techniques commonly used for aluminum alloys, where the presence of an oxide film protects the metal from further oxidation, can not be used for magnesium alloys.

An alternative method of producing powders is to produce rapidly solidified ribbon by free jet melt spinning or planar flow casting and then reducing the ribbon into powder using a series of high speed mechanical comminution processes[4].

2.1 Melt Spinning

Two melt spinning techniques, free jet melt spinning (jet casting) and planar flow casting (PFC), have been used to produce rapidly solidified Mg-Al-Zn-RE alloy ribbon. In jet casting, molten metal is forced through an orifice onto the surface of a rotating water cooled wheel from which a continuous ribbon is chilled. Although jet casting can produce a wide range of alloys, the process is restricted to the production of ribbon less than about 5 mm wide. This potentially serious production deficiency has been solved by the development of the planar flow casting process. The planar flow casting process positions the casting nozzle in close proximity to the rotating chill block and employs a rectangular slot orifice. The molten pool of metal is constrained into a stable shape by balancing surface tension forces with the applied liquid metal pressure, to achieve a dynamically stable melt pool on the rotating substrate. By using planar flow casting techniques, magnesium alloy ribbon about 50 mm in width has been produced in large quantities. While casting magnesium alloys a suitably protective shielding atmosphere has been used to prevent oxidation of the liquid metal surface and air entrapment under the liquid film.

2.2 Comminution

After casting, the ribbon is typically comminuted into -35 to -60 mesh U. S. sieve size powders. In general, the final product consists of irregularly shaped flat platelets with a thickness equal to the original ribbon thickness. The particles obtained by the process have a uniform microstructure irrespective of the particulate size, whereas the rapidly solidified powder obtained by atomization processes has a wide range of particle sizes and the highest cooling rate is usually limited to the finest particle size. As a result, the production yields for high cooling rate (10^5-10^6°C/S) atomized particles are low. Furthermore, if one intends to separate the fine particles which have high cooling rate, there are potentially serious safety hazards that one will have to deal with. For example, large amounts of -200 mesh particles are hazardous, as air suspensions are easily maintained and any source of ignition can result in a violent explosion. In a ribbon comminution process, however, one does not necessarily have to deal with large volumes of fine particles.

2.3 Consolidation

Consolidated bodies were produced from powder/particulate by using conventional powder metallurgy techniques. The powders were either vacuum outgassed in a can and then sealed under vacuum or vacuum hot pressed at 200 - 300°C for 1 - 24 hours depending upon the size of can. The can may be extruded directly or compacted and machined to remove the can before extrusion or forging. Prior to extrusion the cans were soaked at the extrusion temperature for about 1/2 - 4 hours. The cans or compacts were extruded at temperatures of about 200-300°C to rectangular or round bars at extrusion ratios ranging from 14:1 to 18:1. A significant temperature rise occurs by the extensive deformation produced in extrusion[5]. Round bars with dimensions up to 3.6 m in length x 0.06 m in diameter have been extruded in a production facility without much difficulty (Figure 1). The microstructure obtained after consolidation depends upon the composition of the alloy and the consolidation conditions. Excessive time at high temperatures can cause the fine grains and precipitates to coarsen beyond the optimal size, leading to a deterioration of the properties. The stable dispersoids inherent in RS magnesium alloys pin the grain boundaries and prevent their coarsening during high temperature consolidation. As a result, the consolidated body has a fine grain size. Characterization of as extruded Mg-Al-Zn-RE alloys will be discussed in Section 3.1.

3. RAPIDLY SOLIDIFIED ALLOYS

Three alloys (1)Mg-5Al-5Zn-6Y (in wt%) (EA65 RS), (2)Mg-5Al-5Zn-1Si-5Pr (EA55A RS) and (3)Mg-5Al-5Zn-5Nd (EA55B RS) have particularly attractive combinations of strength, ductility and corrosion resistance[6]. These alloys have strengths and ductilities that are equivalent to those of high strength aluminum alloys and significantly exceed those of conventional magnesium alloys and are 5 times as corrosion resistant as the most resistant conventional magnesium alloy. The high strengths and low densities of these alloys result in their specific strengths greater than those of existing low density alloys. The compressive strength of RS magnesium alloys is equal to or greater than the tensile strength.

Fig. 1 A photograph of 3.6 m (in length) x 0.06 m (in diameter) extrusions.

3.1 <u>Microstructure Characterization</u>

Figure 2 shows the typical microstructure of a rapidly solidified alloy EA65. The microstructure consists of uniform distribution of fine dispersoids within the magnesium matrix. The diameter of the dispersoids in EA65 RS is approximately 20 nm in both as-cast and as-extruded conditions. The fact that the dispersoids do not coarsen during consolidation demonstrates the stability of dispersoids at elevated temperatures. Figure 2(b) shows an X-ray spectrum from the matrix and Figure 2(c) shows a spectrum from the dispersoids. It can be seen that the dispersoids contain a significant proportion of Al and Y which are not detectable in the matrix. Quantitative analysis, involving foil thickness measurement and Monte Carlo simulation, shows that the composition of dispersoids is close to Al_2Y[7]. Microdiffraction studies have verified that the dispersoids are indeed Al_2Y [Figure 2(d)]. The formation of Al_2Y in this alloy system is quite interesting because Al_2Y is not an equilibrium phase in Mg-Al-Zn-Y alloy. Similar results have been obtained in

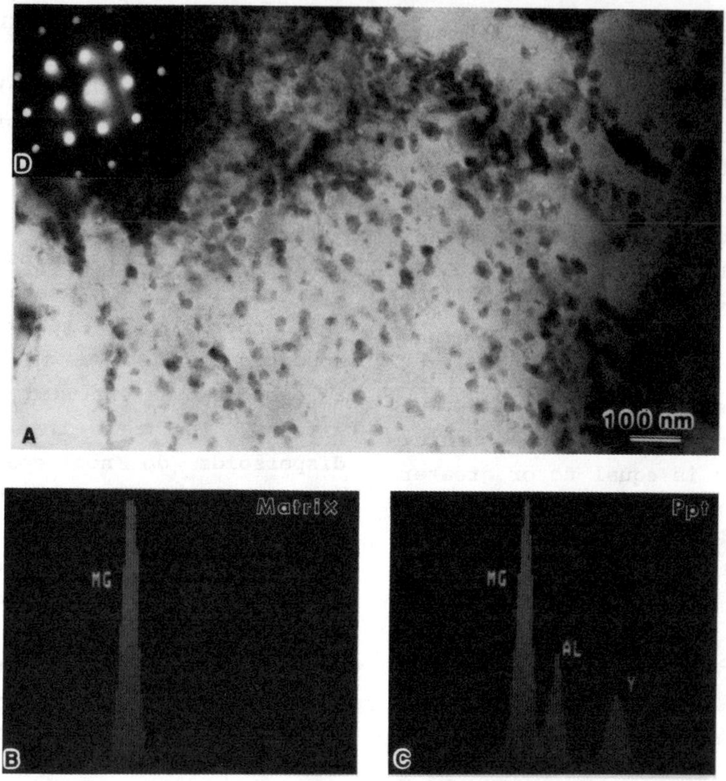

Fig. 2 STEM analysis of as-extruded EA65 RS including (a) bright field micrograph showing general microstructure , and EDS analysis of (b) Mg matrix and (c) precipitate, and microdiffraction of Al_2Y phase.

EA55B RS. The major microstructural constituent in this alloy is Al_2Nd. Unlike EA65 RS, however, a small volume fraction of plate-like $Mg_{17}Al_{12}$ was observed within the matrix of EA55B RS.

3.2 <u>Mechanical Properties</u>

Table 1 shows room temperature mechanical properties of 25.4 mm diameter extrusions of three alloys extruded at different temperatures. For samples extruded at a low temperature, yield strengths ranging from 435 MPa (63 ksi) to 476 MPa (69 ksi) with ductilities ranging from 5% to 14% were achieved. The high strength of these RS Mg alloys is due to the fine grain and dispersoid size retained in these alloys after extrusion. The fracture toughness of these high strength alloys is poor. Increasing the extrusion temperature, decreases the

Table 1 Room temperature mechanical properties of 25.4 mm diameter extrusions of Mg-Al-Zn-RE alloys extruded at a moderate speed and at different temperatures.

Alloy	Extrusion Temp.	Y. S. (MPa)	U. T. S. (MPa)	El. (%)	K_q (MPa$\sqrt{m}$)
EA65 RS	Low	456	513	5.0	7.0
	High	430	473	12.3	7.0
EA55A RS	Low	476	516	5.0	8.0
	High	349	423	13.3	10.0
EA55B RS	Low	435	476	13.8	7.0
	High	361	439	11.5	15.0

yield strength but increases the ductility, and fracture toughness (K_q). Further improvement in the fracture toughness can be achieved by solution heat-treatment which dissolves the brittle and hard second phase particle of $Mg_{17}Al_{12}$(3). As discussed in the previous section, because of the extensive deformation produced in extrusion, there is a temperature rise during extrusion. This may result in the loss of strength at the end of extrusions (Table 2). By reducing the extrusion speed to minimize the heat build up in the material, uniform properties can be obtained throughout the entire extrusion (Table 2). The mechanical properties of Mg-Al-Zn-RE alloys strongly depend on the temperature and strain rate (Figures 3-4). At a constant temperature increasing the strain rate increases the tensile strength. Moreover, the strain rate dependence of strength increases with increasing temperature. Testing at a high temperature and at a low strain rate tends to improve the ductility.

Table 2 Room temperature tensile properties of EA55B RS extrusions extruded at a low temperature and at a moderate and low speed.

Diameter (mm)	Extrusion Speed	Section	Y. S. (MPa)	U. T.S. (MPa)	El. (%)
25.4	Moderate	Nose	460	510	9
		Tail	422	466	15
	Low	Nose	460	506	12
		Tail	460	510	13
60	Moderate	Nose	406	444	13
		Tail	388	431	19

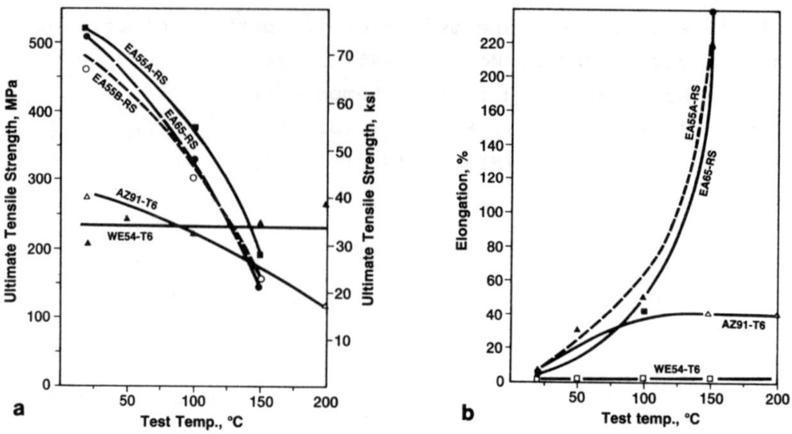

Fig. 3 A comparison of elevated temperature properties of as-extruded RS Mg-Al-Zn-RE alloys with commercial ingot cast magnesium alloys (a) tensile strength (b) ductility.

Unlike conventional magnesium alloys, elongations more than 100% have been achieved at test temperatures as low as 150°C and at strain rates less than $1x10^{-3}/S$ in the as-extruded bar. This is due to the fine grain and dispersoid size obtained by RSP.

3.3 Formability

Magnesium alloys, like other alloys with hexagonal crystal structures, have a limited number of slip systems, and are much more workable at elevated temperatures than at room temperature. For conventionally processed magnesium alloys, the temperature range between the minimum forming temperature required to avoid cracking and the maximum temperature needed to avoid softening is very narrow. However, the combination of low flow stress and high ductility in the RS Mg-Al-Zn-RE alloys extrusion at temperatures as low as 150°C, makes them useful in near-net shape forming such as hot forging. Figure 5 shows extruded bars of EA55B alloy forged at 160°C at a low rate and at 180°C at a moderate rate. Large cracks occurred when the sample was forged at the moderate rate [Figure 5(b)]. Decreasing the ram speed eliminated the crack in the sample and improved the formability [Figure 5(a)]. When forged in a close die at a low rate, a complex part may be formed in a single step and with outstanding precision of

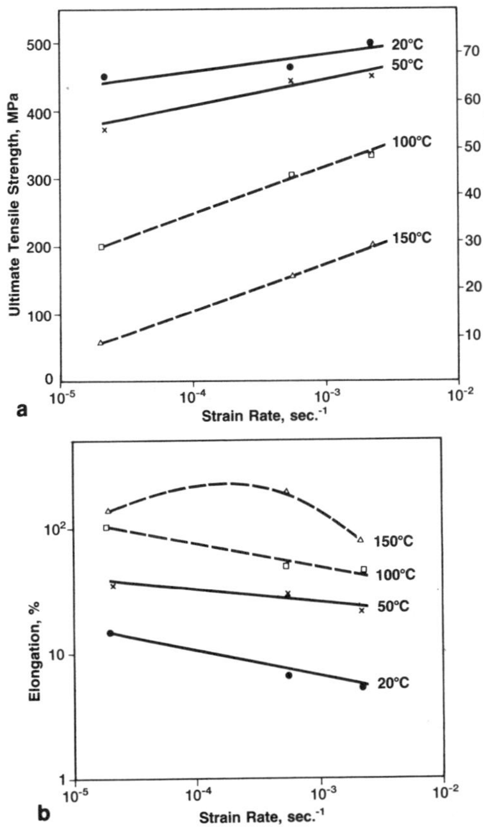

Fig. 4 Effect of strain rate on the tensile properties of EA55B RS extrusions (a) tensile strength (b) ductility.

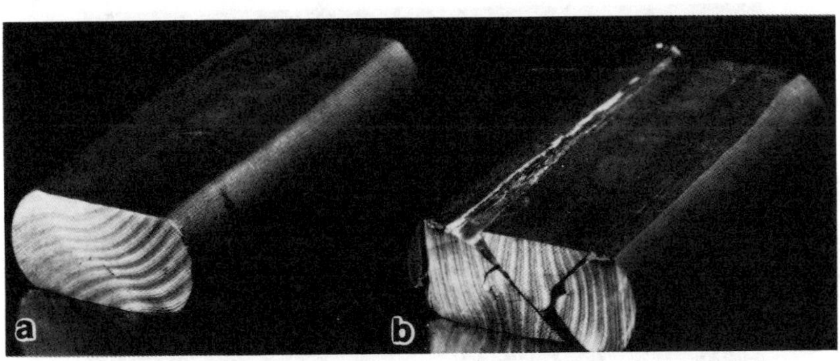

Fig. 5 Photographs of forgings forged at (a) 160°C at a low rate (b) 180C at a moderate rate, illustrating the effect forging speed on the formability of RS magnesium alloys.

Table 3 Tensile properties of as-forged EA55B RS extrusions.

Extrusion Speed	Forging Speed	Temp. (^{o}C)	Y. S. (MPa)	U. T. S. (MPa)	El. (%)
Low	Low	190	483	540	6.8
	Moderate	230	467	537	7.3
Moderate	Low	150	444	499	10.2
	Moderate	220	451	516	13.0

shape and no cracks. It is to be noted that under the same forging condition severe cracks have been found in the commercial alloy, such as ZK60A. The mechanical properties of the forged sample are about the same as the as-extruded sample [Table 3].

3.4 <u>Corrosion Behavior</u>

Superior corrosion resistance of RS Mg-Al-Zn-RE alloys over commercially available corrosion resistant Mg alloy AZ91HP is demonstrated by the results of a 5 % salt spray test shown in Figure 6. The improvement in the corrosion resistance of these RS Mg-Al-Zn-RE alloys is due to the combined effect of refined microstructure and inertness of second phase particles which reduce the tendency to pitting corrosion.

4. CHILL CAST ALLOYS

Three different Mg-Al-Zn-RE alloys (1)Mg-5Al-5Zn-6Y (in wt%) (EA65), (2) Mg-5Al-5Zn-5Nd (EA55B), (3)Mg-8Al-5Nd-6Zn (AZ86) were cast into copper molds which produced 6-25 mm in diameter by 150 mm in length fingers.

4.1 <u>Tensile Properties</u>

The room temperature tensile properties of chill cast Mg-Al-Zn-Nd (Y) alloys are shown in Table 4. Additions

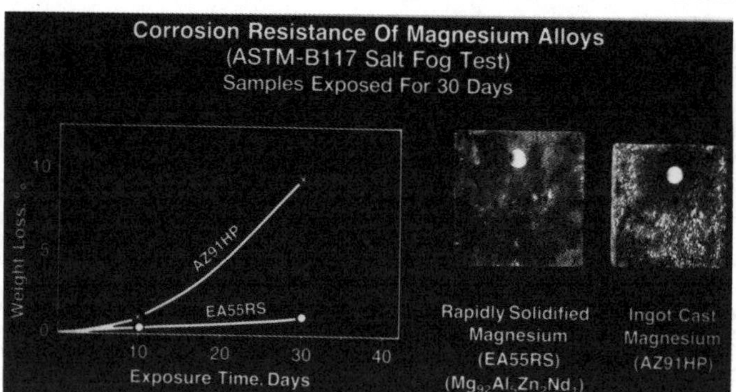

Fig. 6 A comparison of corrosion behavior of as-extruded EA55B RS with ingot cast magnesium alloy AZ91HP.

of Al to EA55B tend to increase the yield strength and to decrease the ductility of the alloy. Yield strength comparable to that of the commercially available high strength casting alloy, WE54, has been achieved in alloy AZ86. As in Mg-Al-Zn and Mg-Y-Nd-Zr alloys, maximum strength in the Mg-Al-Zn-RE alloys is developed in the solution heat-treated and aged (T6) condition. The room and elevated temperature tensile properties of Mg-Al-Zn-Nd (Y) alloys in T6 condition are comparable to those of WE54 as shown in Figure 7.

4.2 Corrosion Behavior

The corrosion resistance of chill cast AZ86 alloy is superior to commercial magnesium casting alloy WE54 in both the as-cast and T6 condition, as reflected by % weight loss in salt fog testing (Figure 8). Superior pitting corrosion resistance of this alloy is due to the alloy chemistry together with the inertness of second phase particles such as Al_2Nd in the corrosive environment.

5. CONCLUSION

Advanced Mg-Al-Zn-RE alloys processed by conventional casting technology have superior corrosion resistance compared to commercial

Table 4 Room temperature tensile properties of as-cast Mg-Al-Zn-RE alloys.

Alloy	Y. S. (MPa)	U. T. S. (MPa)	El. (%)
EA65	142	169	5.7
EA55B	142	253	7.0
AZ86	165	230	2.9
WE54	164	212	2.6

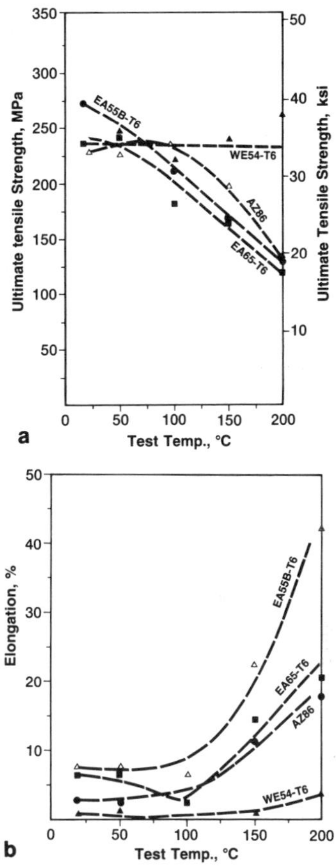

Fig. 7 A comparison of tensile properties of chill cast Mg-Al-Zn-RE alloys with commercial alloy WE54 (a) tensile strength (b) ductility.

magnesium alloys in saline environment. RSP Mg-Al-Zn-RE alloys possess an excellent combination of room temperature mechanical properties and corrosion resistance in both extrusions and forgings. The alloys have excellent formability at elevated temperatures which are critical for near-net shape forging. Parts made from these alloys are suitable for aerospace application where weight savings is critical.

6. ACKNOWLEDGMENTS

The authors would like to thank Messrs. J. Misson, E. Rainal, P. Jeges, R. Costa, D. Timan, R. Fehrman, and Ms. M. Rodriguez for their experimental assistance.

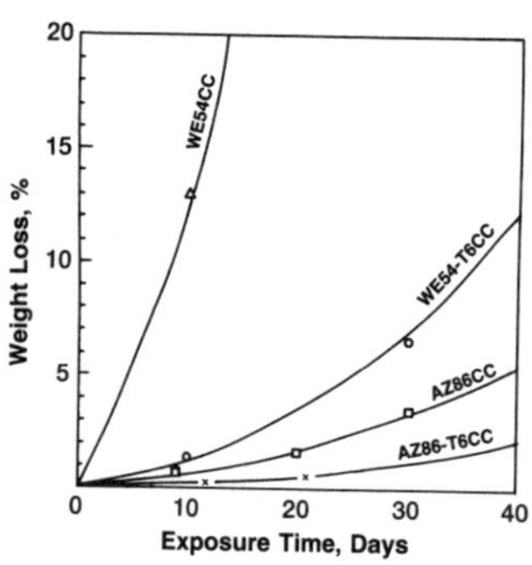

Fig. 8 A comparison of corrosion behavior of chill cast AZ86 with commercial alloy WE54 in the as-cast and T6 condition.

7. REFERENCES

(1) S. K. Das, et al., U. S. Patent No. 4,675,157, High Strength Rapidly Solidified Magnesium Base Metal Alloys, June, 1987.

(2) C. F. Chang, S. K. Das and D. Raybould, Rapidly Solidified Materials, Amercian Society for Metals, 1986, pp 129-135.

(3) C. F. Chang, S. K. Das and D. Raybould, Processing of Structural Metals by Rapid Solidification, ASM International, 1987, pp 409-415.

(4) S. K. Das et al., U. S. Patent No. 4,718,475, Apparatus for Casting High Strength Rapidly Solidified Magnesium Base Metals Alloys, Jan., 1988.

(5) D. Raybould and C. F. Cline, Rapidly Solidified Crystalline Alloys, The Metallurgical Society of AIME, 1985, pp 111-133.

(6) C. F. Chang, S. K. Das, D. Raybould and A. Brown, Metal Powder Report, 41, 302-308 (1986).

(7) N. J. Kim and S. K. Das, "Characterization of Microstructures in Rapidly Solidified Alloys", in Charaterization of Advanced Materials, eds. E. Hennecke and W. Altergott, International Metallographic Society, in press.

2nd International SAMPE Metals Conference
August 2-4, 1988

A REVIEW OF BERYLLIDES FOR VERY HIGH TEMPERATURE SERVICE

J. M. Marder
A. J. Stonehouse
Brush Wellman Inc.
Elmore, Ohio 43416

Abstract

With the renewed emphasis on very high operating temperatures of aircraft and engine structural materials, the potential application of beryllide intermetallic compounds is undergoing serious evaluation. The beryllides, as a group, present a unique combination of high temperature strength and low density, which may be translatable into significant improvements in high speed aircraft performance. The objective of this paper is to review the current state of beryllide technology. Binary intermetallic compounds of beryllium are fabricated, under laboratory conditions, as powder metallurgy products. The typical fabrication processes used for these materials are reviewed with an emphasis upon the characteristics of the consolidated compounds, including density, microstructure, high temperature strength and oxidation resistance. Directions for future development are suggested.

1. INTRODUCTION

Over the past 30 years, intermetallic compounds as a class and beryllides, in particular, were considered as candidate materials for very high temperature applications.(1) Intermetallic compounds were surveyed to determine those which had the greatest potential for meeting the strength and density requirements for a high temperature aerospace material. The density requirement was considered as being as important as high temperature strength in airborne or spaceborne applications.(2) After a review of silicides and aluminides among others, beryllides were selected as the intermetallic compound showing the greatest potential in terms of high temperature strength, density and oxidation resistance. Indeed, by the beginning of the 1960's, the problems which prevented widespread use of beryllides were confined to their low and intermediate properties rather than high temperature properties. In this paper, the current understanding of beryllide fabrication, properties, limitations, and outlook for future development will be reviewed.

2. FABRICATION

Beryllides are prepared from elemental powders to produce the required composition. The powders are mixed, and then cold pressed to approximately 80% of theoretical density. The compacts are then placed in covered beryllium oxide crucibles in a vacuum furnace and heated to react the constituent powders. One of the problems generally encountered at this stage is volatilization of beryllium under the vacuum and high temperature generally necessary for the reaction. A lack of control over the loss of one of the constituent elements can lead off-stoichiometry compositions which cannot achieve the desired property levels. This is of particular importance when a small change in chemistry can result in a transition from one compound to a less desirable one. After the reaction step, the resulting cake of intermetallic material is comminuted into powder for fabrication.

The powdered intermetallic beryllide may be treated in the same fashion as any powder metallurgy product with proper regard for reactivity, and the high temperature which is required to induce consolidation into a solid body. Typically, cold pressing followed by vacuum hot pressing could be used as a consolidation technique, achieving densities of greater than 95% of the theoretical value for most of the compounds studied. As in the reaction step, the loss or volatilization of the component elements can limit the control over the chemistry, and therefore over the properties of these compounds. Typical vacuum hot pressing temperatures range from 2500°F to 3130°F, depending upon the beryllide under investigation. With the advent of hot isostatic pressing as a consolidation technique, it would be expected that greater densities and better control of chemistry would be achieved if suitable can materials and HIP parameters can be developed.

3. PHYSICAL CHARACTERISTICS

3.1 Crystal Structure

With the notable exceptions of aluminum and silicon, beryllium forms intermetallic compounds with just about every other element evaluated to date. In general, more than one compound can be formed, depending upon the exact stoichiometry. Table 1 is a compendium of the physical characteristics of beryllides which, because of their high melting point and their strength at high temperature, as well as their oxidation resistance at high temperature, might suggest themselves as being useful in the severe environments which may be imposed by advanced aircraft and engine designs. The temperature range of 2000-3000°F is taken to be the range of interest in future applications. The beryllide forming elements of interest are the transition metals Nb, Ta, Mo, Ti, Zr, and Hf. Typically the MBe_2, MBe_{12}, and MBe_{13} intermetallics, which are of greatest interest, are found in these systems.

Table 1. High-Temperature Oxidation Resistant Beryllides

		Melting Point (°F)	X-Ray Density (g/cc)
Nb-Be	$NbBe_{12}$	3070	2.92
	Nb_2Be_{19}	~3100	3.17(?)
	Nb_2Be_{17}	3100	3.28
Ta-Be	$TaBe_{12}$	3360	4.18
	$TaBe_{17}$	3610	5.05
Mo-Be	$MoBe_{12}$	~3100	3.03
Ti-Be	$TiBe_{12}$	2900	2.26
	Ti_2Be_{17}	2970	2.46
Zr-Be	$ZrBe_{13}$	3500	2.72
	Zr_2Be_{17}	3600	3.08
Hf-Be	$HfBe_{13}$	2900	3.93
	Hf_2Be_{21}	<3180	4.25
	Hf_2Be_{17}	<3180	4.78

In general, the group VB and VIB metals (V, Nb, Ta, Mo, and W) produce MBe_{12} structures, as does titanium, whereas the lanthanides, actinides, Y, Zr, and Hf form MBe_{13}. Exceptions are magnesium and calcium which also form MBe_{13}.

3.2. Density and Melting Point

The density of the beryllides makes them particularly attractive in aerospace applications. The density is naturally dependent upon the other metals involved, and ranges from 2.26 g/cc for $TiBe_{12}$ to 5.05 g/cc for $TaBe_{17}$. This may be compared with the density of pure beryllium, 1.85 g/cc, and the density of typical nickel-based superalloys, 8.28 g/cc. The density of titanium aluminide is 3.35 g/cc, and may be compared with that of titanium beryllide, with a much higher temperature capability and a density of 2.26 g/cc for $TiBe_{12}$. The specific properties at high temperature will therefore be much greater for the beryllides than for comparable aluminides, or any other high temperature material. The density of the beryllides changes rapidly as the composition deviates from the exact intermetallic composition, because of the low density of beryllium. This makes the processing steps, and particularly those precautions taken to reduce the loss of high vapor pressure elements during fabrication, all the more crucial.

The melting points of several beryllide compounds are also given in Table 1. It can be seen that the variation from one compound to another is substantial, with values in the zirconium-beryllide system up to 2000°C. As a rough comparison, Ti_3Al melts at approximately 1200°C, whereas $TiBe_{12}$ melts at 1593°C a 400°C advantage. A consideration in the processing and eventual application of beryllides, however, must take into account the reactivity of beryllium, particularly at the high potential use temperatures.

3.3 Oxidation Behavior

The utility of many high temperature materials can be limited by their interaction with the environment. In general, the beryllides exhibit excellent oxidation resistance at high temperature.

Paine, et al.(3) have published the results of an extensive survey of the oxidation behavior of the transition metal beryllides. With the exception of $ZrBe_{13}$, the MBe_{13} compounds are not resistant at temperatures as high as 2300°F although good resistance below this temperature may be recorded. The MBe_{13} compounds tested in this connection included YBe_{13}, $LaBe_{13}$, $CeBe_{13}$, $GdBe_{13}$, $ThBe_{13}$, and UBe_{13}. The majority of the MBe_{12} and M_2Be_{17} phases are quite oxidation resistant.

From a nuclear standpoint it is unfortunate that the actinide MBe_{13} compounds are not oxidation resistant in the 2300-2500°F temperature range. The MBe_{13} series is, however, isomorphous and a continuous solid solution may be formed between $ZrBe_{13}$ and UBe_{13}. The resultant single phase material has been found to be oxidation resistant at 2500°F up to at least the equimolar $ZrBe_{13}$-UBe_{13} composition which contains 42 wt % uranium level. A more typical fuel element loading would be a 10 wt % uranium level. The (U, Zr)Be_{13}-10 wt % U material exhibits a penetration of 0.5 mils after 100 hours of exposure to 2500°F air and 3.0 mils upon 100 hours of exposure to 2900°F air, an oxidation behavior essentially equivalent to $ZrBe_{13}$. Other lanthanides and actinides would be expected to form similar solid solutions. These have been prepared and tested with the expected results in the case of $GdBe_{13}$ and $HfBe_{13}$.(3)

3.4 Mechanical Properties

Characteristic of intermetallic compounds in general, the beryllides are hard, strong materials. The room temperature hardness of several beryllides are given in Table 2. For orientation in the hardness spectrum, aluminum is about 40 on this scale, tungsten about 400, BeO 1300, and titanium carbide 3200. The beryllides are, therefore, much harder than common metallic materials, but not among the extremely hard materials.

Table 2. Room Temperature Hardness of the Beryllides

Compound	VHN - 2-1/2 Kg Load
Nb_2Be_{17}	1000
Nb_2Be_{19}	1050
$NbBe_{12}$	500
Ta_2Be_{17}	1120
$TaBe_{12}$	720
Zr_2Be_{17}	1130
$ZrBe_{13}$	1000
$MoBe_{12}$	950

The strength of Ta_2Be_{17} as a function of temperature is illustrated in Figure 1. This modulus

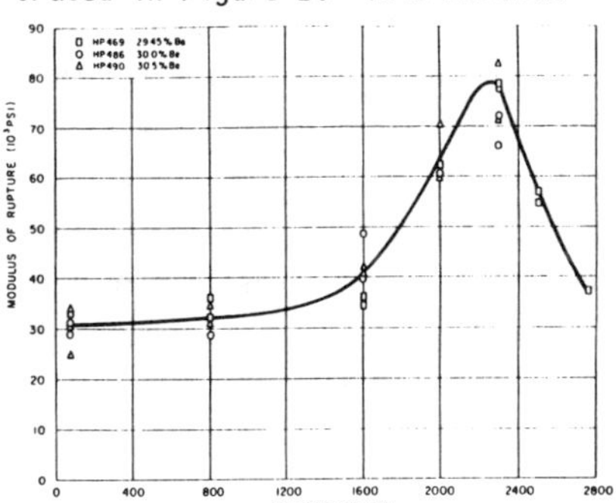

FIG. 1—Strength of Ta_2Be_{17}.

of rupture curve is quite typical of the beryllide family in that a gradual increase in strength with temperature is noted up to 1600°F, after which a sharp increase in strength is observed, the peak strength occurring in the 2300 to 2500°F region. In this case, the effect of minor variations in stoichiometry was under examination, 29.8% Be being the stoichiometric composition for Ta_2Be_{17}. It was concluded that the minor variations in stoichiometry tested here had little effect upon the observed strength of the material, the results being indicative of the range of values encountered in modulus of rupture testing.

Typical modulus of rupture values for other beryllides are shown in Table 3. The highest values are recorded for Nb_2Be_{17}, Nb_2Be_{19}, and Ta_2Be_{17} materials. The niobium compounds are particularly attractive when their low density (3.28 and 3.17 g/cc respectively) is remembered.

Table 3. Modulus of Rupture of Niobium, Tantalum and Zirconium Beryllides

	Room Temp	2300°F	2500°F	2750°F
	(psi)			
Nb_2Be_{17}	31,000	70,000	63,000	36,000
Nb_2Be_{19}	30,000	70,000	65,000	27,000
$NbBe_{12}$	22,000	45,000	40,000	22,000
Ta_2Be_{17}	30,000	78,000	56,000	35,000
$TaBe_{12}$	31,000	53,000	43,000	26,000
Zr_2Be_{17}	25,000	40,000	40,000	35,000
$ZrBe_{13}$	25,000	40,000	37,000	25,000

The relatively low strength values of the zirconium compounds are attributed in part to the relatively large grain sizes obtained in such bodies to date. The average grain size of the hot-pressed zirconium beryllides lies in the region of 25-50 microns while the niobium compounds have a grain size of 10-15 microns when fabricated by the techniques used here. Where such comparison is possible, distinct advantage is seen for fine-grained bodies even at the high temperatures.

The data in Figure 1 are for vacuum hot-pressed Ta_2Be_{17}. This fabrication technique was used for practically all of the materials described in this paper. The more standard ceramic fabrication technique of cold pressing followed by vacuum sintering was used to obtain the material for the modulus of rupture versus temperature curve for Nb_2Be_{17} shown in Figure 2. These data compare

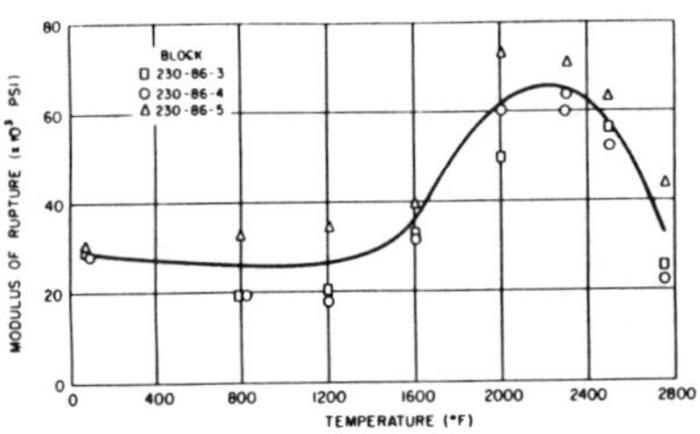

FIG. 2—Strength of cold pressed and sintered Nb_2Be_{17}.

favorably with similar data for vacuum hot-pressed Nb_2Be_{17}, and again show the characteristic shape of the strength curves for the beryllides. The modulus of rupture versus temperature curves for the zirconium beryllides differ from these two curves in that a substantial decrease in strength as compared to room-temperature and high-temeprature values is recorded in the region of 1200 to 1800°F when testing in air. This decrease in strength is associated with anomalous oxidation behavior which is discussed by Paine, et al.[4] The modulus of rupture test used to gather these data has been described elsewhere and this description will not be repeated here.[5] The test involved single point loading at the rate of 120 pounds/minute on bars 3 in. long x 1/2 in. x 1/4 in. with 2-1/2 in. between the edges of the supporting anvil.

Tensile testing of the beryllides has been very limited to date. A few tests have been made using a 3-3/4 in. long by 5/8 in. diameter tapered specimen with a gage diameter of about 1/4 in.[6] Stainless steel grips were used with a small induction coil serving as the heater for the high-temperature tests. The tensile strength of Nb_2Be_{17} as a function of temperature is shown in Figure 3. The tensile strength ranged from 50 to 60% of modulus of the rupture as expected.

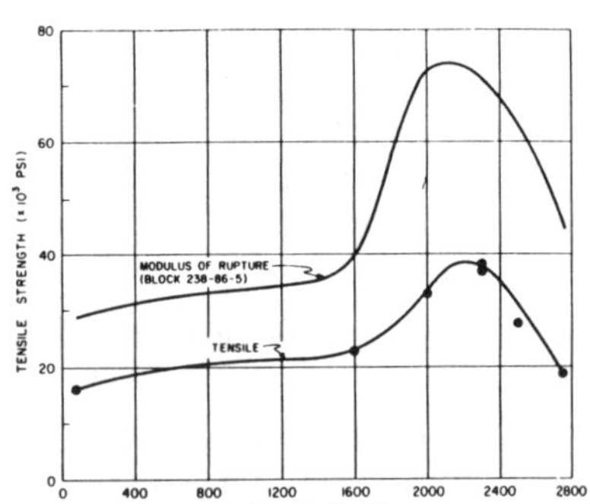

FIG. 3—Tensile strength of cold pressed and sintered Nb_2Be_{17}.

Static Young's moduli values determined in the course of the modulus of rupture test are 43-47 x 10^6 psi at room temperature, 40 x 10^6 psi at 1600°F, and 20-25 x 10^6 psi at 2500°F for the beryllides under discussion, with little variation among individuals of the family. Dynamic measurements of Young's modulus have indicated anomalies in the behavior of the niobium compounds, and a minimum is recorded at about 1400°F after which a rapid rise occurs to peak at 2400°F in the case of Nb_2Be_{17}.(5,6) The values for Nb_2Be_{17} are 43 x 10^6 psi at room temperature, 41 x 10^6 psi at 1400°F, 47 x 10^6 psi at 2000°F, 53 x 10^6 psi at 2300°F, and 46 x 10^6 psi at 2500°F.

Compressive strength determinations have indicated 190-200,000 psi at room temperature, 130-150,000 psi at 1600°F, and 70-80,000 psi at 2500°F for $ZrBe_{13}$ and $NbBe_{12}$.

Little capacity for plastic strain is exhibited by the beryllides under discussion between room temperature and 2000°F, but significant plastic deformation has been recorded at higher temperatures. Calculations from the deflection of the modulus of rupture bars indicate that a maximum of about 0.1% tensile elongation occurs in the outermost fibers of such bars at room temperature prior to fracture. At 2500°F, elongations between 1 and 2.6% may be calculated for the niobium and the tantalum compounds, but the zirconium materials remain at about 0.1%. Ta_2Be_{17} and Nb_2Be_{17} shown about 5 and 7% respectively at 2750°F, while Zr_2Be_{17} shows only 0.1% and $ZrBe_{13}$ only 0.6% at this temperature. Tensile elongations for Nb_2Be_{17} in the tensile test showed a maximum of 2%. This difference is believed to be due to the design of the specimen which in effect provided a very small gauge section making it necessary to calculate strain from reduction in area.

3.5 Thermal Properties

The thermal properties of the transitional element beryllides are quite good, generally being intermediate of the more familiar ceramic and metallic high-temperature materials.

The thermal expansion coefficients of a number of the members of the beryllide family are presented in Table 4. These coefficients are of the same order of magnitude as those exhibited by metals and are higher than the expansion coefficients of refractory oxides. The expansion versus temperature curves for these materials are quite similar with only small differences between individuals. The difference between the expansion of $ZrBe_{13}$ and that of

Zr_2Be_{17} is as great as the difference between $ZrBe_{13}$ and any of the other compounds shown here. This difference may give rise to some of the anomalous behavior exhibited by the zirconium beryllides as discussed by Paine, et al.[7]

Table 4. Mean Linear Thermal Expansion Coefficients for Several Beryllides (in./in./°F x 10^{-6})

	80-600	80-1200	80-1800 (°F)	80-2400	80-2700
Nb_2Be_{17}	6.3	7.3	7.9	8.5	8.8
$NbBe_{12}$	6.7	7.5	8.3	9.0	9.3
Ta_2Be_{17}	6.2	7.1	7.8	8.4	8.7
$TaBe_{12}$	5.8	6.9	7.6	8.1	8.4
Zr_2Be_{17}	5.6	7.0	7.5	8.1	8.4
$ZrBe_{13}$	6.5	7.7	8.6	9.4	9.8
Hf_2Be_{21}	6.5	7.7	8.2	8.8	9.1
$MoBe_{12}$	6.7	7.8	8.4	9.0	9.3

The thermal conductivity of the niobium and tantalum beryllides along with $ZrBe_{13}$ are given in Table 5. These values plot as essentially straight lines with slight increase in conductivity with temperature. $ZrBe_{13}$ is again an exception as the conductivity of this compound decreases with temperature across the range of 1200 to 2600°F. The conductivity of BeO in the units used here is about 9.7 at 2500°F. The beryllides, therefore, have a thermal conductivity at elevated temperature of about twice that of the best of the common ceramic materials.

Table 5. Thermal Conductivity of Several Beryllides (Btu · hr^{-1} · ft^{-2} · ft · $°F^{-1}$)

	1200	1600	2000 (°F)	2400	2600
Nb_2Be_{17}	18.1	18.5	18.9	19.3	19.5
$NbBe_{12}$	17.7	17.9	18.2	18.5	18.6
Ta_2Be_{17}	16.9	17.5	18.2	18.9	18.2
$TaBe_{12}$	16.7	18.3	19.5	21.0	22.3
$ZrBe_{13}$	23.1	22.0	21.0	20.0	19.5

Specific heat values are shown in Table 6. Compared to many other materials, these values are quite high as a result of the beryllium content of the compounds. The electrical conductivity of the materials as a function of temperature is shown in Figure 4.

Table 6. Specific Heat of Beryllides (Btu/lb°F)

	1000	1500	2000 (°F)	2500	2800
$NbBe_{12}$	0.385	0.400	0.414	0.429	0.437
Ta_2Be_{17}	0.213	0.227	0.240	0.253	0.261
$TaBe_{12}$	0.271	0.280	0.288	0.297	0.302
$MoBe_{12}$	0.381	0.402	0.423	0.444	0.457
$ZrBe_{13}$	0.381	0.406	0.431	0.456	0.471

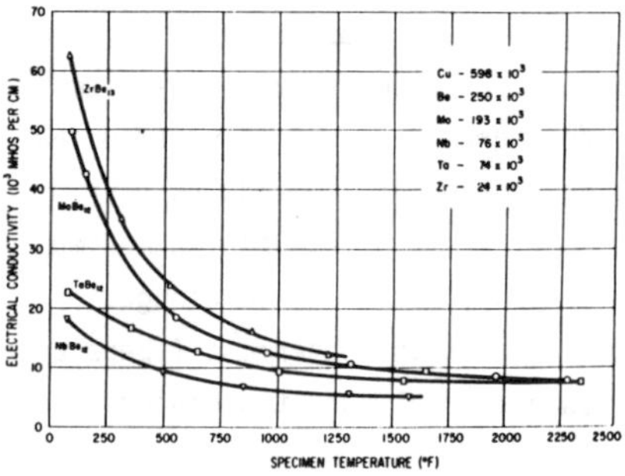

FIG. 4—Electrical conductivity of beryllides as a function of temperature.

4. DISCUSSION

Exploratory work has shown that transition element beryllides have properties which qualify them for consideration in high-temperature structures. This includes non-nuclear applications as well as the self-evident nuclear reactor usage. The temperature span of particular interest is 2000-2700°F for long time service while short time service may be extended to 3000°F. Ta_2Be_{17}, e.g., shows a penetration of only about 3 mils in 10 hours at this temperature, a performance which is superior to refractory metals by several orders of magnitude. A tremendous saving in weight is, of course, potentially possible when a beryllide replaces a refractory metal component in a design due to the very favorable strength-to-weight ratio for the materials.

While the beryllides are not as refractory as many ceramics, they possess a stability toward surrounding atmospheres approaching that of the refractory oxides due to the layer of beryllium oxide, an excellent ceramic material in its own right, which is formed when the beryllide surface is exposed to high temperature oxidizing conditions. While the surface of the compounds are thus essentially a refractory oxide surface in service, the beryllide matrix possesses properties quite unlike that of the refractory oxide, most notably high strength at elevated temperature. Significant further increases in the strength of beryllides at temperature are believed possible with the development of techniques for attaining and maintaining very fine-grained bodies, an approach which has been indicated by the work completed to date. Further gains in mechanical behavior may be possible through the use of metal working techniques which have been demonstrated to be feasible.(6) The present beryllide materials are typically not high-purity, single-phase materials. The work to date has shown that the range of homogeneity for any of the beryllide phases under discussion is quite narrow. This means that a nominal $ZrBe_{13}$ compound, for example, will normally contain 2-5% Zr_2Be_{17} or similar quantities of free beryllium. The materials discussed in this paper were always controlled so that the exact composition was on the M_2Be_{17} side of the MBe_{12} or MBe_{13} phase in preparations of these latter compound types and slightly on the MBe_{12} or MBe_{13} side in the case of M_2Be_{17} preparations. Preparations have been made which are single phase by X-ray, but only with extreme care. Chemical impurities include BeO in the 0.7 to 1.5% range, Fe 500 to 800 ppm, Si 200 to 600 ppm, and Al 300 to 800 ppm.

The ductility picture with beryllides is at once the greatest difficulty with the materials and perhaps the greatest hope. Being intermetallic compounds, the beryllides might be dismissed briefly as hard, brittle phases. Brittle behavior is certainly characteristic presently at low temperatures, but significant plastic deformation has been shown above 2000°F. Lewis(7) reports evidence of slip in $NbBe_{12}$ at -196°C during microhardness determinations. While the structure of the compounds is quite complex, Lewis does postulate a possible slip mechanism for the plastic deformation of $NbBe_{12}$. Purity control to a greater degree than practiced to date and structure modification through ternary "alloying" agents may dramatically alter the present plastic deformation picture.

The study of deformation of the beryllides and means to achieve increased capacity for such deformation at temperatures below the 2000-2300°F range should be an extremely interesting area for further research on these materials.

5. REFERENCES

1. Paine, R. M., Stonehouse, A. J., and Beaver, W. W., "An Investigation of Intermetallic Compounds for Very High Temperature Applications," WADC TR 59-29, Part I, 1959.
2. Stonehouse, A. J., Paine, R. M., and Beaver, W. W., "Physical and Mechanical Properties of Beryllides," paper presented at NACE symposium on Aerospace Metals, March 9-13, 1964.
3. Paine, R. M., Stonehouse, A. J., and Beaver, W. W., "High Temperature Oxidation Resistance of the Beryllides," paper presented at NACE Symposium on Aerospace Metals, March 9-13, 1964.
4. Paine, R. M., Stonehouse, A. J., and Beaver, W. W., "Oxidation of the Beryllides at Intermediate Temperatures-Anomalous Behavior and a Solution," Symposium on "Compounds of Interest in Nuclear Reactor Technology," sponsored by The Metallurgical Society of AIME, at the University of Colorado, August 3-5, 1964.
5. Stonehouse, A. J., Paine, R. M., and Beaver, W. W., "Mechanical Properties of Some Transition Element Beryllides," Mechanical Properties of Intermetallic Compounds, Edited by J. H. Westbrook, John Wiley and Sons, Inc., 1960, Chapter 13.
6. Truesdale, R. S., Lympany, B. B., Bielawski, C. A., Grala, E. M., and Beaver W. W., "Investigation of the Effects of Processing Variables and Fabrication Tech-

niques Upon the Properties of Intermetallic Compounds," ASD-TDR-62-476, September, 1961.

7. Booker, J., Paine, R. M., and Stonehouse, A. J., "Studies on the Vapor Pressure and Oxidation of Certain Beryllides," Symposium on "Compounds of Interest in Nuclear Reactor Technology," sponsored by The Metallurgical Society of AIME, at the University of Colorado, August 3-5, 1964.

6. BIOGRAPHY

Dr. JAMES M. MARDER

James M. Marder received his Bachelor of Science in Metallurgy from Polytechnic Institute of Brooklyn in 1968; a Master of Science in Metallurgy from the University of Connecticut in 1976 and a Ph.D. in Materials Science from Case Western Reserve University in 1981. In 1981, he joined the staff of Brush Wellman where his efforts have been directed toward the development and processing of beryllium. Major areas of emphasis have been the development of improved optical grades of beryllium, and the investigation of net-shape processing techniques, including hot isostatic pressing. He is currently the supervisor of the Beryllium Metal Research Group.

A. J. STONEHOUSE

A. J. Stonehouse graduated from Iowa State University in 1949 with Bachelor of Science in Chemical Technology. He joined The Brush Beryllium Company in Research and Development in 1949. His work has included research on the extractive and process metallurgy of beryllium, zirconium and uranium, as well as determination of chemical and physical properties of beryllium and its alloys. He and his group conducted the original research which established the beryllides of a number of transition element metals as a class of materials with excellent high temperature strength and oxidation resistance. About ten U.S. patents have been awarded and a number of technical papers pre-

2nd International SAMPE Metals Conference
August 2-4, 1988

THE INFLUENCE OF PROCESSING VARIABLES AND DEFECTS ON THE PERFORMANCE OF ADHESIVELY BONDED JOINTS

T.S. Srivatsan
The University of Akron
Akron, Ohio 44325

T.A. Place, Raju Mantena, and R.F. Gibson
University of Idaho
Moscow, Idaho 83843

T.S. Sudarshan
Materials Modification Inc.
Falls Church, Virginia 22044

ABSTRACT

In this paper, the role of heterogeneities and defects caused by the improper curing and inadequate degassing of the adhesive on the performance of double-lap adhesive joints bonded with a structural epoxy adhesive is non-destructively analyzed through damping capacity measurements. The damping measurements were made for the flexural mode of vibration using an impulse-frequency response technique. Strength of the adhesive joint was determined for the different types and degree (size and volume) of defects. The damping characteristics were found to correlate well with strength; the shear strength of the adhesive joint decreasing and damping increasing with increasing severity of the defect. The micromechanisms governing the performance of adhesively bonded joints are discussed and include the influence of several concurrent factors involving curing time, curing temperature, adhesive degassing time and specimen geometry. The role of processing variables on the intrinsic severity of the flaws is rationalized.

1. INTRODUCTION

Bonding metals and fiber reinforced composites with the aid of adhesives has been extensively practiced for many years in the aerospace, transportation and other high technology industries. Through weight savings, manufacturing cost effectiveness, reduction or minimization of local stress concentration effects and the provision of aerodynamically smooth surfaces, adhesive bonding offers significant advantages and is a desirable alternative to conventional joining techniques using mechanical fasteners [1,2].

Use of adhesive bonding aids in the effective absorption of vibrations during service of a component coupled with smooth transfer of stresses over the entire bond area in contrast to the stress concentrations that occur with welding, riveting and fastening with screws [3]. Besides, bonding with the aid of adhesives simplifies construction and eliminates the harmful effects of high temperatures associated with welding and brazing, resulting in improved appearance. Additionally, joining of cold rolled and heat sensitive materials, elimination of galvanic corrosion by the separation of dissimilar materials, and reduced part count with concomitant weight savings are some of the attractive features that favor the use of adhesive bonding for cost-effective design [4].

The essential requirements needed to achieve a good adhesive bond are [5]:

(a) good wettability of the substrate to be bonded,

(b) absence of weak boundary layers, both in the adhesive and in the substrate and

(c) avoidance of stress concentrations which can lead to disbonding.

These requirements have to be met in such a way that not only does the adhesive bond exhibit the necessary bond strength but that this bond strength is maintained for the service life of the application. It is the durability or permanence of a bonded joint in service which is an important consideration in the acceptance of an adhesive joint for a particular end use. Ideally, the most reliable information on the durability characteristics of an adhesive or an adherend-adhesive system needs to be developed by subjecting the test samples to conditions comparable to those of the glued product in service. However, such long-term tests are impractical. Consequently, over the years strenuous efforts were made to develop non-destructive test methods for the detection of areas having inadequate strength.

Analyses of adhesively bonded structures have frequently revealed the presence of flaws or defects. The actual defect types which occur in bonded structures are [6]:

(a) complete voids, disbonds or porosity,

(b) poor adhesion, i.e., a weak bond between the adhesive and one or both adherends,

(c) poor cohesive strength, i.e., a weak adhesive layer.

Voids or large gas bubbles in the adhesive are caused either by a lack of adhesive or by the presence of foreign matter on the adherends, or even in the adhesives. Porosity of the adhesive is similar to voiding except that the size of the bubbles is much smaller. It is usually caused by volatiles or gases trapped in the adhesive. Porosity is a major problem in composite adherends and occurs when the adsorbed moisture vaporizes during the cure cycle to produce bubbles in the adhesive.

Because adhesive bonding is a surface phenomenon, surface preparation is one of the keys to successful bonding. The durability of an adhesive bond is very much dependent on the surface treatment of the adherends. Dis-bonds or zero-volume bonds (areas where no void exists but also no adhesion is present), occurs due to: (i) the presence of a contaminant, such as grease, on the adherend, (ii) a poorly prepared adherend surface, and (iii) environmental degradation after manufacture. The surfaces of a dis-bond are generally in close proximity, or are touching, but are incapable of transferring load from the adherend to the adhesive.

Improper curing, that is, too low a curing temperature or inadequate time at the required curing temperature or improper mixing of the adhesive components are factors responsible for variation in physical properties of an adhesive due to insufficient cross-linking of the polymer. While destructive tests provide information on the cohesive properties of the adhesive, the nondestructive tests are preferred for they aid in the evaluation of the adhesive-adherend interface. Failure of adhesively bonded joints often results from catastrophic crack propagation originating from inherent flaws or impurities on the adhesive-adherend interface or defects [7].

This paper presents a rational assessment of the influence of induced porosity and bond flaws caused by improper curing of the adhesive on the performance of double-lap adhesively bonded joints. The analysis is based on correlation of the results of the non-destructive tests which provides information on the dynamic response of the adhesive joint with results from destructive tests which provide a measure of the strength of the adhesive joint. The role of processing variables, namely, curing temperature, curing time, and adhesive degassing time on the severity of the flaws is discussed and the intrinsic micromechanisms governing the performance of adhesively bonded joints are rationalized.

2. EXPERIMENTAL PROCEDURES

In order to accomplish the objective of understanding the effects of flaws, impurities and defects on adhesive joint performance, a double-lap specimen configuration with provisions for clevis pins was chosen (Figure 1). The material chosen for the adherend was hot rolled structural steel having a fine grain size, while the adhesive chosen was a mixture of EPON 815 epoxy and Hardener V-40. The specimen design and configuration is such that the adhesive is only along the clevis walls of the adherend and is subjected to pure shear during static loading.

The prime difficulty in bonding metals is not so much one of making a strong joint as one of keeping it that way. Solid metal surfaces have a texture which is characterized by waviness, roughness and indentations such as cracks and pits. Surface identations may interfere with wetting if they retain trapped air or selectively concentrate solvents from the adhesive. Besides being highly textured, solid surfaces in the normal state are often impure. Most often, the surface of a metal is covered with a layer which has quite different properties from the bulk. The layer may contain grease, processing lubricants, oxidation products or incidental contaminants. Hence, the strength of an adhesive joint is limited by forces between this layer and the adhesive, or between this layer and the metal or by the cohesive strength of the layer; whichever is the weakest.

2.1 Adhesive Preparation and Curing:

The adhesive used for fabricating the joints was prepared in accordance with manufacturers instructions, and due regard was given to handling precautions for the epoxy resin material used. EPON 815, a plasticised liquid epoxy resin was mixed with Hardener V-40, in the proportion of 100 parts of EPON and 80 parts of Hardener by weight. The adhesive was prepared at ambient-room temperature in laboratory air environment (relative humidity 65 percent). The resin and hardener were thoroughly mixed for 20 seconds and subsequently degassed under vacuum at a temperature of 100 C for a period of seven minutes. The viscous adhesive was then poured into the female adherend positioned in a jig. The male adherend was lowered into the double-lap female adherend to form the required joint.

The EPON 815 epoxy resin normally cures at room temperature. In order to study the effects of cure time on adhesive joint performance, an accelerated cure was performed by placing the jig holding the composite specimens in an oven at 100 C. The cure time is the amount of time the specimen is in the oven and the cure temperature is the temperature in the oven. Cure schedules with cure time settings of 60 minutes, 120 minutes, 180 minutes and 240 minutes at the cure temperature setting of 100 C were used. A fast cool-down rate was simulated by turning the oven off, leaving the oven door open and allowing the jig and specimen assembly to cool to ambient temperature. In order to minimize any environmental effect on bond stability the cured composite specimens were stored in a low humidity and room temperature environment prior to testing. This was done by storing the cured specimens in a desiccator.

The degree of porosity in the adhesive was controlled by varying the adhesive degassing time prior to curing. By degassing the adhesive under vacuum for time periods less than the time prescribed (seven minutes) for complete degassing, air and product gas resulting from the reaction between the EPON epoxy and the hardener are trapped in the adhesive mixture. Degassing time of 0 minutes (no degassing of the adhesive mixture), 2 minutes and 7 minutes were chosen. In order to minimize the depletion of induced pores and fine voids, the porosity specimens were allowed to cure at ambient temperature (26 C) for a period of 7 days, instead of the elevated temperature (1 hour at 100 C) cure used for preparing the other category specimens.

2.2 Mechanical Testing:

The shear strength of the adhesive joint was determined by uniaxial loading of the composite specimen (two steel adherends held together by a double-lap adhesive joint) in a MTS servohydraulic testing machine. Failure was taken to be the point at which the 'male' adherend separated from the 'female' adherend due to failure of the adhesive layer. To obtain the dynamic mechanical properties (damping and frequency) of the double-lap adhesive joint, non-destructive tests based on the impulse-frequency response technique were performed for the flexural vibration set-up. Details of the impulse-frequency response technique and the testing procedure can be found elsewhere [8,9].

3. RESULTS AND DISCUSSION

3.1 The Influence of Curing on Adhesive Joint Performance

The influence of curing on adhesive joint performance was established through evaluation of composite specimens with the adhesive cured over a range spanning the under-cured to the fully cured condition. The dynamic mechanical properties in the flexural mode were obtained using the impulse frequency response technique. Care was taken to avoid prolonged delay in testing following accelerated curing for the different time periods in order to minimize contributions that could result from the self-correction of the poorly cured regions. The influence of cure time on both dynamic properties and shear strength of the adhesive joint,for a fixed cure temperature, is shown in Figures 2 and 3. It is observed that at cure times of 120 minutes, 180 minutes and 240 minutes the damping of the double-lap adhesive joint is fairly constant while it is higher by about 35 percent for the composite specimens cured for only 60 minutes. The marginal change in damping suggests that the adhesive achieves a stable and favorable molecular network structure at time periods beyond 120 minutes coupled with perhaps complete bonding of the adhesive to the steel adherend. The increased damping (higher loss factor) (Figure 2) Resulting from under-curing (60 minute cure) is reflected by the lower shear strength (Figure 3).

The higher damping at shorter cure times is rationalized on the basis of the fact that the high temperature, short-time cure schedule represents a highly unsteady condition favoring the build-up of large thermal stresses. However, it is most likely that these stresses are relieved during the slow cool-down rate and is an appealing rationale for the adhesive developing a more stable and favorable molecular structure [10]. In the case of bonding steel specimens shrinkage of the adhesive is a function of its thickness and adhesive shrinkage will put the adhesive into tension. The ultimate tensile strength will be the same, but the residual tensile stress adds to the applied stress. Consequently, a lower applied force is needed to reach the ultimate tensile strength. A change in strength results in a concomitant change in other properties [11]. A plausible reason to account for the observed change in dynamic properties is a variation in the degree of shrinkage of the adhesive. Other reasons for the observed dynamic and static response could be due to a combination of the chemical nature of the adhesive mixture and the physical characteristics of the adhesive in relation to the adherend. For the formation of a joint of sound integrity, controlled wetting and penetration are both essential. For these conditions to be fulfilled, an appropriate viscosity level of the adhesive is imperative [12]. Attainment of this viscosity level is contingent upon: (a) the components of the adhesive, (b) the cure temperature, and (c) the cure time. Since components of the adhesive and adhesive cure temperature were held constant in the preparation of the composite specimens, the over-riding or controlling factor is the cure time and it was observed to have only a marginal influence on adhesive joint performance.

The shear strength of the double-lap adhesive joint specimen was determined by uniaxial loading of the composite specimen in a servohydraulic testing machine. It is observed that the shear strength is lowest for the under-cured composite specimen, whose adhesive was cured for only 60 minutes at the cure temperature of 100 C. A decrease in shear strength of the adhesive joint results in a concomitant increase in damping (loss factor). For the specimens cured for 120 minutes and 180 minutes, the strength of the adhesive joint is comparable or better than the perfectly cured specimen. This trend confirms the observation that the adhesive achieves a stable molecular network structure at times beyond 120 minutes. A comparison of the shear stress-shear strain curves for the different cure times is made in Figure 4.

3.2 The Influence of Porosity on Adhesive Joint Performance:

Traditionally, adhesive bonded structures have been analysed and modeled under the assumption that the adhesive is uniform and flawless throughout. The effects of porosity have historically been accounted for by coupon tests of porous bonds correlated with some form of ultrasonic non-destructive inspection. Porosity is caused by the presence of volatiles and entrapped air in the adhesive (Figure 5). In this study, the degree of porosity in the adhesive was controlled by varying the adhesive degassing time of the liquid adhesive. By degassing the adhesive for times less than the time prescribed for complete degassing, air and product gas resulting from a reaction between the EPON 815 epoxy and the hardener are trapped in the adhesive mixture. The volume fraction of air and reaction gas pockets (pores) is likely to be greater when the adhesive mixture is not degassed. To minimize the depletion of pores and voids during curing at elevated temperature following degassing, the composite specimens were cured at ambient temperature (26 C) for a time period of 7 days.

To have a better understanding of the effects of porosity on performance of adhesive bonded joints, it is necessary to characterize the behavior of adhesive joints having no imperfections (i.e., pores and voids). In comparison with a imperfection free specimen, that is, one whose adhesive mixture was totally degassed, the composite specimens whose adhesive was not degassed revealed different dynamic mechanical properties of the adhesive joint. The change in damping (loss factor) for the different adhesive degassing times was significant while the change in resonant frequency due to the presence of pores and voids was only marginal as shown in Figure 6.

The impulse-frequency response technique successfully detects and magnifies the presence of pores and voids as is evident by the large change in damping. In order to have a better understanding of the effects of porosity on adhesively-bonded joints, characterization of the load transfer through porous adhesive bonds is essential and is achieved through an analysis of the stress-strain curve in shear. Results of the destructive tests reveal a marked degradation in shear strength of the adhesive joint with short degassing times. The increased volume fraction of pores and/or voids in the undegassed adhesive mixture, coupled with contributions from changes in reaction kinetics, results in the adhesive joint having inferior strength (Figure 7) and exhibiting increased damping (Figure 6). While the degradation in strength is 75 percent, the concomitants change in damping is as high as 450 percent.

Location of porosity within the bonded joint influences joint strength. It seems plausible that the fine pores at an adhesive-adherend interface act as debond area and consequently, degrade the strength of the adhesive joint. Porosity is also present in the interior of the adhesive joint where virtually no load is transferred. Porosity in the interior locations does not possibly have the same effect as porosity on the adhesive-adherend interface. Degassing the adhesive mixture for even short periods of time reduces the total amount of entrapped air in the adhesive mixture and completes the reaction kinetics of the adhesive mixture. As a result, contributions from these factors are minimized resulting in only a marginal change in damping and frequency. A comparison of the shear stress-shear strain curves for the different levels of porosity is shown in Figure 8.

Porosity, as with any flaw, is critical in bonded joints primarily because the local imperfections have a widespread effect. As the adhesive bond is weaker the load that can be transferred safely is drastically reduced. The adhesive joints were observed to fail partly by interfacial separation and partly by tensile fracture of the adhesive. The separated joints (separation of the 'male' adherend from the 'female' adherend) were examined in a low power stereomicroscope. A few fine microscopic pores were observed on the surface and embedded in the epoxy of the specimens that were not degassed, while none were observed in specimens in which the adhesive mixture was degassed. The 0 minute degassed specimens failed predominantly by interfacial shear, while the specimens in which the adhesive was degassed for 2 minutes and 7 minutes, failure was a combination of interfacial shear and tensile fracture of the adhesive. Examination of the fracture surface of the adhesive in a scanning electron microscope revealed the fractures to be brittle, with evidence of extensive mirror and hackle (river patterns) regions. High magnification examination revealed the presence of few microscopic pores in the specimen whose adhesive mixture was not degassed (0 minute degassing), and an isolated group of pores in the interfaces of the specimens in which the adhesive was degassed for 2 minutes and 7 minutes.

Failure of the adhesive joints by interfacial separation confirms the argument that the pores at an interface are detrimental to strength. The reaction kinetics between the EPON 815 epoxy and the hardener is different when the adhesive mixture is not degassed under vacuum when compared to the strong reaction accompanied by the evolution of gas bubbles that occurs when degassed under vacuum. The change in reaction kinetics of the adhesive mixture is another key factor that contributes to the observed differences in damping and strength.

4. CONCLUDING REMARKS:

This paper has discussed the influence of flaws, imperfections and defects caused by improper curing and incomplete degassing of the adhesive mixture on the dynamic mechanical properties and shear load transfer capability of adhesively bonded joints.

The impulse-frequency response technique successfully detects the presence of fine microscopic flaws as is reflected by an increase in damping (loss factor) with an increase in size and number of the flaws.

The observed change in damping with cure time at the accelerated cure temperature is attributed to several competing mechanistic effects involving: (a) shrinkage of the adhesive, (b) chemical and physical characteristics of the adhesive relative to the adherend, and (c) thermal stresses.

Fine microscopic pores were induced by not degassing the adhesive mixture. The pores coupled with changes in reaction kinetics of the adhesive mixture causes a large change in damping (loss factor). The internal pores affect damping while those at the interface cause a marked degradation in strength, besides contributing to damping changes. Results of this study has revealed the key to understanding the sensitivity of adhesive bonds to porosity caused by incomplete degassing and flaws caused by improper curing.

ACKNOWLEDGEMENTS

The investigation reported herein was conducted under a contract from the U.S. Navy, Naval Surface Weapons Center, Contract No. N60921-86-C-0210. Mr. C.W. Anderson was the U.S. Navy program monitor. The authors thank Mrs. Sandra Collins for her assistance in the preparation of this manuscript.

REFERENCES:

1. Petric, E.M., Adhesives Age, Vol. 23, 1980, pp. 14-23.

2. Pitrone, L.R., and Brown, S.R., "Environmental Durability of Adhesively Bonded Joints," presented at the ASTM Symposium on Adhesive Bonded Joints: Testing, Analysis and Design, September 1986.

3. Villalobos, M.M., and Czarnock, P., J. Adhesion, Vol. 19, 1986, pp. 79-87.

4. Guyott, C.C.H., Cawley, P., and Adams, R.D., J. Adhesion, Vol. 20, 1986, pp. 129-159.

5. Davison, A.P., I. Mechanical Engineers, C50/86, 1986, pp. 231-234.

6. Adams, R.D. and W.C. Wake: "Structural Adhesive Joints in Engineering" Elsevier Applied Science Publishers, 1984, pp. 133-137.

7. Dwight, D.W., Sancaktar, E., and Brinson, H.F., Polymer Science and Technology Series, Vol. 12A, L.H. Lee, (editor) Plenum Press, New York, 1980, pp. 141-164.

8. Srivatsan, T.S., "Portable Impulse Measurements to Non-Destructively Predict the Integrity of Adhesive Joints," Final Report for the Naval Surface Weapons Center, Contract No. N60921-86-C-0210, April 1987.

9. Suarez, S.A., and Gibson, R.F., J. of Testing and Evaluation, Vol. 15, No. 2, 1987, pp. 114-121.

10. Sancaktar, E., JoZavi, H., and Klein, R.M., J. Adhesion, Vol. 15, 1983, pp. 241-251.

11. Hylands, R.W., and Sidwell, E.H., J. Adhesion, Vol. 11, 1980, pp. 203-218.

12. Ebewele, R.O., River, B.H., and Koutsky, J.A.: J. Adhesion, Vol. 14, 1982, pp. 189-217.

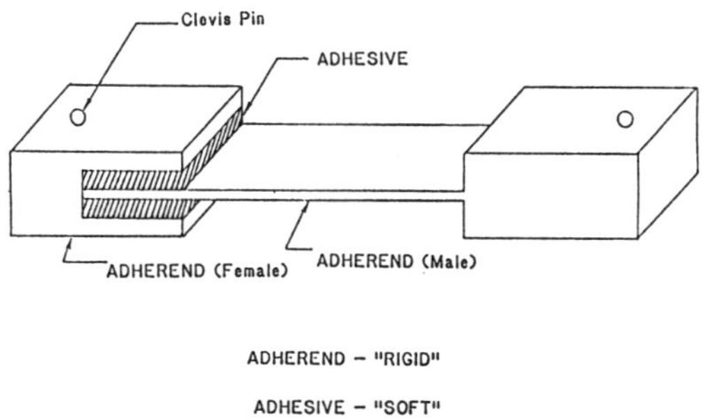

Figure 1. Schematic of the composite test specimen

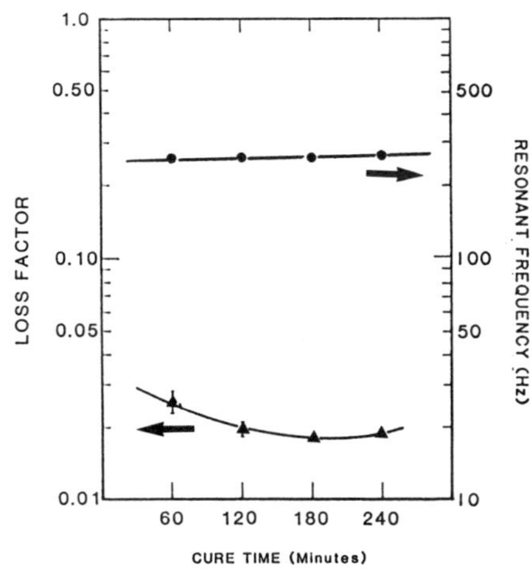

Figure 2. Effect of accelerated cure time (minutes) on adhesive joint damping and first mode resonant frequency.

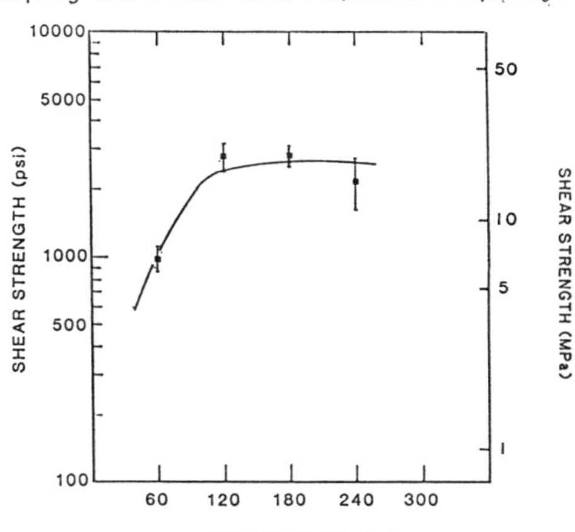

Figure 3. Variation of adhesive joint shear strength with adhesive cure time.

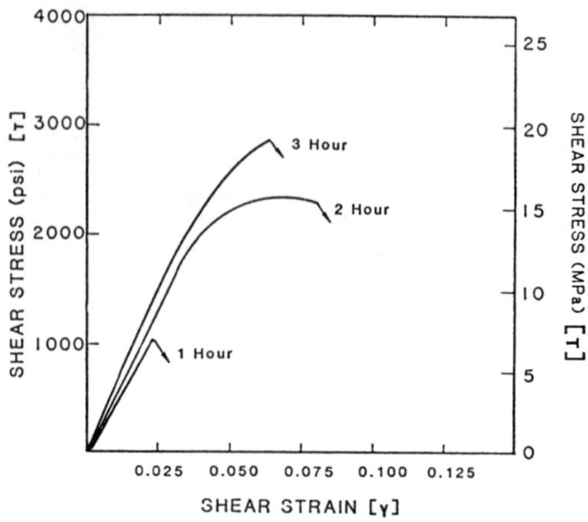

Figure 4. Comparison of representative shear stress-shear strain cures for the different cure times.

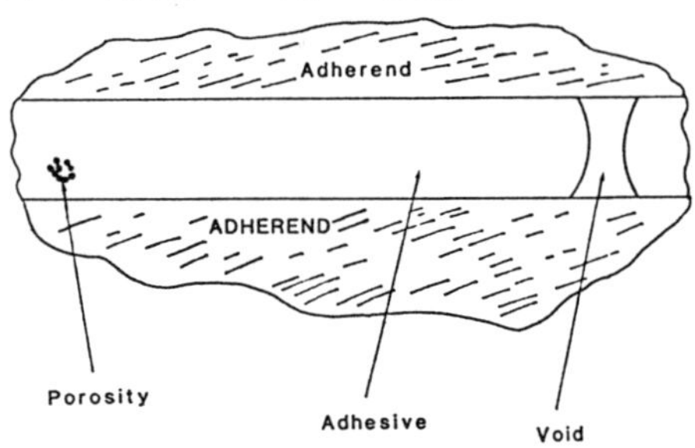

Figure 5. Typical defects: porosity and voids, caused by entrapped air in the adhesive.

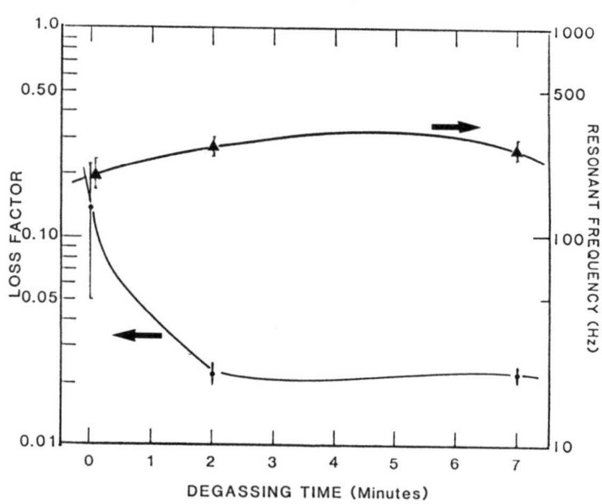

Figure 6. Effect of adhesive degassing time on joint damping and first mode resonant frequency (Hz).

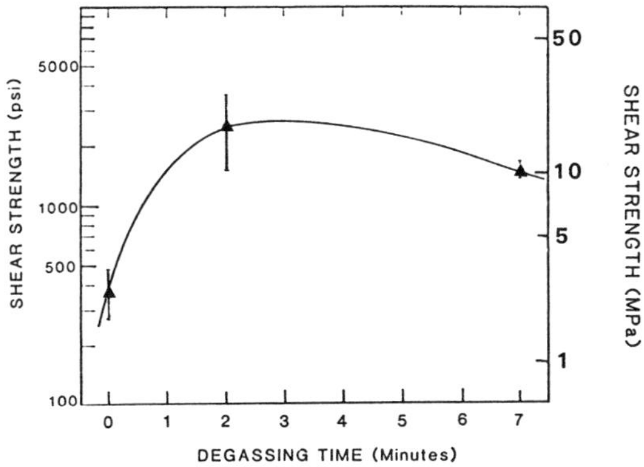

Figure 7. Variation of adhesive joint shear strength with adhesive degassing time (minutes).

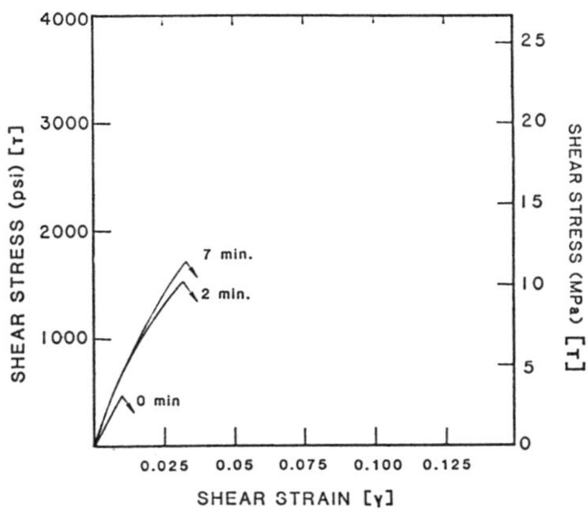

Figure 8. Comparison of representative shear stress-shear strain curves for the different adhesive degassing times.

T. S. Srivatsan has a BS (1980) degree in Mechanical Engineering, an MS(1981) Aerospace Engineering and a PhD (1984) in Mechanical Engineering from Georgia Institute of Technology, Atlanta, GA. He is a member of ASME, ASM International and the Metallurgical Society of AIME. He has published over 45 papers in the areas of structure-property relationships in materials, microstructural analysis, deformation and fracture characteristics of advanced aluminum alloys, fracture mechanics and computational techniques. He is a Key Reader for Metallurgical Transactions. He is currently a Assistant Professor of Mechanical Engineering at the University of Akron.

T. Alan Place: has a B.Sc. in Metallurgy and Materials Science from the University of Nottingham, an M.Eng., degree in Materials Science and Engineering from McMaster University in Hamilton, Ontario, and a PhD in Metallurgy from the University of British Columbia. He is a member of ASM International, ASTM, ISTFA, IMS and the Institute of Metals. He has published 45 papers in the areas of microstructure-property relations, deformation, fracture ans non-destructive evaluation of advanced materials. He has been employed by Rolls Royce Aero-Engine Division, and has taught and performed research at six colleges and universities in four countries.

Raju Mantena: received a BS degree in Mechanical Engineering from Andhra University in 1973, and a Post-Graduate Diploma in Design Engineering from IIT, Delhi in 1975. He was employed as Development Engineer for eight years by Bharat Heavy Electricals Ltd., where he was responsible for the establishment of the company's corporate research and development laboratory complex. Raju received his MSMS (1985) degree in Mechanical Engineering from the University of Idaho, and is currently pursuing research in the area of non-destructive evaluation of advanced materials while working on his doctoral degree at the University of Idaho. He is the co-author of 10 publications.

Ronald F. Gibson: received his BS (1965) degree in Mechanical Engineering from the University of Florida, and was employed as Development Engineer by the Nuclear Division of Union Carbide Corporation. He received a MSME degree (1971) from the University of Tennessee, and a PhD (1975) in Mechanics from the University of Minnesota. Since that time, he has served on the faculties of Iowa State University, the University of Idaho, and the University of Florida. He is currently a Professor of Mechanical Engineering at the University of Idaho.

T. S. Sudarshan: has a B.Tech (1976) in Metallurgy from IIT, Madras, an MS (1978) in Materials Engineering from Virginia Polytechnic Institute and State University, and a PhD (1984) in Materials Engineering Science from Virginia Polytechnic Institute and State University. He is a member of ASM International, Metallurgical Society of AIME, AEP, NACE and ADPA. He has published over 50 papers in the areas of fatigue and fracture behavior of materials, environmental effects, corrosion, failure analysis tribology and non-destructive testing. He has edited a handbook on Surface Modification Technologies and is the key reviewer of Technical Books for Journal of Metals. He is the principal investigator on programs U. S. Army, U. S. Navy and is currently the Technical Director of Materials Modification Inc (MMI).

2nd International SAMPE Metals Conference
August 2-4, 1988

JOINING OF RAPIDLY SOLIDIFIED POWDER METALLURGY AL-FE-CE ALLOYS

W. A. Baeslack III
Department of Welding Engineering
The Ohio State University
Columbus, Ohio 43210

Abstract

The joining of rapidly-solidified powder metallurgy (RS/PM) Al-Fe-Ce alloys by capacitor-discharge and inertia-friction welding has been investigated from a metallurgical perspective. Capacitor-discharge welding was effective in the joining of 6.4 mm diameter Al-8Fe-4Ce rod. The application of pressure simultaneously with fusion zone solidification suppressed the formation of fusion zone porosity in this high hydrogen PM product. In addition, extremely rapid cooling rates promoted RS fusion zone microstructures and prevented heat-affected zone (HAZ) formation, thereby providing joint efficiencies up to 100%. High joint efficiencies were also obtained in 12.5 mm diameter Al-9Fe-4Ce and Al-9Fe-7Ce rod using the solid-state inertia-friction welding process. Welds produced at low axial force were characterized by defects along the weld interface and a relatively large heat and deformation zone (HDZ). Increased axial force significantly reduced HDZ width and minimized defect formation, resulting in optimum joint efficiencies exceeding 90% with satisfactory ductilities.

1. INTRODUCTION

Al-Fe-Ce alloys offer significant advantages over conventional aluminum alloys for elevated-temperature applications up to 350°C (1). These dispersion-strengthened alloys are basically hypereutectic Al-Fe compositions with ternary additions of Ce for enhanced dispersoid stability. Through air or flue-gas atomization, extremely fine, irregular-shaped powder particles are produced which vary in size from about 5 to 40 microns in diameter (Fig. 1a). The rapid cooling rates (10^3 to 10^6 ^{o}C/s) experienced by these particles suppresses formation of the equilibrium primary intermetallic structure and instead promotes solidification to supersaturated alpha aluminum. As

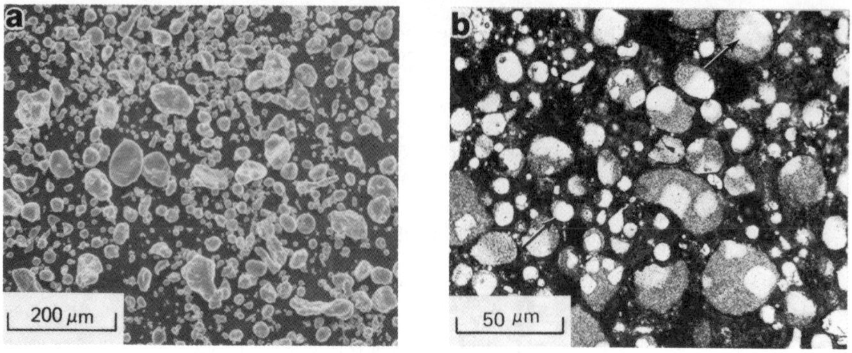

Figure 1
SEM (a) and light (b) micrographs of Al-9Fe-4Ce flue-gas atomized powder. Arrows in (b) indicate featureless Zone A regions.

shown in Fig. 1b, the finest particles exhibit a featureless microstructure designated as "Zone A," while the coarser particles exhibit a dark-etching "Zone B" microstructure. Generally, both of these microstructures solidify in a cellular mode, with the Zone B contrast resulting from a coarser cellular structure.

Subsequent powder consolidation and thermomechanical processing results in a unique microstructure comprised of fine dispersoids in a matrix of submicron alpha aluminum grains. As shown in Fig. 2a for an Al-9Fe-4Ce alloy, the extruded product exhibits good homogeneity with a light texture in the extrusion direction. The SEM micrograph in Fig. 2b reveals a variety of fine, acicular intermetallics in the Al-9Fe-4Ce alloy. TEM and X-ray diffraction studies have identified these dispersoids to include Al_3Fe, Al_6Fe, $Al_{10}Fe_2Ce$, Al_8Fe_4Ce and their metastable precursor phases (2).

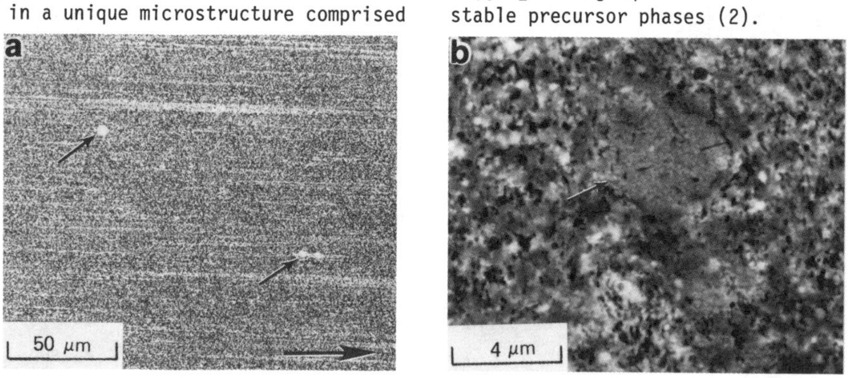

Figure 2
Light (a) and SEM (b) micrographs of Al-9Fe-4Ce base metal microstructure. Large arrow in (a) indicates extrusion direction, small arrows indicate undeformed powder particles.

The efficient utilization of Al-Fe-Ce alloys in structural applications will require their joining. Unfortunately, the application of conventional welding processes has been only marginally successful with the Al-Fe-Ce alloys. Gas tungsten-arc (GTA) welding studies by Metzger (3) on thin-sheet Al-10Fe-5Ce found poor weld quality to result principally from the high inherent hydrogen content of these flue-gas atomized alloys (1-10 ml/100 gm) and the associated evolution of gross fusion zone porosity (Fig. 3a). In addition, the moderate cooling rates experienced by the GTA weld fusion zone allowed the formation of a coarse, primary intermetallic solidification structure (Fig. 3a). Although the addition of filler metal was ineffective in reducing porosity (Fig. 3b), the combination of a preweld vacuum heat treatment (400°C/100 hrs), a direct current, electrode negative welding arc and the addition of 5356 filler metal significantly improved weld integrity (Fig. 3c). However, poor filler metal/base metal mixing

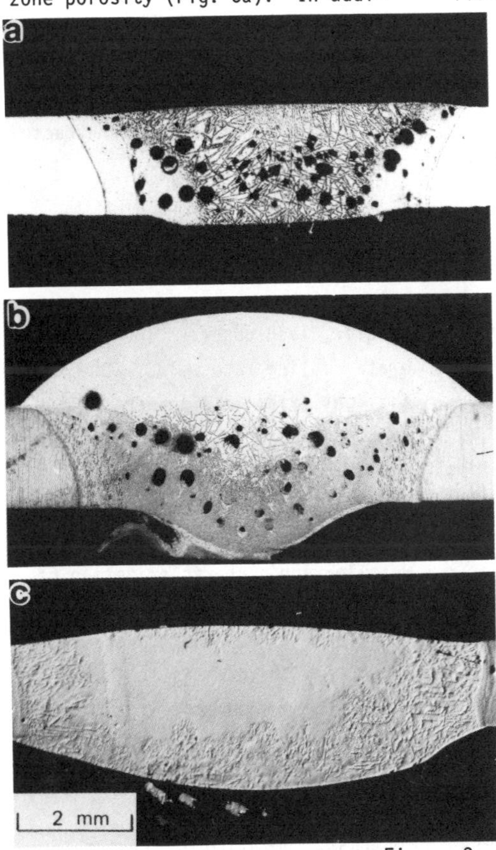

Autogenous, no preweld vacuum heat treatment

ER 4043 filler metal, AC, no preweld vacuum heat treatment

ER 5356 filler metal, DCEN, 400°C/100 hr preweld vacuum heat treatment

Figure 3
Light macrographs of GTAW welds in 1.6 mm thick Al-10Fe-5Ce sheet (3).

prevented elimination of the coarse, primary intermetallic solidification structure near the weld fusion boundary. This region served as the preferential fracture site in transverse tensile testing which determined a 55% joint efficiency and negligible ductility (3).

Considering the physical metallurgy of the Al-Fe-Ce alloy system, it is apparent that achieving high joint efficiencies requires the application of joining techniques which can either recreate or retain the metastable base metal microstructure across the joint while precluding the formation of defects. In this work, capacitor-discharge and inertia-friction welding processes were evaluated in an attempt to meet these requirements.

2. CAPACITOR-DISCHARGE WELDING

Recent work by Devletian (4) has demonstrated the effectiveness of capacitor-discharge (CD) welding in producing defect-free joints in SiC/Al metal-matrix composites. In addition to suppressing the formation of fusion zone porosity in these high hydrogen, PM materials, the process also provided rapid solidification rates in the melt zone and minimized HAZ formation. Based on these advantages, the CD welding process was employed to join 6.4 mm diameter Al-8.4Fe-3.6Ce (Al-8Fe-4Ce) extruded rod.

The "initial-gap" CD welding process involves the impacting of specimen surfaces under gravity with the simultaneous discharge of a capacitor bank, causing arcing and subsequent melting. Upon impact, the arc is extinguished, molten metal is expelled and rapid solidification occurs. In the present work, welding parameters (Table 1) were controlled to provide both high and moderate cooling rate welds (about 10^7 and 10^5 $^oC/s$, respectively). Representative welds were sectioned axially and either mounted for metallographic characterization and DPH hardness testing, or three-point bend tested in both the as-welded condition and after removal of the flash/notch at the outer weld periphery. SEM was used for fractographic analysis.

Visual examination of the CD welds revealed minimal flash at the outer weld periphery, with a somewhat greater flash observed in the moderate cooling rate welds. Due to a slight axial misalignment of the specimens during welding, the welds exhibited a notch or ledge at the outer weld periphery.

Metallographic characterization of the CD welds revealed narrow melt zones about 50 to 100 microns in width. In the high cooling rate welds, the fusion zone width was relatively constant across the entire specimen, tapering slightly toward the outer periphery. In contrast, the width of the moderate cooling rate weld fusion zone

Table 1. Hardness and bend ductility of Al-8Fe-4Ce base metal and capacitor-discharge weld specimens.

Specimen*	DPH** (50 gm load)	Bend Ductility***	Fracture Initiation Site
Base Metal	132-137	>20%	-
HC Weld/AW	286-341	3.3%	outer periphery notch
HC Weld/Mach.	"	>16%	FZ weld defect
MC Weld/AW	148-161	3.7%	outer periphery notch
MC Weld/Mach.	"	>17%	base metal

*HC weld: 0.4 mm tip dia., 0.5 mm tip length, 60^0 conical tip, 80 μF, 100V, head + 5.68 kg drop wt., 50 mm drop ht.
MC weld: 0.8 mm tip dia., 1.0 mm tip length, 60^0 conical tip, 80 μF, 100V, head + 1.93 kg drop wt., 25 mm drop ht.
**Hardness at center of fusion zone
***Flat surface in tension

varied continually across the specimen cross section.

Figure 4 illustrates the high (a) and moderate (b) cooling rate CD weld fusion zone microstructures at increased magnification. As shown, high cooling rates promoted solidification to a featureless Zone A microstructure similar to that shown earlier in the finest powder particles. Although a fine region of partial melting (PMR) was evidenced at the fusion boundary as dispersoids in a Zone A matrix, microstructure analysis found no evidence of a HAZ. Microhardness traverses across the weld showed an extremely high fusion zone hardness as compared to the base metal (DPH 300 versus 135) and no evidence of base metal softening adjacent to the fusion boundary. Moderate weld cooling rates promoted solidification to a coarser Zone B solidification structure, as shown in Fig. 4b. Despite experiencing

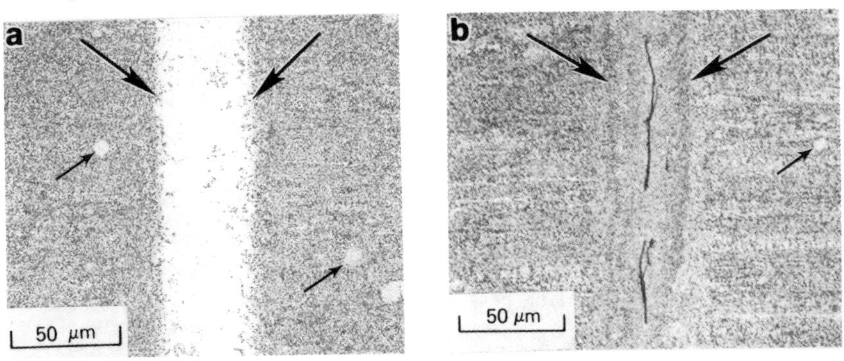

Figure 4
Light micrographs of high (a) and moderate (b) cooling rate capacitor-discharge welds in Al-8Fe-4Ce. Large arrows indicate fusion boundaries, small arrows indicate undeformed powder particles in unaffected base metal. Note linear defects along the center of fusion zone in (b).

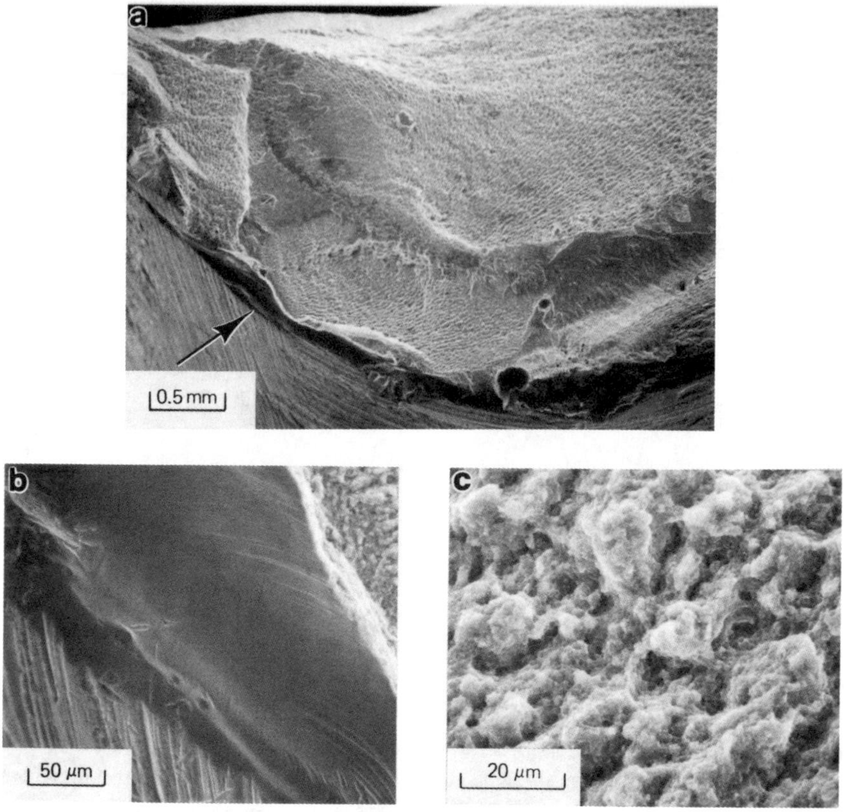

Figure 5
SEM fractographs of moderate cooling rate capacitor-discharge weld bend specimen in Al-8Fe-4Ce tested in as-welded condition: (a) fracture surface at low magnification, arrow indicates fracture initiation site at outer periphery; (b) liquated surface of outer periphery "ledge;" (c) fracture surface of unaffected base metal.

slower fusion zone cooling rates, these welds also exhibited a hardness superior to that of the base metal (DPH 175 versus 135) and showed no evidence of HAZ softening. Occasional defects were observed in both the high and moderate cooling rate welds in the form of small pores and linear discontinuities (Fig. 4b), and were attributed to the evolution of hydrogen-induced porosity during the final stages of solidification. The near elimination of porosity in the CD welds, as compared to that previously shown in GTA welds, was attributed to the influence of high cooling rates in suppressing pore nucleation and growth, and the high welding pressures forging closed the pores and possibly shrinkage cavities during the final stages of solidification.

Three-point bend tests performed on the axially-sectioned specimens showed a significant reduction in bend ductility versus the base metal (Table 1). As shown in Fig. 5 for a moderate cooling rate weld, analysis of the fractured specimens showed fracture to initiate at the outer weld periphery notch and propagate across the unaffected base metal. Removal of the flash/ledge by machining markedly improved bend ductilities to near base metal values and in the moderate cooling rate welds promoted fracture exclusively in the unaffected base metal. Occasionally, fracture in the high cooling rate welds initiated at solidification-related defects in the fusion zone, as shown in Fig. 6. Such fracture initiation did not, however, significantly reduce bend ductility.

In summary, CD welding was found to be effective in producing high quality welds in the Al-8Fe-4Ce alloy. Based on hardness and bend test results, it is anticipated that welding parameter optimization and improved specimen alignment could readily provide 100% joint efficiencies in the as-welded condition.

3. INERTIA-FRICTION WELDING

Inertia-friction (IF) welding is a solid-state welding process which offers an important alternative to fusion welding for joining materials in which melting and solidification must be avoided. The process involves a combination of frictional heating at faying surfaces and the application of axial pressure to produce a metallurgical bond. The absence of melting during IF welding can alleviate weldability problems including the formation of porosity and solidification cracking. Also, the expulsion of contaminated surface layers and heat-and-deformation affected metal at the weld interface during the final forging and upset stage of the process allows final bonding between nascent subsurface

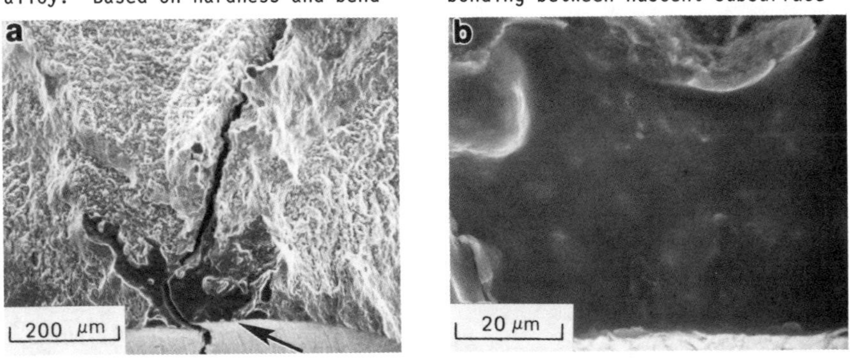

Figure 6
SEM fractographs of high cooling rate capacitor-discharge weld bend specimen in Al-8Fe-4Ce after removal of outer periphery notch: (a) a solidification-related defect near specimen center (arrow); (b) liquated surface of defect at increased magnification.

materials. Based on these inherent characteristics, this process offers the capability to "retain" to a large degree the base metal microstructure across the joint, making it highly suitable for the joining of RS/PM Al-Fe-Ce alloys with a high residual gas content.

In the present study, 12.5 mm diameter rod of Al-8.6Fe-3.8Ce (Al-9Fe-4Ce) and Al-8.6Fe-6.9Ce (Al-9Fe-7Ce) were joined axially using a conventional IF welding system. Through preliminary testing, welding parameters including the moment of inertia, rotational speed and axial force were optimized. In addition to joints produced using these optimized parameters, welds were also produced at lower and higher axial forces for comparative purposes (Table 2). Following visual analysis, representative welds were sectioned axially and metallographically prepared for microstructure analysis using light and electron microscopy and DPH hardness testing. In addition, round tensile specimens were machined from the center of the welded specimens (4 mm gage dia., 16.5 mm gage length) and tested at an extension rate of 0.0042 mm/s. Fractography of the fractured specimens was performed using SEM.

As shown in Fig. 7a, visual analysis of the welds produced at all three axial forces showed good symmetry and continuity of flash around the weld circumference. As expected, an increase in axial force promoted an increase in axial displacement (Table 2) and the quantity of flash (Fig. 7a). A consistently greater axial displacement in the Al-9Fe-4Ce versus the Al-9Fe-7Ce alloy was attributed to the lower compressive yield strength of the Al-9Fe-4Ce.

Macroscopic characterization of the as-polished specimens revealed two basic defect types: 1) voids which resulted from the nonuniform deformation and insufficient extrusion of metal out of the weld interface

Table 2. Inertia-friction weld axial displacements and tensile properties

Alloy	Axial Force kN	Axial Displacement mm	Tensile Strength MPa	Joint Efficiency %	% Elong.	Fracture Location*
Al-9Fe-4Ce	4.4	2.5	234	56	1.9	IR
"	6.5	5.8	262	62	2.9	IR
"	8.7	9.6	386	92	3.0	IR/HDZ
Al-9Fe-7Ce	4.4	1.8	262	56	1.9	IR
"	6.5	4.6	386	82	2.7	IR
"	8.7	8.9	427	91	3.0	IR/HDZ

*IR - Interface Region, HDZ - Heat-and-Deformation Zone

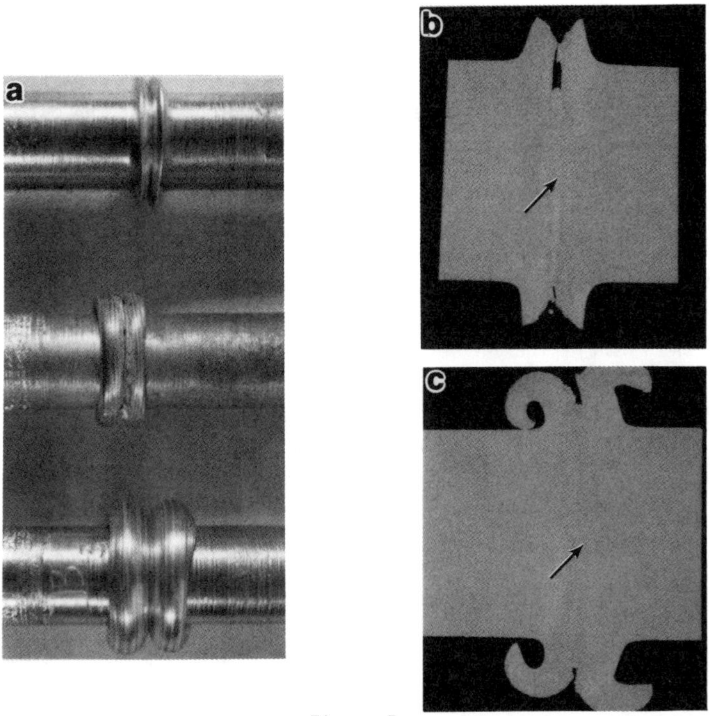

Figure 7
Inertia-friction welds in Al-9Fe-4Ce: (a) as-welded specimens produced at low, medium and high axial force (top to bottom); (b,c) axial sections of welds produced at low and high axial forces, respectively. Arrows in (b) and (c) indicate the weld interface region.

region (Fig. 7b) and 2) fine, linear lack-of-bonding defects which may have been associated with dispersed oxides. The number and size of these defects decreased with increased axial force, and were only occasionally observed in welds produced at the highest axial force (Fig. 7c). Although the base alloys exhibited high hydrogen contents of 2-3 ml/ 100 gm, no evidence of porosity or blistering was observed. Macroscopic examination of the welds after etching with Keller's reagent revealed heat-and-deformation zones (HDZ's) which were observed to decrease in width with increased axial force. This decrease was attributed to the greater extrusion of heat-and-deformation affected metal out of the weld interface as flash.

Figure 8 illustrates microstructures of IF welds produced at low (a) and high (b) axial forces. Evidence of a radially outward flow of metal which is characteristic of IF weld ing is apparent in both microstructures, with the original base metal texture changing 90^0 to an orientation parallel to the weld interface. In the vicinity of the weld interface, elevated temperatures promoted coarsening of the dispersoid structure, as indicated by the light-etching region and the adjacent dark-etching transition region in Fig. 8a. The use of high axial force essentially eliminated the most severely coarsened light-etching region and also reduced the width of the dark-etching transition region (Fig. 8b). As indicated in Fig. 8c, microhardness traverses across the weld interfaces correlated well with the observed microstructures. The dispersoid-coarsened, light-etching structure in the low-axial force weld exhibited a significant decrease in hardness versus the base metal, with the dark-etching region showing a range of hardnesses. Elimination of the light-etching region in the high axial force weld and a reduction in the width of the transition region resulted in a minimal drop in hardness across the weld zone.

Tensile properties for the welded specimens are shown in Table 2. Increased axial force was found to promote a significant increase in tensile strength, with an increase in joint efficiency from 56% for welds produced at low axial force to over 90% for high axial force welds. Tensile ductilities of the specimens also increased with increased axial force, although elongation values were still well below those of the unaffected base metal.

Fractographic analysis of the tensile specimen surfaces showed fracture in low and medium axial force welds to occur along the weld interface through the low hardness, dispersoid-coarsened structure. The presence of voids along the weld interface may also have contributed to property degradation in these welds. Higher weld strengths associated with welds produced at increased axial force promoted a shear mode of failure across the weld interface into the unaffected base metal (Fig. 9a). All of the weld fracture surfaces exhibited a distinct spiral appearance emanating from the center of the rod, which reflected the effects of combined torsional and axial forging forces on metal flow experienced at the weld interface during IF welding (Fig. 9b). In addition, on a microscopic scale all fracture was observed to involve microvoid formation around the fine dispersoids and the associated ductile failure of the alpha aluminum matrix (Fig. 9c).

In summary, the IF welding process has been shown to be effective for the joining of RS/PM Al-Fe-Ce alloys. Increased axial force was found to improve weld integrity and tensile properties by eliminating weld interface defects and minimizing microstructural coarsening across the weld zone.

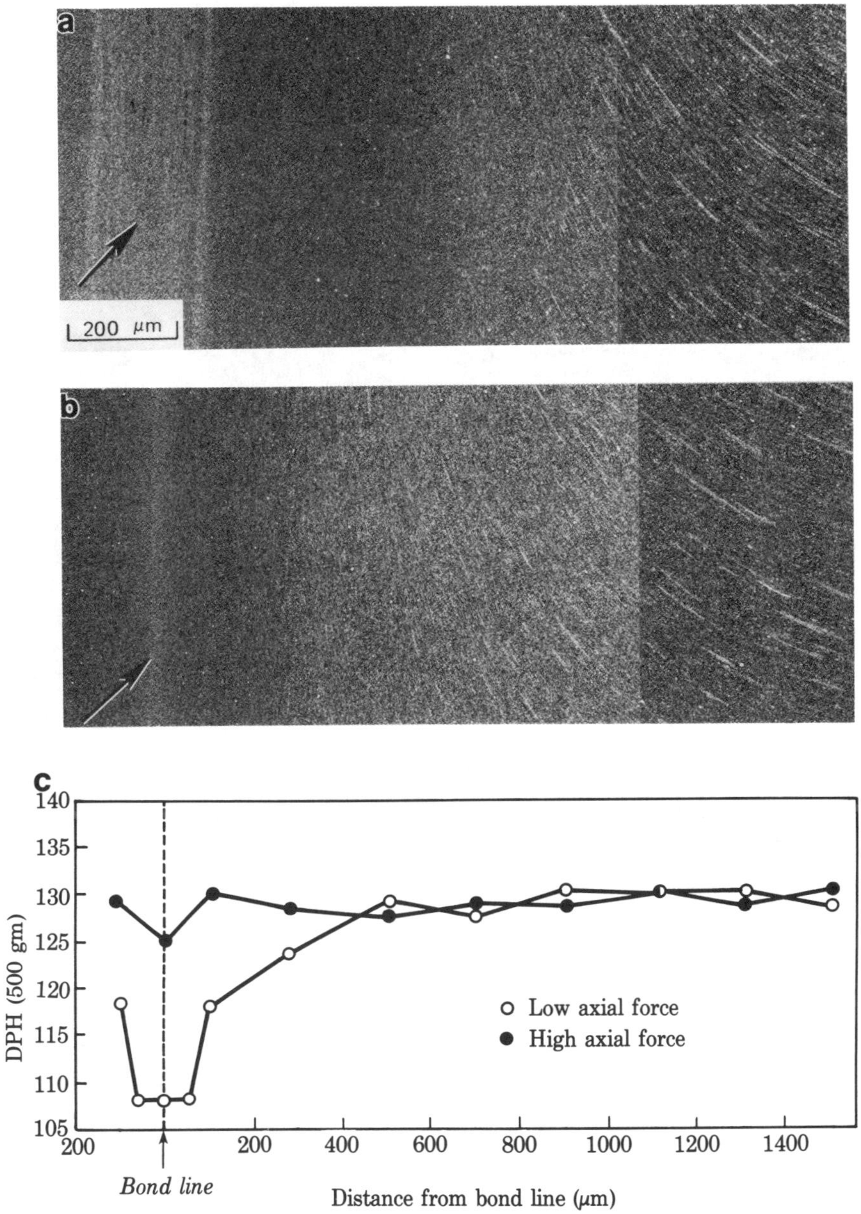

Figure 8
Microstructures (a,b) and DPH hardness traverse (c) near center of inertia-friction welds in Al-9Fe-4Ce produced at (a) low and (b) high axial force. Arrows indicate weld interface.

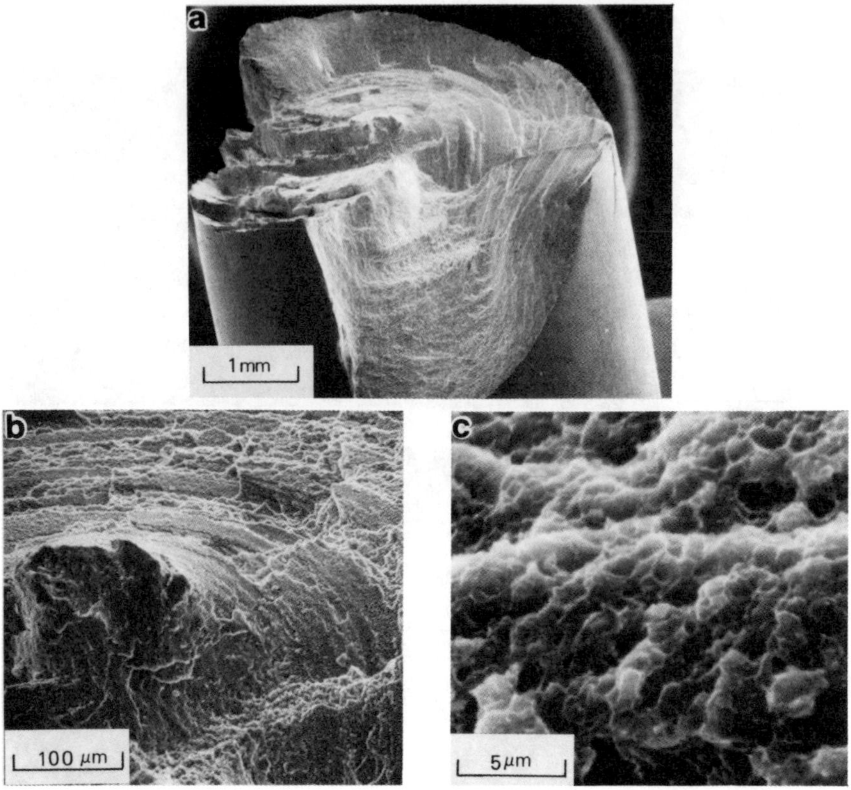

Figure 9
SEM fractographs of tensile tested Al-9Fe-7Ce inertia-friction weld produced at high axial force: (a) macroscopic fracture surface; (b,c) interface fracture surface at increased magnification.

4. REFERENCES

1. Langenbeck, S. L., et al., Rapidly Solidified Powder Aluminum Alloys, ASTM-STO-890, American Society for Testing Materials, 1986, pp 410-422.
2. Angers, L. M., Ph.D. Dissertation, Northwestern University (1985).
3. Metzger, G. M., AFWAL-TR-87-4037 (1987).
4. Devletian, J. H., Weld. J. 67 33 (1987).

5. ACKNOWLEDGEMENTS

The author is indebted to Dr. G. J. Hildeman of ALCOA for supplying the Al-Fe-Ce alloys. Appreciation is also expressed to Prof. J. H. Devletian of Oregon Graduate Center, Ms. K. Hagey of Westinghouse, Dr. G. M. Metzger of AFWAL, Mr. D. E. Kruzar of MTI, Inc., and Messrs. K. H. Hou and S. Krishnaswamy of OSU for their technical contributions. This work was supported by the Army Research Office under Contract # DAAG-29-84-0176.

2nd International SAMPE Metals Conference
August 2-4, 1988

MACHINE VISION FOR CONTROL OF GAS TUNGSTEN ARC WELDING

Robert A. Anderson
Manufacturing Technology Laboratory
General Electric Aircraft Engines
Cincinnati, Ohio

Abstract

A gas tungsten arc welding robot system using machine vision control to weld aircraft engine components is described. The equipment, capabilities, and operational theory of the system are discussed. Both vision controlled robot motion for weld joint tracking and vision-based weld process control are used to overcome typical production conditions such as non-repeatable joint locations, joint width variations, and part thickness variations. This state of the art robotic system is currently being used to weld aircraft engine hardware in a production environment. Improvements in productivity and quality have been realized.

1. INTRODUCTION

Advanced manufacturing processes and techniques, along with advances in materials, are of critical importance in the aerospace industry. Automation and particularly the use of industrial robotics is one method of increasing productivity and quality in today's competitive environment. Although new technological advancements in the areas of composites, near net shape processes, and ceramic bonding have somewhat reduced the need for welding in the aerospace industry, gas tungsten arc welding (GTAW) will continue to be an important and necessary requirement for manufacturing in the aerospace and other industries.

Conventional arc welding robots are preprogrammed using a teach pendant or in some cases, an off-line programming facility. The teaching of robot locations are programmed using a first part or an expected joint location. The desired path is stored in the robot controller to be repeated for subsequent parts. This preprogramming method requires repeatable weld joint alignment and often results in unacceptable quality due to variations in joint fitup and thermal distortion. Dissimilar thicknesses at the joint may also cause the weld pool to be asymmetric about the joint. Similarly, nominal welding parameters may not repeatedly produce high quality welds due to local and part to part variations in metal thickness, joint width variations and heatsink differences in the part and fixturing.

It is desirable, therefore, to monitor and control robot motion and the welding process in real-time much as a human welder does in using his sight for manual welding. Sensors and closed-loop control systems which can emulate human adjustments could be added to an industrial robot to ensure repeatable, high quality welding, and overcome the difficultes presented by conventional robots. Machine vision, as well as other control schemes have been investigated extensively for use in gas tungsten arc welding (1). The use of an integrated through-torch vision sensor for coaxial viewing of the weld joint and weld pool is well suited for GTAW (2).

The subject of this paper is to report on a GTAW closed-loop robot system using machine vision control to weld aircraft engine components. The development of a prototype of this system has been reported previously (3). The integrated robot system uses a customized vision and control system to automatically alter the robot path and welding parameters using visual feedback information received from the weld.

2. SYSTEM DESCRIPTION

The system equipment, illustrated in Figure 1, includes both custom modules and standard components. Cell control is provided by the Adaptive Control Processor. The system uses a 5 axis articulated arm robot which has been integrated with a 2 axis welding positioner. The welding equipment includes a fast switching 150 ampere welding power supply, a customized cold wire feeder, a standard automatic voltage control unit, and a custom weld interface control. The system uses an integrated weld torch/vision sensor. The control and vision system includes both standard commercially available computer boards as well as custom computer boards. Most system communications are performed via fiber optic lines which increases immunity to electromagnetic interference. The system also incorporates the use of EMI hardened wire troughs and enclosures.

The robot system is used for production welding of sheet metal and castings from 0.020" to 0.100" thickness for single pass applications. Welding is performed in the downhand or flat position. The system is used primarily for butt joints but has also been used for edge and lap joints. Stainless steel, titanium, and nickel-base alloys can be welded.

2.1 System Architecture

The system consists of three main hardware sub-systems; the Adaptive Control Processor (ACP) and Feature Extractor module, the P50 robot and positioner, and the various welding equipment components. Figure 2 illustrates the computer architecture of the system. The heart of the entire system is the ACP which runs the menu-driven user interface, performs the weld process control, and controls all system communications. The ACP cabinet contains a single board computer, associated boards and electronics, the Feature Extractor, a CRT/keyboard, and weld image monitor. The Feature Extractor is a multi-board custom vision system which uses bit-sliced computer technology. The bit-sliced architecture allows for custom intruction sets and large word lengths which make it possible to execute multiple instructions simultaneously. The vision system is, therefore, capable of the very fast computations necessary for real-time process control. The ACP sends information to the P50 robot control to perform weld joint tracking. A second single board computer, servo control boards for positioner integration, and other associated boards have been added to the P50 controller to achieve closed-loop, coordinated arm and positioner motion control. The seven axes of coordinated motion enable the system to perform complex welds involving intricate three-dimensional shapes while keeping the weld in the flat position.

Both the ACP and the P50 Controller send and receive signals to and from the Weld Interface. Both RS 232 protocol and 24 volt logic signals are used. Weld sequence and pulsing signals are sent to the welding power supply. Wirefeed sequence and speed signals are sent to the customized wirefeeder which feeds the wire in a reciprocating or dabbing manner synchronized with current pulsing. Signals are sent and received from a commercially available automatic voltage control unit to control arc length.

2.2 Weld Torch Sensor

Vision information of the weld area is detected using a specially designed weld torch sensor. The sensor provides direct images of the weld pool and weld joint as

System Configuration

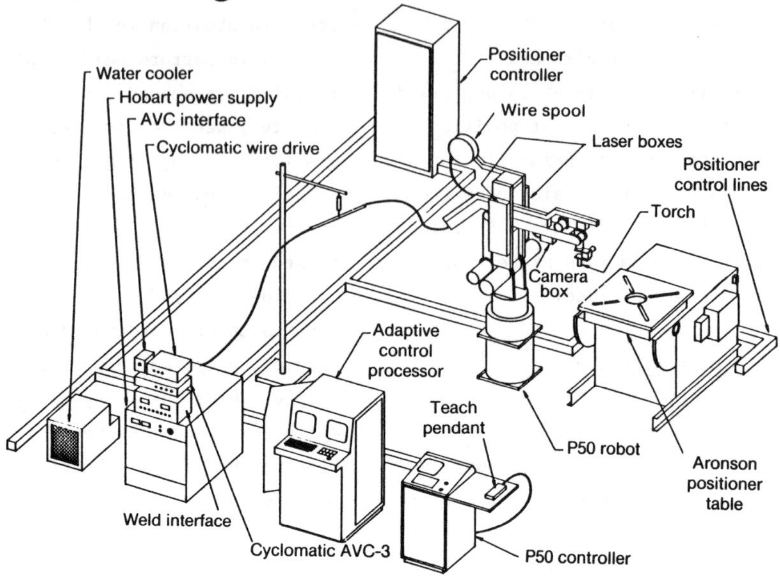

FIGURE 1

Computer Architecture Diagram

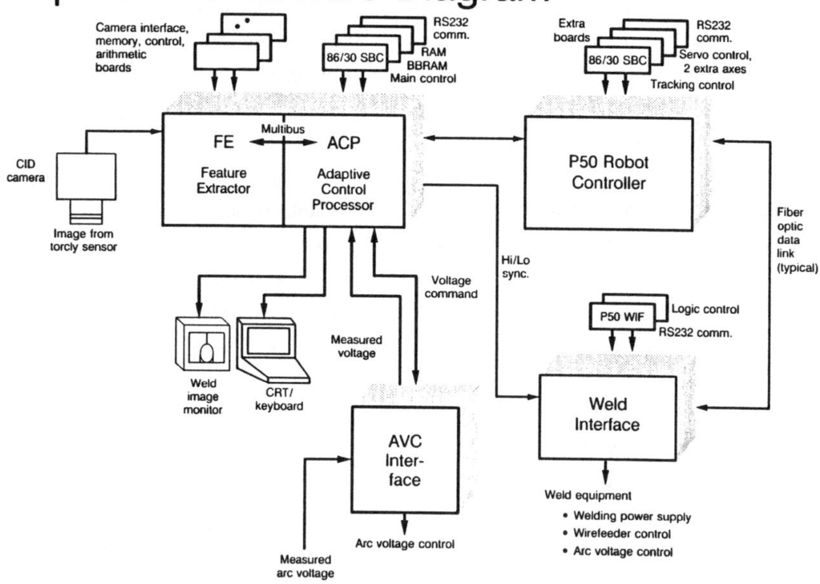

FIGURE 2

viewed axially along the torch electrode. The concept of using a vision sensor integrated with a torch to provide an axial view of the weld was first proposed by Richardson et. al. (4). The main components of the vision sensor are shown in Figure 3. The two-dimensional weld image is transmitted through several lenses in the optical torch, through a coherent fiber optic image guide, into a solid state, CID camera. The 256 x 256 pixel image with 128 levels of gray is transmitted from the camera into the vision computer for image processing and analysis. Image illumination is provided by four defocused, low power helium neon lasers. The laser light is transmitted to the optical torch by two bifurcated random fiber optic bundles. The intense arc light in the immediate vicinity of the electrode is masked in the axial view by the electrode itself. Further filtering is achieved by a narrow band pass filter.

Figure 4 shows a photograph of a weld image. The photo shows the detection of the weld pool width and weld joint as indicated by the white lines. The photo also shows a weld tack immediately in front of the weld pool, cold wire addition, and weld pool asymmetry about the joint. The weld joint in this image is a tight butt joint with no gap, only a surface indication of 0.004" in width.

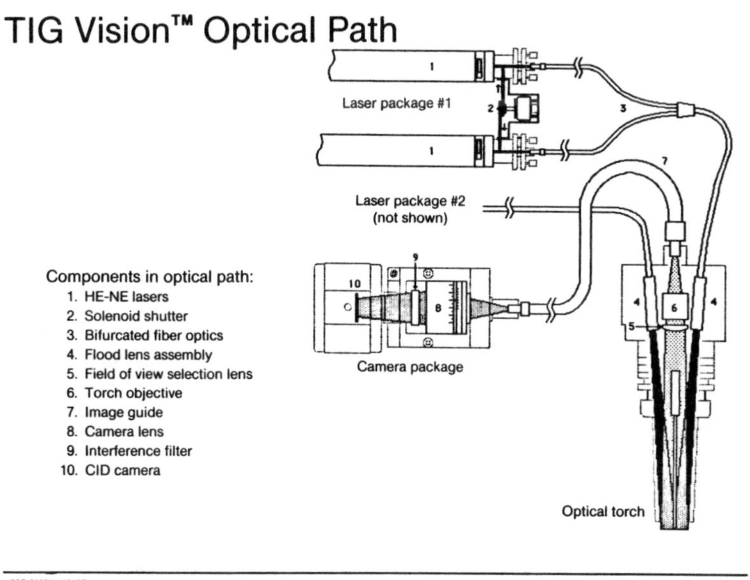

FIGURE 3

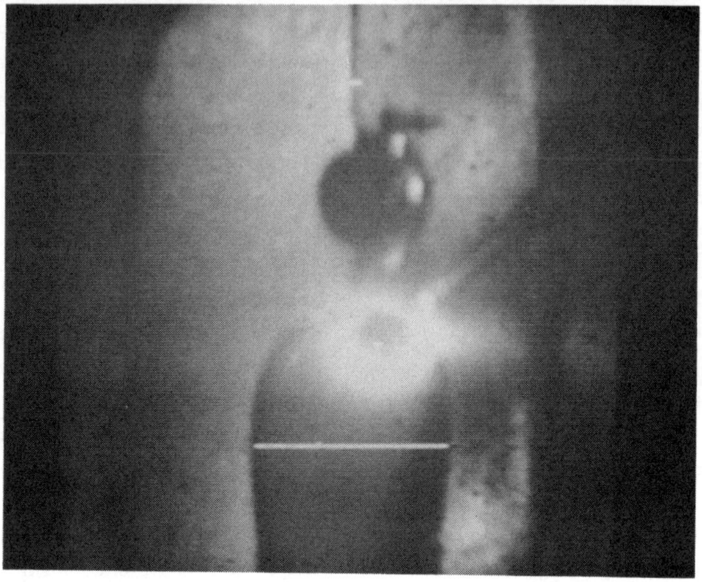

FIGURE 4: Weld Image

3. MOTION CONTROL

Robot motion control involves real-time detection of the weld joint location and weld pool center, and the calculation of robot steering commands. These commands are used to position the center of the pool over the weld joint. To be able to perform weld joint tracking with coordinated positioning, calculations in three coordinate systems are necessary; 1) camera or vision coordinates, 2) robot coordinates, and 3) positioner faceplate coordinates.

The first task required for real-time motion control is joint detection. A pre-defined portion of the image is analyzed to find the location and width of the joint. The optics of the vision sensing system must have sufficient resolution and illumination for the system to be able to detect the weld joint under a variety of production conditions. Vision algorithms must be sufficiently robust to perform under a variety of surface finishes and joint widths. Chemically cleaned surfaces, a number of mechanically cleaned surfaces, and joints with virtually zero gaps can be detected. Different joint detection algorithms are used when the type of joint to be weld, butt or edge, is specified.

Measurements of the joint location and pool offset from the electrode are transformed from vision coordinates to robot coordinates. The errors between actual joint location and the taught path location in each orthangonal direction of the robot coordinate system are used to calculate robot velocity commands. These commands alter the motion of the robot so it follows the actual path instead of the programmed path. The motion algorithms incorporate a delay to account for the fact that the joint is being detected ahead of the weld pool. Motion about the welding electrode is also controlled in such a manner that the vision coordinates of the torch and wirefeed guide maintain a correct orientation with respect to the joint.

For welding three-dimensional joints, the part is rotated with two degrees of freedom by the positioner such that the weld surface remains flat. Robot and positioner motion is coordinated so that the speed of the weld torch relative to a moving workpiece surface is constant. Robot and positioner faceplate coordinate systems must be related and transformations between the two must be accomplished in order to achieve coordinated motion.

An automatic joint location correlation algorithm is included as part of the joint detection routines. In each control cycle the current joint location is compared to a history of joint locations; a low correlation indicates errant joint detection which may occur at weld tacks. Upon recognition of errant joint data the robot follows the taught path plus an offset from an accumulated integration of velocity corrections.

In addition to real-time tracking, the robot system also has the capability to perform pre-scan or two-pass joint tracking. In applications in which the weld joint pulls together so tightly during tacking or welding so that no joint can be detected, the joint can be pre-scanned. Actual joint locations are stored in memory during a first pass to be retreived during tacking or welding so that the previously tracked path is followed. The robot system also has the capability to automatically center the welding electrode on the joint before welding begins. In practice, the upper limit of joint tracking speed is well beyond the GTAW process speed limitation.

4. PROCESS CONTROL

The objective of the system's process control is to make full penetration GTA welds with proper reinforcement and drop-through on sheet metal and castings. The system must be capable of producing high quality welds under a variety of conditions including different alloys, various surface finishes, different metal thickness at the joint or from part to part, varying joint gaps that range from a tight butt joint with zero gap to a gap equal to one times the thickness of the metal.

The robot system is capable of achieving this objective by performing three closed-loop control functions; 1) weld pool geometry control, 2) arc gap control, and 3) wirefeed speed control. The weld pool geometry and wirefeed speed controls use vision feedback, while the arc gap control uses arc voltage feedback. These controls will be described in the following sub-sections.

4.1 Weld Pool Geometry Control

It must be assumed that in order to perform process control using top side weld pool geometry as a feedback variable, that this geometry is a satisfactory indicator of weld quality. Unlike some iron-base alloys which display a wide variance in penetration from one heat to another due to minor element effects, as cited extensively in the literature, experience has shown that for the alloys to be welded by the robot system, top side pool geometry is an excellent indicator of ultimate weld quality.

In addition to being satisfactory indicators of weld quality, weld pool geometrical variables must also have enough dynamic response to be adequate feedback control discriminants. A third criteria that geometrical features used for process control must possess is the ability to be detectable by the vision system with high reliability under a variety of conditions and disturbances. There are several geometrical attributes which can be analyzed or detected in a two-dimensional image of a GTA weld pool; these include area, width, length and symmetry, among others. It has been determined that both pool area and width meet all three criteria. Production conditions and disturbances include various alloys with significantly varying physical and optical characteristics, various surface finishes with differing reflectivities, the presences of disturbances such as oxides floating on the weld pool, glare due to specular reflection of the laser illumination, disturbances due to the addition of filler wire and current pulsing. Under all of these conditions image analysis algoritms must be fine tuned and control routines must have different gains and time constants. The robot system has been designed to allow for this application specific fine tuning.

It has been determined that with some alloys weld pool area is the best control discriminant; and with other alloys, weld pool width is more desirable (5). For example, some combinations of metal thickness and joint width in most stainless steel alloys investigated, display an increase in weld pool length but little or no increase in pool width. In these cases pool area is used as a control discriminant. In other alloys there is a good correlation between weld pool width and penetration. In titanium for example, the weld pool reflectivity and size make it necessary to use pool width as a discriminant since the entire periphery of the pool can not be reliably detected.

A direct variable must be regulated in response to weld pool geometry in order to control dependent weld quality variables such as penetration, bead width and reinforcement. The most significant direct variable in the GTAW process is heat input. Heat input can be changed by regulating welding current or welding speed. Weld current is used as a direct control variable for the regulation of pool area or width for two reasons. First, weld current has a more direct and predictable influence on the heat input to the weld. And second, the mathematics and dynamics required to provide for a continuously varying torch speed are currently too involved for practical use.

The pool geometry dynamics can be represented by a linear transfer function model. A block diagram of the control scheme is shown in Figure 5. The pool geometry algorithm determines the pool location, pool center, offset of the pool center from the electrode, and pool width or area. The measured width or area is compared to a desired width or area and multiplied by a normalization factor, a_1 or a_2. This factor is than passed into a proportional, integral, derivative dynamic compensation routine. The function of this algorithm is to minimize the effects of disturbances on the pool geometry, stabilize the control loop, and provide for robust dynamic performance. The dynamic compensator determines a correction to the nominal welding current. The control loop is completed by the regulation of the weld pool size and visual analysis of a new pool image. The control loop update rate for the pool geometry control is 30 hertz. In summary, weld pool geometry is controlled by operator selection of width or area values. The current is automatically and continuously regulated to create a weld pool of the desired area or width

Pool Geometry Control Block Diagram

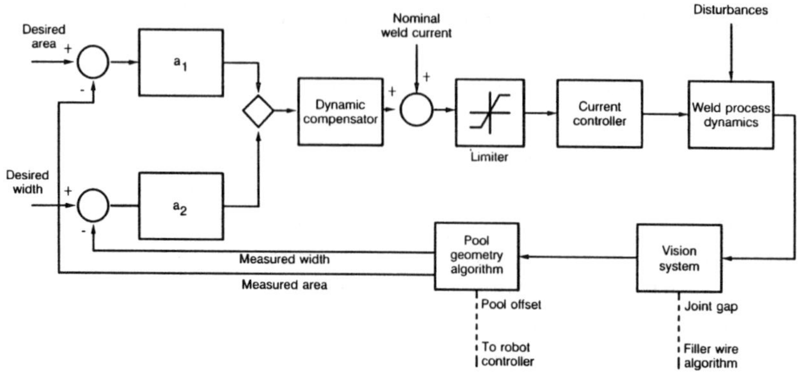

FIGURE 5

regardless of part thickness, joint gap, heat sinking or other variables.

4.2 Arc Gap Control

The regulation of a constant arc gap or arc length is critical to assuring accurate placement of the wire into the weld pool. If the arc gap is too high the molten wire can accumulate into a large molten ball before detaching from the solid wire. This could result in adhesion of the wire droplet to the electrode or in large disturbances when the droplet enters the weld pool. If the arc gap is too low the wire can miss the weld pool and jam into the solid base material. Further, a constant arc length aids in regulation of the desired heat input to the weld and minimizes arc disturbances.

Arc voltage provides an indirect measurement of the arc gap and can be used to maintain a constant arc gap. The fundamental governing relationship used in the control is that the arc voltage is proportional to the arc gap at a constant current. However, since current changes automatically, the control routine must compensate for these current changes. A true adaptive control algorithm is used to make this compensation in which the control gains are automatically changed in response to current changes.

The robot system has two methods for controlling arc gap. A

commercially available automatic voltage control (AVC) with a one inch linear slide mounted at the robot wrist can be used in applications with large voltage gradients. These gradients could be due to rapid changes in part geometry or positioner/robot movement. The AVC unit accepts a current compenated remote voltage signal from the system's process control computer. An alternate technique for controlling the arc gap is to use robot arm motion. In applications which have small voltage gradients or that have tight access requirements, it may be desirable to use robot motion to control the arc gap.

A block diagram of the arc gap control is shown in Figure 6. An error between the reference current and the actual current command is first calculated. Beta2[n] is a piecewise 1st order function which simulates the 2nd order current to voltage relationship. Beta1 is a slope which simulates the arc gap to arc voltage relationship. The top part of Figure 6 shows that this function results in a voltage value to be added to a set voltage which compensates for current changes. The bottom portion of the block diagram shows that the output of the Beta function can be input to a proportional, integral dynamic compensator which computes z-axis velocity corrections. These commands modify the z-axis position of the robot to maintain a constant arc gap.

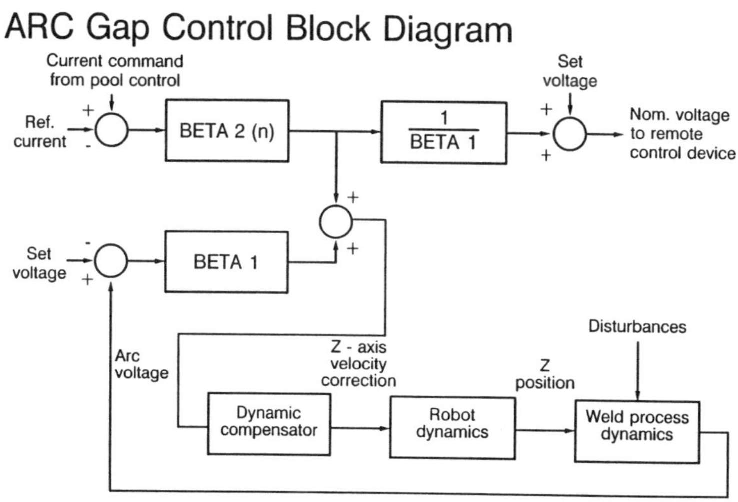

FIGURE 6

4.3 Wirefeed Speed Control

The third process control function of the robot system is that of wirefeed speed control. This function controls the volume of wire added to the weld and hence, allows the system to accomodate for large variances in joint gaps and for the presence of tacks. By sensing the width of the joint and performing a proportional feedback control on wirefeed speed, consistent weld reinforcement and drop-through can be maintained. The presence of a weld tack is detected by the autocorrelation routine described in Section 3. The autocorrelation status is input to the wirefeed control algorithm which reduces the speed of the wire when welding over a tack.

As mentioned earlier, the wire is fed in a reciprocating manner. This minimizes the chances of the molten wire accumulating into a ball before it detaches from the solid wire. Reciprocation can be programmed at varying frequencies, duty cycles (ratio of forward feed time to reverse feed time), and desired net or average forward feed speed. The required actual forward wirefeed speed necessary to result in the desired nominal forward speed is calculated and sent to the wirefeed device.

5. MANUFACTURING APPLICATIONS

One robot system is currently being used for production welding with hourly personnel. A second system is being used to investigate and develop new applications. Several different types of parts have been welded with the most notable being large titanium ducts in the range of 0.020" to 0.060" thick. The system is shown welding one such duct in Figure 7. The photograpgh of the image shown in Figure 4 was taken during the welding of this part. This part presents problems for conventional automated welding due to local and part-to-part thickness variations in the skin, non-repeatable joint locations due to trimming and fixturing, variations in joint gaps, and dissimilar thicknesses at the joint. The part is robot tacked and welded in about one fourth the time required for manual welding. Rework has also been cut significantly over mechanized welding.

Other types of parts such as stainless steel manifolds, sheet metal to casting welds in stainless steel and nickel-base engine structural components, and edge welds on struts have been welded with the robot system.

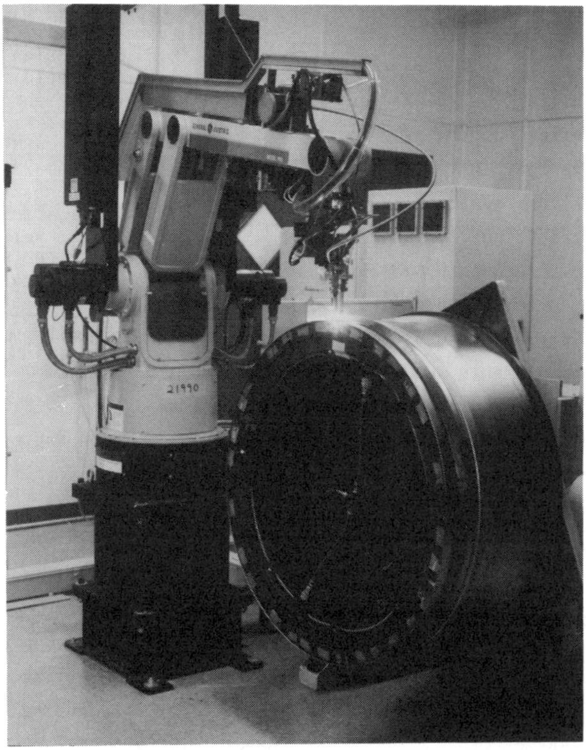

FIGURE 7: Robot System Welding Duct

6. CONCLUSIONS

Closed-loop control of arc welding has been the subject of extensive research for over two decades. Many ideas and technologies have been developed but only a very few have been successfully applied to large scale production. The vision controlled robot system described herein is one such successful system. This system is one of the first integrated, closed-loop GTAW robot systems to make a successful transition from the laboratory to production.

The robot system has proved to be a flexible, production worthy tool which has increased welding productivity, improved weld quality, and reduced rework. The development of this system has demonstrated that although the GTAW process is extremely complex, it can be automatically controlled for use in a production environment where conventional automation has not succeeded. This system is expected to be a building block for the continued development of advanced welding automation in the years ahead.

7. ACKNOWLEDGEMENTS

The author gratefully acknowledges the diligent effort of the many people who have contributed to the development of this system. Important contributions to the early research were made by members of the General Electric Corporate Research and Development Center. Many system improvements and software features were implemented by several key employees at the Manufacturing Technology Laboratory, General Electric Aircraft Engines, Evendale, Ohio. Equipment productization was performed by a dedicated team at the Robotics and Vision Systems Department of General Electric.

8. REFERENCES

(1) Richardson, R.W., "Review of The State-Of-The-Art of Adaptive Control For The Gas Tungsten And Plasma Arc Processes," Center For Welding Research Technical Report 529501-82-5.

(2) Richardson, R.W., Anderson, R.A., Farson, D., Gutow, D.A., "Co-Axial Arc Weld Pool Viewing For Process Monitoring and Control," Welding Journal, 63 (3), March 1984, pp 43-50.

(3) Beheti, R.S., Haefner, K.B., Sweet, L.M., "Operational Performance of Vision-Based Arc Welding Robot Control System," ASME, Winter Meeting, Dec. 1984.

(4) Richardson, R.W., Gutow D.A., Rao, S.H., "A Vision Based System for Arc Weld Pool Size Control, Measurement and Control for Batch Manufacturing D.E. Hardt, ed., ASME, New York, 1982, pp.65-75.

(5) Nied, H.A., "Two-Dimensional Gas Tungsten Arc Welding Process Model," CRD Report 84CRD092, April 1984.

9. BIOGRAPHY

Mr. Anderson received B.S. and M.S. degrees in Welding Engineering from The Ohio State University in 1981 and 1982 respectively. Following graduate studies in vision aided robotic welding, Mr. Anderson worked for Automatix in robotic welding product development and applications. Starting at General Electric, Manufacturing Technology Laboratory, in 1984, Mr. Anderson has worked on the development and applications of the GTAW robot system, as well as other welding activities. Mr. Anderson is a member of the American Welding Society (AWS). He is a co-recipient of the Charles H. Jennings Memorial Award presented by the AWS for the best technical paper during 1984. The Industrial Research IR-100 Award was presented to a GE team, including Mr. Anderson, in 1984, for their work on the GTAW robot system.

2nd International SAMPE Metals Conference
August 2-4, 1988

ADAPTIVE GAS TUNGSTEN ARC WELDING SYSTEMS

Mr. Harvey R. Castner
Edison Welding Institute
1100 Kinnear Road
Columbus, Ohio 43212

ABSTRACT

The production of aerospace components often involves arc welding of precision weldments that require the highest levels of weld quality and reliability. The production of these weldments necessitates "precision" control of the welding operation.

Automated or robotic welding of precision weldments can produce the accuracy and control that are necessary to meet the high quality requirements and, at the same time achieve effective levels of productivity. However, this type of welding of precision components has been limited because:

- The robot may not have sufficient accuracy.
- Programming is time consuming.
- Tolerances of the weld joint are not always adequate.
- Manual operator intervention during welding is impractical and imprecise.

This paper describes work at Edison Welding Institute that is directed toward the development of advanced sensors and adaptive control techniques for precision welding.

1.0 INTRODUCTION

Manufacturing companies must continually improve not only the quality of their products but must also increase productivity in order to remain competitive. Automation is one of the techniques that can be used to accomplish these goals. Therefore, industries have turned to automation to achieve greater control of welding operations and to reduce their reliance on the skill of manual welders.

While automated welding is not new, robots and computer controlled welding systems have become practical only recently. These systems offer a flexible way to economically produce high-quality parts and can integrate welding into computerized manufacturing systems.

Automation has been applied most rapidly in those industries with medium- or high-production volumes. However, aerospace companies, manufacturers of nuclear components and electrical products are also interested in automating their welding operations. These industries produce "precision weldments;" parts that have exacting quality requirements, high unit costs and low production volumes.

Since precision weldments require the highest levels of weld quality, they also necessitate "precision" control of the welding operation. Precision components are typically welded with either the gas tungsten arc (GTAW) or plasma arc (PAW) processes. Although the quality of these processes is high, productivity is low.

The complex shapes, low production volumes and the high degree of accuracy required for precision weldments make automation or the use of robots very difficult. Robots offer the flexibility which is an advantage for low production volumes. However, robot welding of low volume, precision components is limited because:

- The robot may not have sufficient accuracy.
- Programming is time consuming.
- Tolerances of the weld joint are not always adequate.
- Manual operator intervention during welding is impractical and imprecise.

The use of sensors to monitor the welding process and to provide adaptive control can solve many of these problems with inaccurate parts, work cell location and off-line programming errors. Laser-based weld joint tracking systems and coaxial vision systems can provide information on variations in the weld path and joint tolerances. Weld penetration sensors can overcome deviations in material composition and welding parameters. While these sensor techniques are available, there are only a limited number of companies who are applying them.

This paper describes an adaptive GTA welding system concept for precision weldments. The concept integrates the following components into an intelligent welding system:

- Coaxial and laser-based weld vision systems

- Sensors for the control of weld penetration
- Off-line programming techniques
- Welding process data bases.

This paper will describe a number of vision and penetration sensors that are suitable for use in the adaptive GTA system. A number of these sensors are under development at Edison Welding Institute. Others have been developed by robot or sensor manufacturers and are commercially available. Not all of these sensors will be required for all applications. Individual sensors must be selected to meet the needs of the parts being welded. This advanced GTA welding system will solve many of the remaining technical problems so that automated and robotic welding can be applied to precision weldments.

2.0 SYSTEM DESCRIPTION

2.1 System Overview

The adaptive GTA welding system will include feedback control of all aspects of the welding operation in order to produce the highest weld quality. The system will integrate welding equipment, a robot and a part positioner with real-time sensors to monitor and adaptively control all welding process variables.

The adaptive welding system for precision weldments may be a stand-alone work cell, it may be a part of a flexible manufacturing system, or it may be interfaced to a computer integrated manufacturing system.

The block diagram in Figure 1 shows the major system components.

- System controller
- Robot or programmable welding torch manipulator
- Advanced welding equipment
- Coordinated part positioner
- Penetration control system
- Weld vision system
- Process monitors
- NDE system.

A welding robot or a programmable manipulator will accurately control torch position, torch angles, oscillation parameters and torch travel speed. The system control computer will interface with all system components; generate, control programs and monitor welding and control functions.

A microprocessor or computer controlled welding power source and wire feeder will accurately control welding power and wire feed even during changes in input line voltage.

One or more sensors will monitor weld penetration to insure proper fusion. Weld vision systems will accurately track the weld joint and control heat input and joint fill. Other vision systems could monitor the completed weld bead to insure proper bead placement, proper bead size and proper fusion.

Welding parameter monitors will produce a permanent and accurate record of all welding variables. Finally, an NDE system would confirm weld quality in real time.

2.2 Control of Welding Parameters

The adaptive arc welding system will include computer or microprocessor controlled welding equipment that is capable of producing pulsed output current in a variety of waveforms and with extremely short pulse durations. The ability to pulse welding current improves the control of heat input and penetration. Pulsation at high frequencies can also improve the stiffness of the arc and enhance arc stability.

The system control computer will provide the programmed welding parameters to the welding equipment for each weld pass. Once the welding parameters are set, the feedback control within the power source and wire feeder will maintain these parameters within the desired limits. The system controller may determine that changes are required in the welding parameters in order to compensate for changing joint conditions, fitup or material variations.

Welding parameters will also be monitored during welding by the process monitoring system to insure that they do not exceed proper limits.

2.3 Path Control

To produce extremely high-quality welds, dimensional variations in the weld joint preparation and joint location tolerance variations must be accommodated by the automatic welding system.

The sources of weld joint mislocation and joint size variations may be due to: manufacturing tolerances during the preparation of the parts, fixturing and part location tolerances, torch position repeatability and thermal distortion that may occur during welding. In order to control the effects of these dimensional variations, an automated means is required to: locate and measure the weld joint, position and control the alignment of the torch in the joint, and adaptively control welding parameters in real time.

Various joint interrogation systems have been developed to address one or all of the above functions.

Sensors for weld joint interrogation include: mechanical probes, inductive probes, arc sensors, thermal sensors, ultrasonic sensors, laser-based vision systems and video imaging systems. Weld joint interrogation systems must be flexible enough to accommodate a wide variety of joint designs and must also be extremely accurate.

Optical joint sensing systems have the greatest potential for obtaining the maximum amount of information. These systems can examine the weld joint as well as the weld pool and the completed weld bead.

The adaptive GTA welding system will observe the weld joint, weld pool and solidified weld bead with two joint sensing systems, the laser-based system and the coaxial weld vision system.

2.3.1 Laser Based Joint Sensing Systems

A laser-based optical preview sensor is shown in Figure 2. This sensor uses a laser beam and detector array to determine the size and shape of the weld joint and the distance from the detector to the workpiece. The laser beam scans across the joint to gather multiple distance measurements and generates the joint profile.

The laser-sensing system allows multiple joint designs to be measured because it has the ability to recognize joint features including: the top edge of the preparation, root opening, root land height, sidewall angles and edges of bevels. The laser system can also calculate the location of some hidden features as well as the area of the joint. Joint area information is therefore available to control the welding parameters and joint fill.

2.3.2 Coaxial Weld Vision System

The coaxial weld vision system, which is shown in Figure 3, utilizes a specially designed GTA welding torch that contains a lens aligned with the tungsten electrode. The lens projects a view of the weld area onto a video camera that is mounted inside the torch. Alternative designs use an optical fiber to transmit the image to a remote video camera. This torch design uses the tungsten electrode to shield the camera from the majority of the arc light. An advantage of this technique is that the sensing point is located within the welding torch itself which reduces joint access problems.

The camera receives a view of the weld joint, the weld pool and the solidified weld bead. The weld scene is normally illuminated only by light from the arc itself. However, external illumination can be added if necessary to enhance the video image. Incorporation of a laser into the system adds the height measurement function required for joint profiling.

The video image is digitized and then analyzed by an image processing computer. A small portion of the weld image just ahead of the weld pool is analyzed to extract joint edge locations from which the joint centerline can be calculated for tracking control. The system can be taught to recognize the specific weld joint type. Measurement of the joint gap can be used to adjust welding current or filler wire feed. The molten weld pool and the solidified weld bead are also within the field of view. These features are used to center the weld bead on the joint and weld bead size is maintained by adaptively controlling welding current and wire feed speed.

This system has been applied to a number of automated and robotic welding applications and is under continuing development at OSU and EWI.

2.4 Penetration Control

Penetration is very important because it controls fusion with the base material. Effective control of penetration ensures that proper fusion is maintained yet prevents excessive heat input and melt-through. The use of in-process penetration feedback control represents an effective way of maintaining the desired weld penetration despite the large number of factors that affect it.

Weld penetration is influenced by: material composition, joint design, fitup, joint location and welding parameters such as current, voltage, travel speed, electrode diameter, shielding gas composition and torch-to-workpiece orientation.

Several feedback penetration control techniques have been developed. All of these techniques involve the use of a sensor to provide real-time information which is directly or indirectly related to weld penetration. This information is then compared to a predetermined value and one or more of the welding variables (current, voltage or travel speed) is adjusted to maintain penetration. The adaptive GTA welding system will use one or more of the following sensors to control weld penetration. The sensor will be selected to match the needs of the specific application.

2.4.1 Backside Light Sensing

The sensing of light intensity emitted from the backside of the weld is an effective penetration control technique. This approach typically uses a photo-detector to monitor the radiation from the backside of the weld pool. The amount of radiation is a direct measure of weld penetration and the output of the detector is used to adjust welding current or travel speed to maintain constant penetration.

While this technique is very effective, it requires access to the backside of the joint and can only be used in those situations where the backside of the weld joint is accessible.

2.4.2 Pool Viewing

EWI is developing two penetration control methods for use from the face of the weld. The coaxial vision system can be used to view the weld pool with either an infrared or visible light video camera and then use a computerized image processor to monitor the weld pool size or temperature distribution. A control algorithm relates the temperature distribution or desired pool size to the welding parameters in order to control penetration.

The primary advantage of the pool viewing technique is that the sensor does not contact the workpiece and does not intrude into the weld area. A disadvantage of the technique is that pool width is an indirect measure of weld penetration. A further disadvantage of the thermal sensor system relates to the varying thermal emissivity of the weld zone due to surface conditions that make accurate temperature measurements difficult.

2.4.3 Pool Oscillation Sensing

Another penetration control technique for GTA welding currently under development at OSU and EWI relies on detecting the oscillation frequency of the molten pool surface. As shown in Figure 4, the oscillation frequency of the weld pool for both partial and full penetration welds is related to pool size and can be sensed through the arc. After a short pulse of welding current is applied to excite the pool into oscillation, a computer analyzes the arc voltage and extracts the pool frequency. The welding current is adjusted to maintain a constant weld pool frequency and hence penetration. This technique is an indirect measure of penetration and also requires a very stable power supply to allow detection of the small pool motions.

2.5 Thermal Process Control Model

Feedback control of the cooling rate of an arc welding process is also under development at EWI and OSU.

Two control algorithms have been formulated and tested using a mathematical heat flow simulation model. The input to both of these control algorithms is the width of the weld pool which is obtained by the coaxial weld vision system. The size of the weld pool is directly related to welding parameters and is therefore indicative of the cooling rate of the weld. The output of the control algorithm changes one of three welding parameters to adjust cooling rate.

The operation of both models have been tested using computer simulation and results have been compared to a mathematical heat flow model.

The conventional control algorithm was found to be too sensitive to modeling errors to be of practical use. Therefore, a second, adaptive model was developed. The adaptive model adjusts arc efficiency in response to changes in weld pool dimensions. This adaptive feature improved the performance of the control system.

2.6 Monitoring of Parameters for Quality Control

In addition to the closed loop feedback control that is designed into each component of the adaptive GTA welding system, all welding parameters will also be monitored by a separate set of instruments to ensure that they do not exceed control limits. Sensors will detect welding current, wire feed speed, voltage, torch position, torch angles and travel speed of the welding heads. Data from these sensors will be analyzed in real time to insure that parameters are within expected tolerances. If parameters exceed the tolerance limits, the welding operation will be halted and corrective measures taken. The sensor data will also be recorded to produce a permanent record of all weld passes.

A laser-based vision system could be used to measure the contour of each completed weld bead and to compare the size and shape of the bead to that of a properly placed weld bead. The coaxial weld vision system could also provide a permanent video record of each weld pass.

A real-time NDE system based upon radiographic or ultrasonic techniques is another important part of the completely automatic, adaptive weld system. The NDE system would inspect the completed weld in real time and verify the quality while the part was still in the welding position. This would provide a final control on the proper performance of the welding system.

2.7 Off-Line Robot Programming

One of the present limitations on the productivity of arc welding robots is the need to generate robot programs by manually "teaching" each point of the robot path. Teaching is inefficient because the robot is not productive during teaching. Therefore, off-line programming can increase the productivity of robot welding by generating and testing robot programs while the robot continues to perform useful work. The ability to off-line program the adaptive welding system is a final important ingredient in achieving maximum productivity.

Off-line programming of welding robots has been feasible for a number of years. Nearly all robot systems have software to generate off-line programs. However, the off-line programming of welding robots has not achieved the level of sophistication that makes it useful for many applications. There are still problems that must be solved before off-line programs can be widely used. These problems involve inaccuracies in the off-line programs themselves as well as dimensional errors in the robot work cell.

- Fixtures, positioners and other components in the robot cell are not located accurately with respect to each other or to the robot. This creates location errors that are not accounted for by the off-line program.

- There are limitations in the accuracy of the construction of the computer model of the robot cell that is used to generate the off-line program.

- Finally, there are computational errors that occur during the mathematical manipulation of the off-line models which are used to create and simulate the robot programs.

Edison Welding Institute is supporting work to improve off-line robot programming (OLP) so that it is better suited for production welding operations. This OLP system, shown in Figure 5, includes elements for:

- The generation of robot paths from part geometry
- Simulation of robot motion
- Determination of the need for a weld vision system
- Selection of welding procedures

- Incorporating welding parameters into the robot program
- Design of weld fixtures
- Design of the robot work cell
- Calibration routines for vision systems, torch alignment and welding parameters
- Economic justification of the robot cell
- Determination of economic batch sizes.

Software is being written as a part of this project to perform a number of new functions that are needed by OLP systems. Algorithms are being developed for welding parameter definition, optimum positioning of the weldment in the robot envelope, design of weld fixtures and the generation of sensor data. The system operator will select icons on a graphic display to perform these functions. The software interface between the user, the OLP system and the robot is also being improved.

One portion of this OLP system will determine the welding parameters for each joint. Material type and thickness, welding position, shielding gas and electrode type will be automatically identified for each weld. Other algorithms will then select welding current, voltage and travel speed and this information will be transferred to the robot program.

While the repeatability of robot arms is generally superior to that of the parts being welded, the degree of robot accuracy which is essential for off-line programming is still only available on a few robots. The need for weld sensors or vision systems to compensate for the inaccuracies of the parts, programs and work cell layout has already been discussed. A basic question to be answered by the off-line programming system is: "Does this particular job require a sensor or vision system?"

The OLP system under development will evaluate the need for a welding sensor or vision system based on the accuracy and repeatability of the robot, piece part tolerances, fixture tolerances and welding process tolerances. The accuracy and repeatability of the robot work cell will be compared to that which can be accommodated by the welding process. If the process tolerance is less than that of the work cell, then a sensor will be required to track the weld joint and adjust the robot path during welding. If a weld sensor or vision system is to be used, the off-line programming system must be able to generate the commands that are required by the sensor as well as the robot.

3.0 CONCLUSIONS

Automated or robotic welding of precision weldments can produce the accuracy and control necessary to meet high-quality requirements and, at the same time achieve effective levels of productivity. This paper has described the components of an adaptive GTA welding system that can be applied to both robotic and automated welding of precision parts. This adaptive system will overcome the current limitations of part accuracy, work cell tolerances and errors in off-line programs. The system includes sensors and control algorithms for joint tracking, penetration control and weld fill control.

This advanced system will solve many of the current technical problems and continue to expand the application of automated and robotic welding to precision weldments.

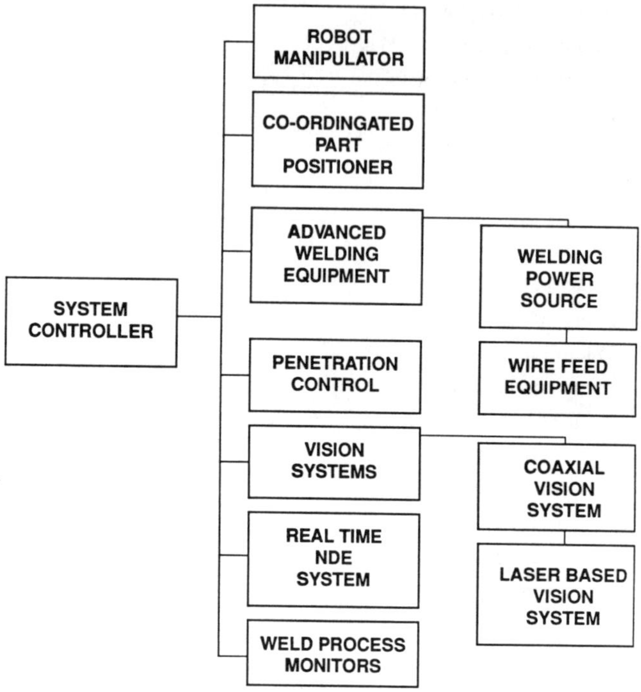

FIGURE 1. ADAPTIVE GTA WELDING SYSTEM

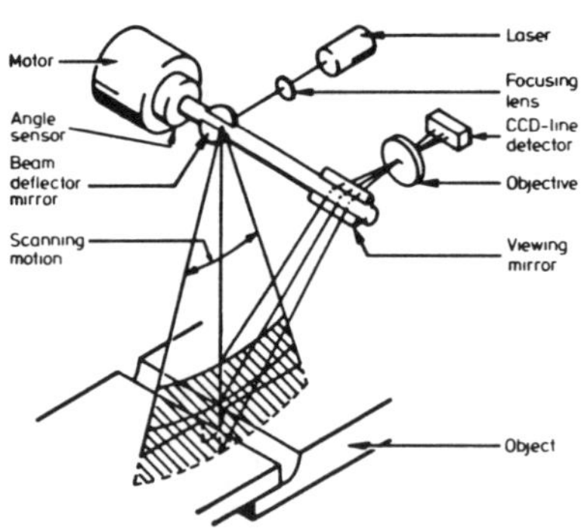

FIGURE 2. LASER OPTICAL JOINT SENSING SYSTEM

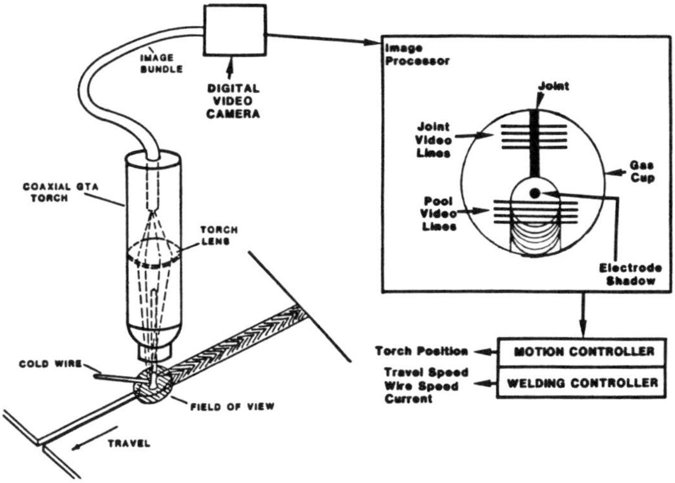

FIGURE 3. COAXIAL WELD VISION SYSTEM

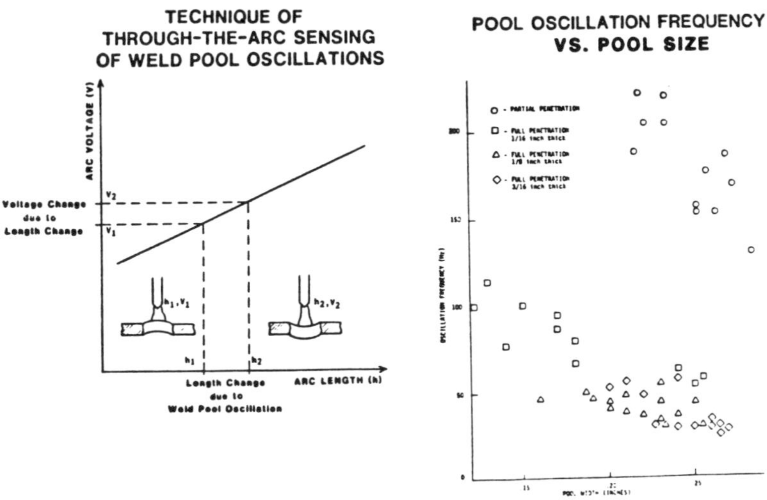

FIGURE 4. PENETRATION CONTROL BY WELD POOL OSCILLATION SENSING

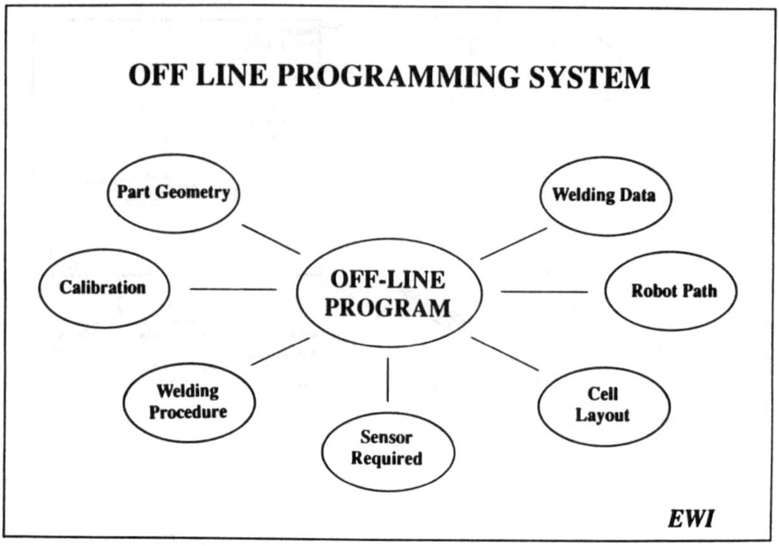

FIGURE 5. OFF-LINE PROGRAMMING SYSTEM

2nd International SAMPE Metals Conference
August 2-4, 1988

ANALYTICAL MODELING OF 3-D CASTING USING THE CAST3 FINITE ELEMENT CODE

Douglas R. Barker, Mark T. Samonds and Sokka M. Doraivelu
Universal Energy Systems, Inc.
4401 Dayton-Xenia Road
Dayton, Ohio 45432

ABSTRACT

The transient, nonlinear heat-conduction code CAST3 has been validated using a series of problems that range from very simple to highly complex geometries and boundary conditions. A set of support programs has been developed that assist the user in pre- and post-processing and in setting up the finite element runs. These programs include interfaces to several commercially available mesh generators and post-processing packages.

1. INTRODUCTION

Successful implementation of advanced process modeling in an industrial setting requires a numerical simulation code that has been carefully validated and is thoroughly supported by a system of ancillary programs. The validation of CAST3 has been carried out using a two pronged approach of developing a set of elementary validation problems as well as seeking out and running complex problems from industry. The ancillary programs that support the numerical simulation code are the only interface that the user encounters and thus they form the entire simulation environment. This environment is broken down into two parts: pre-processing and post-processing as shown in Figure 1.

2. PRE-PROCESSING

Pre-processing is the building of the model that will be input to the numerical simulation code. For the complex geometries that are often encountered in casting problems, pre-processing is a formidable task that can take up to several man-weeks. The first step is the construction of the finite element mesh that defines the geometry of the mold, the casting and, in some cases, the surrounding furnace. Next, the material properties such as thermal conductivity are supplied. Finally, the boundary

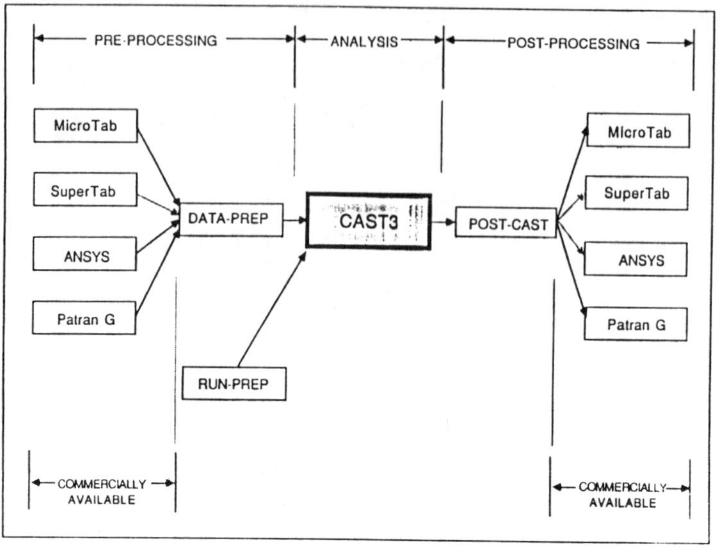

Figure 1. Casting Simulation System

conditions must be specified at the casting/mold interface and at the free surfaces.

The approach to pre-processing that is being implemented in the current project is to use existing software tools where they are appropriate and to develop new tools where they are required to speed up tasks that are particularly man-hour intense or unique to casting simulation.

2.1 <u>Mesh Generation</u>

One area where we can take advantage of existing technology is mesh generation. The extensive use of finite element modeling in structural analysis and product design over the past twenty years has fueled the development of sophisticated geometric modelers and mesh generators. In many vendor industries, these modelers are already in place to support CAD. For these reasons, the current project takes advantage of these existing tools for mesh generation by interfacing to several commercially available pre-processors.

Interfaces have been developed to take the geometry in the form of nodal coordinates and element connectivity from ANSYS, Patran, SuperTab and MicroTab and convert it into a CAST3 style input file.

2.2 <u>Interface and Surface Boundary Conditions</u>

An area where the commercial pre-processors are not suited to the casting problem is that of boundary condition specification.

The interface and surface boundary conditions that are required for casting simulation by CAST3 are quite different from what the popular pre-processors are designed to handle. For this reason, the program DATAPREP has been developed.

DATAPREP takes the geometry that has been read from the pre-processor files and performs several tasks. The most critical of these tasks is the specification of those element faces at which to apply mold/casting interface boundary conditions. This procedure has been fully automated. The DATAPREP code searches the finite element mesh for element faces that are shared by the casting and the mold materials. Whenever such a face is found, a coincident node interface is generated by replicating the nodes, assigning the new nodes to the mold material's element and modifying the element connectivity data. The coincident interface data that will be required by the CAST3 analysis code is then generated.

The DATAPREP code also locates the free surfaces on the exterior of the mold and generates the Neuman boundary condition data that are required there.

The sequencing of the nodes in the element connectivity data is checked by DATAPREP for satisfaction of the sequencing rules that CAST3 uses. Elements whose nodes are not properly sequenced are reordered automatically.

While DATAPREP performs the above operations automatically, some user interaction is required for specifying file names and what type of analysis is to be run by CAST3. Most of the user interaction is performed in a screen-oriented environment in which the user can jump between input blocks using his cursor. In this way, the user can proceed through the program quite quickly and easily, with a reduced risk of errors.

The program also displays key statistics about the model. By showing the user such critical numbers as the number of nodes, elements, coin- cident interfaces, free surface faces, node use frequency and min/max coordinates the program enables the user to spot errors such as unused nodes in the model setup.

2.3 Material Properties

The material properties that are required for the casting analysis are enthalpy as a function of temperature, thermal conductivity as a function of temperature, emissivity and density. DATAPREP can get these properties from two different sources. The first source is DATAPREP's own simple material property files. The user provides the materials' names and the corresponding properties are read in. The second source is a CAST3 style file from a previous problem or from the program MATLPROP.

The program MATLPROP was developed in response to solving industry problems for which ANSYS style files were available. The ANSYS files contain some but not all of the material properties that are required by CAST3. In addition to reformatting the data, MATLPROP converts the specific heat curves from ANSYS into the enthalpy curves that are required by CAST3.

2.4 Setup of Analysis Run

Every time that a finite element program is run, a number of parameters must be supplied by the user. These parameters include such things as whether the analysis is transient or steady state, which solver to use and what the time-stepping scheme is. The program RUNPREP generates the file that contains all this information. Like the DATAPREP program described above, the user interface in RUNPREP is screen oriented. The user steps through a series of screens in which he is given a description of the various parameters that are required. If the user wishes to accept the default values, he can go straight past the screen, without having to go through each and every response. In this way, the user can go in and modify a single parameter, such as the initial timestep, without running the risk of changing some other parameter by mistake.

3. POST-PROCESSING

Post-processing is the analysis and display of the data that has been generated by the numerical simulation code. The CAST3 code produces large files containing the temperature of the nodes at the end of all or some timesteps.

As was the case with pre-processing, there are several commercial programs available that do an excellent job on some of the post-processing tasks that are required for casting simulations. Several supporting programs have been developed that tailor the post-processing to casting.

3.1 Nodal Temperatures

The first thing that a user often wishes to see after a simulation run is a simple 2-D plot of several nodes' temperatures as a function of time. This plot shows at a glance whether or not the run was numerically stable, how the time-stepping scheme behaved and how much the rate of temperature drop has changed over the course of the simulation. In order to provide this plot to the user quickly and easily, the special purpose graphics program TRACE was developed. Graphics are so machine specific that slightly different codes were written for the Compaq 386 and the MicroVax. TRACE asks the user only for the name of the output file and the node numbers to be plotted. The graphics parameters will default to automatic values or can be overridden by the user. If the user does not want to specify particular nodes to be plotted, the program can plot all nodes or can select node numbers at regular intervals.

3.2 <u>Isotherms</u>

The commercial post-processors Patran and MicroTab can be used to provide 3-dimensional contour plots of temperature (isotherms) at a particular timestep. The reformatting of the CAST3 output data that is required is done using the program POSTCAST. These isotherm plots provide a snapshot view of what was happening at a selected moment during the casting.

3.3 <u>Isochrons</u>

The commercial post-processors can also be used to generate 3-dimensional contour plots of more useful parameters than nodal temperature, but some analysis is required first. The most useful plot is that of the time at which the casting passed though a particular temperature; these contour plots of time are commonly referred to as cooling curves or isochrons. If the temperature selected is the temperature at which the alloy is partially solidified, then the plot effectively shows the path that the solidification front took through the casting. This plot will help reveal any problems with necking.

The program COOL scans through the large data files output by CAST3 and generates the data required for the isochrons. This data is then output in a format that can be utilized by Patran or MicroTab for drawing the 3-dimensional contour plots.

4. VALIDATION PROBLEMS

A set of simple validation problems is essential to the development of any numerical simulation code. When changes are made to the numerical code or to the various support programs, the validation set is run to assure that updates are working properly. As new capabilities are added to the simulation system, new validation problems are added that isolate and test the new features. The problems in the validation set should be as simple as possible. Analytical solutions can be made to verify the numerical results for straightforward problems. If an analytical solution is not possible, at least a qualitative behavior can be predicted. Problem simplicity also keeps the solution times low, which encourages more frequent use.

The development of a set of validation problems has been initiated. The first problems in the set are simple geometric shapes that are particularly useful for validating data handling and boundary condition generation. The validation of thermal results for more complex problems requires that actual castings be poured with thermocouples recording the temperature at numerous locations. For this advanced validation work, we are turning to several different sources for experimental data. These sources include previous work by the investigators, problems supplied by industry, simulations documented in the literature and work conducted at universities.

5. BRAKE VALVE PROBLEM

A gravity die casting of a brake valve was simulated with the CAST3 code. The problem involves an aluminum alloy in a cast iron die, as shown in Figure 2. It is a fully three dimensional problem and was modeled with 645 nodes and 374 eight-node brick elements.

The brake valve problem was run using CAST3 on both the Compaq 386 and the MicroVax. Running the problem for 100 timesteps, which takes the simulation out to 4.4 minutes took 1 hour 5 min of CPU on the MicroVax. Results from the two runs were identical and agreed with previous work and thermocouple measurements.(1)

6. FUTURE WORK

As the CAST3 simulation code is further enhanced, especially in the area of fluids flow modeling, the validation effort will continue to expand. New problems will be added to the validation set in order to isolate and test new capabilities in a stepwise manner.

The simulation support programs that have been developed to date have been maintained as several distinct packages. These separate tools will now be tied together into a single menu-driven package that presents a more unified interface to the user.

7. REFERENCES

1. "Efficient Three Dimensional Finite Element Metal Casting

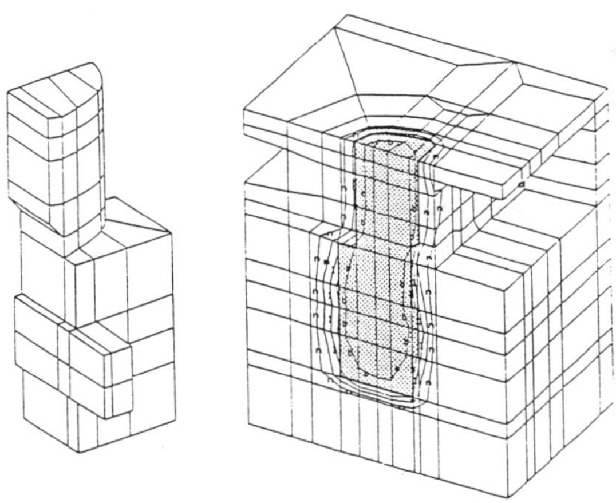

(a) Part Geometry (b) Solidifying Al and Mold

Figure 2. Brake Valve Simulation

Simulation Employing a Quadratic Conductivity Approximation"; M. Samonds, R.W. Lewis, K. Morgan, R. Symberlist; pp. 64-87; Numerical Methods and Thermal Problems, Vol. 4; published by Pineridge Press; 1985.

8. BIOGRAPHIES

Mr. Douglas R. Barker received his B.A. in Mathematics and his M. Engr. in Materials Science from Dartmouth College. He is currently Technical Program Manager for casting process simulation, design and consultation services at Universal Energy Systems.

Dr. Mark T. Samonds received his B.S. from the University of North Carolina in Physics and his Ph.D. from the University of Wales, Swansea in Numerical Methods. He is now the Technical Manager for Finite Element Modeling of heat transfer and fluid flow related to casting and plastics processing at Universal Energy Systems.

Dr. Sokka M. Doraivelu received his B. Tech. in Metallurgy and his M.S. and Ph.D. in Metal Forming from the Indian Institute of Technology, Madras, India. He is now Technical Director of the Processing Science Division of Universal Energy Systems.

2nd International SAMPE Metals Conference
August 2-4, 1988

DEVELOPMENTS AND TRENDS IN ELECTRODEPOSITION

J. W. Dini
Lawrence Livermore National Laboratory
Livermore, California 94550

Abstract

Electrodeposition is a very important technology. Covering inexpensive and widely available base materials with a thin (typically 0.1 to 10 μm) coating with specific properties extends their use to applications which otherwise would have been prohibitively expensive. Furthermore, the ability to provide property control in coatings as thick as 10 mm adds further value to this important technique of surface modification.

In recent years a number of advances in areas such as composite plating, pulse plating and high speed plating have added further credence to the use of electrodeposition as a viable coating technology. This paper will review these branches of the industry and discuss future trends.

1. INTRODUCTION

Electroplating is the deposition of a coating of a metal onto another metal or non-metal by electrolysis (and for the purpose of this paper in aqueous solutions). By use of this technique, the physical, mechanical and chemical properties of a surface can be modified to meet special requirements such as corrosion protection, ornamentation and engineering or functional requirements. Electrodeposition methods feature applicability to complex shapes, lower cost than many other coating processes, better control of deposit properties including internal stress, and the ability to deposit very thick coatings.(1) A variety of properties can be obtained with electrodeposits.(2) This is done by using different plating solutions, by varying electrodeposition parameters or by using additives. Advances in recent years in areas such as composite plating, pulse plating and high speed plating have provided the opportunity to further enhance deposit properties or produce new materials. The purpose of this paper is to review some of these new aspects of the technology.

2. ELECTROCOMPOSITES

The subject of composites is familiar to all members of SAMPE. Sessions on composites are normally available at any SAMPE meeting and the journals of this organization continually print articles on this topic. One aspect of this subject that is perhaps less familiar to SAMPE members is composites produced by using electroplating or electroless deposition. These less common methods of producing composites do not require high temperatures or pressures that tend to damage the fibers and allow for production of a variety of forms without machining.(3)

The American Electroplaters and Surface Finishers Society (AESF) has recently formed a committee on "Composites" in an attempt to bring together people involved with this technology and help promote its understanding and further use by industry. Those interested in this committee can obtain information from myself or Pete Greco, 110 Chapel Pl., Amsterdam, NY 12010.

Publication of reports on electrodeposited composites began nearly 25 years ago, but commercial applications are still relatively limited, despite the low costs associated with fabrication methods and the encouraging available property data.

Codeposited particles of powdered ceramics such as alumina and silicon carbide generally enhance wear resistance. Such particles can contribute up to 30 percent by vol. of the composite. For most applications particle size ranges from less than 1 to 10 µm. Relatively smooth surfaces can be obtained with the preferred small particles. However, deposits containing larger particles can be polished. Applications include metal forming dies, oil-well tubes and textile machinery parts.[4] A summary of electrocomposite categories can be found in Table 1.

The most common method of producing electrodeposited or electroless composites is by addition of dispersed particles with a maximum diameter of 1 or 2 µm to the plating solution. Submicrometer particles are even better when obtainable. An effective technique for keeping the particles dispersed uniformly in solution is important. Mechanical stirring is used in some cases, but in others the solution containing the particles is pumped to the plating solution at high velocity to maintain an effective dispersion.

Electrodeposited coatings typically are not effective for parts requiring wear resistance at high temperatures. However, the use of composite coatings offers promise at high temperatures. A composite containing 20 to 25 percent by weight Cr_2O_3 in a cobalt matrix exhibited excellent resistance to adhesive wear at temperatures of 300 to 700°C, due to the formation in

TABLE 1—ELECTROCOMPOSITE COATINGS CAN BE CATEGORIZED IN SIX CLASSES

APPLICATION	PARTICLES
Corrosion resistance—bright nickel overlaid with thin Cr	$BaSO_4$
Wear resistance—strongly influenced by wear properties of hard codeposited material	Al_2O_3, TiC, WC, SiC, CrC_3
Dry lubricant coatings—soft particles with a low shear strength dispersed in the metal matrix provide good antifriction properties	MoS_2, graphite, $BaSo_4$, PTFE in electroless nickel
Heat-treatable metals/alloys—particles are codeposited and then the composite is heat treated	Cr with Ni, B with Ni, and Cr, Al, and Y with Co
Nuclear coatings—composites for possible use in fuel elements or as reactor materials	UO_2 and Pu with Ni, B with Ni
Mechanical properties—improved strength at room and elevated temperatures	Al_2O_3 in gold, Al_2O_3 and TiO_2 in nickel

air of an oriented oxide layer of Co_3O_4.[5] This electrodeposited composite was much better resisting wear at elevated temperatures than composite coatings of Ni/TiC or Ni–P. Figure 1 shows wear data for a variety of composite coatings as a function of temperature. Irrespective of the particles used, nickel dispersion coatings exhibited more wear than the Co composites from room temperature to 700°C.

An electrodeposited composite coating consisting of a matrix of cobalt containing approximately 30% by volume chromium carbide powder (2–4 μm diameter) is capable of controlling certain forms of wear on aircraft engines at temperatures up to 800°C and has been used on tens of thousands of parts.[6] The wear resistance of the coating principally stems from the formation of a cobalt oxide glaze on the load bearing contact areas during interfacial motion.

A barrel plating process that employed a slurry of chromium, aluminum, and yttrium powders in a cobalt solution was used to coat turbine blades with CoCrAlY.[7] After heat treatment, these coatings performed better than plasma-sprayed CoCrAlY coatings during 600 hours of testing at 1373 K in a burner rig. The composition of the coating after heat treatment was ~20% Cr, 10% Al, 0.5% Y, balance Co, all in wgt. percent.

Another type of electrodeposited composite consists of several alternating layers of dissimilar metals. Plating is done by cyclic modulation of the cathodic current or potential during deposition. Copper-nickel cyclic multilayered alloy deposits were significantly stronger than simple combinations of the component materials taken independently. A room temperature tensile strength of 750 MPa (110,000 psi) was obtained for a 32.5 μm for a (0.0013 in.) thick foil consisting of alternate 0.05 μm (2 μm) thick layers of relatively low strength copper and nickel –245 MPa (36,000 psi) and 423 MPa (60,000 psi), respectively.[8,9]

Researchers at the National Bureau of Standards have shown that electrodeposition technology can be used to produce artificially modulated alloys with sharp interfaces. Copper and nickel or cobalt thicknesses ranging from 2 to 300 nm were continuous over their entire sample length. Optical microscopy showed that even disturbances—due for example to the presence of codeposited impurities—occurring in the planarity of the interfaces, healed out and the planarity of the interface was restored within a few layers.[10] Others have produced compositionally modulated alloys of Cu–Ni with 0.8 nm thick layers,[11] Ag/Pd films with repeat distances as small as 50 nm,[12] and Ni/Ni–P films with wavelengths between 2.1 and 4 nm and an average P content around 12 at.%.[13]

With the knowledge of the cyclic multilayered alloy electrodeposition technique gained from the above studies it should be possible to produce other alloys where beneficial efforts might be realized from the strongly anisotropic structure. One potential future application is the preparation of magnetic alloy films for information storage since magnetic properties are sensitive to both film thickness and composition. Other possible areas of application include the preparation of superconducting films, highly protective coatings for use in corrosive environments, layered structures with optical or improved properties such as toughness and wear resistance, new electroformed alloys and particle codeposition with the aim of producing superalloy properties at high temperatures.[12,14]

3. PULSE PLATING

Pulse plating, which is simply interrupted direct current-interrupted at very high frequencies, is becoming more and more useful as a tool for electrodeposition. Current pulsing is proving to be a powerful method for achieving properties and characteristics that cannot be matched using direct current.

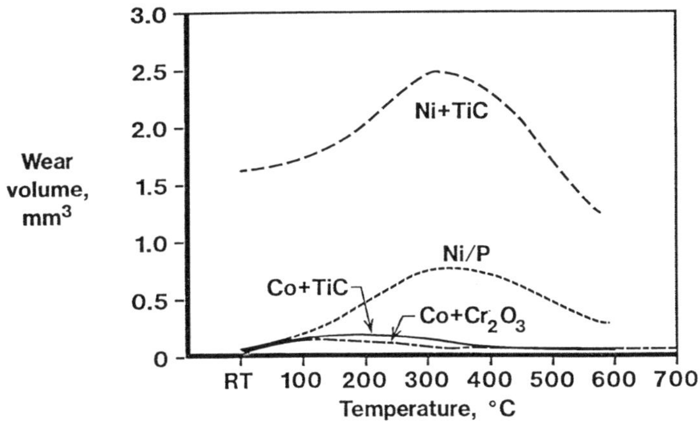

Figure 1—Wear resistance of a variety of electrocomposite coatings as a function of temperature (ref. 5).

Some of the potential benefits of pulse plating are summarized in Table 2. A number of recent symposia[15,17] and a book[18] have appeared on this subject. A few examples of applications of pulse plating will be discussed in the following paragraphs.

The life of pulse plated rhodium electrical contacts was improved 5–10 times by comparison with similar contacts plated with direct current and porosity and contact resistance were reduced.[19] Cracking of palladium deposits was reduced by the use of pulsed reverse current because codeposited hydrogen was anodically oxidized during the reverse cycle.[20]

With direct current plating, cobalt-tungsten deposits containing about 20% W were obtained but they were brittle and readily cracked on heating and deformation. Deposits produced with pulsed current were of similar W content, but crack free.[21]

Chromium deposits are often cracked and this results from thermal decomposition of a meta-stable chromium hydride. Substantial quantities of hydrogen are discharged at the cathode since the metal deposition efficiency is low. Some of the cathodically generated hydrogen atoms are incorporated into the metal lattice to produce a hydride and subsequent thermal decomposition of this hydride results in a highly strained deposit which spontaneously cracks. The adoption of pulsed plating techniques allows hydrogen atoms to combine to form hydrogen molecules rather than form metal hydrides, thus a stable lattice suitable for engineering application is obtained.[22]

4. HIGH SPEED PLATING

High speed plating is finding increased usage in a variety of plating applications and will continue to do so in the future. The plating of wire and steel strip with zinc or tin was one of the earliest applications of fast rate electrodeposition and these activities continue today. Many car companies have now selected electroplated coatings of pure zinc and various zinc alloys over hot dip galvanizing and other materials and treatments such as zinc rich paints and Galvaneal (heat treating to form a Zn–Fe alloy). As a result, in 1987 in the United States, five new zinc electrogalvanizing lines with an annual capacity of 2.4 million tons of steel strip were commissioned. In terms of zinc usage, these lines represent greater than a two fold increase from 45,000 tons in 1987 to nearly 100,000 tons by the early 1990's when they will be at full capacity.[23]

The increased demand in the electronics industry for high reliability along with the increasing cost of gold led to many developments in high speed, controlled deposition. Reel-to-reel units for selective plating on designated areas of a continuous strip or webbing of copper alloy account for many installations. In addition, high speed technology has been adopted for plating lead frames and the contact tabs of printed wiring boards.[24]

Fast rate electrodeposition (FRED) has been developed for a number of applications at Battelle Columbus Laboratories. FRED requires high mass transport of metal ions to the electrode surface (accomplished by turbulent electrolyte flow) and very high current densities to achieve a deposition rate 50 to 100 times faster than conventional plating. To apply high current densities of 3 A/cm^2 (2800 A/ft^2) or more, an electrolyte flow rate of 1.0 to 2.5 m/sec (3.3 to 9 ft/sec) past the cathode surface is required.[25] During one of the studies conducted at Battelle, ten metals were plated at fast rates using solution flow above 1.0 m/sec. The metals included Cr, Co, Cu, Au, Fe, Pb, Ni, Sn, Cd, and Zn and the current densities varied from 0.3 A/cm^2 for gold to 6.9 A/m^2 for iron. The corresponding plating rates ranged from 18 μm/min (0.7 mil/min) for gold to 150 μm/min (5.9 mil/min) for iron. These rates are 30 to 50 times greater than those for plating at customary current densities below 0.1 A/cm^2.[24] Properties of some fast rate and conventional electrodeposits are compared in Table 3. The tensile strength of copper, nickel and zinc plated at high speeds ranged to higher levels and this is attributed to a fine grained structure.

TABLE 2—POTENTIAL BENEFITS FROM PULSE PLATING
• Increase in rate of deposition
• Less danger of hydrogen embrittlement
• Reduction in the need for additives
• Lower level of impurities in deposit
• Finer grain size
• Lower stress in deposit
• Reduced porosity
• Improved uniformity

Data in Table 3 also reveal that electrodeposits produced at conventional rates tend to be denser than their high speed counterparts. However, some of the deposits produced at fast rate exhibited a lower electrical resistivity and in some cases this would be advantageous for electronic applications.[24]

High speed copper plating has been developed for commercial applications such as electroforming of flexible printed circuits[26] and in-mold plating of polymeric materials.[27] A process for electroforming copper circuit paths on a reuseable metal substrate and transferring the circuitry to a flexible polymeric surface coated with adhesive has been advanced to production. The copper circuit paths are electroformed at rates up to 50 μm/min using fast rate electrodeposition technology. The process does not generate spent etching solutions or plating rinse water and costs can be 70 to 80 percent of those associated with the conventional etched circuit process for high volume production. The properties of the flexible circuitry meet or exceed IPC standards.[26]

Electroforming of printing plates for stamps and currency requires 18 to 36 hours to deposit thick (0.9 to 1.8 mm) nickel. A successful investigation

TABLE 3—PROPERTY FOR FAST RATE AND CONVENTIONAL ELECTRODEPOSITS[1]

DEPOSITION MADE	METAL/SOLUTION	TENSILE STRENGTH MPa	ELONGATION %	DENSITY g/cm^3	RESISITIVITY microhm/cm
Fast Rate[2]	Copper/2M fluoborate	294 to 333	6 to 10	8.65 to 8.80	1.69 to 2.34
Conventional[3]	Copper/2M fluoborate	137 to 255	8 to 31	8.91 to 8.93	1.75 to 2.1
Fast Rate[2]	Nickel/2.75M chloride–sulfate	735 to 863	2 to 3	8.65 to 8.69[4]	6.84[4]
Conventional[3]	Nickel/2.75M chloride–sulfate	294 to 677	5 to 35	8.86 to 8.90[4]	7.44 to 8.90[4]
Fast Rate[2]	Zinc/2M sulfate	137 to 152	3 to 5	6.60 to 7.14[5]	7.6 to 8.2[5]
Conventional[3]	Zinc/2M sulfate	49 to 118	1 to 51	7.2 to 7.6[5]	7.5[5]

[1]From W. H. Safranek, ref. 24.
[2]Current densities of 1.5 to 3.1 A/cm^2 for copper, 0.5 to 2.5 A/cm^2 for nickel and 0.9 to 2.5 A/cm^2 zinc.
[3]Current densities of 0.03 to 0.05 A/cm^2 for copper and nickel and 0.015 A/cm^2 for zinc.
[4]Nickel/3.8M chloride–sulfate.
[5]Zinc/1.75M sulfate.

was carried out to accelerate the electrodeposition rate up to four times.[28] When forced solution flow and higher current densities were used without modifying the solution composition, physical properties of the nickel deposits were essentially unchanged (Table 4). Plates with dimensions of 7.5 × 8.6 dm were electroformed by the fast rate process using a current density of 20 A/dm².

Wear tests on cylinders of cast aluminum air conditioning compressors showed that fast rate electrodeposited chromium (deposition rates of 20 μm/min, 0.0008 in./min) performed as well as, or better than, conventional deposits (0.42 μm/min, 0.000017 in./min). The fast rate chromium deposit also had a lower hardness so it was easier to hone than conventional chromium.[29]

The combination of fast rate electrodeposition with interrupted current (FRIC) has been applied successfully to deposit metals and alloys with excellent properties.[25] FRIC offers a number of potential advantages including (1) a high plating rate due to the minimization of metal ion depletion and hydrogen evolution, (2) better control of crystal growth and other deposit properties, and (3) improved control over the ratio of metals and the relative amount of phases and compounds in the coatings. Some potential applications include coatings composed of chromium, cadmium and various alloys for the aircraft and automotive industries.[25]

Many more high speed plating applications can be expected in the near future to supplement those already adopted commercially. An excellent review on this technology can be found in the paper by Safranek.[24]

5. SUMMARIZING REMARKS

In this paper, a number of recent advancements in the field of electrodeposition have been discussed. It was shown that these items; composite plating, pulse plating, and high speed plating offer additional means of improving properties or producing new materials. With the advent of these methods of enhancing the electrodeposition process, continued advances can be expected from this already viable coating technology.

Work was performed under the auspices of the U.S. Department of Energy by Lawrence Livermore National Laboratory under contract No. W-7405-ENG-48.

TABLE 4—PROPERTY DATA FOR NICKEL DEPOSITS USED FOR PRINTING PLATES[1]

	CONVENTIONAL	HIGH SPEED
Ultimate	586 MPa (85,000 psi)	545–620 MPa
Yield	448 MPa (65,000 psi)	390–416 MPa
Tensile Stress	55 MPa (8,000 psi)	—
Hardness R 15T	90	89–92
Elongation	10%	7–13%
Plating Current Density	5 A/dm²	20 A/dm²

[1]From Schaer and Krasley, ref. 28.

References

1. Dini, J.W., "Electrodeposition—A Viable Coating Alternative," *Thin Solid Films* **95**, 123 (1982).

2. *The Properties of Electrodeposited Metals and Alloys,* 2nd Edition, Safranek, W.H., Editor, American Electroplaters and Surface Finishers Society, Orlando, Florida, 1986.

3. Greco, V.P., "Evaluation of Filament-Reinforced Electrocomposites," Proceedings, 31st International SAMPE Symposium, 1807 (April 1986).

4. Safranek, W.H., "Deposit Disparities," *Plating and Surface Finishing* **74,** 16 (Sept. 1987).

5. Thoma, M., "A Cobalt/Chromic Oxide Composite Coating for High Temperature Wear Resistance," *Plating and Surface Finishing* **71,** 51 (Sept. 1984).

6. Kedward, E.C., and Wright, K.W., "The Wear Control of Aircraft Parts Using A Composite Electroplate," *Plating and Surface Finishing* **65,** 38 (Aug. 1978).

7. Honey, F.J., Wride, V., and Kedward, E.C., "Electrodeposits for High-Temperature Corrosion Resistance," *Plating and Surface Finishing* **73,** 42 (Oct. 1986).

8. Tench, D., and White, J., "Enhanced Tensile Strength of Electrodeposited Nickel-Copper Mutlilayer Composites," *Metallurgical Transactions* **15A,** 2039 (Nov. 1984).

9. Ogden, C., "High-Strength Composite Copper-Nickel Electrodeposits," *Plating and Surface Finishing* **73,** 130 (May 1986).

10. Lashmore, D.S., Oberle, R., and Dariel, M. P., "Electrodeposition of Artificially Layered Materials," Proceedings of AESF 3rd International Pulse Plating Symposium, Washington, D.C. (Oct. 28–19, 1986).

11. Yahalom, J., and Zadok, O., "Formation of Composition—Modulated Alloys by Electrodeposition," *Journal of Materials Science* **22,** 499 (1987).

12. Cohen, U., Koch, F.B., and Sard, R., "Electroplating of Cyclic Multilayered Alloy (CMA) Coatings," *Jour. Electrochemical Soc.* **130,** 1987 (1983).

13. Goldman, L.M., Blanpain, B., and Spaepen, F., "Short Wavelength Compositionally Modulated Ni/Ni–P Films Prepared by Electrodeposition," *Jour. Applied Physics* **60** (4), 1374 (Aug. 1986).

14. Safranek, W.H., "Deposit Disparities—Crystal Balling," *Plating and Surface Finishing* **74,** 16 (Dec. 1987).

15. Proceedings of the AES First International Pulse Plating Symposium, Boston, MA (April 19-20, 1979).

16. Proceedings of the AES Second International Pulse Plating Symposium, Rosemont, IL (Oct. 6–7, 1981).

17. Proceedings of the AESF Third International Pulse Plating Symposium, Washington, D.C. (Oct. 28–29, 1986).

18. *Theory and Practice of Pulse Plating,* Puippe, J. C., and Leaman, F., Editors, AESF, Orlando, Florida, 1986.

19. Miuand, W.S., and Fung, Y.S., "Pulsed Current Electrodeposition of Rhodium," *Plating and Surface Finishing* **73,** 58 (March 1986).

20. Grimmand, R.D., and Landolt, D., "Investigation of the Influence of Pulse Parameters on Crack Formation in Palladium Electrodeposits," *Surface and Coating Technology* **31,** 151 (1987).

21. Barnes, C., "Electroplating of Cobalt–Tungsten Alloys," *Trans. Inst. Metal Finishing* **62,** 47 (1985).

22. Faust, C.L., Schaer, G.R., and Semones, D.E., "Electroplating Chromium Directly on Aluminum," *Plating* **48,** 605 (1961).

23. Sard, R., "Advances in Functional Zinc and Zinc Alloy Coatings," *Plating and Surface Finishing* **74,** 30 (Feb. 1987).

24. Safranek, W.H., "High Speed Electroplating," *Plating and Surface Finishing* **69,** 48 (April 1982).

25. El-Shazly, M.F., White, J.L., and Brooman, E.W., "High-Speed Depostion Using Interrupted Current," *Plating and Surface Finishing* **74,** 136 (May 1987).

26. Schaer, G.R., and Wada, T., "Electroformed Flexible Printed Circuitry," *Plating and Surface Finishing* **68,** 52 (July 1981).

27. Schaer, G.R., and Preston, J.R., U.S. Patent 4 **597,** 836 (1986).

28. Schaer, G.R., and Krasley, P., "Electroforming Accelerated by Forced Solution Flow," *Plating and Surface Finishing* **66,** 36 (Dec. 1979).

29. Dufrane, K.F., and Schaer, G.R., "Process Development and Wear Performance of Fast Rate Electrodeposited Chromium on Aluminum Castings," *Jour. Amer. Soc. of Lubrication Engineers* **41** (2), 75 (1984).

2nd International SAMPE Metals Conference
August 2-4, 1988

PLASMA APPLICATIONS FOR SPACE AGE METALS

M. P. Schlienger
Retech, Inc.

With the introduction of more complex alloys and the emphasis on zero defect material plasma melting is receiving considerable attention as related to high performance materials. Some of the early work as related to aerospace materials was accomplished under a Wright-Patterson Air Force Base contract in 1971. This work was carried out through Carnegie-Mellon Institute with the melting portion being conducted in San Rafael, California by the predecessor to Retech, Inc.

It should be pointed out that the results of the 1971 contract were somewhat disappointing because the initial thrust was to demonstrate the titanium casting capability of cold wall plasma melting. In the early stages of the contract it was discovered that plasma melting was primarily a shallow pool melter with pool characteristics somewhat similar to E.B. melting. With the above results the program was redirected to melting in a withdrawal crucible which proved feasible for titanium, zirconium, and nickle-base alloys.

For many years plasma melting was ignored by the Special Metals Industry until 1981 when Frankel Corp. had a plasma consolidation system built to enable the consolidation of a variety of titanium scrap sweetened with virgin material as required. The Frankel unit was a horizontal boat melter as depicted in Slide 1.

Following the Frankel unit "Oregon Plasma" installed a withdrawal type melter in late 1983 (Slide 2) this unit has been in successful operation for approximately four years and has the capability of producing first melt ingots from 14 inches in diameter to 28 inches in diameter and lengths up to 200 inches. Slide 3 (Ingots).

During this same period of time some of the U.S. aircraft engine manufacturers were engaged in active programs related to making powder products via the use of plasma. (Slide 4 Pratt & Whitney - Slide 5 Univ. of Illinois). It was also found that plasma melting had some real advantages in melting of metals and alloys having high vapor pressures that were not compatible to vacuum melting. With its relatively long arc characteristics, it has also proved to be effective in hearth melting of a variety of materials including the following:

(Slide 6)

1. Titanium and Ti Alloys
2. Zirconium and Zirconium Alloys
3. Nickel Base Alloys
4. Niobium and its alloys
5. Chromium
6. Copper Alloys
7. Super Conducting Alloys
8. Memory Metals
9. A variety of other special alloys with wide variations in vapor pressures.

In non-hearth applications, materials such as tantalum, tungsten, molybdenum, iridium and others have been melted. So it is safe to say that due to the pressure regime in which plasma torches operate materials from aluminum through tungsten can be melted. Difficulty of melting and maintenance of furnace hardware is pretty much proportional to melting temperature raised to some appropriate exponential power.

With the advent of plasma hearth melting in 1985 several tests have been made by a variety of people to determine the effects on high and low density inclusion removal. These tests have been very encouraging and I am sure the results have been and will be reported by people more qualified than myself. Slide 7 shows the two torch plasma hearth melter which has been used for a variety of hearth and non-hearth melting tests.

At this time Wyman-Gordon is starting up a large 2250 kW melter at its Worcester plant that will have the capability of melting ingots to 15,000 pounds and also for producing nickel base powder (Slide 8). An R & D hearth melter has also been installed at Pratt & Whitney, East Hartford (Slide 9). In addition, General Electric has been very active in pursuing plasma processing for both ingot and powder applications. It is expected that within the near future the plasma melting capacity in the United States will grow substantially. Primarily because of the ability to melt materials with a wide variety of vapor pressures and better handle a variety of volatile raw materials that cause problems in high vacuum systems.

With the emphasis on RSP powders for the next generation of engine and air frame applications, plasma has some very profound applications. Since powder production usually requires pressure melting, it is a regime in which plasma operates well and especially with its capability as a cold wall process. Like any of the advanced melting processes, plasma melting has its limitations but with the work that has been accomplished by the many parties presently involved, it is becoming more evident that this is a viable process for not only many of our present aerospace applications but also for some of the future requirements.

It is apparent that more use for the non-consumable melting processes and combinations of various processes will become more widespread in the future. Among non-consumable cold wall melting processes, the principle processes for consideration are as follows: (Slide 10)

Electron Beam Melting (Slide 11)

Cold Wall Induction Melting (Slide 12)

Rototrode Melting (Slide 13)

Plasma Arc Melting (Slide 14)

They, of course, all have certain advantages and disadvantages which must be considered for a specific application but they do enable optimizing of many of our aerospace requirements.

To this point this presentation has been somewhat general in nature but I would like to briefly mention some possibilities for future metals and perhaps non-metal applications.

Over the past few years work has been accomplished on a toxic waste reactor for treatment of extremely hazardous waste by the PCR system (patent pending). The PCR stands for Plasma Centrifugal Reactor and the mode of operation is as follows: (Slide 15)

Material is fed into a spinning reactor of a given diameter. The reactor has a cold zone in its outer diameter and a hot zone in the inner diameter confines. The melting mode is skull and the reactor material is water-cooled copper in the hot zone.

As you will note, there is a hole in the center of the reactor for downloading of molten material. We literally open and close material flow through this orifice by centrifugal force.

We feel that such a system may perhaps well be used for a number of metallurgical applications such as:

(Slide 16)

1. Production of titanium sponge.
2. Casting of reactive metals.
3. Consolidation of scrap.
4. Production of powder.
5. Carbothermic reduction.
6. Production of metallic and/or intermetallic alloys.

I am sure that the PCR must first be proven as a hazardous waste fixer but its operation to date has been intriguing and we believe it does have metallurgical merit. Although some critical design aspects must be worked out as related to certain mechanical requirements of the equipment, we do see pending application of the technology. (Slide 17) The original prototype of an 18 inch centrifugal reactor is shown in Slide 17. This equipment has been operating for approximately one year.

There are other aspects of hearth melting that we will also briefly discuss and this is related to non-linear hearths. The reason for hearth configuration other than the classic straight hearth approach is to accommodate material feeding and mounting of non-consumable melting devices such as the plasma torch or Rototrode. One of the disadvantages of these systems when compared to electron beam melting is the structure mass that must be moved or manipulated in a programmed motion. For the above reason, a variety of hearth configurations have been considered to enhance the feeding of material and manipulation of heating devices as related to advanced metal melting systems.

(Slide 18)

1. Rotary hearth design with inside material exit.

(Slide 19)

2. Rotary hearth design with outside material exit.

3. Centrifugal crucible design with center hole exit.

4. Rectangular or around the corner hearth to minimize electrode jamming.

Slide 18 relates to a rotary hearth design with inside material exit. This configuration enables material to be moved below the torch, thus expanding the range of plasma torch-to-hearth coverage and enabling remote feeding from the torch location. The pour lip moves in an arc of perhaps 180° with a receiving crucible or withdrawal crucible mounted on the central axis of the hearth. (Patents pending).

Slide 19 relates to a rotary hearth design with a water-cooled center post to prevent feed material from short circuiting to the crucible. In this configuration, material is removed from the hearth at pour lips located on the outer diameter. Such a system could be oscillated as above mentioned, or continuously rotated with multiple pour lips to feed the secondary mold or crucible. (Patents pending).

Slide 20 relates to around-the-corner, statically-mounted hearth design which is being applied to one of the more recent plasma systems being constructed. This geometry also incorporates a bulkhead to prevent feed material from short circuiting to the secondary crucible; in this case, a withdrawal crucible. The torches are mounted 90 degrees from each other on a fixed radius resulting in manipulation patterns that minimize overlapping of torch top structures; maximizing room for torch mounting and auxiliary equipment. I should point out that this geometry can also be applied to electron beam melting but is probably not as due to the high flexibility of the magnetically positioned electron beam.

Slide 21 relates to the overall arrangement of a furnace utilizing an around-the-corner hearth.

I am sure there are many pros and cons relating to what has been presented but for given applications these arrangements can be effective.

In conclusion, I would like to summarize some of the important characteristics of plasma melting that are applicable to the special metals industry. These characteristics are as follows;

(Slide 22)

1. Non-consumable melting tool.
2. Can be operated over a large range of pressures.
3. Used in cold wall or refractory lined furnaces.
4. Wide range of plasma gas chemistry that can be applied.
5. Relatively low voltage operation 100 to 600 volts.
6. Long directional arc with appropriate gas flow.
7. Expensive gases (helium) can be recycled.
8. Mobility of arc permits hearth melting.
9. Clean up capability.
10. Excellent for difficult alloy combinations.

Slide 1 - Frankel Plasma Melter

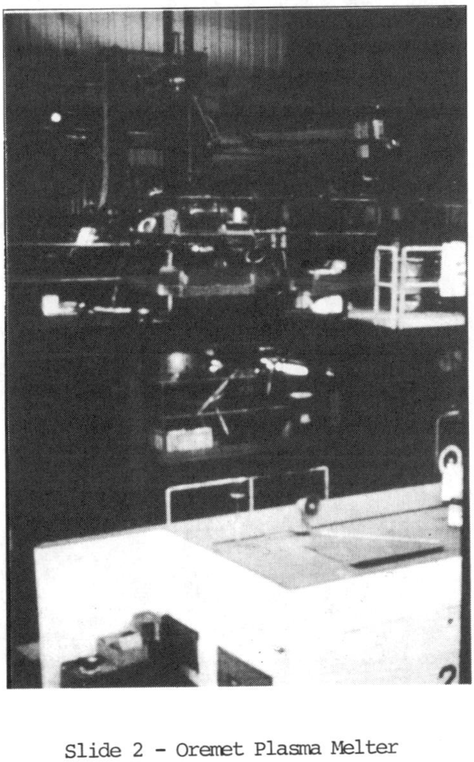

Slide 2 - Oremet Plasma Melter

Slide 3 - Oremet Plasma Melted Ingots

Slide 4 - P&W Plasma Powder Process

Slide 5 - Univ. of Illinois
Plasma Powder Process

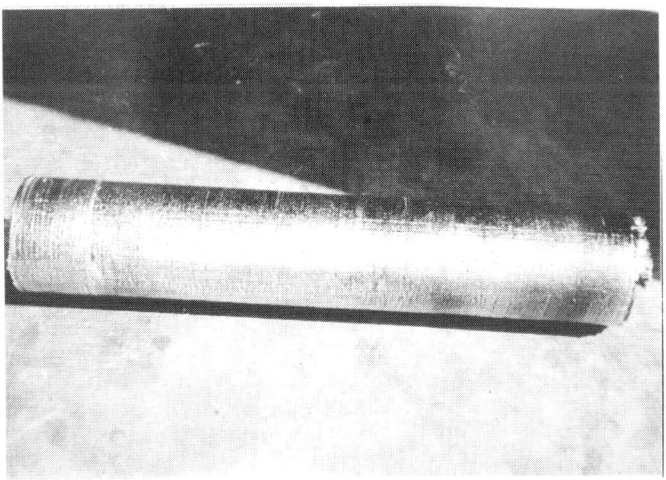

Slide 6 - Hearth Melted Metals
and Alloys

Slide 7 - Retech's Plasma Hearth Melter

Slide 8 - Not Available at
this time.

Slide 9 - P&W Plasma Hearther Melter

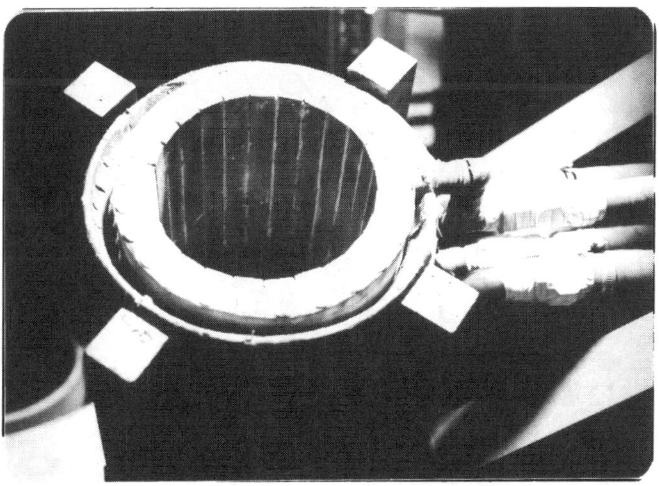

Slide 10 - Cold Wall Melting Processes

Slide 11 - Axel Johnson Electron
Beam Melter

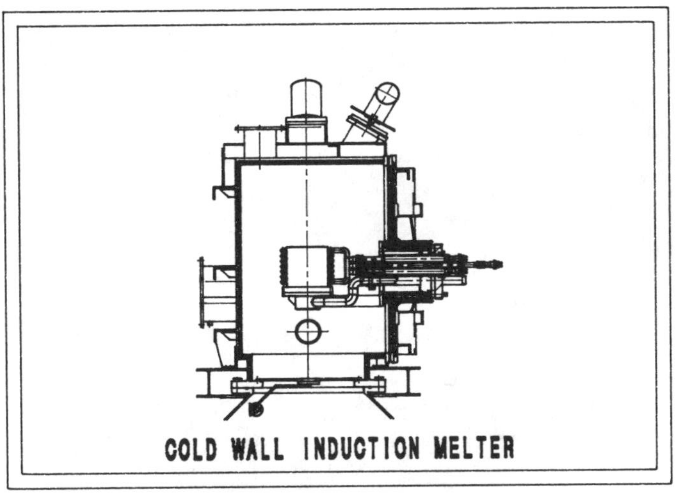

Slide 12 - Crucible Research Cold
Wall Induction Melter

Figure 13 - Rototrode Melting Equipment

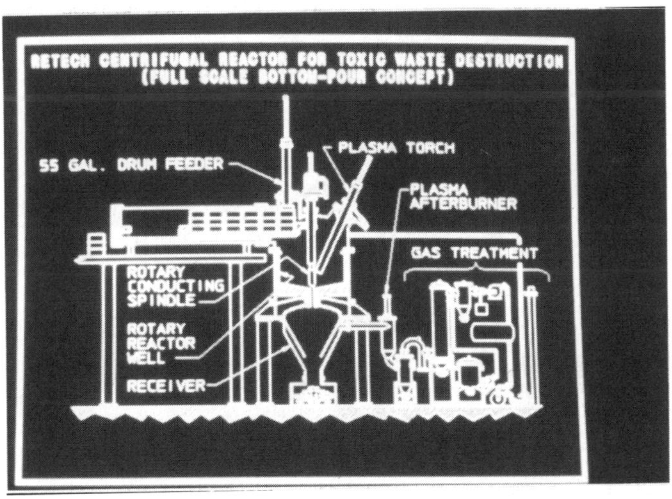

Figure 15 - Centrifugal Reactor

Slide 17 - Centrifugal Reactor
R & D Unit

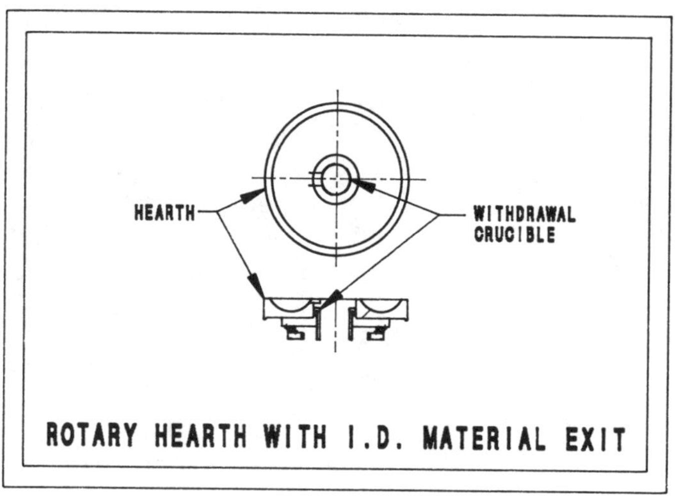

Slide 18 - Rotary Hearth with
I.D. Material Exit

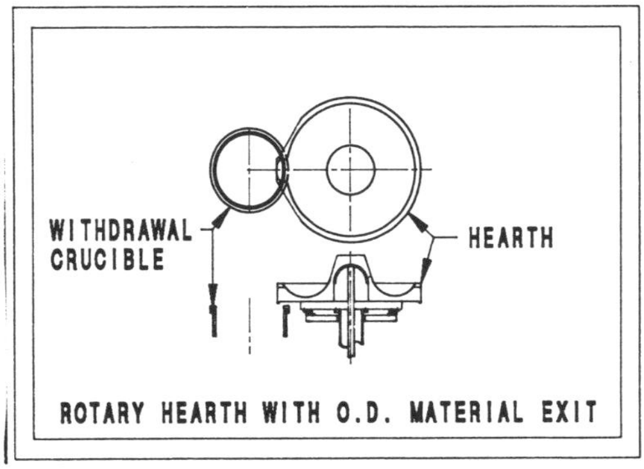

Slide 19 - Rotary Hearth with
O.D. Material Exit

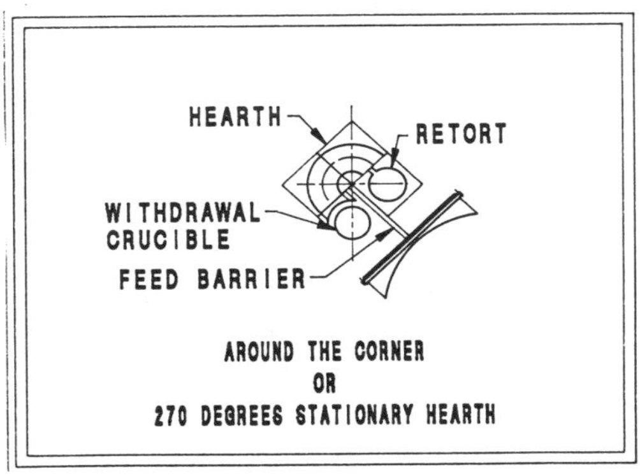

Slide 20 - Around-the-Corner or
270 Degrees Stationary Hearth

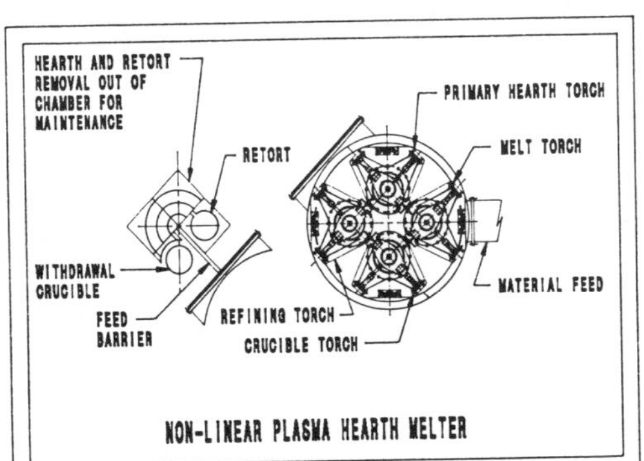

Slide 21 - Overall Arrangement of Non-
Linear Plasma Hearth Melter

2nd International SAMPE Metals Conference
August 2-4, 1988

PROCESSING AND MECHANICAL PROPERTIES OF LIGHTWEIGHT STRUCTURAL COMPOSITES

John J. Lewandowski
Department of Materials Science
and Engineering
Case Western Reserve University
Cleveland, Ohio 44106

Abstract

The properties of discontinuous metal matrix composites depend sensitively on a variety of processing parameters, including choice of reinforcement and type (e.g. plate, particulate, whisker), volume percent reinforcement, size and size distribution of reinforcement, as well as matrix characteristics and properties. Techniques available to encorporate reinforcement into potential matrices include casting, powder metallurgy, as well as spray deposition techniques. In addition, post-processing treatments such as thermomechanical treatment and heat treatment may be particularly important in producing optimum properties. This paper will review recent work which has focussed on structure-property relationships in aluminum alloy metal matrix systems.

1. INTRODUCTION

Considerable interest has lately been generated by the relatively low material and fabrication costs and approximately isotropic properties of discontinuously reinforced metal matix composites (MMC's). This is in part due due to recent advances in the manufacture of reinforcement materials,(1) although most recently there have been additional advances in processing techniques utilized to encorporate reinforcement into the metal matrices.(2-5) In general, reinforced composites may be classified into three types: a) continuously reinforced with long, continuous fibers of relativley large diameter (e.g. 100's μms) high strength material, b) reinforcement by whiskers (i.e. defect-free materials) of diameter ~ 1μm and discontinuous in nature, c) discontinuous reinforcement by particulates, ranging in size from ~ 1μm - <30μm. This review is based on the class of discontinuous composites covering the Al-SiC system, and will review recent works which illustrate the importance of the matrix microstructure on subsequent properties.

Although it was recognized 20 years ago that the mechanical properties of aluminum alloys discontinuously reinforced with whiskers could provide mechanical properties that were competitive with those of continuously reinforced materials,(7) the prohibitive cost of the whiskers prevented the use of such composites in a variety of structural applications. However, the recent development of relatively inexpensive techniques for producing SiC whiskers from rice hulls has renewed interest in discontinuous composites utilizing SiC whiskers. Recent reviews have summarized some of the current work on composites utilizing these whiskers(8,9). This paper will summarize some of those works in light of the more recent work on systems based on the Al-SiC whisker system. Subsequent discussion will focus on composites based on reinforcement using SiC particulate. The focus in the paper will be on works designed to determine the effects of microstructural changes on subsequent properties. The information contained in this paper has been obtained from published papers and from other sources of public information.

2. REINFORCEMENT CHARACTER

SiC whiskers derived from rice hulls are typically β-SiC with diameters ranging from 0.2 - 1.0μm and lengths up to 50μm. Although strengths of these whiskers may exceed three times the strength of SiC fibers(6), the commercially produced SiC product can also contain debris that can be as high as 50%, as noted in recent work(1). The whiskers themselves also may contain growth faults(10). SiC in the form of particulate can also be obtained in sizes ranging from ~1μm up to 30μm, while the particles may be highly irregular with respect to size and shape(11), although recent attempts at providing more regular sized and more nearly spherical particles is being pursued.(12) Encorporation of the reinforcement into the aluminum alloy matrix is often acheived via powder metallurgy techniques (i.e. PM) and its variations, and typically involves blending, cold-compaction of the SiC with the matrix alloy powder,degassing, and subsequent hot pressing (HP) of the green body. Hot pressing is often performed at temperatures exceeding the matrix solidus temperature to minimize the amount of reinforcement damage as well as enhance the bonding of the reinforcement to the matrix. Detailed Transmission Electron Microscopy(TEM) investigations of the interfaces in whisker reinforced materials processed via this technique have revealed the presence of amorphous layers at the reinforcement surface(13), as well as the formation of aluminum carbide in the as-processed material. Considerable reinforcement fracture has been observed in discontinuous composites processed via sub-solidus processing(11,15), although the particle matrix interfaces in sub-solidus processed materials were free of aluminum carbide. Various liquid metal infiltration techniques have also been utilized, in addition to proprietary casting techniques (4,14).

Subsequent processing of these materials includes extrusion which may cause additional reinforcement breakage, alignment of the reinforcement and "banding" of the reinforcement into regions that are SiC-rich and SiC-lean. Subsequent hot-rolling or cross-rolling has been utilized in an attempt to produce a more homogeneous dispersion of reinforcement.

While considerable work has investigated the variety of techniques for encorporating the reinforcement into the matrix, recent work has also focussed on the development of new techniques to disperse ceramic reinforcements into a variety of matrices. These consist of spray deposition techniques(5), in-situ formation of reinforcement in the molten metal(2), and squeeze casting(16) techniques which may be utilized on composites that are produced using casting techniques(4,14) where the composite is supplied for remelting. Use of the composite material requires subsequent solution treatment and aging treatments to produce a material with optimum performance for the properties of interest. While the effects of heat treatment on the properties of unreinforced aluminum alloy matrices have been extensively studied and recently reviewed(17), only recently have more controlled studies been initiated to investigate the effects of changes in aging conditions and heat treatment on the mechanical behavior of discontinuously reinforced materials.

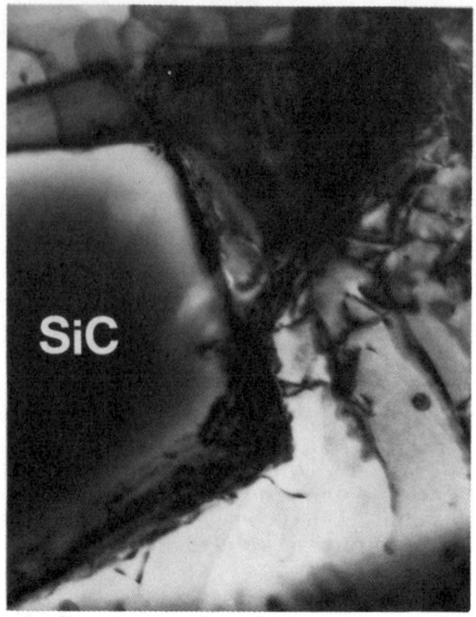

Figure 1

3. MICROSTRUCTURAL DEVELOPMENT

Although it would be convenient if the heat treatments designed to produce optimum properties in monolithic aluminum alloys would similarly produce optimum properties in composites based on the same matrices, recent works (9,14) have indicated that the prescribed treatments for the monolithic alloys may not be suitable for the composite. A variety of recent works have begun to investigate the aging response and mechanical properties of composites, where identical studies have been conducted on the monolithic matrix for comparison(11,15,18-22). In general, it has been found that accelerated aging at both room temperature and at elevated temperature is obtained in composites which were solution treated and quenched. The source of this accelerated aging response in many of the Al-SiC systems has been attributed to enhanced diffusivity in the as-quenched product due to a high dislocation density in the matrix. It is proposed that the high dislocation density results from the difference in thermal expansion of the matrix and the reinforcement(23).Figure 1 provides a TEM image of the interface and matrix region in an extruded high strength aluminum alloy reinforced with SiC particles.(11) The matrix exhibits a high dislocation density in the as-extruded composite as well as in the as-quenched condition. Although TEM work has been conducted on aluminum MMC's produced via a variety of techniques (e.g. PM(11,13,15, 19,20), spray cast(5), hot pressed(24)), containing either whisker or particulate reinforcement, the details of the precipitation sequences have been shown to be sensitively affected by the processing technique. For example, precipitation at the particle/matrix interfaces was observed in the PM composites produced by sub-solidus consolidation(11,15,19) for certain heat treatments, while no evidence of precipitation at interfaces was obtained in composites produced by spray deposition(5) or in whisker composites produced via super-solidus processing. Although work is continuing in this area, it is likely that relatively subtle changes in processing techniques may produce significant changes in the details of the aging response and nature of precipitation reactions in these materials. Thus, generalizations as to the effects of reinforcement on aging response must be cautioned.

4. MECHANICAL PROPERTIES

As indicated in the recent reviews of composites based on aluminum alloy matrices(8,9), much of the data on these materials has focussed on properties such as stiffness, strength, and ductility. Fewer works have focussed on the fracture properties in general, or on the role of microstructure in particular. The remainder of the paper will review some of the more recent work on the effects of microstructure on the micromechanisms of fracture in these materials and will cover areas such as toughness and fatigue.

Recent works which have investigated the toughness of aluminum based MMC's have revealed that fracture is typically macroscopically brittle, while the fracture surfaces exhibit micro-void coalescence(MVC)(11,15,24-28). Figure 2 presents the fracture surface of an aluminum alloy composite reinforced with SiC particles.(15) In general, it has been observed that the toughness of the composite may be considerably lower than that of the unreinforced matrix, although recent results indicate that the matrix microstructure may have a significant effect on the magnitude of the difference between the properties of the monolithic material and that of the composite.(15) In that work,(15) the effects of particle clustering and changes in matrix

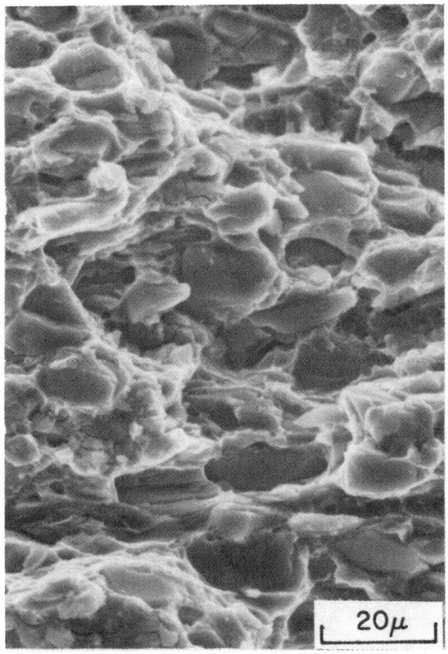

Figure 2

microstructure on the uniaxial tensile properties and various measures of toughness were investigated. The degree of SiC particle clustering was varied by choosing different sizes of reinforcement, while the degree of clustering was quantified by using recently available quantitative techniques to characterize the spatial distribution of SiC. (29) Figure 3 presents the results of short-rod fracture toughness specimens tested for different matrix aging conditions and for the different particle size distributions.(15) The microstructures tested consisted of specimens heat treated to either underaged(UA), peakaged(PA), or overaged(OA) conditions, while the reinforcement level was 20% and either consisted of an average particle size of 5μm or 16μm. Trends similar to those exhibited in Figure 3 were similar trend was observed in blunt notched specimens,(19) where the notched toughnesses were similarly higher in the UA material.

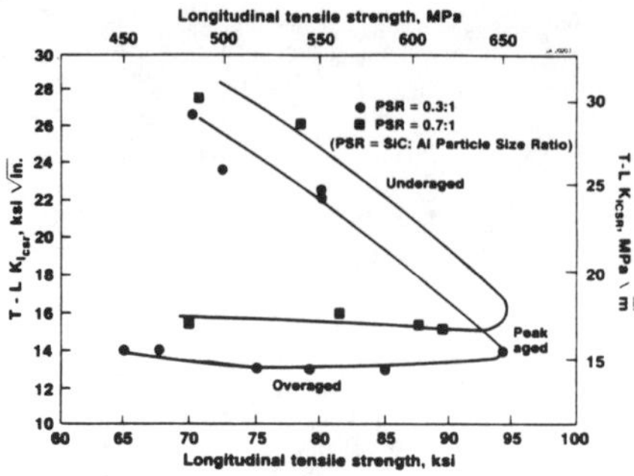

Figure 3

path of already fractured SiC particles.

Interestingly, the toughness of the UA material was significantly higher than that of the OA material, despite nearly identical uniaxial tensile properties(11,15,19). It was shown by TEM and by quantitative fractography that this change in properties accompanied a change in fracture mode from predominantly SiC fracture in the UA material to failure near the SiC/matrix interface in the OA material. Quantitative fractography revealed that large SiC particles preferentially fractured in the UA material. In an attempt to determine the details of fracture initiation and propagation, a specimen was designed to permit in-situ observations of fracture initiation and propagation. The details of the specimen are described elsewhere(15), while the intent was to slow the crack sufficiently to enable in-situ monitoring of fracture initiation and growth. It was additionally observed that the composite manufacturing process often fractured many of the SiC particles. Although fractured SiC particles were observed on the fracture surfaces, it was not clear if those particles had fractured during testing. Thus, an additional interest in utilizing the in-situ fracture studies was to determine if fracture preferentially follows the path of already fractured SiC particles.

The in-situ fracture studies on a 7XXX series (i.e. Al-Zn-Mg-Cu) composite revealed that the previously unfractured SiC particles preferentially fractured in the UA material, followed by subsequent failure in the matrix. Despite a population of already fractured particles (i.e. due to processing) fracture was not observed to follow those already fractured particles. The OA materials, in contrast, exhibited both matrix failure and failure near the SiC/matrix interfaces. Figure 4 shows a typical crack tip region in the UA material, while Figure 5 shows failure at and near the interfaces in the OA material. Work is continuing on specimens aged to the other conditions shown in Figure 3 to determine the micromechanisms of crack initiation and propagation. From that work, (15) however, it was demonstrated that the properties of the matrix microstructure may dominate the fracture behavior of the composite. Although the degree of particle clustering was varied by the selection of different reinforcement sizes, this did not appear to have a major effect on the properties obtained in that work.

Other recent work(26,27) compared the toughnesses of whisker reinforced to particulate reinforced material, where both materials had the same levels of reinforcement. In both cases, the particulate material exhibited a higher toughness. Unfortunately, it is difficult to directly compare the results obtained on these materials since the strengths were lower in the particulate reinforced materials.

Figure 4

Figure 5

As with the monotonic fracture behavior, the fatigue properties of a variety of MMC's have been investigated (26,30,31), although only recently has there been emphasis placed on determining microstructural effects on fatigue crack initiation, threshold, and crack growth behavior.(32,33) While fracture under monotonic conditions in particulate reinforced materials often occurs via fracture of the SiC (cf. Fig. 2) or failure near the particle/matrix interfaces, recent work has shown that at low ΔK levels, crack growth occurs predominantly in the matrix, regardless of the aging condition of the matrix.(34) Figure 6 shows the fracture surface obtained at low ΔK in an underaged 7XXX aluminum alloy. No evidence of failure of the SiC particles is obtained until higher levels of ΔK are approached during the test. At the higher ΔK levels, static modes of fracture (e.g. SiC fracture in underaged material) begin to predominate.

Previous work(30) on the smooth-specimen fatigue endurance properties of 6061 heat treated to the peak strength condition revealed improvements in the fatigue endurance properties in whisker reinforced material in comparison to particulate reinforced material. More recent work has focused on systematic studies of the effects of controlled changes in matrix microstructure on both the fatigue threshold as well as crack growth. It was shown(33) that a whisker reinforced composite exhibited fatigue thresholds roughly two to three times that of the unreinforced matrix. Other recent work(32) on particulate reinforced materials showed that fatigue thresholds in the composites were somewhat lower than those of the unreinforced matrix. This was attributed as due to the lower levels of fatigue crack closure in the

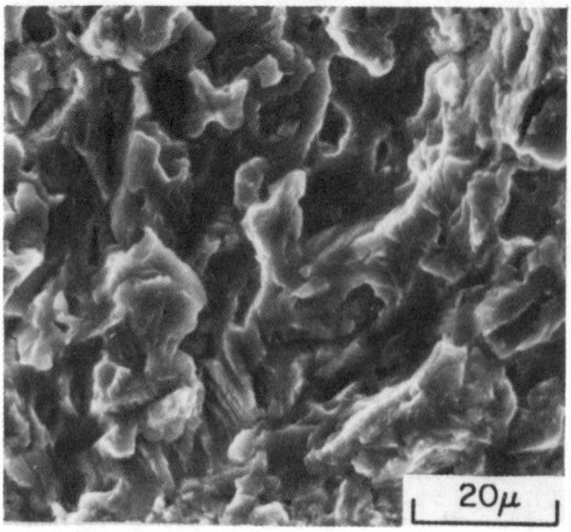

Figure 6

composites.(32) At higher stress intensities, growth rates in the composites were somewhat lower than that of the unreinforced alloy, which was attributed to crack tip shielding effects occuring in the wake of the crack.(32)

5. SUMMARY

Considerable potential exists for the utilization of MMC's in a variety of structural applications. Property improvements in the areas of fatigue and fracture toughness are continuing to be made while significant progress has already been made with the economic feasibility of these materials. Continued research into the micromechanisms of fracture initiation and propagation and the role that microstructure has on controlling these events remains an important area for additional research.

6. ACKNOWLEDGEMENTS

The support of an AIME Research Initiation Grant for the preparation of this document is gratefully acknowledged, while thanks are extended to W.H. Hunt at ALCOA for supply of materials and discussions on collaborative work reported herein.

7. REFERENCES

1. Divecha, A.D., et al. J. Met. 33, 12 (1981).
2. Westwood, A.R.C. and Winzer S.R., Advancing Materials Research, National Academy Press, 1985, pp. 225-243.
3. Mortensen, A., et al. J. Met. 40, 12 (1988).
4. Schuster, D., Presented at the 9th Meeting for Discontinuous Composites, Park City, Utah, (1987).
5. Warner, T.J. et al., Composite Interfaces, Pergamon Press, 1988.
6. Yajima, S. et al., J. Mat. Sci., 16, 3033, (1964).
7. Parrat, N.J., Powder Met, 14, 152, (1964).
8. Divecha, A.D. and Fishman, S.G., SAMPE Q., 40, (1981).
9. Nair, S.V. et al, Intl. Met. Rev. 30, 275, (1985).
10. Arsenault, R.J. and Fisher, R.M., Scr. Met., 17, 67, (1983).
11. Lewandowski, J.J. et al, Powder Metallurgy Composites, TMS-AIME,(1987) in press.
12. Norton Company, private communication, (1987).
13. Nutt, S.R. and Carpenter, R.W., Mat. Sci. Eng., 75, 169, (1985).
14. Skibo, M.D., Rep.SAND 81-8212, Sandia Labs, Livermore, Calif., June 1981.
15. Lewandowski, J.J. et al, Mat. Sci. Eng. (1988) in press.
16. Verma, S.K. and Dorcic, J.L., Adv. Mat. and Proc, 5, 48, (1988).
17. Vasudevan, A.K. and Doherty R.D., Acta Met., 35, 1193, (1987).
18. Flom, Y., Ph.D. Thesis, Univ. of Maryland, (1988).
19. Lewandowski, J.J. et al., Composite Interfaces, Pergamon Press, 1988, in press
20. Christman, T. and Suresh, S., Acta Met., 1988, in press.
21. Nieh, T.G. and Karlak, R.F, Scripta Met., 18, 25, (1988).
22. Harrigan, et al, 112th AIME Meeting, Atlanta, March, 1983.

23. Vogelsang, M. et al, Met. Trans. A, 17A, 379, (1986).
24. Flom, Y. and Arsenault, R. J., J. Met., July, 31, (1986).
25. Divecha, A.D. and Fishman, S.G., Proc. 3rd Int. Conf. on Mech. Behav. Matl's., Pergamon Press, 1980, 351.
26. Logsdon, W.A. and Liaw, P.K., Westinghouse R and D, Paper 83-1D7-NODEM-P1, Pgh. Pa., Dec, (1983).
27. Crowe, C.R. and Gray, R.A., Failure Mechanisms in High Performance Materials, Cambridge University Press, 1984, pp. 157-166.
28. Shang, J.K. and Ritchie, R.O., Mat. Sci. Eng., (1988), in press.
29. Wray, P.J. et al, Metallography, 16, 39, (1983).
30. Crowe, C.R. and Hasson, D.F., Strenth of Metals and Alloys, Pergamon Press, 1983, p. 859.
31. Tien, J.K. and Purushothaman, S., Mat. Sci. Eng., 34, 247, (1978).
32. Shang, J.K. and Ritchie, R.O., Met. Trans. A, (1988), in press.
33. Christman, T. and Suresh, S., Mat. Sci. Eng. (1988), in press.
34. Liu, C. and Lewandowski, J.J., Unpublished results, (1988).

8. BIOGRAPHY

John J. Lewandowski received his B.S., M.S., and Ph.D. degrees in Metallurgical Engineering and Materials Science from Carnegie-Mellon University in Pittsburgh, Pa. He subsequently spent two years as a N.A.T.O. Postdoctoral Fellow at the Department of Metallurgy and Materials Science, University of Cambridge, England. In 1986, he became an Assistant Professor in the Materials Science and Engineering Dept. at Case Western Reserve University,Cleveland, Ohio.

2nd International SAMPE Metals Conference
August 2-4, 1988

VARIABLE-POLARITY PLASMA-ARC WELDING OF ALUMINUM-LITHIUM ALLOY 2090

Paul T. Wood
United McGill Corporation
Columbus, OH 43216

W. A. Baeslack III
Department of Welding Engineering
The Ohio State University
Columbus, OH 43210

Abstract

The process and metallurgical aspects of variable-polarity plasma-arc welding (VPPAW) aluminum-lithium alloy 2090 have been investigated. Results found that with proper inert gas shielding, preweld cleaning and welding parameter selection, alloy 2090 could be effectively joined with this advanced process. Metallurgical structures and mechanical properties were determined for welds produced with different filler metal/postweld heat-treatment combinations. Optimum mechanical properties were obtained for a 5356 filler metal weld subjected to a postweld solutionizing and aging treatment. This combination resulted in a joint efficiency of 68 percent. Weld defects included occasional porosity and heat-affected zone microfissuring.

1. INTRODUCTION AND BACKGROUND

The addition of lithium to aluminum alloys reduces density and increases elastic modulus, resulting in strong, lightweight materials that provide the aerospace industry with an alternative to composites. Alloy 2090 is a promising aluminum-lithium alloy currently experiencing increased commercial utilization by the aerospace industry. The nominal chemical composition and typical mechanical properties of this alloy are summarized in Table 1 (1). Alloy 2090 is a heat-treatable alloy in which strength is provided by a complex precipitation reaction involving the coprecipitation of T_1-prime, T_2-prime and delta-prime(1). Zirconium is added as a dispersoid-forming element which retards subgrain boundary migration and

Table 1: Characteristics of Aluminum-Lithium Alloy 2090 - T8E41

Alloying Elements
2.7% Copper
2.2% Lithium
0.12% Zirconium

Yield Strength: 559 MPa [81 KSI]
Ultimate Tensile Strength: 593 MPa [86 KSI]
Elongation: 9%
Density: 2.594 gm/cm^3 [0.0937 lbs./in^3]
Fracture Toughness (KIc): 1034 MPa (m)$^{0.5}$ [30 KSI (in.)$^{0.5}$]

coalescence and thus suppresses recrystallization (1). In addition, zirconium would be expected to promote grain refinement in the weld fusion zone.

Variable-polarity plasma-arc welding (VPPAW) is a relatively new variation of the keyhole plasma-arc welding process which offers several advantages over gas tungsten-arc welding (GTAW) and is, therefore, experiencing increasing use in the aerospace field. Advantages over GTAW include: complete joint penetration, less required joint preparation, improved weld quality, decreased distortion and lower weld costs (2).

Plasma-arc welding produces coalescence of metals by heating with a constricted arc which promotes high energy densities and plasma gas velocities. These arc characteristics permit deep penetration welding to be performed in the keyhole mode. In this mode of welding, a cavity is formed through the base metal thickness at the leading edge of the weld pool. As this cavity progresses along the joint, metal melted by the arc plasma is forced to flow around the plasma stream, along the molten sides to the rear edge of the pool, where the weld is formed. Filler metal, when used, is added to the leading edge of the keyhole. An auxiliary source of shielding gas is used to protect the weld from atmospheric contamination. The process is schematically shown in Figure 1.

Previous attempts to plasma-arc keyhole weld aluminum were generally unsuccessful due to inherent problems with the use of both AC and DC welding currents. Unlike AC welding, the DC VPPAW machine produces cyclic changes of electrode-positive (DCEP) and electrode-negative (DCEN) polarities with independent control of current amplitude and duration. This allows an optimum balance of DCEP for cathodic cleaning and DCEN for penetration

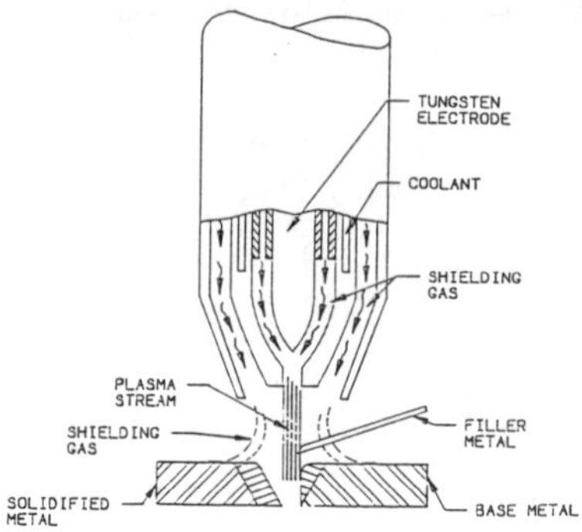

Figure 1: Schematic Representation of VPPAW process

to be achieved, resulting in quality welds in aluminum. Research found that adequate cathodic cleaning did not require the amount of DCEP time associated with AC, while as much DCEN time as possible was needed for penetration (2). It was also determined that an additional spike of current added to the DCEP cycle provided additional cathodic cleaning action with minimal heat buildup and deterioration in the electrode. A typical waveform representative of these conditions is shown in Figure 2.

2. EXPERIMENTAL

VPPAW was performed using a commercial Hobart 300 amp VPPAW system, which included a VP power

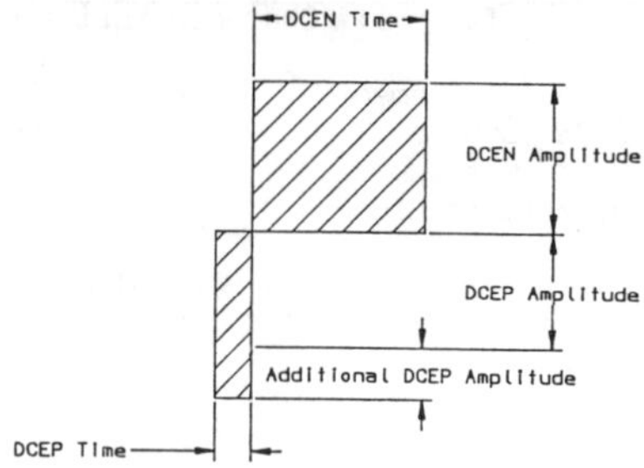

Figure 2: Typical VPPAW Waveform

source, plasma console, wire feeder and fixturing including auxiliary shielding.

The material evaluated in this study was 1.27 cm (0.5 inch) thick Alcoa aluminum-lithium alloy 2090-TE841. Filler metals used were 0.158 cm (0.0625 inch) diameter 2319 and 5356 aluminum alloys. Nominal chemical compositions of these materials are given in Table 2.

Alloy 2090 plate was provided with saw cut edges. Various cleaning procedures evaluated included: acetone degreasing, chemical cleaning, mechanical scraping and dry milling.

A parametric study evaluated process parameter effects on producing 1.27 cm (0.5 inch) thick butt welds in the

Table 2: Nominal Chemical Compositions

Alloy	Al	Li	Cu	Mg	Mn	Other
2090	Bal.	2.2	2.7	-	-	0.12 Zr
2319	Bal.	-	6.2	-	0.3	0.15 Ti
5356	Bal.	-	-	5.0	0.12	0.12 Cr, 0.13Ti

vertical-up position. All welds were performed in the keyhole mode in a single pass. Selected process variables shown in Table 3 were determined from previous studies and were maintained constant throughout the investigation (2).

Transverse-oriented weld specimens were mounted in epoxy and rough polished on 240, 320, 400 and 600 grit paper followed by 1 micron diamond polishing compound. Final polishing was accomplished using 0.3 micron magnesium oxide.

The specimens were then etched using Keller's reagent. Optical microscopy was used to characterize the weld macrostructures and microstructures for the various filler metal/postweld heat-treatment combinations. Rockwell B hardness traverses were then performed across the same cross sections.

Prior to removal of the tensile specimens, all welds were radiographically inspected. Flat, traverse-weld oriented tensile specimens with a 0.953 cm [0.375 inch] nominal thickness and 1.27 cm [0.5 inch] by 3.8 cm [1.5 inch] gage were used for mechanical property characterization. Two different postweld heat-treatment conditions were evaluated: aged and solutionized plus aged. The aging treatment was conducted at 190°C [375°F] for 8 hours, followed by an air cool. The solutionizing treatment was performed at 537°C [1000°F] for 2 hours, followed by a tap water quench and an 8 hour age at 190°C [375°F].

3. RESULTS AND DISCUSSION

3.1 Parametric Evaluation

The process application study revealed that both front and back-side shielding of the weld was necessary to avoid formation of excessive oxide on the molten pool surface and permit the

Table 3: VPPAW Variables Maintained Constant

Wire Feed Speed: 2.33-2.36 cm/sec [55-56 ipm]
Wire Diameter: 0.158 cm [0.0625 inches]
Travel Speed: 0.2 - 0.30 cm/sec [5-7 ipm]
Electrode Type: 2% Thoriated
Electrode Diameter: 0.39 or 0.32 cm [0.155 or 0.125 inches]
Electrode Extension: 0.81 - 0.84 cm [0.320 - 0.330 inches]
Electrode Tip Geometry: Blunt
Standoff: 0.48 - 0.64 cm [0.1875 - 0.25 inches]
Tip Type: Flat
Orifice Diameter: 0.39 or 0.32 cm [0.155 or 0.125 inches]
Torch Angle: 5° Push
Shield Gas: Helium/19 - 24 liter/min. [40-50 CFH]
Plasma Gas: Argon
Pilot Arc Current: 28 amps (continuous)
Voltage: 30 volts

production of acceptable welds. Initially, welds were produced using procedures recommended for aluminum alloy 2219. Due to the lower thermal conductivity of alloy 2090 it was necessary to use a lower DCEN current. The DCEP amplitude and time were also increased in an effort to further optimize the cathodic cleaning action of the VPPAW process. The resulting welds exhibited poor quality due to atmospheric contamination of the molten weld pool, which produced excessive oxidation. In an effort to reduce contamination effects on process performance, an inert gas purge was placed on the weld backside and a gas trailer shield was added on the weld face. This generally resulted in acceptable quality welds, however, consistency was difficult to obtain along the entire weld length. Therefore, to eliminate contamination problems as a variable, the final weld trials were performed inside a purge chamber. A comparison between the variables recommended for alloy 2219 and those determined to be optimum for alloy 2090 in the purge box are shown in Table 4. The cross-section macrostructure, top surface and underbead of a representative 2090/filler metal VPPAW weld is shown in Figure 3.

3.2 Cleaning Requirements

Several different cleaning methods were evaluated during the course of this study, including: mechanical scraping, chemical cleaning, acetone degreasing and dry milling. The dry milling operation, followed by an acetone wipe immediately prior to welding produced the best results and was used in all the final interation welding trials.

3.3 Metallurgical Evaluation

Light microscopy was used to identify the microstructural characteristics of the 2090 welds produced with each filler metal/postweld heat-treatment combination.

The resulting microstructural regions in the as-welded,

Table 4: VPPAW Procedures - 2219 vs. 2090

Alloy	2219	2090
Filler Metal:	0.158 cm [0.0625 in.] dia 2319	0.158 cm [0.0625 in.] dia 2319/5356
Wire Feed Speed:	1.9 cm/sec [45 ipm]	2.37 cm/sec [56 ipm]
DCEN Current:	230 amps	160 amps
DCEN Time:	19 ms	19 ms
Add.DCEP Current:	70 amps	90 amps
DCEP Time:	4 ms	5 ms
Plasma Gas:	Argon 2.4 liter/min [5 CFH]	Argon 2.4 liter/min [5 CFH]
Shielding Gas:	Helium 28.3 liter/min [60 CFH]	Helium 18.9 liter/min [40 CFH]
Electrode:	0.39 cm [0.156 in] - EWTh2	0.31 cm [0.125 in] - EWTh2
Travel Speed:	0.275 cm/sec [6.5 ipm]	0.254 cm/sec [6 ipm]

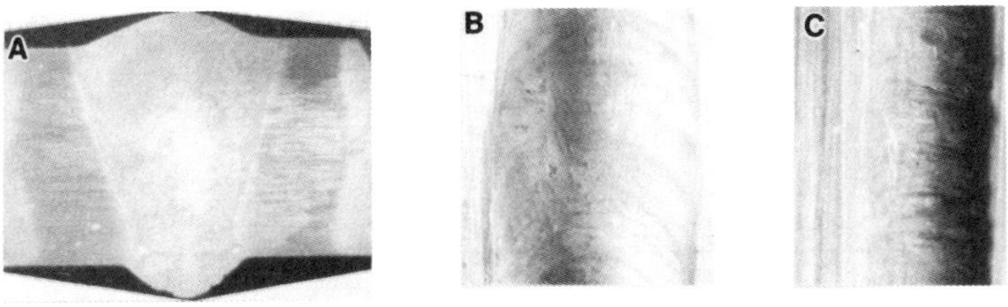

Figure 3: Cross-section macrostructure (A), top surface (B) and underbead (C) 2090/2319 VPPAW weldment.

autogeneous weld are shown in Figure 4. These regions were very similar to a typical 2xxx series heat-treatable aluminum alloy weld, and included a fusion zone, partially-melted, resolutionized, overaged and unaffected base metal regions. The fusion zone was characterized by a relatively coarse cellular-dendritic microstructure with eutectic constituents appearing along the grain and substructure boundaries and some fine precipitates within the alpha dendrites. The grain structure varied throughout the fusion zone, exhibiting an equiaxed morphology at the center and a columnar morphology near the fusion boundary.

A partially-melted region was evident at the fusion line, which exhibited liquation of eutectic precipitate phases at the original base metal grain boundaries. This eutectic liquid surrounded the grains and prevented grain growth, as evidenced by the small grain size in this region. In the resolutionized region, the peak welding temperatures were sufficiently high and the cooling rates sufficiently rapid to produce a supersaturated alpha aluminum solid solution. In the overaged region, lower peak temperatures and slower heating and cooling rates allowed the precipitates to coarsen and agglomerate, as evidenced by the dark-etching regions. The unaffected base metal was characterized by elongated grains of alpha aluminum with second phases at the grain and subgrain boundaries.

The aged, autogeneous weld appeared similar to the as-welded specimen. Although not visible via light microscopy, aging promoted growth of precipitates in the fusion, partially-melted and resolutionized regions. The overaged region showed precipitate coarsening as revealed by the large, dark-etching regions.

Microstructural regions present in the solutionized and aged weld

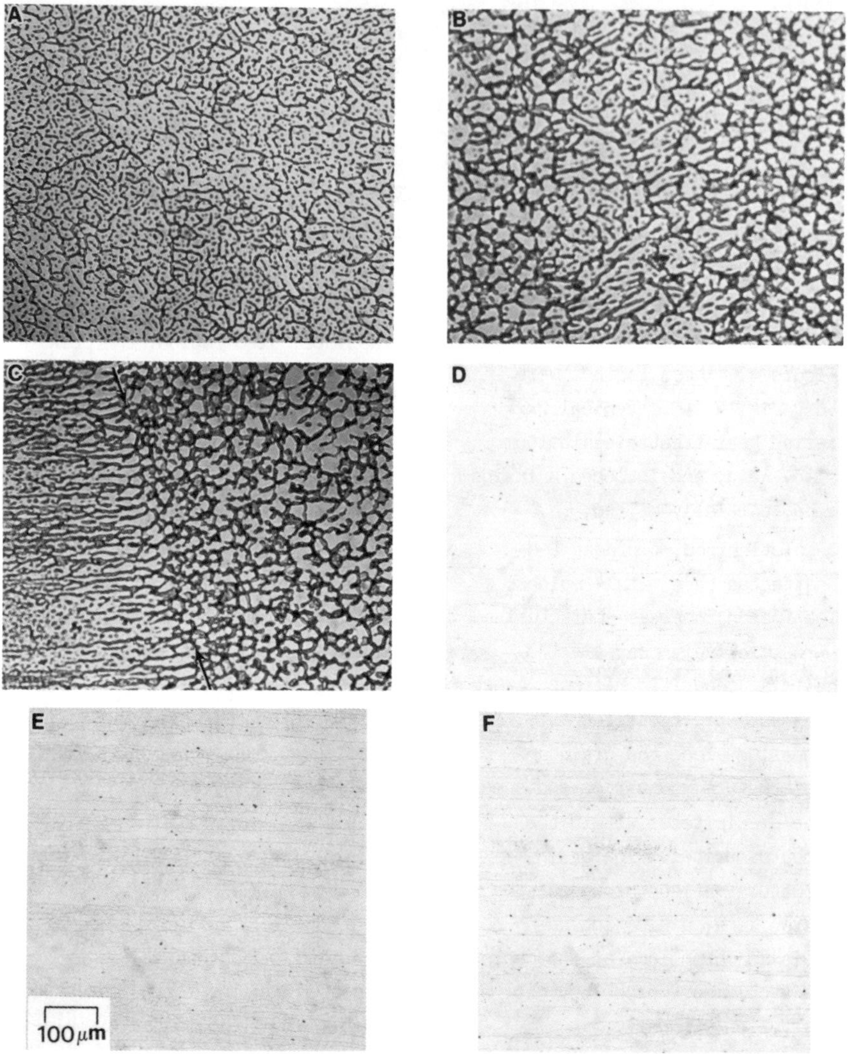

Figure 4: As-Welded Fusion and Heat-affected Zone of Autogeneous VPPAW Weld in Alloy 2090: a) Fusion Zone Center, b) Fusion Zone Near Fusion Boundary, c) Fusion Line, d) Resolutionized Heat-Affected Zone Region, e) Overaged Heat-Affected Zone Region and f) Unaffected Base Metal

differed more significantly from the as-welded condition, exhibiting less interdendritic segregation and second phases. There was little distinguishable difference between the original resolutionized, overaged and unaffected base metal regions. Despite the solutionizing treatment, second phases were still observable at the alpha grain boundaries.

The fusion zone microstructures of the as-welded, postweld aged and solutionized plus aged 5356 filler metal welds are shown in Figure 5. The fusion zones were microstructurally similar to the autogeneous welds subjected to the same heat-treatment. This was attributed to the high base metal dilution in the fusion zone of approximately 83 percent. In all three conditions the 5356 fusion zone appeared coarser than the corresponding autogeneous weld. The only other difference was in the types of eutectic constituents and precipitates, which were not resolvable with optical microscopy.

The 2319 filler metal fusion zone was characterized by regions and structures very similar to those in the 5356 filler metal fusion zone.

3.4 Weld Defects

Light microscopy, radiography and visual inspection were used to identify and characterize defects in the fusion and heat-affected zones. In the final weld trials, the defects observed included occasional porosity and heat-affected zone microfissuring. Porosity was present in minute levels and was spherical in shape, so it was determined not to be detrimental to mechanical properties. HAZ microfissures initiated at the fusion line along grain boundaries where the eutectic precipitates had liquated, as shown in Figure 6. These cracks extended well into the HAZ where peak welding temperatures were not high enough to cause eutectic melting, thereby indicating a solid-state mechanism for crack propagation into this area.

3.5 Mechanical Property Characterization

Figure 7 shows the results of hardness surveys on the autogeneous, 2319 and 5356 filler metal welds in the as-welded and postweld heat-treated conditions. The fusion and heat-affected zone hardness were very similar for all three as-welded samples. The 5356 weld fusion zone responded to the postweld aging treatment more effectively than the autogeneous or 2319 filler metal welds. At and beyond the fusion line, the hardnesses were essentially the

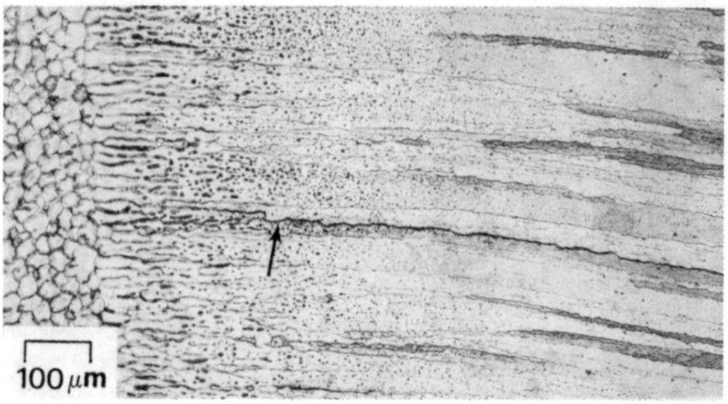

Figure 6: HAZ Microfissure in VPPA Welded 2090

same for all three welds. Similar results were noted for the solutionized and aged welds.

Correlating hardness results with the mechanical properties given in Table 6, reveals that most of the specimens failed in the region of lowest hardness. Although the autogeneous weld exhibited the highest as-welded joint efficiency (48%), welds produced with 5356 filler metal showed the highest postweld aged and solutionized and aged joint efficiencies of 52 and 68 percent, respectively. Low elongation levels were observed primarily in the aged welds due to strain localization in the low strength HAZ region. The 2090 VPPAW weld mechanical properties were very similar to those obtained in previous 2090 welding studies which evaluated the GMAW and EB welding processes, as can be seen in Table 6 for the different postweld heat-treatments.

4. CONCLUSIONS

The results of this study determined that:

1. VPPAW can successfully produce full-penetration, single-pass, square-butt welds, vertical-up in 1.27 cm [0.5 inch] thickness alloy 2090.
2. Adequate cleaning and inert gas shielding techniques are required to prevent excessive oxide formation, which is prohibitive to producing quality welds.
3. Microstructural characteristics of VPPAW welds in 2090 as revealed by light microscopy appeared very similar to those of typical 2xxx aluminum alloy.
4. Weld defects observed included porosity and heat-affected zone microfissuring, neither of which was shown to influence tensile properties.
5. Joint efficiencies of approximately 45%, 49% and 67% were obtained for the as-welded,

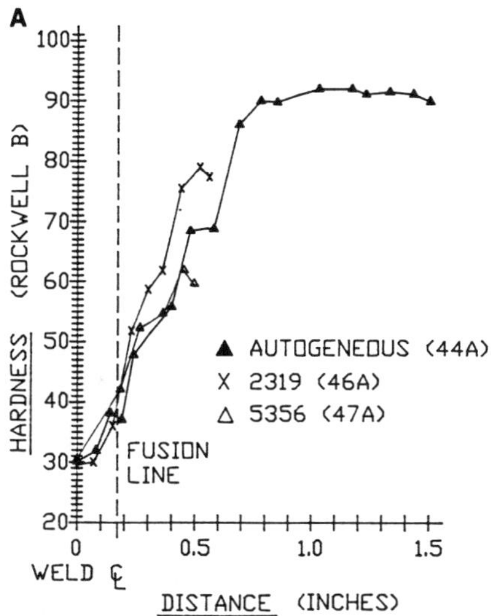

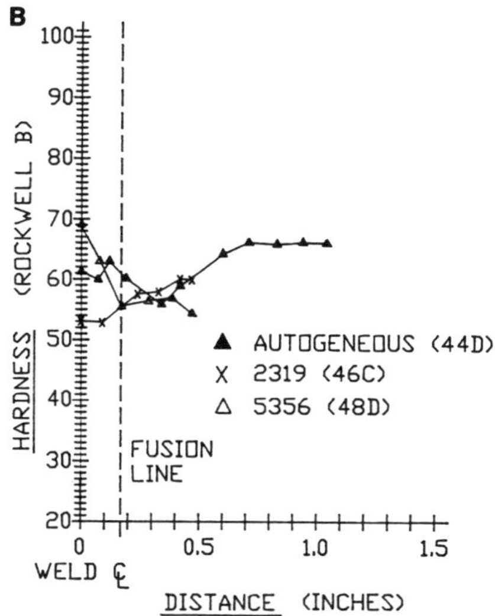

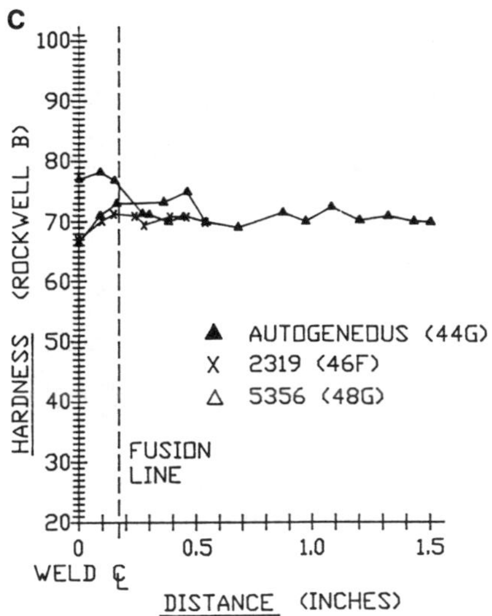

Figure 7: Hardness Versus Distance From Weld Centerline for A) As-Welded, B) Postweld Aged and C) Postweld Solutionized and Aged Conditions

postweld aged and postweld solutionized plus aged conditions, respectively.

6. The 5356 filler metal weld responded the greatest to postweld heat-treatment versus the autogeneous or 2319 filler metal welds.

5. ACKNOWLEDGMENTS

The authors express appreciation to R. P. Martukanitz of ALCOA Laboratories for providing the aluminum-lithium alloy 2090 material and Hobart Corp. for providing the variable-polarity plasma-arc welding equipment. We would also like to acknowledge Gary Reed and Steve Barhorst of Hobart Corp. for their technical support in operation of the welding system and optimization of the process variables.

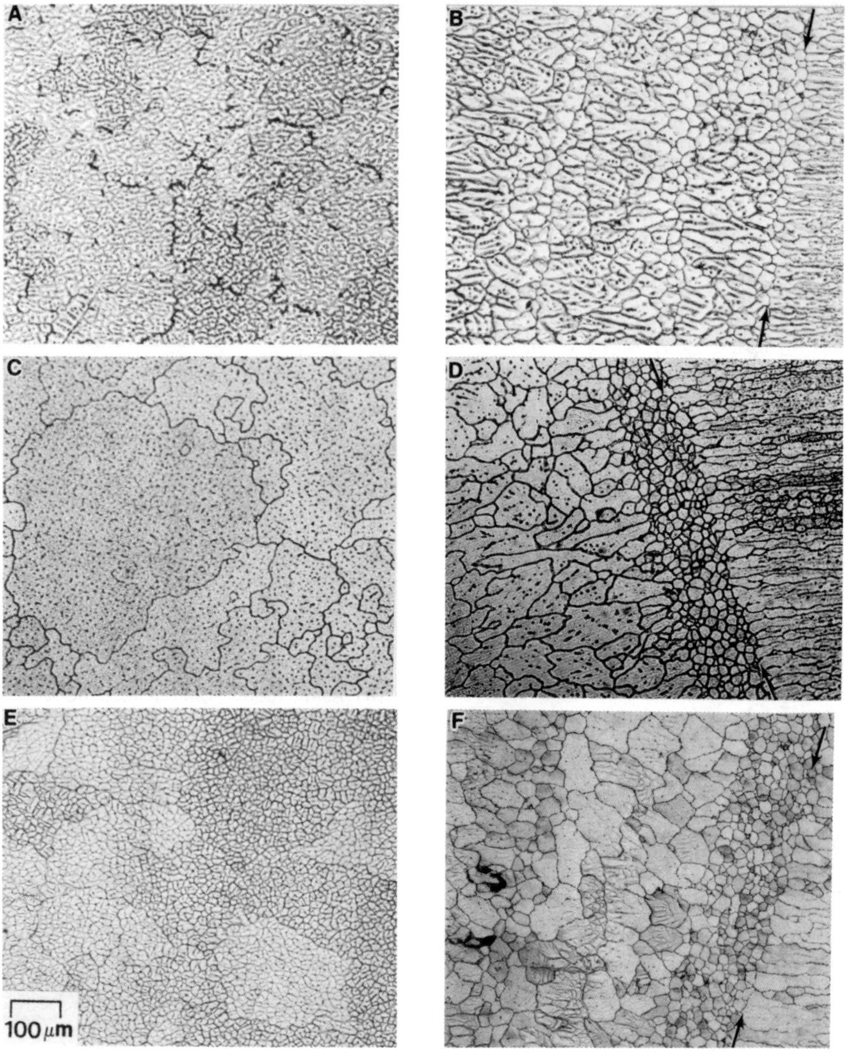

Figure 5: Fusion Zone of VPPAW Weld in Alloy 2090 Produced With 5356 Filler Metal: a,b) As-Welded, c,d) Aged, e,f) Solutionized and Aged

Table 6: Mechanical Properties of VPPAW, GMAW and EB Welds in Alloy 2090

Filler Metal	Ultimate Strength MPa	[KSI]	Elongation (% in 2.54cm)	Failure Location	Joint Efficiency(%)
VPPAW(1)					
As-Welded					
Autogeneous	282.8	[41.0]	4.9 (2)	FZ/FB	48
2319	245.9	[35.6]	7.1	FZ/FB/HAZ	41
5356	236.1	[34.2]	3.8	FZ	40
Postweld Aged					
Autogeneous	280.3	[40.6]	1.6	HAZ/FZ	47
2319	285.9	[41.4]	1.7	HAZ/FZ	48
5356	305.5	[44.3]	1.4	HAZ	52
Postweld Solutionized and Aged					
Autogeneous	390.3	[56.6]	7.2	FZ	66
2319	399.6	[57.9]	7.0	HAZ/FZ	67
5356	405.5	[58.8]	7.1	FZ/HAZ	68
GMAW (Ref.2)					
As-Welded					
2319	231.7	[33.6]	5.2 (3)	-	39
Postweld Aged					
2319	258.6	[37.5]	0	-	43
Postweld Solutionized and Aged					
2319	386.9	[56.1]	0	-	65
Electron Beam (Ref.2)					
As-Welded					
Autogeneous	322.7	[46.8]	0 (2)	-	54
Postweld Solutionized and Aged					
Autogeneous	413.1	[59.9]	0	-	70

Note: (1) Average of duplicate specimens
(2) % in 2.54 cm [1 in.]
(3) % in 5.1 cm [2 in.]

6. REFERENCES

1. Martukanitz, R.P., et al, "Microstructural and Microhardness Charac. of a GMAW Al-Li Alloy", Alcoa Lab., 1986.

2. Tomsic, S.M., and Barhorst, S., Weld. J. 63(2) 25-32 (1984)

2nd International SAMPE Metals Conference
August 2-4, 1988

THE EFFECT OF ELEVATED TEMPERATURE EXPOSURE ON THE PROPERTIES OF SiC/Ti6Al4V COMPOSITES

by F. E. Wawner
University of Virginia
Charlottesville, VA
and
W. J. Whatley
General Dynamics
Convair Division
San Diego, CA

Abstract

One of the most promising reinforcing fibers for use in titanium alloy matrices is the vapor deposited SiC fiber. These fibers are very strong and stiff and maintain these properties well at very high temperatures. It was the goal of the current research to evaluate the effectiveness of two types of surface coatings to protect these fibers from matrix attack. Several criteria were used to evaluate the effectiveness of the coatings in protecting the fiber. The first measure was the kinetics of the formation of the reaction zone. The kinetics of the reaction between the matrix and both of the fibers were described by a parabolic growth process, which was controlled by an effective activation energy. Further evaluation included monitoring degradation of both fiber and composite properties with annealing time.

1. INTRODUCTION

In recent years fiber reinforced titanium alloys have been of interest for use in gas turbine engines and, for structural members in aircraft. The high specific properties of the titanium alloys and their excellent retention of properties at elevated temperatures make them attractive candidates for use as matrix materials. The extreme reactivity of the titanium alloys at the temperatures required for forming the composites has slowed the development of these systems. Attempts have been made to apply surface coatings to allow the fibers to withstand the forming procedures and subsequent thermal exposures without loss of strength.

Perhaps the most promising fiber for use in titanium matrix composites is the vapor deposited SiC fiber. These fibers are very stiff and strong and maintain these properties at high temperatures. This study evaluated the effectiveness of two surface coatings in protecting the fibers from matrix attack.

2. EXPERIMENTAL METHODS

Two types of vapor deposited SiC fibers and the composites made from them were studied. Both fibers and the composites were fabricated by AVCO Specialty Materials. Nominal tensile strengths of the fibers is 4.1 GPa (600 ksi) with a tensile modulus of 427 GPA (62 msi). The major difference between the fibers is the surface coating applied to protect against matrix attack.

Chemical vapor deposition (CVD) processes are used to apply the coatings to the fibers. Both coatings are carbon rich SiC with varying composition through the thickness. Fiber I has a coating approximately 2 μm thick. The coating was applied in a single process. Fiber II has a coating which is approximately 4μm thick. Essentially the same process was used to coat this fiber, but it was coated twice. Figure 1 depicts a schematic representation of the fiber and an Auger trace of Si/C ratio in the SCS layer. Standard processing techniques were used to fabricate composites with fiber loadings from 28 to 35 volume percent from both fibers.

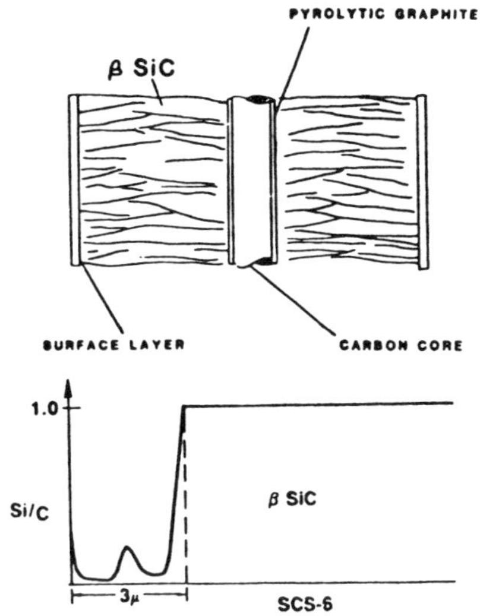

Figure 1. Schematic of fiber section and Auger trace of Si/C ratio in SCS layers.

Mechanical properties of the virgin fibers were determined by testing at several gauge lengths. Statistical parameters describing the distribution of fiber strength were determined by extensive testing at 2.54 cm gauge length. Fiber testing was performed using a floor model Instron tensile test machine equipped with pneumatic jaws. To insure failure of the fiber in the gauge, the jaws were padded with soft aluminum foil. The foil padding was replaced after 8 to 12 tests.

Composites were annealed in evacuated quartz tubes to prevent oxidation of the matrix. Encapsulated samples were placed in a preheated furnace maintained to within ±5 C of the set point. At the end of the prescribed heat treatment, the samples were water quenched.

Two sets of annealing experiments were performed. The first was to simulate a superplastic forming operation. The second was to determine the reaction kinetics between the fiber and matrix at several temperatures.

During the first set of experiments, the composites were annealed at 900 C (1652 F) for 1 to seven hours. The matrix was then etched from the fibers using a 10% bromine in methanol solution. Fibers were carefully washed, and tested using a 2.54 cm gauge length. A minimum of thirty samples were tested from each annealed composite to determine the distribution of fiber strengths.

Kinetics of fiber matrix interaction were determined by applying a matrix of annealing conditions. Thickness of the resulting fiber/matrix reaction zone was measured, and used to determine reaction kinetics. Details of the experimental procedures and the results are discussed in Reference 1.

3. RESULTS AND DISCUSSION

Strengths of the virgin fibers as a function of gauge length are shown in Figure 2. Fiber 1 demonstrated strengths of 4.3 GPa (623 ksi) at gauge lengths up to 2.54 cm while fiber II had strengths of only 3.3 GPa (478 ksi). Both fibers exhibit lower strengths at increased gauge lengths. At 10 cm Fiber I and Fiber II had strengths of 3.4 and 2.8 (493 and 406 ksi) respectively. Weibull parameters determined at the 2.54 cm gauge length predicts a more rapid decline in strength than measured experimentally. This is somewhat anomalous since the Weibull distribution function fits the data for the 2.54 cm gauge length quite well.

As shown in Figure 3, growth of the reaction zone on Fiber I followed parabolic reaction kinetics at all temperatures. The activation energy of the fiber-matrix reaction was determined using the reaction constants for the temperatures evaluated. Measured reaction zone thickness and the reaction zone thickness calculated using the kinetic model (1) are shown in Table 1. An Arrhenius plot for the reaction constants of both Fiber I and Fiber II composites is shown in Figure 4. The activation energy for both composites is approximately 65 Kcal/mole.

At 900°C (1652°F) the Fiber II composites exhibited two step kinetics. For exposures up to 25 hours, the kinetics were parabolic with a reaction constant of 0.0068 $\mu m/sec^{\frac{1}{2}}$. At 25 hours there is an increase in the reaction rate. At exposure times greater than 25 hours parabolic kinetics apply with a reaction constant of 0.014 $\mu/sec^{\frac{1}{2}}$.

Strengths of Fiber I, the single coated fiber, degraded drastically

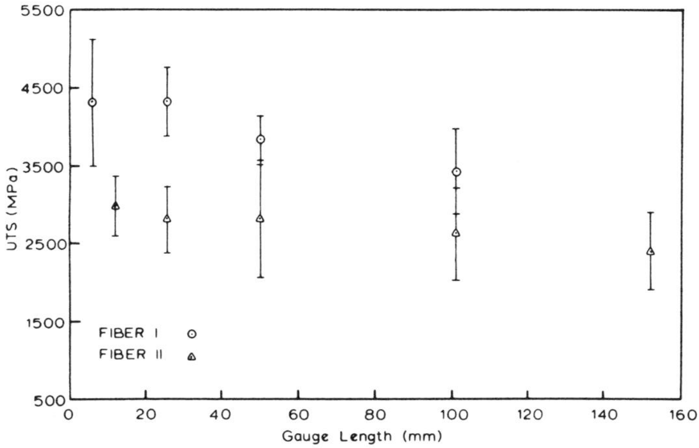

Figure 2 Strengths of SCS fibers tested as manufactured.

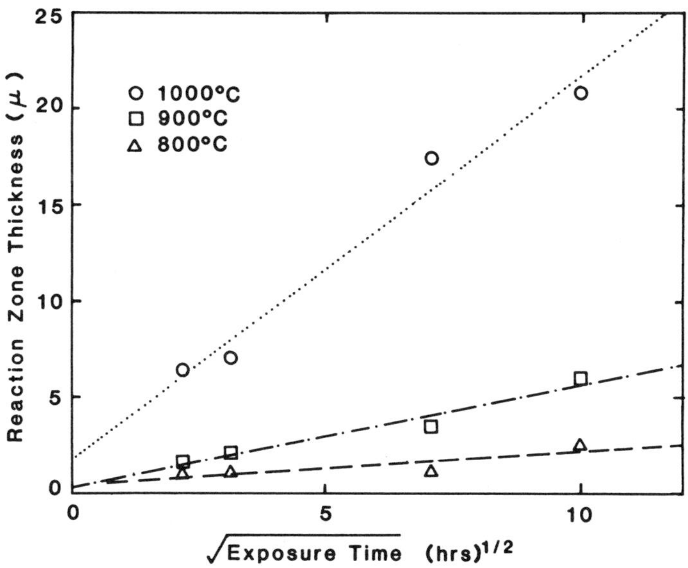

Figure 3. Reaction kinetics of Fiber I were parabolic at all temperatures tested.

Table 1. Results of strength versus exposure experiment for Fiber I composites.

REACTION CONSTANT (μm/sec .5) 0.0089
TEMPERATURE (C) 900
FIBER MODULUS (GPa) 414

EXPOSURE TIME(hours)	REACTION ZONE Measured(μm)	REACTION ZONE CALC.(μm)	FIBER STR. (MPa)	FIB.STR. CALC.* (MPA)	FIB. STR. CALC.** (MPa)
AS FAB	0.48		3169		2098
1		0.5	1688	1987	
4		1.1	1185	1408	
5	1.59				1152
7		1.4		1221	
10	2.1	1.7		1118	1000
20		2.4		938	
50	3.5	3.8		745	780
100	6.0	5.3		628	593

* Calculated using calculated zone thickness values.
**Calculated using measured zone thickness values.

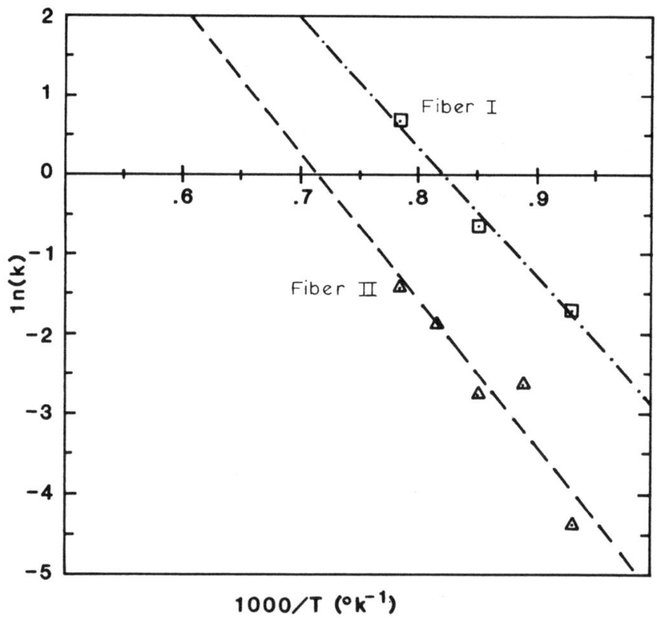

Figure 4. Arrhenius plot for determination of the activation energy for the both fibers tested.

during the simulated processing exposure. Results of the strength versus exposure time experiment are shown in Table 1. Processing the fibers into a composite degrades the strength from 4.3 to 3.2 GPa. A heat treatment analogous to a superplastic forming cycle further degrades the fiber strength to 1.2 GPa.

Comparison of the measured strength data versus reaction zone thickness is shown in Figure 5. Although there are only three data points, they correlate well with the theory of fiber strength reduction proposed by Ochiai[2,3]. The rate of strength reduction is quite accurately predicted. Absolute values of predicted fiber strengths are displaced due to assumptions of material properties.

Results of the simulated superplastic forming cycle on Fiber II are shown in Figure 6. It is evident that the double coating on these fibers prevents tensile strength degradation. Table 2 summarizes the strength and reaction zone data for this fiber. Examination of the microstructure of the reaction zone and fiber coating explain this anomalous behavior. As seen in Figure 7, the reaction zone on the Fiber I composites is adherent and tightly bonded to the protective coating. The uneven attack of the protective coating is also evident. This uneven attack results in pitting of the fiber. These pits act as stress risers, drastically reducing the strength of the fiber. Evidence of the fiber pitting is seen in micrographs of the fracture surface of a transverse tension

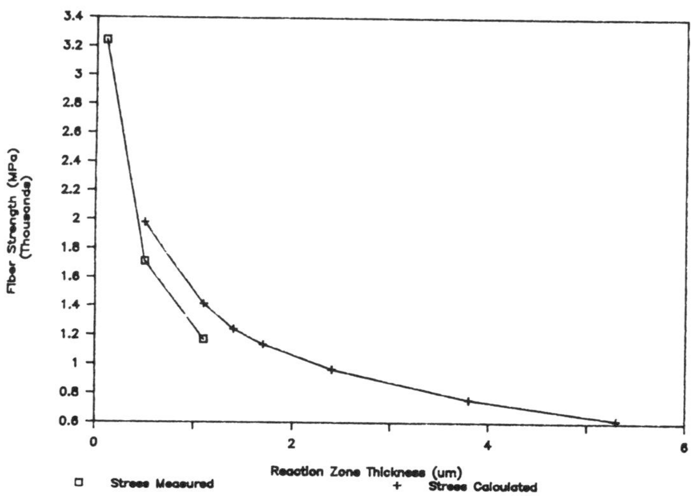

Figure 5. Fiber strength versus reaction zone thickness for Fiber I.

Table 2. Summary of strength versus reaction zone data for Fiber II composites.

REACTION CONSTANT (μm/sec .5) 0.0089
TEMPERATURE (C) 900
FIBER MODULUS (GPa) 414

EXPOSURE TIME(HOURS)	REACTION ZONE MEASURED (μm)	REACTION ZONE CALC. (μm)	FIBER STR. (MPa)	FIB.STR. CALC.* (MPa)	FIB. STR. CALC** (MPa)
AS FAB	0.48		2673		2098
1		0.5	2571	1982	
4		1.1	2675	1408	
7		1.4	2867	1221	
10	1.8	1.7		1118	1083
20	2.3	2.4		938	973
50	5.3	3.8		745	635
100	7.0	5.3		628	552

* Calculated using calculated zone thickness values.
**Calculated using measured zone thickness values.

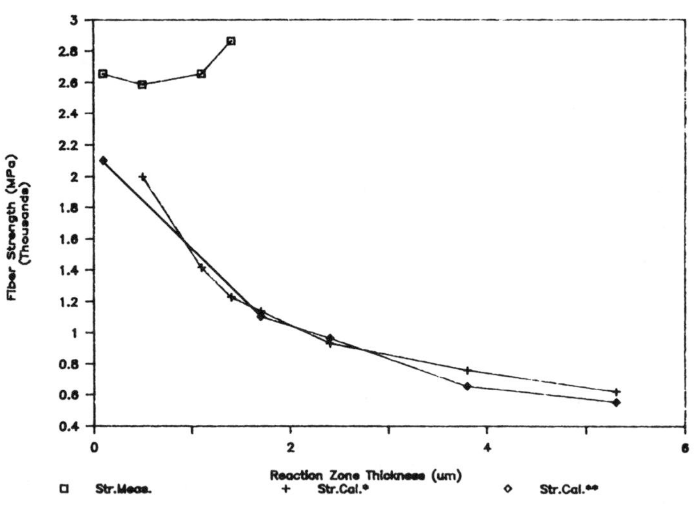

Figure 6. Fiber strength versus reaction zone thickness for Fiber II.

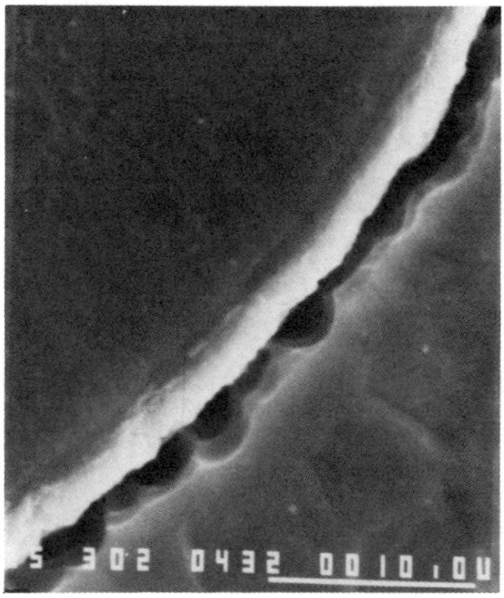

Figure 7. Typical microstructure of reaction zone in Fiber I composites after 10 hours of exposure at 900°C.

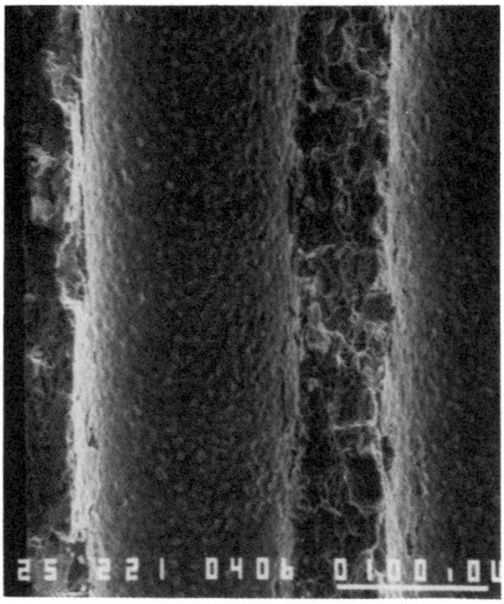

Figure 8. Fracture surface of transverse tensile specimen of a Fiber I composite.

test (Figure 8). Note the sharp points on the reaction surface left after fiber pull out.

Microstructure of the reaction zone around Fiber II is quite different. A circumferential crack is apparent in the protective coating of the fibers (Figure 9). The crack is apparent in the samples annealed for 10 and 20 hours. This crack is probably the boundary between the first and second deposited layers in the double coated fibers. At times over 20 hours the fiber-matrix reaction has proceeded past the point where the crack forms.

Figure 10 shows the fracture surface of a Fiber II composite transverse tensile specimen. The poor fiber-matrix bonding is evident in the smooth surface where the fiber pulled out. Closer examination shows that the failure occurred in the protective layer.

Axial tensile tests on composites produced from the fibers showed that samples made from Fiber I gave 1650MPa (239 ksi) in the as-fabricated condition which decreased to 940 MPa (136 ksi) after heat treating at 900°C for 7 hours. Samples produced from Fiber II had a slightly lower as-fabricated axial strength (1380 MPa - 200ksi) but were not degraded at all by the simulated superplastic forming heat treatment. This is consistent with the results on the individual fibers and attests to the protective nature of the double coating.

Transverse tensile strengths for the samples subjected to the simulated superplastic forming heat treatment were 554MPa (80ksi) and 374MPa (54ksi) for composites made

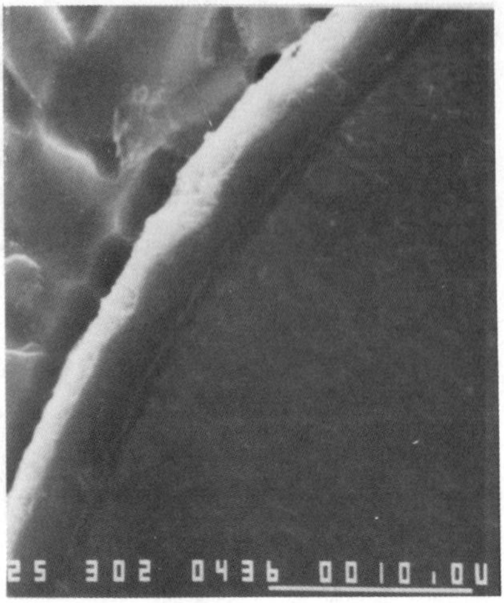

Figure 9. Typical microstructure of reaction zone in Fiber II composites after 10 hours of exposure at 900°C.

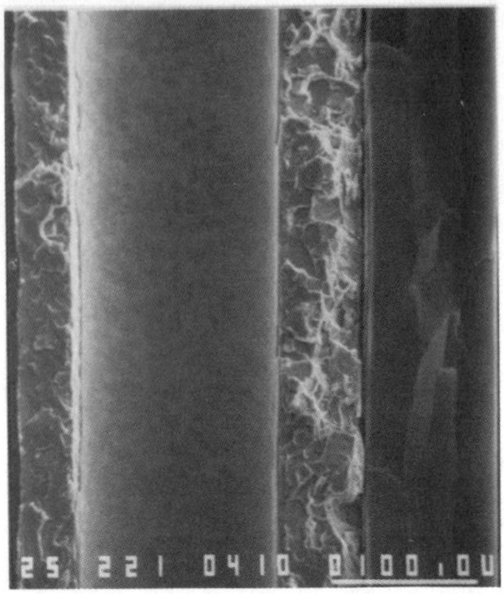

Figure 10. Fracture surface of transverse tensile specimen of a Fiber II composite showing smooth fracture surface and poor fiber/matrix bonding.

from Fiber I and Fiber II, respectively. The lower transverse value for the composite containing the double coated fiber can be attributed to the lack of a bond between the two surface layers. This was demonstrated by flexing a double coated fiber into a small loop and observing the outer layer spall away.

The presence of the circumferential boundary within the protective coating explains several of the anomalous observations from this fiber. Two step reaction kinetics are explained due to the change in concentration of reacting species when the reaction proceeded past the boundary. Low composite transverse strengths are attributable to the poor fiber-matrix bonding.

4. CONCLUSIONS

The single coated fiber degraded excessively with heat treatments required for fabrication of parts from the composite. This system is not an effective protection system for titanium matrix composites.

Application of a double thickness coating protected the fiber from degradation of strength at the expense of composite transverse properties. The poor adhesion of the thick protective coating precluded good fiber-matrix bonding. This resulted in a low transverse strength composite.

Use of a sacrificial coating for fiber protection was not successful in the two fibers tested. Fiber I had a well adhered coating which allowed good fiber-matrix bonding. Unfortunately the coating was not thick enough to prevent severe fiber strength degradation. The thicker coating on Fiber II prevented strength degradation, but did not allow adequate composite transverse properties.

Successful coatings should be thin enough to allow good bonding to the fibers. Further, they should ideally act as diffusion barriers and not as sacrificial coatings.

5. REFERENCES

(1) Whatley, W.J. and Wawner, F.E., J. Mater. Sci. letters, 4, 173 (1985).

(2) Ochiai, S. and Murakami, Y., J. Mater. Sci., 14, 831 (1979).

(3) Ochiai, S. and Murakami, Z. Metallkd., 72, 827 (1981).

6. BIOGRAPHY

Dr. Frank E. Wawner is a Research Professor in the Department of Materials Science, School of Engineering and Applied Science, University of Virginia. He received a Ph.D. degree in Materials Science in 1971 from the University of Virginia. Dr. Wawner has been active in composite materials research, including chemical vapor deposition of boron and SiC fibers, mechanical property, micro-structure, fracture, and processing characterization of metal matrix composites, since 1961. He has published over 100 technical papers.

Dr. Walter J. Whatley is a Research Engineer for General Dynamics, Convair Division. He received his Ph.D. degree in Materials Science from the University of Virginia in 1985. Dr. Whatley is presently working the area of composite materials development.

2nd International SAMPE Metals Conference
August 2-4, 1988

INTERACTION ZONE FORMED BETWEEN TiC AND TITANIUM ALUMINIDE

M. H. Loretto, Dept. of Metallurgy and Material
University of Birmingham, Edgbaston, B152TT, UK
D. G. Konitzer, Alcoa Laboratories
Alcoa Center, PA 15069, USA

Abstract

Samples of Ti-24Al-11Nb containing a dispersion of TiC have been exposed to various thermal treatments. The reaction product between the two phases was identified as Ti_3AlC with some substitution of Nb for the Al. Attempts to determine the kinetics of the reaction using image analysis were unsuccessful due to the low contrast between the reaction zone and the TiC.

1. INTRODUCTION

Interaction between the matrix and reinforcement has been considered to be detrimental to the mechanical properties of composite materials (1). Most of the theories and the resultant calculations have been performed for continuously reinforced composites such as boron in Ti or Al (1-4). These theories show that the degradation of the properties is a function of the extent of reaction between the matrix and reinforcement. The extent of reaction determines the trade-off between debonding of the reaction zone from the fiber and fracture of the reaction zone. The point where the fracture mechanism change from one type to the other is determined by the properties of both the products of the reaction and the bonding strength of the reaction zone to the fiber.

In order to determine the properties of the reaction zone, identification of the composition and structure of the phases in the reaction zone is necessary. While it may be possible to predict these phases from thermodynamic calculations in very simple systems, it would be impractical when the matrix is a complex alloy. Furthermore, thermodynamic calculations give no information on the kinetics of the reaction, thus there would be no way to estimate whether the maximum reaction thickness would be exceeded for a given thermal exposure. Therefore, the reaction products and their extent must be determined empirically.

In this paper a characterization of the reaction between the titanium aluminide alloy Ti-24Al-11Nb and TiC is reported. This characterization includes a determination of the reaction products and the kinetics of their growth at elevated temperatures.

2. PROCEDURE

Small specimens of the alloy Ti-24Al-11Nb containing a dispersion of TiC were heat treated at temperatures of 1100 and 1200°C for times of 1 to 5 hours. These specimens were examined using optical microscopy (OM), scanning electron microscopy (SEM), in secondary and backscattering modes, and analytical transmission electron microscopy (AEM). The specimens for OM and SEM were sectioned, mechanically polished and examined in the unetched and etched conditions. Specimens for AEM were prepared by dimpling followed by ion beam thinning. Chemical analysis was performed using energy dispersive X-ray spectroscopy, both with a standard Be window detector and with a thin window detector, and electron energy loss spectroscopy on microscopes operating at either 120 or 400 kV.

The heat treatment of the material was carried out in an air atmosphere furnace. The size of the sample was such that no contamination of the center was expected based on the calculation of the diffusivity of oxygen at the heat treatment temperature. All specimens for observation were obtained from the center of the sample.

3. RESULTS

The reaction zone was visible in the as-received condition, which represents a heat treatment of 2 hours at 1000°C, Figure 1. The microstructures, both unetched and etched, for two thermal treatments, 1100°C - 2 hours and 1200°C - 1 hour, are shown in Figure 2. Porosity is evident in the TiC phase in both of these figures; this porosity was also present in the as-received TiC. It is difficult to determine the extent of reaction zone in either condition from the optical microscope images, therefore, SEM analysis was carried out. The SEM images recorded using backscattered electron imaging are shown in Figure 3 for the above two heat treatment conditions. The matrix in the 1100°C heat treatment

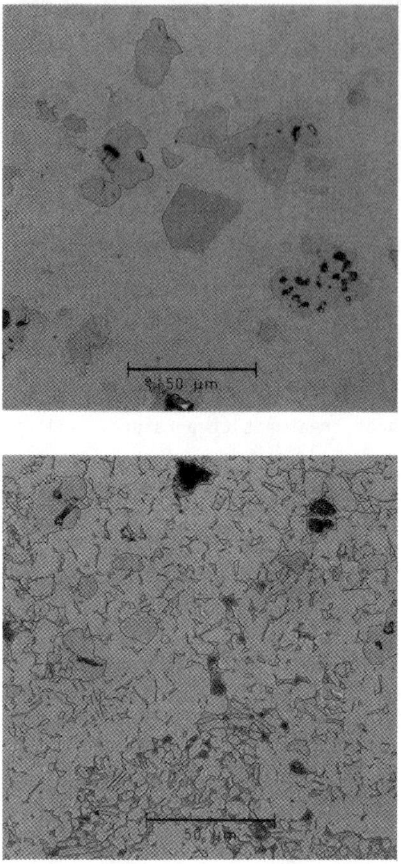

Figure 1: Optical micrographs in the unetched and etched condition for the as-received Ti-24Al-11Nb plus TiC.

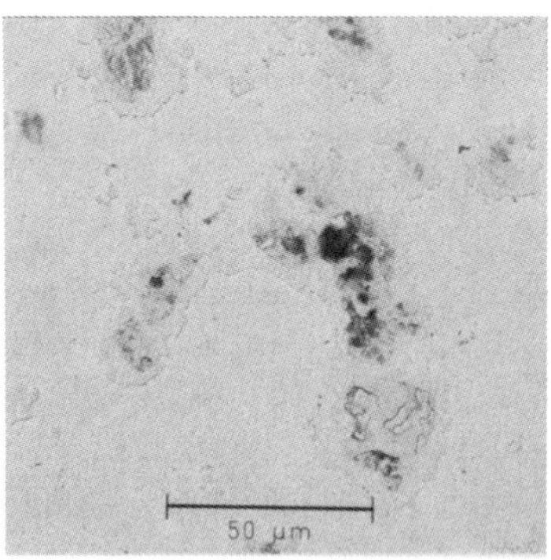

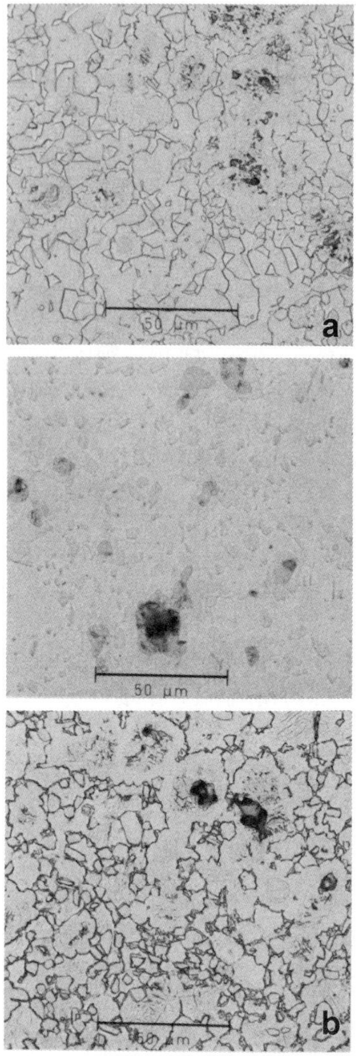

Figure 2: Optical Micrographs in the unetched and etched condition for heat treatment at a) 1100°C 2 hours and b) 1200°C 1 hour.

consists of a mixture of alpha-2 and beta phases with the alpha-2 appearing dark in the backscattered image and the beta lighter because of the increased Nb concentration of this phase. The reaction zone,

Figure 3: SEM micrographs obtained using backscattered electron imaging for heat-treatment at a) 1100°C 2 hours and b) 1200°C 1 hour.

which is the phase with the rounded boundaries of about the same intensity as the alpha-2 phase, can be seen to surround the somewhat darker TiC phase in this figure. The intensity of the reaction zone indicates that it consists of material which has a lower average atomic number than the TiC phase. The matrix in the 1200°C heat treatment consists entirely of the transformed beta phase, i.e., fine laths of the alpha-2 phase. Also in this structure, the reaction product phase is uniformly distributed throughout the matrix. There is a larger volume fraction of reaction product in this structure compared to the 1100°C treatment, presumably because the higher temperature has resulted in an increased rate of reaction.

Attempts were made to quantify the extent of reaction between the matrix and reinforcement using image analysis on the micrographs shown in Figure 3. This imaging condition was used because of the higher contrast available using the backscattered imaging compared to the optical microscopy. However, the contrast was still insufficient for a reliable analysis to be performed and the tests were abandoned.

Identification of the reaction zone products was carried out using analytical electron microscopy. A bright field image showing the TiC, reaction zone, and matrix is shown in Figure 4. The reaction zone shows a finer grain size compared to the matrix and the TiC and contains a large number of faults across its width. Diffraction analysis of the reaction product using convergent beam electron diffraction showed it to be cubic with a lattice parameter of 0.415 nm. The symmetries evident in the 001 and 111 zone axes show the only consistent point group to be m_3m while the allowed reflections

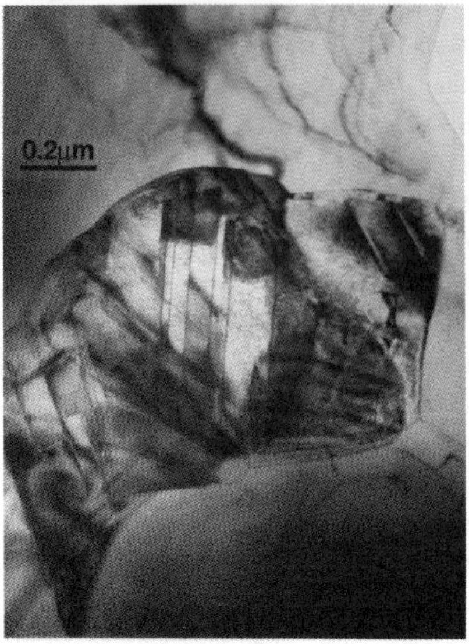

Figure 4: Bright field TEM micrograph of the matrix (top), reaction zone (middle) and TiC (bottom).

indicate the space group to be Pm_3m. EDS analysis of the reaction zone, shown in Figure 5, indicates a decrease in the amount of Nb present compared to the matrix and a change in the Ti:(Al+Nb) ratio to approximately 3:1. Carbon is also present in the reaction zone as shown in the EELS spectrum shown in Figure 6.

4. DISCUSSION

The reaction of the TiC with the Ti-24Al-11Nb was present in the as-received specimens and became larger/thicker as the thermal exposure increased. Comparing the as-received, 1100°C and the 1200°C heat treatment indicates that, as

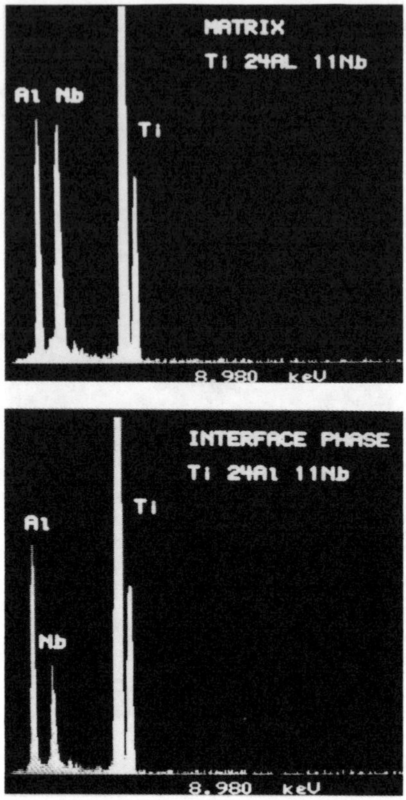

Figure 5: EDS spectra obtained from the matrix and reaction zone.

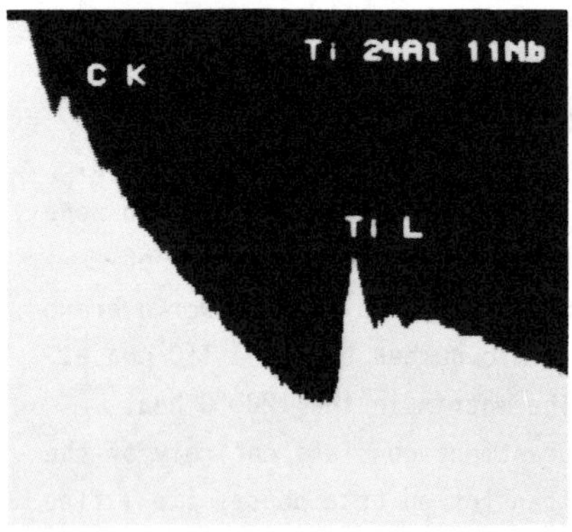

Figure 6: EELS spectrum obtained from the reaction zone.

expected, the kinetics are much faster at the higher temperature. The thickness of the reaction zone increased from approximately 1 micron in the as-received specimens to approximately 6 microns in the 1200°C treatment. The kinetics of the reaction for this system are very different from those reported for the reaction of TiC with Ti-6Al-4V (5).

The reaction zone consists of a cubic phase with a lattice parameter of 0.415 nm, a space group Pm_3m and a composition of approximately 3:1 for Ti:(Al+Nb) and containing some carbon. This evidence is consistent with the reaction zone being Ti_3AlC with some Nb substituting for the Al. Examination of the phase diagram for the Ti-Al-C system indicates that the reaction between Ti_3Al and TiC should result in the formation of the phase Ti_3AlC (6). The phase diagram for the quaternary system Ti-Al-Nb-C is unavailable, however the system Ti-Al-Ta-C does not show that there is some solubility in the Ti_3AlC for Ta (6). Therefore, it does not seem unreasonable for the substitution of Nb to occur in the present case.

5. SUMMARY

The reaction zone between Ti-24Al-11Nb and TiC consists of a fine grained region of the phase Ti_3AlC with some Nb substituting for the Al. The reaction kinetics could not be measured. However, the reaction was fast enough as to make use of the TiC phase in Ti-24Al-11Nb impractical.

6. ACKNOWLEDGMENT

The authors wish to acknowledge the experimental assistance of Mr. R. B. Tallarico and Mr. G. D. Gray of Alcoa and Mr. A. J. Burbery and Mr. T. Doel of the University of Birmingham.

7. REFERENCES

1. Ochiai, S. and Murakami, Y., J. of Mater. Sci., 14, 1979, p. 831.

2. Ochiai, S., Osamura, K. and Murakami, Y, Z. Metallkd, 74, 1983, p. 68.

3. Ochiai, S. and Murakami, Y., Z. Metallkd, 73, 1982, p. 229.

4. Rhodes, C. and Spurling, R., "Recent Advances in Composites in the United States and Japan," ASTM-STP 864, J. R. Vinson and M. Taya, eds. American Society for Testing and Materials, Philadelphia, 1985, p. 585.

5. Konitzer, D. and Loretto, M., submitted to Acta Met.

6. Sridharan, S. and Nowotny, H., Z. Metallkd, 74, 1983, p. 400.

8. BIOGRAPHY

M. H. Loretto is a Professor in the Department of Metallurgy and Materials at the University of

Birmingham. His research interests are in the areas of electron microscopy and defect structures in materials. He has authored two books on the topic of characterization of defect structures using electron microscopy techniques.

D. G. Konitzer is a Staff Engineer at the Alcoa Laboratories. He received his Ph.D in Metallurgical Engineering from the University of Illinois at Urbana-Champaign in 1984. He currently works on the development of high temperature materials.

2nd International SAMPE Metals Conference
August 2-4, 1988

FRACTOGRAPHY OF TITANIUM ALUMINIDE METAL MATRIX COMPOSITES

Mary Lee Gambone
and
Kenneth R. Bain
Allison Gas Turbine Division,
General Motors Corporation

Abstract

Tensile and low cycle fatigue tests have been performed on continuous SiC (SCS-6)/Ti_3Al metal matrix composite. The composite consisted of 8-ply unidirectionally reinforced titanium aluminide of two compositions, Ti-24Al-11Nb and Ti-25Al-10Nb-3V-1Mo. (atomic percent) Fractographic analysis has been performed on the failed specimens. Characteristic features of overload and fatigue fracture of the matrix metal are identified. Elements of fracture structure which can be related to the relative strength and endurance of this material have been studied. Features such as fiber to fiber contact matrix-matrix separation, and adherence of fiber coating to the matrix are analyzed in terms of their effect on tensile strength and fatigue behavior.

1. INTRODUCTION

Titanium aluminide matrix composites are being developed and proposed for use in high temperature and stress applications in aerospace structures and power systems. The allure of this material system arises from the excellent properties of the matrix alloy, Ti_3Al, and the stable nature of the composite it forms with SiC fibers. (SCS-6, manufactured by Textron-SMD) Alpha-two titanium aluminide intermetallics, particularly Ti-24Al-11Nb and Ti-25Al-10Nb-3V-1Mo, (atomic percent) maintain adequate strength and chemical stability to 650°C (1200°F), and, when bonded to SCS-6 fibers exhibit a minimal fiber/matrix reaction. This composite demonstrates the high stiffness and good fatigue behavior of other titanium matrix composites with the advantage of a much higher temperature capability.

The study of the mechanical behavior of this emerging material system remains in its infancy. A great deal may be learned, however, by examining the tensile and fatigue fractures of composite specimens. Both unidirectionally reinforced 8-ply SCS-6/Ti-24Al-11Nb and SCS-6/Ti-25Al-10Nb-3V-1Mo composite were tested in this study. The nominal fiber volume fraction of the material was 30 percent. The objective of this work is to illustrate some of the characteristics of the behavior of this composite system through fractography.

2. RESULTS AND DISCUSSION

2.1 Tensile Tests

Tensile tests were conducted at 24°C (75°F) and 650°C (1200°F) on specimens which were loaded with the fiber axis at 0° to the load axis (longitudinal) and 90° to the load axis. (transverse) Extensometers were used to monitor strain throughout each test until failure to obtain a measure of total strain at failure.

At 24°C the longitudinal composite specimens failed with an ultimate tensile strength that is approximately 80 percent of the value calculated by the rule-of-mixtures. (1) This temperature is below the ductile-brittle transition temperature for alpha-two, which is between 316°C (600°F) and 427°C (800°F), (2) and both matrices showed brittle cleavage fracture. The cleavage facets are equal in size to the alpha-two plates present in the microstructure. A typical fracture is illustrated in figure 1. The matrix of the transverse specimens tested at this temperature also showed cleavage fracture. (figure 2) The fracture path in the fiber in transverse specimens is between the fiber and the fiber coating, not between the coating and the matrix.

The matrix of the specimens tested at 650°C fractured in a ductile manner. Figure 3 shows a transverse fracture surface from a test at the higher temperature. Ductile dimpling in the matrix and a greater separation of the matrix from the fibers is indicative of the increased ductility.

A range of tensile strengths were measured in longitudinal specimens tested at 650°C. Measured strengths were normalized to the rule-of-mixtures strength expected for the volume fraction of fiber in each specimen. Figure 4 shows the variation in normalized strength with volume fraction and total strain at failure. Specimens with a higher volume fraction fiber appeared to perform better than those with a lower volume fraction fiber. This is probably due to processing variations. The measure of total strain at failure seems to indicate how strong a composite specimen is regardless of variations in volume fraction. A well composited specimen with a strength approaching the rule-of-mixtures strength might strain in excess of one percent before fracture while inferior composite will support less strain.

In figure 5 the fracture surface of a weak and a relatively strong composite specimen are compared. Typical of the weaker composite is contact between adjacent fibers due to 'fiber swim' (movement of fibers out of alignment during the consolidation process) and inadequate flow of the matrix between the fibers. Separation between matrix layers along the bondline is also evident in the weaker composite. This does not occur in the composite specimens with the highest volume fraction fiber.

2.2 Low Cycle Fatigue Tests

Longitudinal specimens were loaded in low cycle fatigue at 650°C. Composite of both super and conventional alpha-two titanium aluminide were tested. Tests were performed in strain control at identical strain ranges and a frequency of 20 cpm. An R-ratio of 0.1 was used to avoid cycling the specimens in compression. All the failures occurred in the gage.

Fractographic examination of the failed specimens showed that all the crack initiation sites occurred at the surface of the material. All specimens tested possessed more than one initiation site. Initiations were located primarily at corners and edges where damaged fibers were exposed at the surface; however, occasionally fatigue cracks also initiated along the width of the gage. Figure 6 illustrates several characteristic fatigue crack initiations in this material. The matrix fracture at the initiation sites is flat and featureless and the fiber fractures in the same plane as the matrix. There is complete coincidence between the fiber and matrix in the region of the initiation and fatigue crack growth.

Figure 7 shows the change in fracture morphology from fatigue crack propagation to overload. as the fatigue crack grows away from the initiation site the matrix fractures in cleavage and the fibers fracture along several planes. The surface is rougher than at the initiation, but still appears brittle. Fatigue striations may sometimes be found in this region of crack growth. Some typical stria are presented in figure 8. The spacing of these stria and their distance from the initiation indicates that the approximate crack growth rate at this point was 2.5×10^{-5} cm/cycle and it is estimated that initiation occurred very early in the fatigue life of the specimen.

The fracture surface of the overload area is similar to a tensile fracture at the same temperature. The matrix fails by ductile dimpling and the fibers pull away from the matrix and pull out from the fracture surface.
There was a wide range of fatigue lives measured in these composite specimens. However, a relatively low tensile strength did not neces-

sarily produce a correspondingly low fatigue life. That material which demonstrated a low total strain at failure had a greater degree of scatter in fatigue life than the stronger material tested at the same condition. The same features which were indicative of poor tensile performance were present in the low fatigue life specimens. These included fiber to fiber contact and debonding in the matrix. It was also observed that the initiation site might be less distinct in the poor quality composite. Hysteresis loops recorded during cycling may illustrate the quality of the bond between fiber and matrix. There is a distinct change in slope on loading in those specimens where fiber-matrix debonding was prominent.

3. CONCLUSIONS

The fractographic examination of these SCS-6/titanium aluminide composites led to the following conclusions:

1. Poor bonding, which is manifest in defects such as fiber to fiber contact, will reduce the ultimate tensile strength and total strain at failure of a 0° composite specimen.

2. Failure in transverse tensile specimens occurs between the fiber and the fiber coating. The fiber coating adheres well to the matrix. This failure mode persists at elevated temperatures and so is independent of the matrix ductility.

3. Regions of fatigue crack initiation, growth, and overload can be distinguished in the fracture surface of 0° composite specimens tested at elevated temperature.

4. Low cycle fatigue crack initiations occur at specimen surfaces, not internally. The cracks grow across the fibers (perpendicular to the load axis) and not along the the fiber matrix interface.

5. Low total strain at tensile failure indicates a potential for large scatter in strain controlled fatigue life, but not necessarily reduced lives.

4. REFERENCES

1. Jones, R. M., Mechanics of Composite Materials, Scripta Book Company, 1975, pp. 127-130.

2. O'Connell, T. E., "Production of Titanium Aluminide Products," Final Report for 16 May 1978 to 30 Sept. 1982, United Technologies Corporation, AFWAL/MLLN, WPAFB, Ohio, p. 129.

5. BIOGRAPHIES

Mary Lee Gambone earned a B. S. in metallurgical engineering from Purdue University and an M. S. in metallurgy from the Massachusetts Institute of Technology. She has been working at Allison Gas Turbine Division for four years in the area of advanced material behavior. Her responsibilities include the development of mechanical test procedures for metal matrix composites and coordinating characterization of these materials.

Dr. Kenneth R. Bain received a B.S. in mechanical engineering from General Motors Institute and an M.S. and Ph.D. in metallurgy from the Massachusetts Institute of Technology. He has worked at Allison Gas Turbine Division of G.M.C. for 13 years where as a development engineer he has been responsible for programs to characterize the behavior of engineering materials.

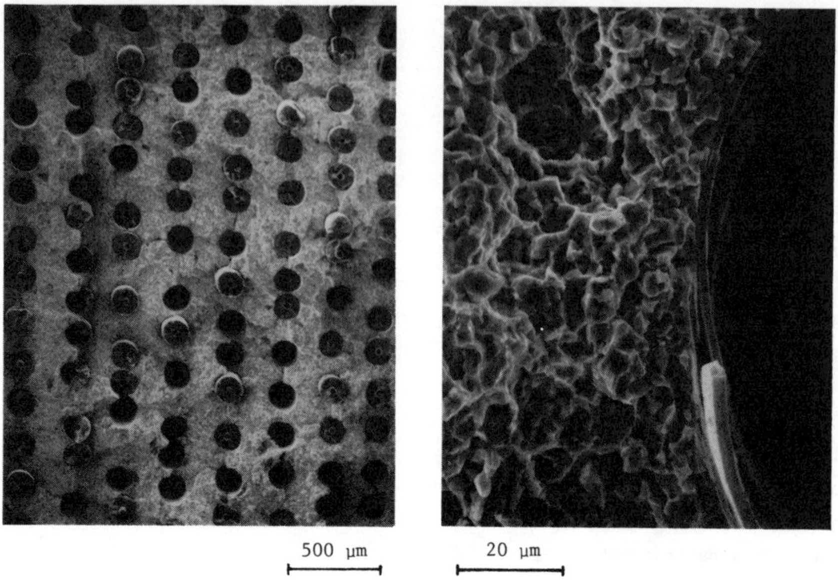

Figure 1: SEM fractograph of a typical room temperature tensile failure of a longitudinal SCS-6/Ti_3Al composite specimen.

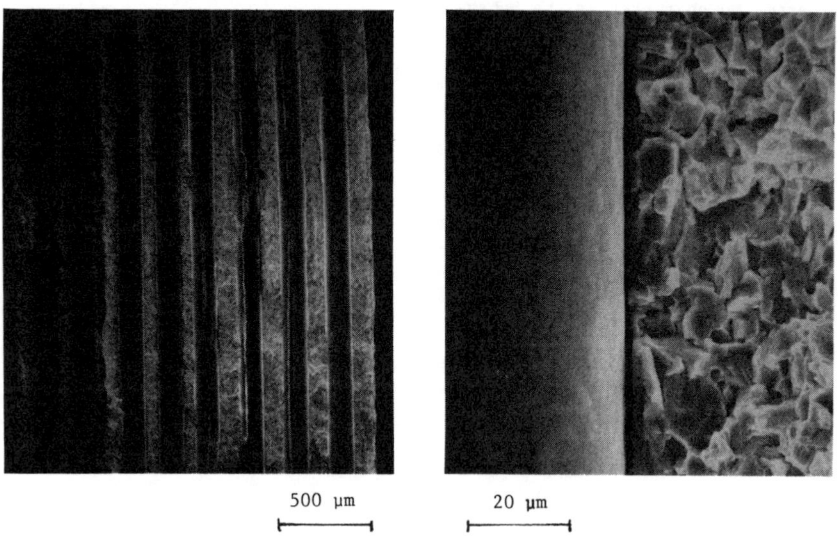

Figure 2: A typical room temperature tensile failure of a transverse SCS-6/Ti_3Al composite specimen. Note the cleavage facets in the matrix and the lack of separation between matrix and fiber.

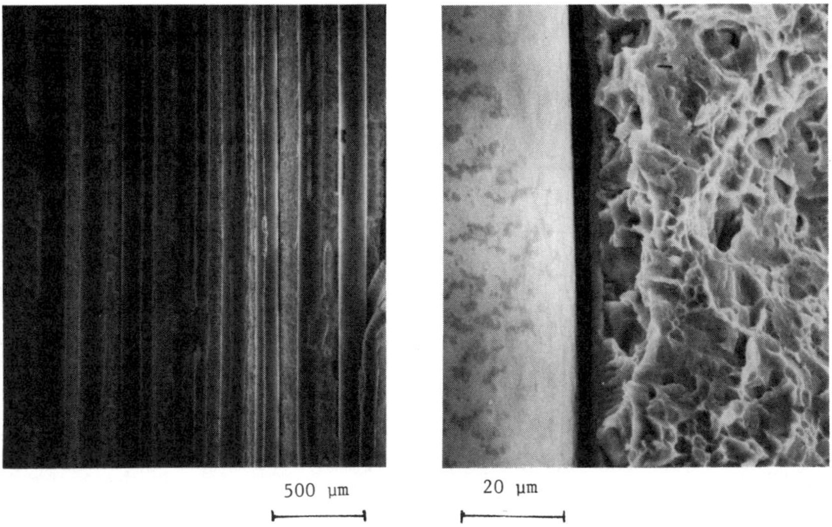

Figure 3: A typical 650°C tensile failure of a transverse SCS-6/Ti_3Al composite specimen. Note the ductile dimpling in the matrix and the separation of the matrix and fiber coating from the fiber.

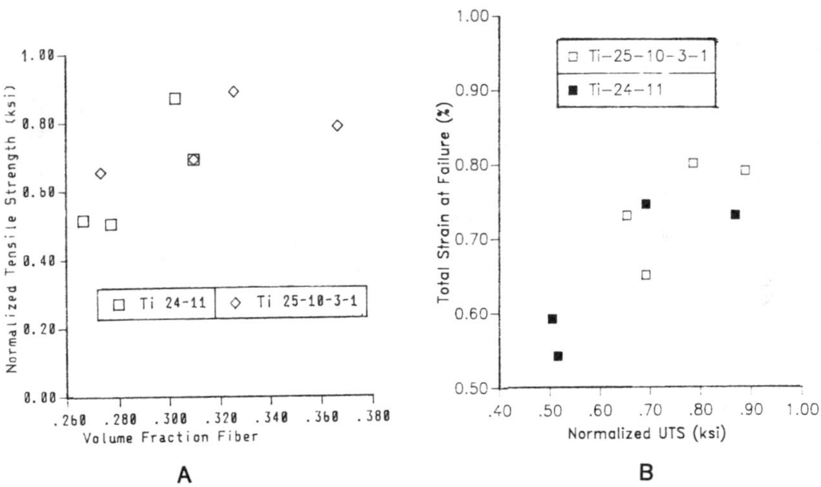

Figure 4: Normalized tensile strength as a function of volume fraction (A) and compared to total strain at failure (B) for 650°C tensile tests of longitudinal SCS-6/Ti_3Al composite specimens.

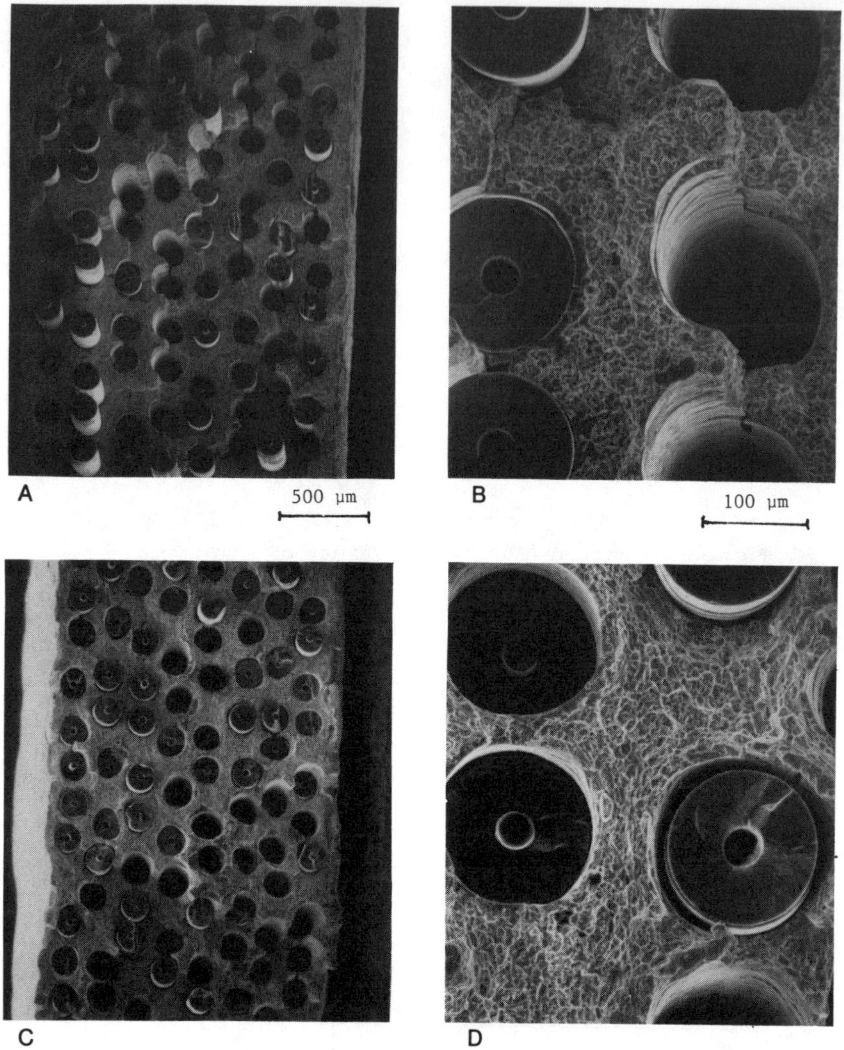

Figure 5: A 650°C tensile failures of longitudinal SCS-6/Ti_3Al composite. Specimen 657-4 in (A) and (B) failed at a normalized UTS of 0.50. Speciman 677-1 in (C) and (D) failed at a normalized UTS of 0.89. Note the greater degree of matrix-matrix separation and fiber in contact in (A) and (B) as opposed to (C) and (D).

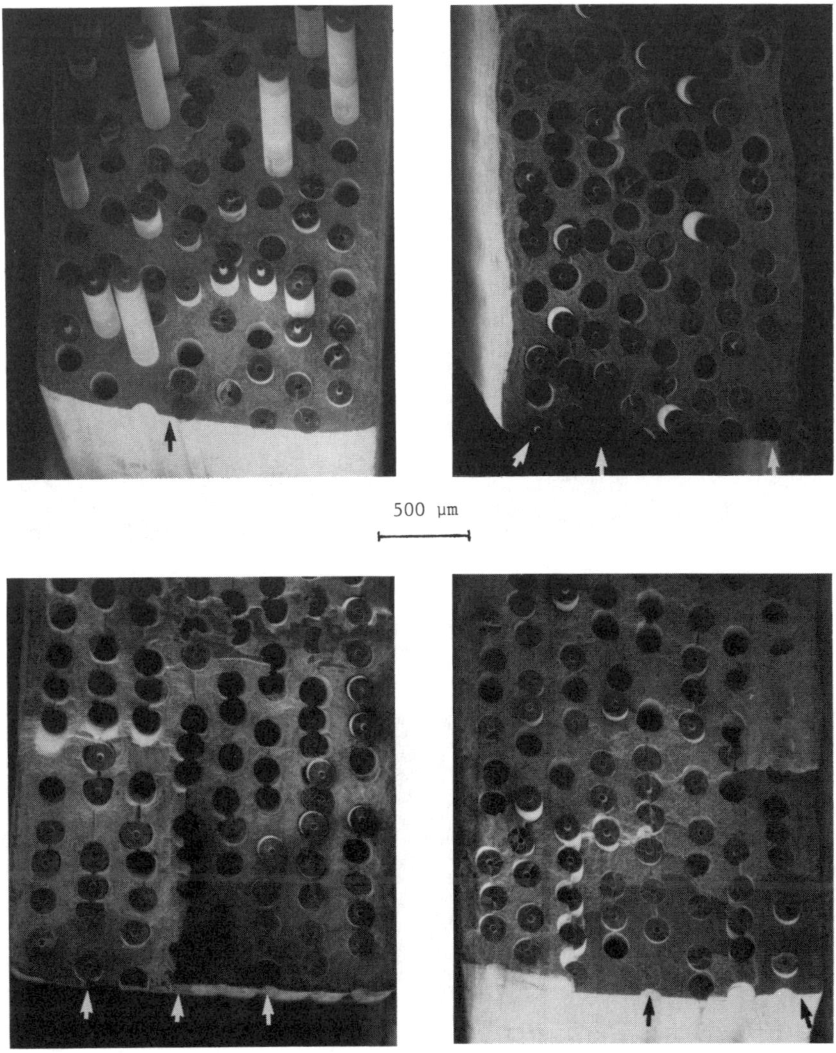

Figure 6: Several low cycle fatigue crack initiation sites. All specimens were tested at 650°C, R=0.10, and 20 cpm. Initiation sites are marked with arrows. Initiation of fatigue cracks occurs primarily at edges and damaged fibers.

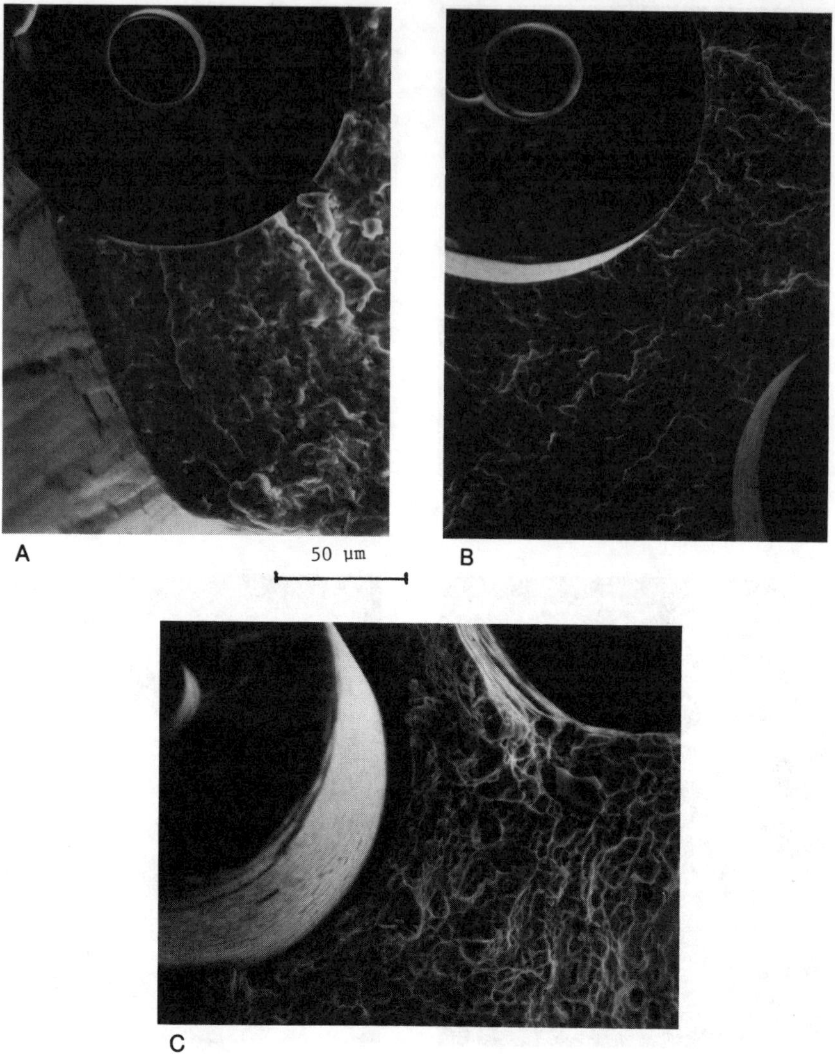

Figure 7: An SCS-6/Ti_3Al composite LCF specimen which illustrates the changes in fracture morphology from (A) fatigue crack initiation, through (B) fatigue crack growth, and (C) tensile overload.

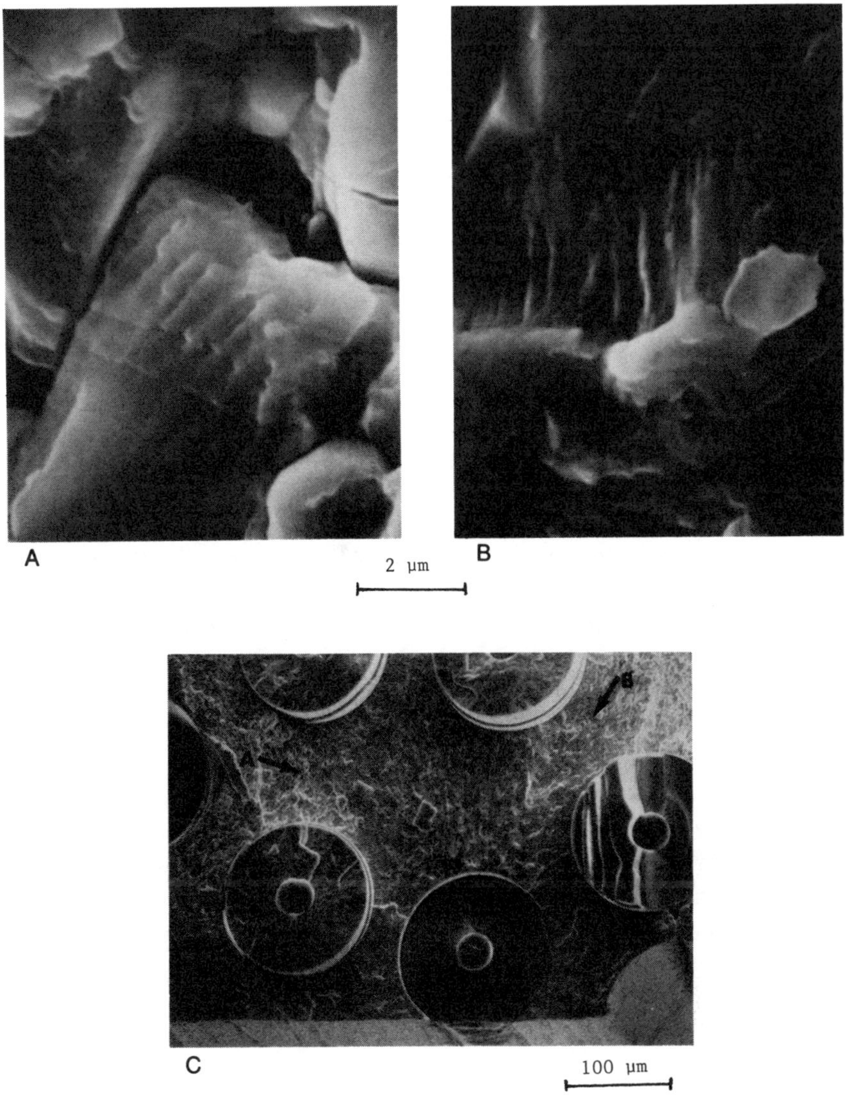

Figure 8: A 650°C LCF specimen of SCS-6/Ti_3Al composite. View (C) shows the initiation and crack growth region where the fatigue stria found in (A) and (B) were located.

2nd International SAMPE Metals Conference
August 2-4, 1988

THE COMMERCIALIZATION OF METAL-MATRIX COMPOSITES PRODUCED BY INGOT METALLURGY

D. M. Schuster, M. D. Skibo and F. W. Yep
Dural Aluminum Composites Corporation
10326 Roselle Street
San Diego, California 92121

Abstract

Metal matrix composites (MMC's) are among the most technologically promising materials of the future. They offer properties that no other material can match, having the high strength, stiffness and wear resistance of ceramics, and the formability, lightness and toughness of aluminum. Unfortunately, these advantages have been offset in the past by the extremely high cost of producing MMC's. Recently, an inexpensive ingot-metallurgy process which allows ceramic particles (SiC or Al_2O_3) to be included into all aluminum product forms (castings, extrusions, forgings, rolled sheets, and drawn wire) has been developed. Discontinuous ceramic platelet grinding media and ingot-grade aluminum are used as the raw materials. The process yields a ceramic particle reinforced aluminum composite material called DURAL™ which is available in the form of foundry ingots, extrusion billets, and rolling blanks. The growing demand for DURAL™ has encouraged the company to scale up to produce large quantities of the material.

1. INTRODUCTION

After some 20 years of research and development on metal matrix composites, or MMC's, it is clear that these versatile materials are well suited for countless high-performance applications in the aerospace industry, among others. By combining the high strength, stiffness, and wear resistance of ceramics with the toughness and formability of metals, MMC's provide mechanical properties markedly superior to those of unreinforced alloys of comparable density. Particularly in aerospace applications, appreciable enhancements in mechanical property values per unit mass are invaluable, in terms of both improved performance and improved fuel efficiency.

But the price of such progress can be too high to pay. The reason that MMC's have not been widely used in fabricating the structural and functional components of aircraft and other vehicles is their hitherto prohibitive cost--or, more precisely, their inadequate cost-effectiveness.

Ceramic particle reinforced light-metal alloys, such as silicon carbide-aluminum or alumina-aluminum, have been made primarily by processes so inherently costly that the large-scale commercialization of these materials has been retarded. Furthermore, MMC's made by other methods cannot be easily melted and cast to net shape--a great disadvantage.

Thus there has been a long-standing need for the development of a particle-reinforced aluminum MMC made inexpensively enough that it can compete with steel, not just in performance but also in price. Realistically, the best way to accomplish that is through ingot metallurgy. This objective has been reached by Dural Aluminum Composites Corporation in San Diego. Dural has developed and patented a unique but simple process for incorporating microscopic ceramic particles into molten aluminum, directly. The particles are thoroughly wet by the melt without unwanted chemical reactions and with a minimum of dissolved gases and included oxides. The process also produces a highly uniform distribution of particles in the melt, as shown in Figure 1. The mechanical properties of the cast material are isotropic.

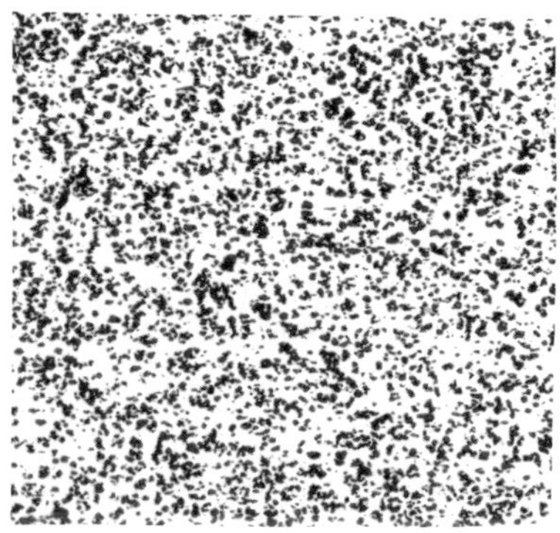

Figure 1: Typical microstructure of a 20 vol. % SiC-Al (AA6061) extrusion. (X50)

The material, called DURAL™ MMC, has all the desirable properties that we have come to expect of MMC's; such as, exceptional strength, stiffness, wear resistance, and resistance to fatigue-crack growth.

DURAL is a trademark of Dural Aluminum Composites Corporation.

It also has excellent high temperature properties (to about 600 degrees F) and a thermal expansion coefficient that can be matched to those of numerous alloys by varying the ceramic content. The latter is typically in the range of 10% to 20% by volume and presently consists of either silicon carbide or aluminum oxide; other reinforcing agents are under development.

Most significant, however, are two factors that set DURAL™ MMC apart from all other currently available MMC's:

1. DURAL™ is inexpensive. Except for steel, all other materials with comparable properties cost much more--some, like titanium, up to several times more.

2. DURAL™ can be easily remelted and cast to net shape, without compromising the material's physical and chemical properties.

2. THE COST FACTOR

DURAL™ MMC is made from inexpensive raw materials; ingot-grade aluminum and commercially available grinding media. Because ingot metallurgical processes are inexpensive, the cost of producing DURAL™ is very low.

In currently available pilot-scale quantities, the price is U.S.$3 per pound, down from U.S.$250 per pound four years ago. Further substantial price cuts will be made when full-scale industrial production begins in 1989. Major new production facilities for this purpose, with an annual production capacity of between 25 and 50 million pounds, are being built by the Alcan Aluminum Corporation, Dural's parent company. Economics of scale will allow the price of DURAL™ MMC to drop to a level marginally higher than that of unreinforced aluminum. All standard sizes of foundry ingots, extrusion billets, and rolling blanks will be available, in a wide variety of formulations.

The ease of fabrication of DURAL™ MMC is another economic plus. Using standard equipment and methods, it can be cast, extruded, forged, rolled, and superplastically formed about as easily as aluminum. Hardly any retooling is required.

3. DURAL™ MMC CASTINGS

Unlike other MMC's, DURAL™ is manufactured as foundry ingots that can be remelted and cast to net shape; for some examples, see Figures 2 and 3.

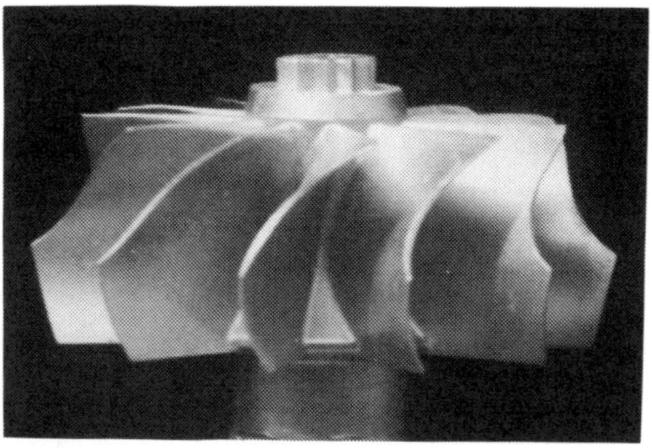

Figure 2: An investment-cast prototype turbocharger impeller made from 10 vol. % SiC-Al (A357). The part was cast to net shape by Dural Aluminum Composites Corporation.

Figure 3: Prototype missile guidance bays, investment-cast from 20 vol. % Sic-Al (A357).

The scrap can also be remelted. The casting parameters are nearly identical to those used for the corresponding unreinforced aluminum alloys, such as A356 and A357. The only unusual requirement is that the melt must be stirred to prevent settling of the ceramic particles. This is easily accomplished with a simple mechanical stirrer that swings in and out of the crucible.

4. DURAL™ MMC WROUGHT PRODUCTS

DURAL™ MMC extrusion billets are produced from such well-known alloys as 6061 and 2014. The process used is a patent-protected, direct-chill (DC) casting technique developed by Dural Aluminum Composites Corporation. It represents a breakthrough that will allow the mass production of these billets at very low cost. Currently, DC billets of 7-inch diameter and up to 60-inches in length are available. Some examples of extruded shapes are shown in Figure 4.

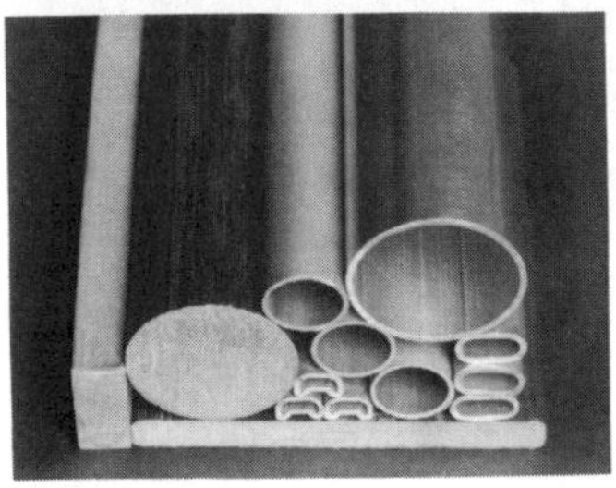

Figure 4: Selected extruded shapes made from various wrought aluminum alloys with SiC contents of up to 20 vol. %.

The faster casting speed of the DC method improves not only the production efficiency but also the physical properties of the material, providing a finer grain structure in the aluminum and a more homogeneous particle distribution in the matrix. Some examples of the T6 tensile properties of near-production extrusions are shown in Table 1.

Table 1:

Examples of T6 Tensile Properties of Near-Production Extrusions

Shape	Volume % SiC	Tensile Properties UTS (ksi)	0.2% YS (ksi)	% Elong.	Young's Modulus ($x10^6$ psi)
Bar	0	47.4	44.8	15.8	9.7
	10	50.9	46.5	7.5	11.4
	20	54.7	59.8	2.3	14.1
I-Beam	0	43.5	39.0	10.5	9.6
	10	49.4	46.5	3.8	11.7
	20	51.1	48.6	2.3	14.0

Note:

Longitudinal properties, I-beam tensile specimen was flat bar, 0.08 in. thick, from web.

5. APPLICATIONS

Although DURAL™ MMC is denser than aluminum by only a few percent, it is 20% to 40% stronger, 40% to 60% stiffer, and an order of magnitude more wear-resistant. It can replace unreinforced aluminum in a wide variety of aerospace and other applications, where its slightly higher cost per pound will be offset by the considerably lower mass of material needed for a given set of performance specifications.

This effect is even more dramatic when DURAL™ MMC replaces titanium or steel, as, e.g., in compressor blades and engine connecting rods (see Figures 5 and 6). Here the cost penalty is offset first by the much lower mass, and second by the greater fuel efficiency that results. Depending on the circumstances, the savings thus realized might be compounded by reducing the masses of other system components, owing to the smaller inertial or vibrational loads from the lighter DURAL™ components. This, of course, would mean still greater fuel efficiency.

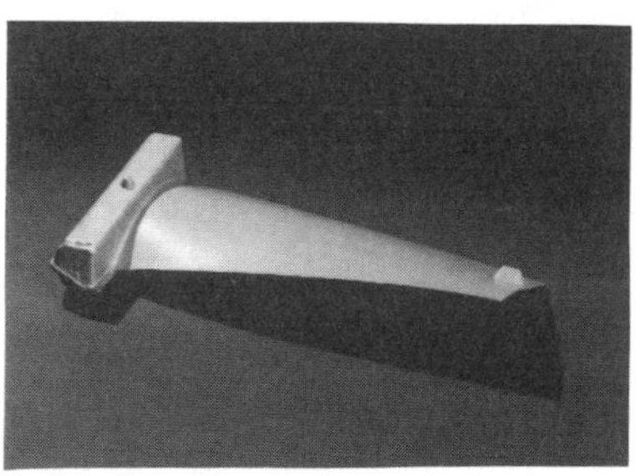

Figure 5: Prototype compressor blade forged from a 20 vol. % SiC-Al (2014) extrusion. It duplicates a blade forged from titanium.

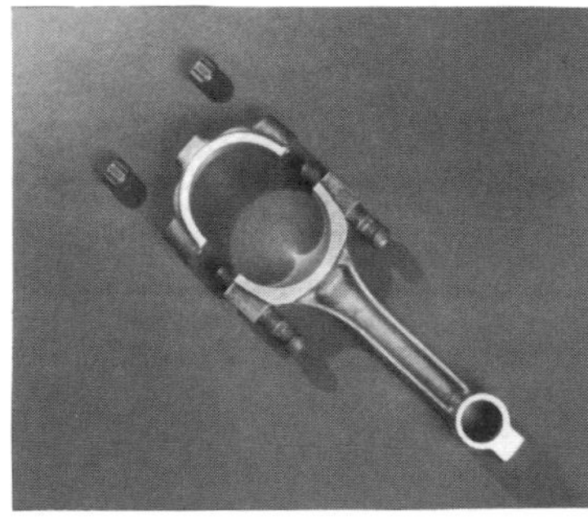

Figure 6: Prototype connecting rod forged from a 15 vol. % SiC-Al (6061) extrusion. Engine performance is significantly improved by substituting this material for steel.

The notion of replacing steel, in some applications, with a much lighter material that has a considerably higher specific strength and a much higher specific stiffness--and that is much more easily worked and formed and much more corrosion resistant--has given design engineers food for thought for many years.

The additional notion that such a material will soon cost not much more than the common structural materials is, however, brand new. With the strong commitment to commercialize these new MMC's being made, we believe that the prospect is very good indeed.

2nd International SAMPE Metals Conference
August 2-4, 1988

V I D P
A NEW SYSTEM CONCEPT FOR VACUUM MELTING OF SUPERALLOYS, FERROUS AND NON-FERROUS ALLOYS

Otto W.Stenzel and Hans-J.Kemmer
LEYBOLD AG, Hanau/FRG

ABSTRACT

This paper introduces a new vacuum melting concept, the Vacuum Induction Degassing and Pouring Furnace, in short called VIDP, which shows some significant advantages in comparison to conventional vacuum induction melting practice.

The vacuum induction melting process in general is gaining importance again with the currently growing demand for cleaner metals. The VIDP concept is a logical development from the standard VIM process which consequently resulted in a highly modular and flexible vacuum melting system with the following characteristics :

- vacuum chamber size for a given molten melt volume is reduced to minimum
- small chamber surface reduces the total amount of desorbed moisture from the furnace walls and enhances vacuum quality
- the outer lining and coil area are pumped separately to prevent any negative influence on the melt from lining moisture etc.
- argon consumption for an inert gas atmosphere, if required, is reduced drastically
- inside the vacuum chamber, risk of leaks is reduced by eliminating all flexible power cables, water hoses and hydraulic lines
- process and handling steps like charging, deslagging, sampling, refining and pouring are straight forward operations with minmal time requirements.

0. INTRODUCTION

The new Vacuum Induction Melting Furnace Concept called VIDP is - in comparison to the well known chamber type furnaces - a bell-type furnace with connected casting chamber.

1. BASIC VIDP CONCEPT

In comparison to chamber type furnaces the vacuum tight shell is close to the coil and forms part of the furnace body.

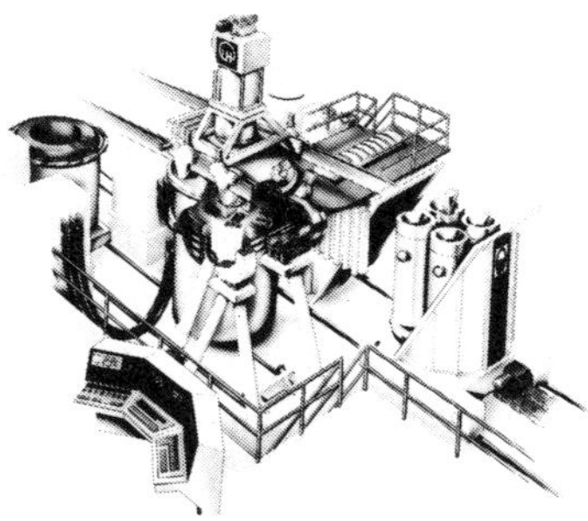

Fig.1: Total view of a VIDP production unit

The melt chamber volume and the chamber surface are drastically smaller than those of a standard VIM. When pouring the shell is tilted together with the coil and the crucible.
The furnace is tilted around a pouring tunnel.The melt is poured into fixed launder-tundish inside the pouring tunnel, which transfers the liquid metal into the casting chamber. The launder-tundish is brought in through a vacuum lock valve shortly before pouring. For high degrees of purity, the launder can be fitted with slag barriers and additional ceramic filters. In the lock-in position the launder-tundish is electrically preheated to maintain an adequate temperature of around 1000°C.

Casting is done in a separate chamber, which can be designed to incorporate electrode/barstick molds, ladles or CC equipment.
At the top of the furnace lid a charging valve is fixed . To its flange a traveling charging chamber can be connected.
All hydraulic cylinders for tilting and the flexible water cooled power cables are outside the vacuum chamber.

The low furnace body with the coil and the crucible is connected to the lid by a quick actuated bayonet connection. This furnace body can be moved aside for relining or can be quickly exchanged by a newly relined second furnace body.
The rear space of the coil is evacuated separately.
By this, the desorbed residual moisture of crucible lining will not come in contact with the melt surface resulting in excellent pressure rise values during the process.

The standard VIM process is certainly possible, i.e.

- melting, alloying, degassing and pouring of the melt into steel or graphite molds.

Additional applications are for instance :

- melting, alloying and liquid melt-transfer to a continuous caster,

- melting, alloying and liquid melt transfer to metall powder atomization plant tower,
- melting, alloying and liquid melt transfer to a small induction heated crucible for investment casting,
- the possibility of melt transfer to both sides of the pouring tunnel allows to connect two casting techniques to only one melting furnace (see Fig. 2).

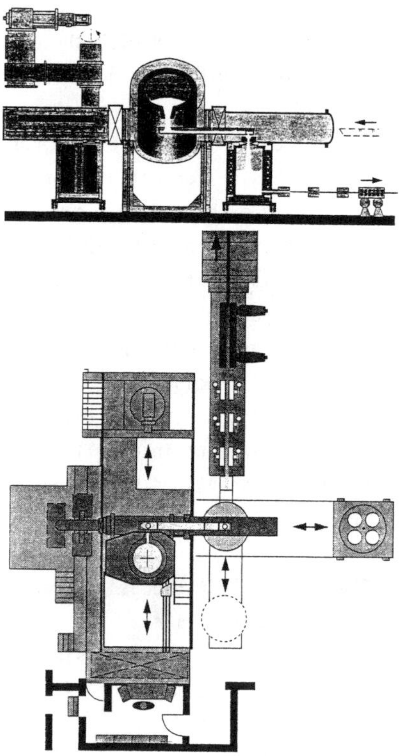

Fig.2: VIDP with combined ingot/ electrode casting and HCC (Horizontal Continuous Casting)

2. ELIMINATION OF GAS SOURCES

The final gas content (especially oxygen) of the VIM melt not only depends on the initial gas concentration and on possible refining reactions (also including oxygen pick-up by melt-crucible reactions), but also on whether a sufficiently low partial pressure of oxygen and other gases can be maintained in the VIM-chamber during melting.

Beside air leaks there are different sources of gas in the VIM-system as listed below :

- Process gas	from	melt
- Outgassing	from	crucible
- Water leak	from	cooling water connections
- Oil leak	from	hydraulic systems
- Air leak	from	seals, welding seams
- Desorption	from	chamber walls

The total gas flow rate in a VIM system which must be pumped away by the vacuum pumping system is the sum of the individual gas flow rates generated by above sources.

Detailed measurements on VIM furnaces resulted in the statements : Besides the gas generated from the melt itself, the gases from the crucible refractory, from the wall desorption and from potential water and oil leaks in the system play a dominant role. At these points the plant manufacturer has to orientate his efforts for improving the technology by new process and equipment developments.

Significant plant/process improvements are for instance :

- selection of suitable refractories
- reduction resp. elimination of any water and moisture
- minimization of the desorption surface in the vacuum chamber
- avoidance of outgassing materials applied for such equipment which is inside the chamber such as hoses, insulation cables,etc.
- minimization of the chamber volume itself
- avoidance of flexible cooling water and hydraulic lines inside the vacuum chamber with their potential danger of leakage
- suitable sealing elements and feed-throughs to obtain low leak rates
- improved on-line control of the process atmosphere.

3. METHODS FOR CLEANER METAL

"Superclean superalloys" are produced today by process combinations such as VIM/ESR/VAR, VIM/VAR/VAR, VIM/EBCHR etc. However it is still the VIM process and the quality controls exercized on this process which largely determines the final material properties.

Besides critical refractory selection - a matter of cost too - the application of modern pouring systems involving filter technique leads to better and cleaner metal. The application of multi-compartment prefired tundish/launder systems which promote non-metallic inclusion removal, has improved electrode quality and improved remelt stability.
Filtering of molten metal by means of a ceramic reticulated foam filter, removes impurities.
The performance of cermic filters (material Al_2O_3 or ZrO_2) with regard to casting rate and removal of all types of non-metallic particles, liquid inclusions, gases and trace elements is of great promise. Test results with typical superalloys have demonstrated filter efficientcies between 70 and 90 %.

The performance of a filter not only depends on the material and design of the filter but also on the alloy, its casting parameters and mainly the temperature conditions of (preheated) filter and liquid metal.

Preheating before and heating of the filter while pouring is of key importance. A typical example of a VIM-launder/tundish arrangement with slag barriers and filter for liquid metal transfer and purification is shown in Fig. 3.

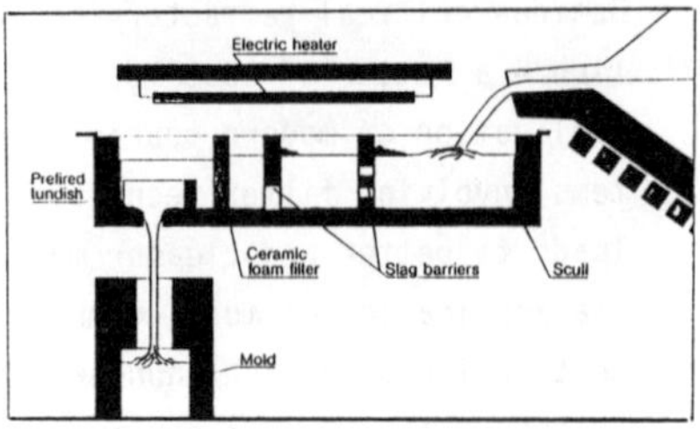

Fig.3: VIM transfer launder-tundish with slag barriers and filter

The tasks of a VIM-launder/tundish are as follows :

- transfer of liquid metal from melt-area (crucible spout) to casting area (mold-top)
- centric filling of any mold with compact metal stream and constant pouring speed
- catching slag particles by slag barriers
- generating laminar metal flow to allow floating up of inclusions
- filtration of smallest non-metallic particles and liquid inclusions.

Errosion of launder/tundish refractory as well as chemical reactions are largely avoided when using the correct refractory material. The adsorbtion of non-metallic inclusions by the refractory surface combined with a low velocity of melt flow is an advantage of a long launder-tundish. The area after the pouring-in-point of the melt must be sufficiently long. Prefired, wear proof launder-tundishes are used.

The knowledge about the importance of various gas sources and about the positive effects of well designed launder-tundishes led consequently to the new VIDP design developed by Leybold AG.

4. APPLICATION OF VIDP

This patented VIDP-technique opens the way for economical production under controlled atmosphere of all high grade metals and alloys commonly processed in chamber-type VIM furnaces.
The VIDP concept is composed of modules, starts with a basic version and can be extended to a fully equipped version.

In 1988 four VIDP furnaces between 2 t and 6 t for various applications have been ordered. The first plant for metal powder production with two atomization stations will start production in summer 1988.

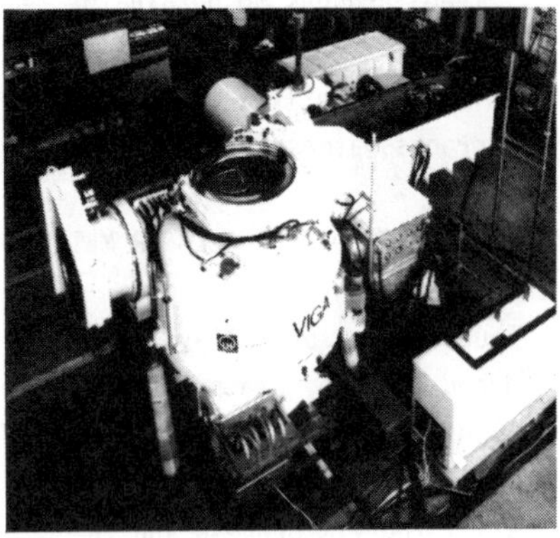

Fig.4: VIDP during preassembly

The flexibility of casting technique, the small furnace volume, the small furnace surface, the reduced risk of leakage, the simple exchange of furnace body with coil and crucible, the possibility of easy cleaning of furnace lid and the reduced space requirements give special advantages in serveral applications.

The VIDP with these features has its special advantages in :

- vacuum operation,
- inertgas operation,
- overpressure operation.

For the production of superalloys or high quality ferrous alloys, which are casted to electrodes, forging ingots or barsticks, the VIDP has its advantages in cleaner atmosphere by reduced gas desorption of chamber walls, reduced lining moisture and reduced leakages.

The production of oxygen-free-copper with an oxygen content below 1 ppm bears the opportunity to adapt casting in stationary molds and continuous casting or powder atomization for flexible production reason.

The production of Al-Li alloys will be done under an extra clean inert-gas atmosphere. Large amounts of scrap can be remelted without Li-loss.

VIDP concept advantages in Al-Li production are :

- no rubber hoses of flexible power cables are inside the furnace, which will be chemically attacked,- small chamber volume for refilling with inertgas,
- easy cleaning of inner furnace surfaces,
- nobody has to enter a furnace chamber for crucible exchange,
- liquid melt will be transfered under protective atmosphere to a continuous caster.

Remelting of low contaminated nuclear steel parts will be done under atmospheric pressure. Charging, melting, deslagging and casting create contaminated fume and dust.

All requirements as :
a closed furnace with filter system, a lock-in charger, a separate casting chamber, a slag lock-out are ideally met. All fittings and armatures, which have to be maintained, are outside the nuclear contaminated area.

The production of the new generation of High-N_2-Steels is possible as this steel must be alloyed and solidified under increased pressure. Only small chamber dimensions allow to use technically acceptable wall thickness and gas consumptions.

5. FLEXIBILITY IN CASTING

The most traditional way of casting is the pouring within a chamber type VIM into molds of various sizes arranged on a movable support.

In order to raise productivity pouring is done via transfer launder into a separate casting/mold chamber which allows the next melting sequence to start shortly after pouring.

Variable pouring/casting techniques can be applied in combination with the VIDP-furnace. Two or more casting techniques can be connected to only one VIDP-furnace. Some examples are shown in Fig. 2, 5 and 6.

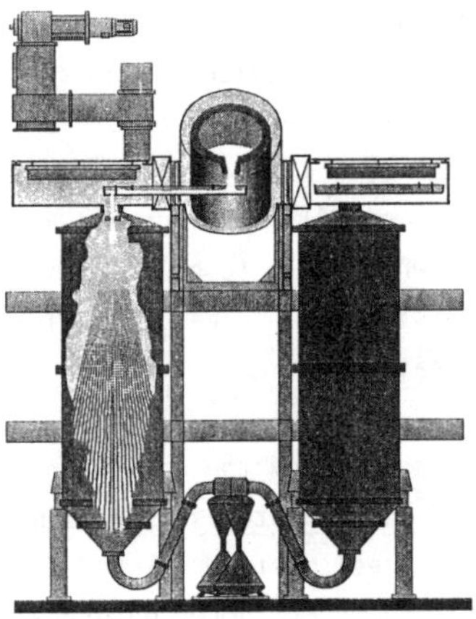

Fig.5 : VIDP with powder atomization process two-station

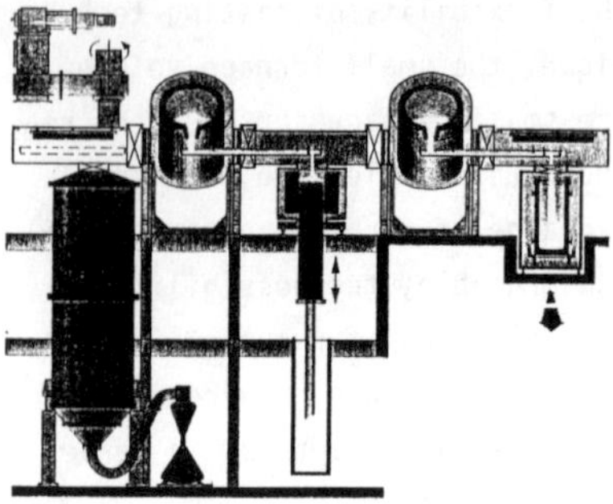

Fig.6: VIDP-duplex system with combined ingot casting/ vertical continuous casting/powder atomization process

It is explained in these firgures, how for instance a horizontal/vertical casting unit and a mold chamber for ingot casting are connected to the same furnace. Because of the reduced furnace height, the reduced space requirements, low inertgas consumption and the possibility of transferring the liquid melt in both directions of the pouring tunnel, the VIDP is especially suited for the connection to metal powder production plants with one or two atomization towers.

A further development in this field is the combination of a vacuum induction melting unit with a horizontal or vertical continuous caster operating under inertgas. This direct way for the production of sheets, strips, bars and wires has obvious advantages.

Fig. 6 shows the basic concept of a VIDP-furnace combined with a horizontal continuous caster and an ingot casting unit. Casting can be done under protective atmosphere without the influence of air.

SUMMARY

The new VIDP furnace design permits a number of new applications, which may give some innovative impulses in ferrous and non-ferrous metal vacuum melting and casting.

BIOGRAPHY OF AUTHORS

Otto W.Stenzel
Born June 21st, 1941 in Aussig/ Germany

- Febr. 1968 - Degree in electrical engineering from the University of Hannover/FRG
- 1968 - 1977 - Assistant engineer at the Institute of Electro-Heat of the University of Hannover/FRG
- 1976 - Dr.-Ing.-Dissertation at the University of Hannover/FRG
- 1977 - 1983 - Manager of Electric. Engineering Department of Special Metallurgical Furnaces at Leybold AG, Hanau FRG
- since 1983 - Senior Engineer, Development of Systems, Industrial Vacuum Metallurgy Division at Leybold AG, Hanau FRG

Hans-J.Kemmer
Born November 14th, 1959 in Sombor/ Yugoslawia

- Febr. 1972 - Degree in Mechanical Engineering from the University Darmstadt FRG
- 1972 - 1976 - Project Engineer for energy, power and heat stations
- since 1976 - Product Manager Vac. Induction Furnaces Industrial Vac. Metallurgy Division at Leybold AG, Hanau/ FRG

2nd International SAMPE Metals Conference
August 2-4, 1988

ENHANCEMENT OF SUPERPLASTIC FORMABILITY IN A HIGH STRENGTH ALUMINUM ALLOY

S.P. Agrawal, G. R. Turk, and R. Vastava
Northrop Corporation
Aircraft Division
Hawthorne, California

ABSTRACT

A 7475 aluminum alloy was developed for superplastic forming (SPF). By lowering the Fe and Si contents in this alloy significantly below their normal levels and optimizing the thermomechanical processing to produce sheet, over 2000% thickness strain to failure was obtained. The microstructure, elevated-temperature uniaxial and biaxial tension, and cavitation behavior of the alloy were determined. In addition, a contitutive model was used to form a generic structural shape from which mechanical test specimens were removed and post-SPF characteristics were evaluated. The constitutive model included both material strain hardening and strain rate hardening effects, and was verified by accurately predicting forming cycles which resulted in successful component forming. Stress-life fatigue, stress rupture, and room and elevated temperature tensile tests were conducted on the formed material.

The results presented in this paper are part of ongoing research under a NASA contract, NAS1-17623.

INTRODUCTION

The initial industry and DoD interest in superplastic forming (SPF) of high-strength aluminum alloys was spurred by the cost and weight savings resulting from the development and implementation of SPF titanium technology. Since then, SPF aluminum technology has progressed rapidly. The industry and DoD have worked jointly to make the material and process developments needed to establish SPF of aluminum as an effective manufacturing technology for fabricating complex monolithic parts at significant cost and weight savings. New alloy concepts for enhanced superplasticity and the effect of alloying additions on precipitates, dispersoids, and grain size refinement in thermomechanically treated alloy sheets have been investigated. These efforts have led to a better understanding of cavitation, allowing aluminum alloys to be superplastically formed to high strains in a manufacturing environment without nucleating microstructural cavities.

This paper presents the results of a program in which the SPF behavior of a high-strength 7475 aluminum alloy was enhanced via modification of alloy composition. The baseline 7475 composition was modified by changing the amounts of Fe, Si, Mn, Cr, and Zr. Laboratory ingots with modified chemistries were cast and rolled into plates. Fine grained sheets were produced from these plates using a previously developed thermomechanical treatment.(1) The microstructure, dispersoid characteristics, and elevated-temperature flow (SPF) behavior of these sheets were evaluated. Correlations were then made between the material microstructure (and chemical composition) and the superplastic ductility and strain rate.

To efficiently and reliably form the material selected by using the above approach into a complex shape, the material formability was predicted using a constitutive model. Several trough-shaped pans, representing various magnitudes of SPF strain, were fabricated according to the model. The predicted formability was then compared to the observed parameters. Post-SPF properties of specimens extracted from these pans, which were solution-treated and peak aged (T6), were determined at room temperature and 300 degrees F.

EXPERIMENTAL PROCEDURE

Four two hundred and twenty-five pound ingots were cast by Reynolds Metals Co., using an experimental foundry equipped with a 100 kW induction furnace. The ingot chemical compositions were those of a 7475 alloy (2), except for variations in the Fe, Si, Mn, Cr, and Zr levels, as indicated in Table 1. The low Fe-Si, Mn-rich, and high-Zr ingots were processed using a thermomechanical treatment (TMT) designated A, Figure 1. The baseline chemistry ingot was processed by TMT B, which is the same as TMT A, except that the ingot received an extended homogenization to promote the formation of larger precipitates (3). These TMTs were determined (4) to be optimum for processing these alloys to attain large SPF strains. The specific steps in the TMT are as follows.

All of the ingots were stress relieved for 12 hours at 550F to prevent cracking. The ingots for TMT A were then homogenized at 850F for 8 hours to promote the formation of fine precipitates. The ingot for TMT B was homogenized at 850F for 8 hours followed by 200 hours at 940F to promote the formation of larger precipitates. After homogenization, all of the ingots were scalped to 3.25 inches and conventionally hot rolled to 1.5-inch plates. They were then solution heat treated at 920F for 4 hr to dissolve soluble alloy constituents, and then furnace cooled at 50F/hr. When a temperature of 625F was reached, the plates were hot/warm rolled to 0.28 inch, followed by cold rolling to 0.125-inch sheet. These sheets then received a final recrystallization anneal treatment which was conducted in a preheated air furnace at 930F for 20 minutes, followed by a cold water quench and roller leveling for flatness. The TMT used for each alloy, as well as alloy designation number, are also listed in Table 1.

The grain size and precipitate morphology of the sheets were evaluated using optical and STEM techniques, respectively. Microstructural features were analyzed in the STEM using energy dispersive x-ray spectroscopy (EDXS). SPF behavior was assessed in both uniaxial and biaxial tension. The uniaxial tension tests were conducted at a constant strain rate at 970 degrees F without cavitation suppression. The biaxial tension cone tests were conducted at 970 degrees F, but with argon gas pressure of up to 500 psi to suppress cavitation (5). The level of residual cavitation in the biaxially formed cones was determined from SEM evaluations.

Based on the results of these evaluations, the alloy having the greatest SPF characteristics was selected and scaled up. Scale-up consisted of producing a four hundred pound ingot of the alloy, followed by evaluation of SPF and post-SPF properties. The ingot TMT and SPF evaluations were the same as those just discussed. To determine the effect of superplastic forming on mechanical properties, several pans, shown in Figure 2, were formed at 965 ± 5 F under the influence of hydrostatic gas pressure at a strain rate of $2 \times 10^{-4} sec^{-1}$. The pans were solution treated and aged to the T6 temper prior to excising coupons from the pan bottoms. Test specimens machined from the coupons had SPF thickness strains of approximately 300%.

Post-SPF tensile and stress rupture properties of the material were determined from the coupons machined from the bottom of the pans. Tensile tests were conducted at both room temperature and 300F. Stress rupture properties were determined at 300F.

RESULTS AND DISCUSSIONS

The chemical composition of the first of the four alloys used in this evaluation (Table 1) had a baseline chemistry representing a commercial product having a large number of intermediate sized Cr-bearing dispersoids. The second alloy, with a Mn content of 0.21%, was designed to have an intermediate number of large, rod-like Mn-bearing dispersoids. In the third alloy, the Fe and Si levels were lowered to produce an alloy with significantly fewer natural sites for cavitation during SPF. The fourth alloy had a Zr level of 0.20 percent, and represented an alloy with the finest, most spheroidal and coherent dispersoids.

Microstructural Analyses

Optical grain size measurements and TEM examination of the four 7475 alloys were conducted. Results of optical grain size measurements on sheets of the four alloys are presented in Table 2.

TEM micrographs for the four alloys are shown in Figures 3(a) through 3(d). Average values of particle length, width, and distribution measured on numerous such micrographs are listed in Table 3. Also listed in this table is the average grain size of each alloy. The measurements listed in Tables 2 and 3 agree well, showing that the two techniques give similar results.

The following observations were made from the microstructural analyses.

a. The high-Zr alloy (S57453) had the smallest and the most spheroidal (aspect ratio of 1) dispersoids. The baseline and low Fe-Si alloys had intermediate size dispersoids. The Mn-rich alloy (S57438) had the largest and most rod-like (aspect ratio of 1.9) dispersoids.

b. The TMT having the higher-temperature homogenization, TMT B, did not produce dispersoids which were measurably different in size from those produced by the lower-temperature homogenization, TMT A. This is probably due to a stabilization of the particle size during the 850F/8h homogenization (TMT A), so that the subsequent higher-temperature homogenization has little effect (TMT B).

c. As stated in a) above, the baseline and low Fe-Si alloys had intermediate size dispersoids and the largest number of dispersoids per unit distance of all the alloys. Also, the sheets produced from these two alloys had the smallest grain size of all the sheets. The Zr-rich dispersoids, in spite of their small size and desirable distribution, were not effective in grain refinement. This is believed to be due primarily to their coherency with the matrix, suggesting that in addition to the requirement of a small size and large number of dispersoids, they must be incoherent with the matrix to promote grain refinement in discontinuously recrystallizing alloy systems such as the 7000-series.

Based upon these results, the four alloys were ranked in the following descending order of SPF potential: low Fe-Si (S57439), baseline (S57437), Mn-rich (S57438, High-Zr (S57453).

Elevated-Temperature Flow Behavior

To investigate the elevated-temperature flow behavior of these materials, a series of constant true strain rate tests were conducted in uniaxial tension at 970 degrees F over a strain rate range of 6 x 10^{-3} sec^{-1} to 10^{-5} sec^{-1}. The results indicated that rapid strain-hardening and reduced strains to failure occur at the higher strain rates. Strain to fracture data are shown in Figure 4 as a function of true strain rate. Using high failure strain at high strain rate (10^{-4} sec^{-1} to 10^{-3} sec^{-1}) as a measure of high SPF potential, these alloys can again be arranged in descending order of SPF potential as follows: low Fe-Si alloy, baseline alloy, Mn-rich alloy, Zr-rich alloy (S57439, S57437, S57438, and S57453, respectively).

The foregoing elevated temperature uniaxial tension tests were conducted without any attempt to suppress in-situ cavitation during deformation. Results of the uniaxial tension tests provided a useful basis for comparison with published data. To study the SPF behavior of these materials in the absence of cavitation, (i.e., under suppressed or completely eliminated cavitation conditions) biaxial-tension SPF cone tests were conducted. A discussion of the validity of this test, its usefulness and limitations, was provided in a previous paper.(6)

Results of biaxial-tension SPF cone tests (a four cone die with cavity suppression(of the four alloys are shown on the true stress versus true strain rate plot of Figure 5. Each sheet coupon was tested at 970 degrees F using top and bottom gas pressures to suppress cavitation. It is seen that the low Fe-Si alloy (S57439) has lower flow stress values and higher slopes (m-values) than the other alloys over a large range of strain rates. The behavior of the four alloys also follows the same order of SPF potential as that seen previously in the results of the elevated-temperature uniaxial tests conducted without cavity suppression.

Strain-rate sensitivities ($m = d\log\sigma / d\log\dot{\varepsilon}$) were determined from the slopes of the curves in Figure 5 and are plotted as a function of log $\dot{\varepsilon}$ in Figure 6. The low Fe-Si alloy (S57439) has the highest m-values in the strain rate range of 1.5×10^{-4} sec^{-1} to 1.5×10^{-3} sec^{-1}. The baseline alloy (S57437) shows a similar peak m-value (>0.9) as the low Fe-Si alloy (S57439), but at a much lower strain rate. The four alloys again followed the same order of SPF performance as mentioned previously.

Engineering thickness strains to fracture obtained from these tests are plotted as a function of true strain rate in Figure 7. It is noted that there is an approximate similarity in the shapes of curves shown in Figure 6 and 7 for the corresponding alloys. Also, the strain rates for the rise and decline in m-values (Figure 6) correspond quite well with those for the increase and decrease in strain to failure (Figure 7). This is in agreement with the generally accepted postulate that the extended ductility (i.e., high strain value) in superplastic materials comes about due to an improved postponement of strain localization (i.e., high m-value) (7,8). The largest strain-to-fracture value observed was 2400 percent in the low Fe-Si alloys (S57439) tested at a strain rate of 3×10^{-4} sec^{-1}.

It is apparent from the foregoing results that the low Fe-Si alloy (S57439) shows the best SPF potential over a range of strain rates typical of SPF operations. To determine the advantages over the state-of-the-art material, the behavior of this alloy was compared with that of MD254, which is a commercially available SPF 7475 alloy produced by Reynolds Metals Co. This comparison was made by comparing log σ vs. log $\dot{\varepsilon}$, m vs. log $\dot{\varepsilon}$ and strain-to-fracture vs. log $\dot{\varepsilon}$ plots shown in Figures 8, 9 and 10, respectively. It is seen that the low Fe-Si alloy offers lower flow stresses than MD254 (in the strain rate range of 5×10^{-5} sec^{-1} to 10^{-3} sec^{-1}), and higher m-values and fracture strains (over the strain rate range of 1.5×10^{-4} sec^{-1} to 1.0×10^{-3} sec^{-1}). Furthermore, while the peak m- values and maximum fracture strains are virtually identical for the two alloys, the corresponding strain rate for the low Fe-Si alloy is almost four times greater than that for MD254. This is considered to be a significant improvement and, if consistently reproduced in full scale production material, could have a significant impact on fabrication related cost savings.

Cavitation Behavior

The in-situ cavitation behavior during superplastic deformation of the four 7475 alloys was studied. Biaxial-tension SPF cone specimens were sectioned, cold mounted, polished, and quantitively examined.

The results of the quantitative image analyses of the four 7475 alloy variants are shown in Figure 11. These results are from cones formed at a strain rate of 1×10^{-3} sec^{-1}. As seen in Figure 11, the low Fe-Si alloy (S57439) showed the least amount of cavitation and produced the greatest values of strain for any cavitation level.

The low cavitation in the low Fe-Si alloy is attributed to its lower number of cavities as well as low mean cavity area (also determined by quantitative analysis, although not shown here).

Based on these results, the low Fe-Si alloy was determined to have the best combination of high fracture strain, and low cavitation (i.e., low cavity nucleation rate, as well as low cavity growth rate). This alloy composition was scaled-up and then formed into a generic pan shape using a constitutive model. Post-SPF tensile and stress rupture properties of the material were then determined.

Selected Alloy Scale-up

The low Fe-Si 7475 alloy (heat S57439) was scaled up to a 400 lb ingot (heat S59658). The chemical compositions of the prescale-up and the scale-up alloys were found to be in excellent agreement.

Scanning transmission electron microscopy (STEM) and optical evaluations of the scale-up material (S59658) were conducted and compared with data for the prescale-up material to determine if the size and spacing of the dispersoids in both materials were similar. It was found that the length, aspect ratio, and density of the dispersoids were essentially the same in the two materials. The grain-boundary dispersoids were most commonly soluble $Mg(Cu,Zn)_2$, and dispersoids in the grain interiors were primarily chromium-rich insoluble E-phase. The grain size of the scale-up material, determined optically as well as by STEM, however, was smaller than that of the prescale-up material. Overall, the microstructures in the two materials were considered by be equivalent.

Elevated-Termperature Flow Behavior

Figure 12 shows the true stress vs. true strain rate results of uniaxial tension tests conducted on the scale-up and prescale-up low Fe-Si alloys at 965F without cavity suppression. A lower value of true stress was obtained in the scale-up alloy for a given true strain rate in the entire range tested. The engineering strain to fracture vs. true strain rate results are seen in Figure 13. Both materials achieved maximum engineering strains at approximately the same true strain rate of 1×10^{-4}/sec. However, the prescale-up material achieved a higher value of the elongation to fracture, 1000% vs. 775%, respectively.

Figure 14 shows the engineering strain to fracture vs. true strain rate results for the biaxial tension SPF cone tests conducted on the scale-up and prescale-up low Fe-Si alloys at 970F using cavity suppression. A maximum thickness strain to fracture of 1950% was obtained in the scale-up alloy at a strain rate of 2.1×10^{-4}/sec. This is comparable to the maximum thickness strain to fracture of 2440% achieved in the prescale-up alloy at a strain rate of 2.6×10^{-4}/sec. The true stress vs. true strain rate plots for these materials, Figure 15, show overlapping bands suggesting that the two materials have comparable true stress values for a given true strain rate.

Comparison of the biaxial tension cone test results of the scale-up alloy to those of a commercial SPF alloy, Reynolds' MD254. shows a three-fold increase in the strain rate at which the S59658 material achieved its maximum strain to failure compared with MD254, Figure 16. This is a significant advancement over the present state of the art.

Cavitation Behavior

Microstructural evaluations were

conducted on the scaled-up low Fe-Si alloy (S59658) to examine its cavitation response during SPF. Figure 17 shows the areal fraction cavitation (ϕ)vs. true strain (ε) plots for the scaled-up alloy at a strain rate of $1x10^{-3}$ per sec, along with a comparison to data for the prescale-up alloy. The results indicate that cavitation produced in each material at this strain rate was essentially the same. The mean cavity area for a given true strain, however, was significantly lower in the scaled-up material.

The foregoing observations made on the scaled-up version of the low Fe-Si 7475 alloy (S59658) are very reassuring of success in further scale-up of this alloy to mill scale using TMT A while retaining the excellent SPF properties of the laboratory version. Significant increases in both the maximum strain to fracture and the rate at which the material can be formed are expected over the commercial counterpart, Reynolds' alloy MD254. This projected success will provide a great advancement in the state of the art of SPF aluminum technology and applications.

Pan-Forming and SPF Evaluations

The deformation characteristics of the low Fe-Si 7475 alloy were analyzed and an equation developed to represent superplastic flow behavior. Parameters determined using this equation were subsequently used to superplastically form a generic pan from which test specimens were removed for post-SPF evaluations.

The flow stress,σ, during superplastic deformation can be denoted as follows when the flow stress is a function of the deformation strain.(9)

$$\sigma = B' * \dot{\epsilon}^m \epsilon^q \quad \text{for} \quad \epsilon \leq \epsilon_c \cdot\cdot$$

Eq.1

where B is temperature and grain size dependent constant

m is the strain rate sensitivity exponent, and

q is the strain hardening exponent.

$\dot{\varepsilon}$ is the true strain rate.

ε is the true strain.

ε_c is the critical true strain above which a significant amount of cavitation is observed.

The equation is applicable up to a critical strain corresponding to either a strain value at which steady state is reached or material instability sets in. The constants in equation 1 were determined experimentally from the SPF so that the equation:

$$\sigma = 58\, \epsilon^{0.6} \dot{\epsilon}^{0.6} \quad ...$$

Eq.2

Where σ is in ksi.

Equation 2 predicts the flow stress in the low Fe-Si 7475 at 965F. This equation should accurately predict material flow behavior due to its inclusion of both strain hardening and strain rate hardening. The effect of strain rate hardening is not usually accounted for in analyses such as this, which leads to a certain amount of error in predicting material flow behavior.

A comparison of the flow stress predicted by Equation 2 and the actual flow stress measured in uniaxial tests is shown in Figure 18. Good agreement between the predicted behavior (solid lines) and the experimental results (data points) is seen over a wide range of strain rates and strains.

Further verification of the accuracy of the variable flow stress was obtained by using equation 2 to determine the forming cycle for a generic pan, Figure 19. Figure 20 shows the forming cycle utilized to form this pan. The forming temperature was 965 ± 5 F and the strain rate was $2 \times 10^{-4} sec^{-1}$. Table 4 lists the actual thickness at four locations along a section of the pan as indicated in Figure 19. Good agreement was obtained between the predicted and actual values.

Post-SPF Properties

Using the foregoing model, numerous pans having thickness strains to failure of approximately 300% in the bottom flat were formed from which coupons for tensile and stress rupture testing were removed.

Post-SPF tensile properties were determined at room temperature and 300F on material oriented in both the longitudinal and transverse sheet rolling directions. All of the test coupons were removed from the bottom of several SPF pans and heat treated to the T6 condition. To fully determine the effect of SPF deformation on material properties, similar tests were also conducted on this material in the pre-SPF condition, i.e., after simply exposing the alloy sheet to the temperature/time cycle encountered during SPF (without the SPF strain), and heat treating it to the same T6 condition.

The longitudinal room temperature properties of the formed material (pan) were similar to those of the pre-SPF material (sheet) and greater than those of 7475-T61 nonSPF sheet (10), Table 5. Virtually no effect of SPF strain is observed on the strength or elongation of the material in the longitudinal orientation . A slight drop (approximately 5%) in strength is, however, noted in the transverse orientation.

The same trends were observed in the longitudinal data at 300F (one-half hour exposure at temperature), Table 6. The ultimate, yield, and elongation of the post and pre-SPF materials as well as nonSPF 7475-T61 sheet were essentially equivalent, except that the transverse elongation of the post-SPF material was appreciably greater than that of the pre-SPF material.

Stress rupture test data of the post-SPF low Fe-Si 7475 alloy are shown in Figure 21. Data for pre-SPF material are also included for comparison, along with data from the ASM Handbook [10]. The post-SPF data compare well with the pre-SPF and ASM handbook data.

SUMMARY

1. The superplastic formability of a thermomechanically treated 7475 aluminum alloy can be significantly enhanced by reducing the Fe and Si contents below their normal levels (to .01 wt % each). The enhanced formability is primarily due to a reduction in the constituent phases which are the natural sites for cavitation. A maximum thickness strain to failure of over 2400%, at a strain rate of 2×10^{-4} sec^{-1}, was obtained in biaxial tension (cavitation suppressed) from such an alloy compared to a similar value of strain for normal 7475 SPF material at a strain rate of approximately 6×10^{-5} sec^{-1}.

2. The enhanced formability obtained from the low Fe-Si alloy, which was produced on a laboratory scale, was successfully reproduced in material made on a scaled-up, or near production level.

3. A constitutive model for predicting the pressure-time cycles needed to fabricate a generic structural shape was verified and several pans were successfully made from the low Fe-Si 7475 alloy.

4. The post-SPF service properties of the low Fe-Si 7475 were at least equivalent to the pre-SPF material as well as conventional 7475.

ACKNOWLEDGEMENTS

The authors would like to acknowledge the help of Messrs. M. Arrieta, J.T. Findley, and D.R. Drott of the Metallic Materials and Processes Research and Development Department and M. Gambone of the Microstructural Analysis Department of Northrop Aircraft Division. Help received from Messrs. B.R. Ward, T. J. Watson, and J. Bennetch of

Reynolds Metals Company is also gratefully acknowledged. This work was supported by NASA Langley Research Center under Contract NAS1-17623, under the technical direction of Mr. T.T. Bales.

REFERENCES

1. Ward, B.R. and Ashton, R.F. US Patent No. 4,486,242, December 4, 1984.

2. Mehr, P.L., "ALCOA 7475 Sheet and Plate," ALCOA Green Letter, Third Revised Edition, Alcoa Technical Center, PA, February, 1978

3. Boyd, J.D., Drennen, D.C., Martin, C.J., Price, C.W., Rosenfield, A.R., Williams, D.N., and Thompson, D.S."Research on Synthesis of High-Strength Aluminum Alloys," Technical Report No. AFML-TR-72-199,WPAFB, Ohio, October 1972.

4. Agrawal, S.P. and Ginter,T.J., "Design, Development, and Characterization of Advanced Aluminum Alloys forEnhanced Superplasticity," Third Interim Report, NASA Contract No. NAS1 17623, Northrop Corporation, Aircraft Division, Hawthorne, CA, December, 1985

5. Agrawal, S.P., US Patent No. 4,516,419, May 14, 1985.

6. Agrawal, S.P. and Tuss, J.M., "Superplastic Forming and Post-SPF Mechanical Behavior of an Aluminum Alloy for Airframe Applications," Proceedings of an International Conference held at Cranfield, U.K., 12-15 July 1985; Pearce, R. and Kelly, L. eds., Ashford Press, England, pp. 296-325 (1985).

7. Woodford, D.A., Trans. Quart. ASM, 1969, Vol. 62, p.291.

8. Hosford, W.F., and Caddell, R.M. "Metal Forming: Mechanics and Metallurgy,"Prentice-Hall,Inc., pp. 80-102, 1983.

9. Bird, J.E., Mukherjee, A.K., and Dorn, J.E., Proceedings of International Conference on Quantitative Relations Between Properties and Microstructure, Israel University Press, 1970, p. 255.

10. ASM Handbook, Ninth Ed., V.2, pp. 136-138.

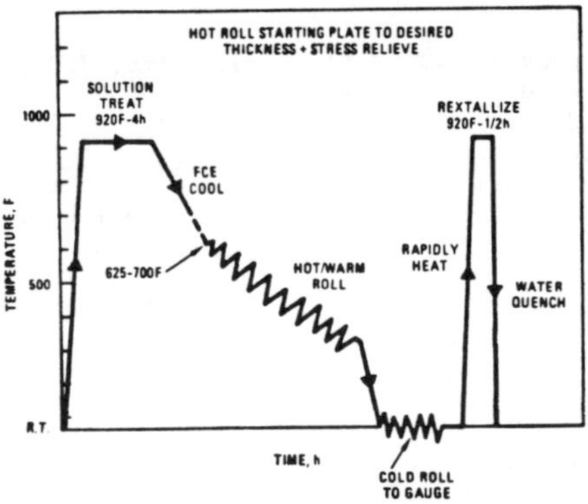

Figure 1. TMT Mill Practice (A) for Producing Fine and Equiaxed Grain Structure in 7475.

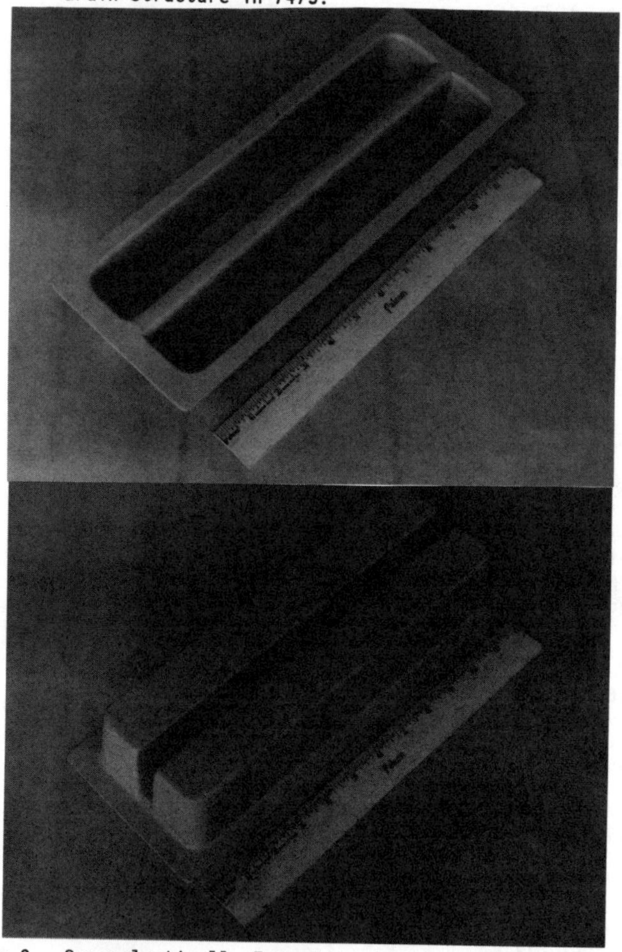

Figure 2. Superplastically Formed Pan Used for Post-SPF Mechanical Property Determination

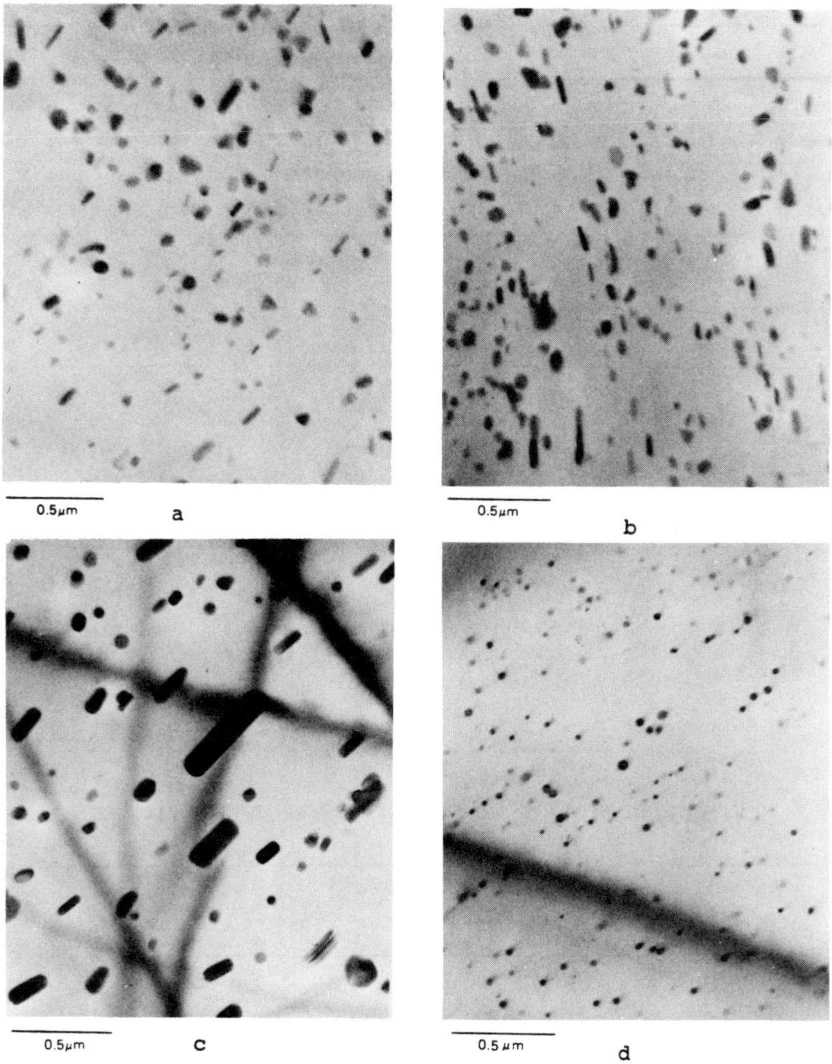

Figure 3. Transmission Electron Micrographs for 7475 Alloys: a) baseline alloy, TMT B (S57437); b) low Fe-Si alloy, TMT A (S57439); c) Mn-Rich alloy, TMT A (S57438); d) High Zr alloy, TMT A (S57453)

TABLE 1. CHEMICAL COMPOSITIONS (IN WEIGHT PERCENT) OF FOUR ALUMINUM ALLOYS

Alloy	Heat Number	TMT	Si	Fe	Cu	Mn	Mg	Cr	Ni	Zn	Ti	Zr	Bal
Baseline	S57437	B	0.03	0.05	1.56	<0.01	2.33	0.20	<0.01	6.12	0.03	<0.01	Al
High Mn	S57438	A	0.03	0.05	1.55	0.21	2.32	0.21	<0.01	6.12	0.02	<0.01	Al
Low Fe-Si	S57439	A	<0.01	0.01	1.55	<0.01	2.36	0.20	<0.01	6.12	0.02	<0.01	Al
High Zr	S57453	A	0.01	0.02	1.51	<0.01	2.30	<0.01	0.01	6.08	0.02	0.20	Al

TABLE 2. GRAIN SIZE MEASUREMENTS OF SHEETS OF FOUR 7475 ALLOYS.

			GRAIN SIZE		
			Longitudinal	Transverse	
ALLOY	TMT	HEAT NO.	µm	µm	Aspect Ratio
Baseline	B	S57437	15.0	12.3	1.2
Low Fe-Si	A	S57439	15.3	13.3	1.1
High-Zr	A	S57453	22.1	16.5	1.3
Mn-rich	A	S57438	16.7	14.7	1.1

TABLE 3. DISPERSOID SIZE AND DISTRIBUTION IN 7475 ALLOYS.

ALLOY/TMT	DISPERSOID SIZE(µm) LENGTH	DISPERSOID SIZE(µm) WIDTH	NUMBER* OF DISPERSOIDS(/µm)	GRAIN SIZE* (µm)
Baseline/B (S57437)	0.08	0.05	2.52	13.5
Low Fe-Si/A (S57439)	0.09	0.06	2.38	14.3
Mn-Rich/A (S57438)	0.15	0.08	1.67	15.7
High Zr/A (S57453)	0.02	0.02	1.99	19.1

*Measured by linear-intercept method

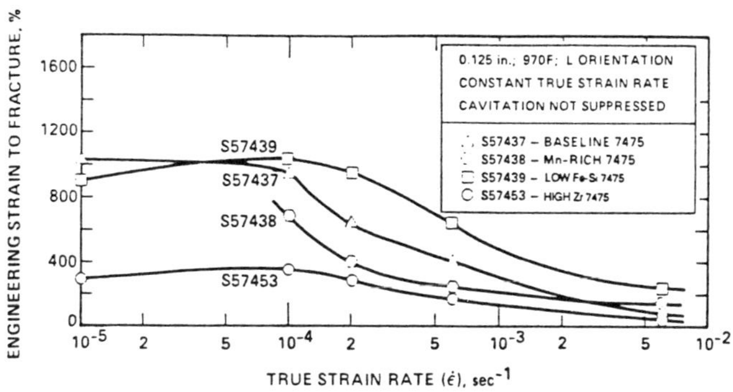

Figure 4. SPF Strain to Fracture vs. Strain Rate For 7475 Alloy Variants in Uniaxial Tension

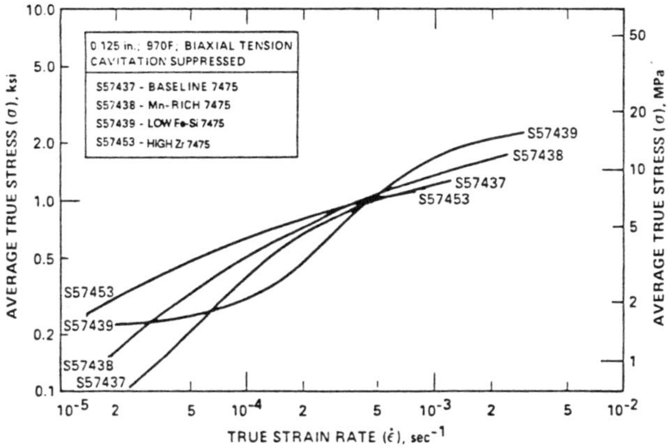

Figure 5. True Stress vs. True Strain Rate for 7475 Alloy Variants

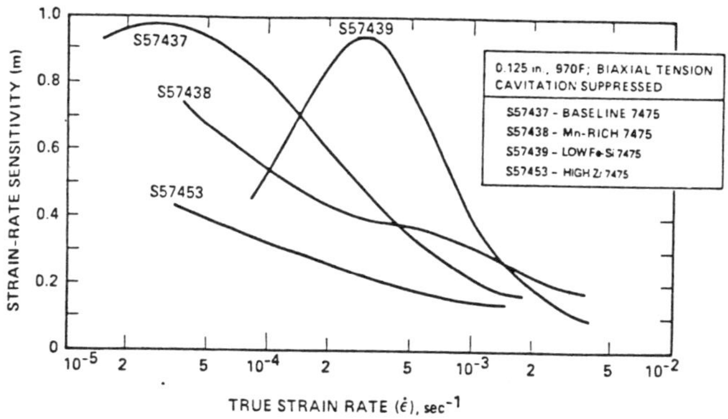

Figure 6. Strain-Rate Sensitivity vs. True Strain Rate for 7475 Alloy Variants

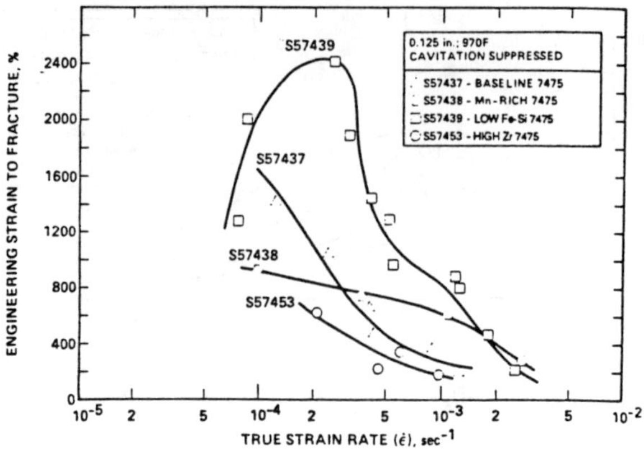

Figure 7. SPF Strain to Fracture vs. Strain Rate for 7475 Alloys in Biaxial Tension

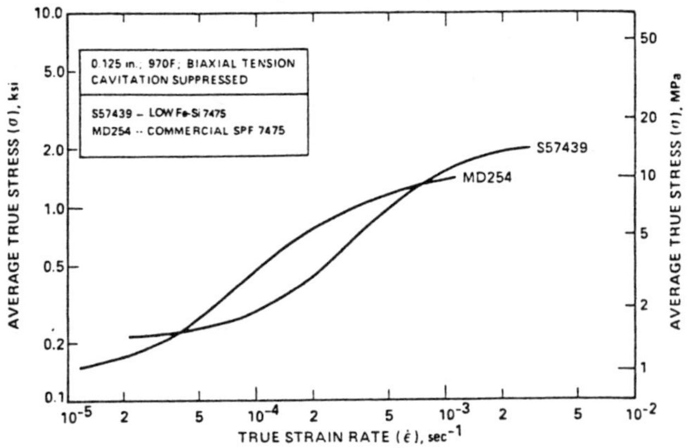

Figure 8. True Stress vs. True Strain Rate for a Commercial (MD254) and a Low Fe-Si 7475 Alloy

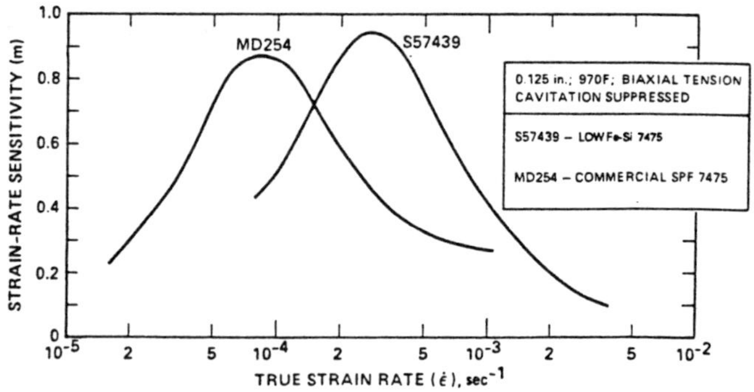

Figure 9. Strain-Rate Sensitivity vs. True Strain Rate for a Commercial (MD254) and a Low Fe-Si 7475 Alloy

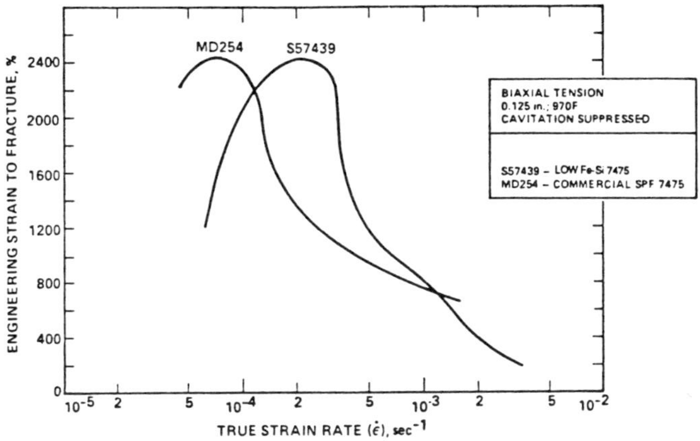

Figure 10. SPF Strain to Fracture vs. Strain Rate for a Commercial (MD254) and a low Fe-Si 7475 Alloy

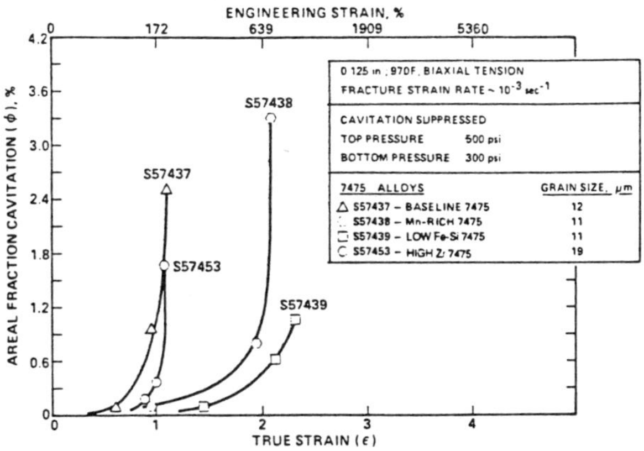

Figure 11. Areal Cavitation vs. SPF Strain for 7475 Alloys

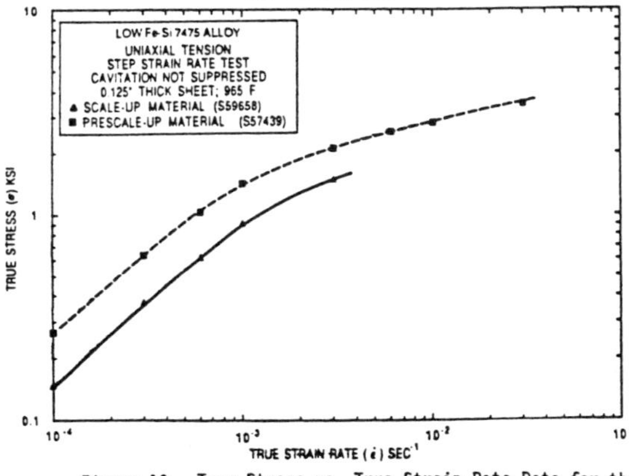

Figure 12. True Stress vs. True Strain Rate Data for the Scale-up and Prescale-up 7475 Alloy Tested in Uniaxial Tension at 965F.

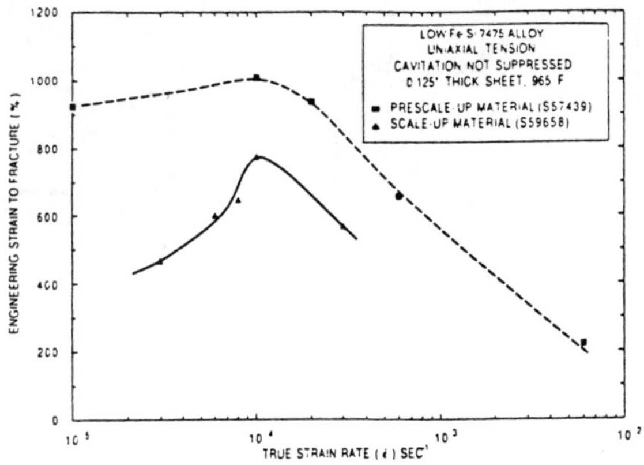

Figure 13. Engineering Strain to Fracture vs. True Strain Rate Data for Scale-up and Prescale-up 7475 Alloys Tested in Uniaxial Tension.

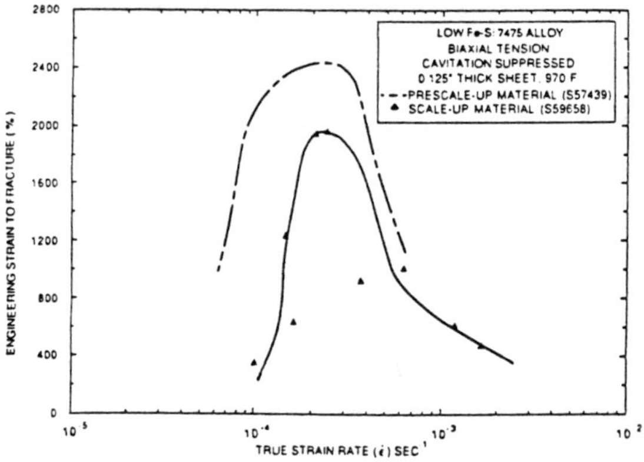

Figure 14. Engineering Strain to Fracture vs. True Strain Rate Data for Scale-up and Prescale-up 7475 Alloy Tested in Biaxial Tension.

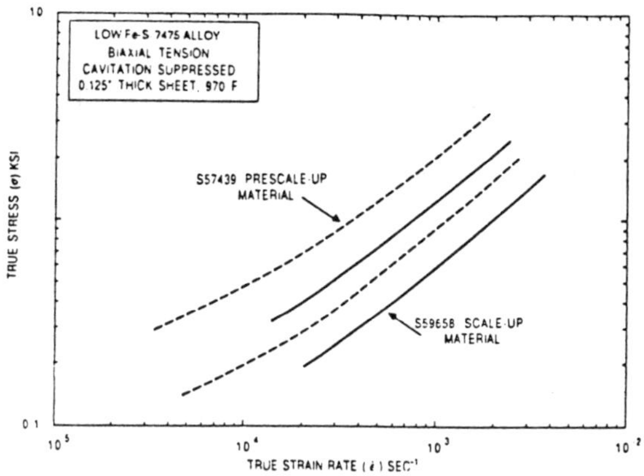

Figure 15. True Stress vs. True Strain Rate Data for the Scale-up and Prescale-up 7475 Alloys Tested in Biaxial Tension at 970F.

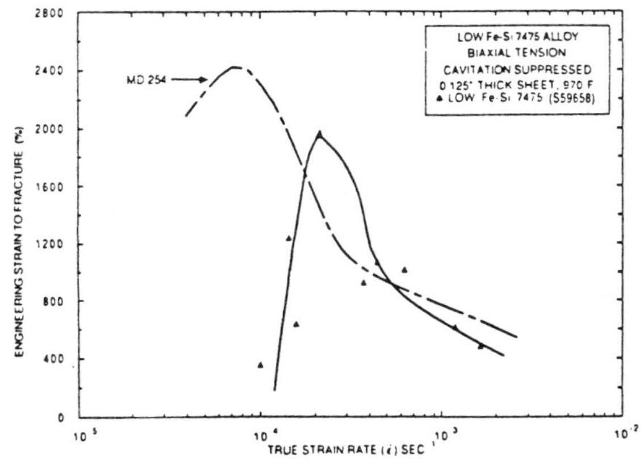

Figure 16. Engineering Strain to Fracture vs. True Strain Rate Data for MD254 and the Scaled-up Low Fe-Si 7475 Alloy Tested in Biaxial Tension at 970F.

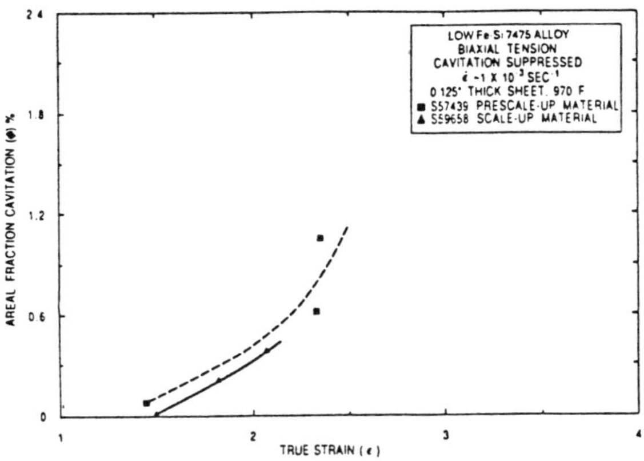

Figure 17. Areal Fraction Cavitation vs. True Strain Data for the Scale-up and Prescale-up Low Fe-Si 7475 Alloy Tested in Biaxial Tension at 970F.

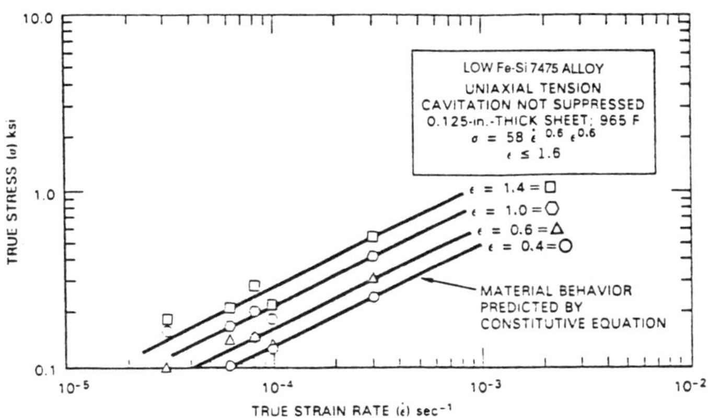

Figure 18. Comparison of Flow Stress Prediction with the Experiments

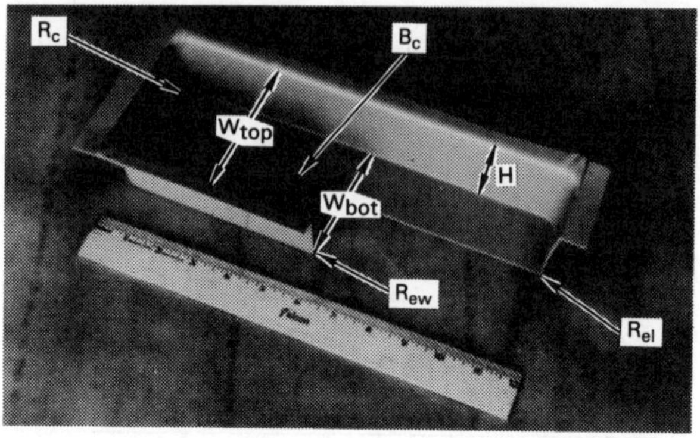

557.10

Figure 19. Pan Which Was Superplastically Formed at a Strain Rate of Approximately $2x10^{-4}sec^{-1}$ and 935F. Thickness Strain of Approximately 60% Obtained in Bottom of Pan.

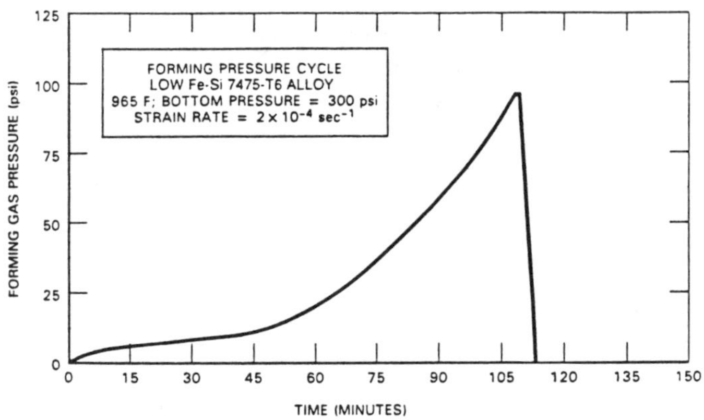

Figure 20. Pressure Time Cycle Used to Form Pans.

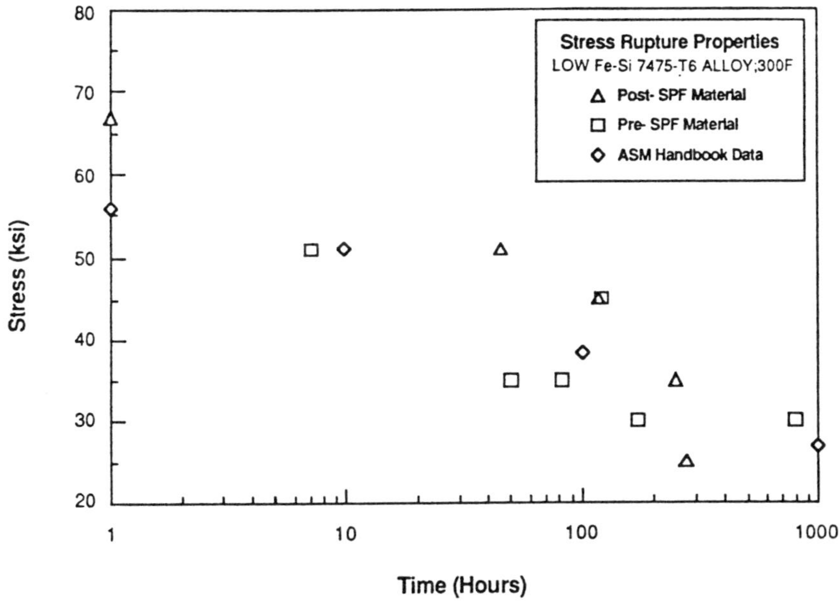

Figure 21. Stress Rupture Test Results for 7475-T6 Obtained at 300°F.

AUTHOR INDEX

Abbaschian, G.J. 102
Agrawal, S.P. 512
Anderson, I.E. 116
Anderson, Robert A. 393
Baeslack III, W.A. 381,458
Bain, Kenneth R. 487
Balsone, S.J. 243
Bania, Paul J. 286
Barice, W.J. 20,28
Barker, Douglas R. 421
Benn, R.C. 188
Bhattacharya, R.S. 254
Bohlen, J.W. 320
Bye, R.L. 345
Castner, Harvey R. 407
Cetel, Alan D. 37
Chanani, G.R. 320
Chang, C.F. 345
Chang, Keh-Minn 76
Chawla, S.K. 225
Das, S.K. 91,345
Dietrich, W. 144
Dini, J.W. 428
Dini, J.W. 428
Domoratsky, Vsevolod A. 202
Doraivelu, Sokka M. 421
Duhl, David N. 37
Elsley, R.K. 298
El-Soudani, S.M. 298
Entrekin, Charles H. 128
Eylon, D. 28
Ferris, D.M. 157
Froes, F.H. 28
Gambone, Mary Lee 487
Gibson, R.F. 368
Gilman, P.S. 91
Gokhale, A.B. 102
Guthmiller, W.A. 298
Hainz, L.C. 157
Harker, Howard R. 128
Henderson, Russell 298
Hirara, L.H. 213
Huron, E.S. 271
Javed, K.R. 102
Kantzos, Peter 334
Kemmer, Hans-J. 504
Kemppinen, A.I. 177
Khobaib, M. 243,254,262
Kim, N.J. 345
Konitzer, D.G. 480
Latanision, R.M. 213
Lewandowski, John J. 450
Lewis, R.E. 102
Loretto, M.H. 480
Mantena, Raju 368
Marder, J.M. 357
Miller, J.A. 271
Mirchandani, P.K. 188
Mitchell, A. 135
Nelson, Howard G. 310
Nikolayev, Vladimir K. 202
Pannen, H. 144
Payer, J.H. 225
Peltier, J.M. 91
Place, T.A. 368
Raffoul, C.N. 254
Rai, A.K. 254
Ranke, H. 144
Raybould, D. 91,345
Samonds, Mark T. 421
Sanin, Anatol F. 202
Sankaran, K.K. 166
Scheglova, Lioudmila L. 202
Schelleng, R.D. 177
Schlienger, M.P. 437
Schuster, D.M. 498
Sheinker, A.A. 320
Sick, G. 144
Sikka, Vinod K. 62
Skibo, M.D. 498
Srinivasan, V. 237
Srivatsan, T.S. 368
Stenzel, Otto W. 504
Stonehouse, A.J. 357
Stumpp, H. 144
Sudarshan, T.S. 368
Telesman, Jack 334
Thorne, J.K. 20
Tripp, D.W. 135
Turk, G.R. 512
Vahldiek, F.W. 262
Vasey-Glandon, V.M. 166
Vastava, R. 512
Vedula, K. 49
Walker-Heckman, E.A. 166
Walsh, J.A. 320
Wawner, F.E. 470
Weber, J.H. 177
Whatley, W.J. 470
Wood, Paul T. 458
Yep, F.W. 498
Young, J.D. 298
Zedalis, M.S. 91

SUBJECT INDEX

2nd International SAMPE Metals and Metals Processing Conference

Volume 2, 1988

APPLICATIONS

Aerospace 1,157,177,310,345
Airfoil 37,437
Fighter aircraft 166
Gas turbine engines 271

ADHESIVE BONDING

Adhesively bonded joints 368

ANALYSIS and MODELING

Analytical modeling of 3-D Castings 421
CAST3 Finite Element Code 421
Failure analysis 298
Fractographic analysis 487

CORROSION/OXIDATION

Corrosion in early stages of condensation 225
Corrosion of MM composites 213
Corrosion/oxidation of surface modified nickel 254
High temperature oxidation behavior of Ti_3Al alloys 262
Hot corrosion 243
Oxidation of ion-implanted nickel-base alloys 237

DEVELOPMENTS, GENERAL

Recent advancements in lightweight metals 1
Superalloy development 76

ELECTROCOMPOSITES 428

FATIGUE

Fatigue behavior of Ti-6Al-4V 320
Fatigue crack growth in a single crystal alloy 334
Fatigue damage mechanism 334

HYDROGEN EMBRITTLEMENT 310

METALS AND ALLOYS

Al-Fe-Ce alloys 381
Al-Fe-V-Si alloys 91
Al-Li alloys 166
Al-Mg-Li alloys 177
Al-Ti based alloys 188
Beryllides 357
Cu-base alloys 116
Ferrous alloys 504
Mg alloys 345
Nb-base alloys 102
Ni-Al alloys 49
Ni-base alloy 37,237,243,437
Nickel-aluminides 62
Surface modified Ni 254
Titanium 28,144,271
Ti_3Al alloys 262
Ti-1100 286
Ti-6Al-4V 28,320
Titanium aluminide 487
7475 Al alloy 512

METAL-MATRIX COMPOSITES

Al alloy MMC's 450
Elevated temperature exposure 470
Graphite/6061-T6 aluminum alloy 213
Corrosion 213
Interaction zone 480
Produced by ingot metallurgy 498
SiC/Ti-6Al-4V composite 470

METALS PROCESSING

Adhesively bonded joints 368
Arc welding 393,407
Cold Hearth Melting 128,144
Electrodeposition 428
Electron beam melting 128,157
HIP processing 202
Joining of powder metallurgy Al-Fe-Ce alloys 381
Melting techniques for Ti 135
Metal Matrix Composites 450
Microstructure-property relationships in Ni-base castings 37
Plasma Melting 437
Processing of nickel aluminides 62

Rapid solidification of Cu-base alloy powders 116
Rapid solidification of Nb-base alloys 102
Superplastic formability of high strength Al 512
Ti alloy castings 20,28
Vacuum Melting 504

WELDING

Arc welding systems 407
Gas tungsten arc welding 393
Plasma-Arc Welding 458
Vision based weld process control 393